The Genetics and Genomics of the Rabbit

The Genetics and Genomics of the Rabbit

Edited by

Luca Fontanesi

CABI

CABI is a trading name of CAB International

CABI
Nosworthy Way
Wallingford
Oxfordshire OX10 8DE
UK

Tel: +44 (0)1491 832111
Fax: +44 (0)1491 833508
E-mail: info@cabi.org
Website: www.cabi.org

CABI
WeWork
One Lincoln St
24th Floor
Boston, MA 02111
USA

Tel: +1 (617)682-9015
E-mail: cabi-nao@cabi.org

A catalogue record for this book is available from the British Library.

Library of Congress Cataloging-in-Publication Data

Names: Fontanesi, Luca, editor.
Title: The genetics and genomics of the rabbit / Luca Fontanesi.
Description: Wallingford, Oxfordshire ; Boston, MA : CAB International, 2021. | Includes bibliographical references and index. | Summary: "Covering all aspects of the genetics and genomics of the rabbit, this book includes domestication, classical genetics, the rabbit genome, genetic maps and markers, coat colour genetics, immunogenetics, transgenic rabbits and biotechnology applications. It also covers the genetics of meat and fibre production, reproduction and disease resistance"-- Provided by publisher.
Identifiers: LCCN 2020054417 (print) | LCCN 2020054418 (ebook) | ISBN 9781780643342 (hardback) | ISBN 9781780643359 (ebook) | ISBN 9781789249484 (epub)
Subjects: LCSH: Rabbits--Genetics. | Rabbits--Genetic engineering.
Classification: LCC SF453 .G38 2021 (print) | LCC SF453 (ebook) | DDC 636.932/2--dc23
LC record available at https://lccn.loc.gov/2020054417
LC ebook record available at https://lccn.loc.gov/2020054418

References to Internet websites (URLs) were accurate at the time of writing.

ISBN: 9 781 78064 3342 (hardback)
9 781 78064 3359 (ePDF)
9 781 78924 948 4 (ePub)

DOI: 10.1079/9781780643342.0000

Commissioning Editor: Alexandra Lainsbury
Editorial Assistant: Emma McCann
Production Editor: Marta Patiño

Typeset by SPi, Pondicherry, India
Printed and bound by CPI Group (UK) Ltd, Croydon, CR0 4YY

Contents

To Giovanna, Davide and Sara

Contributors

Joana Abrantes
InBIO-Research Network in Biodiversity and Evolutionary Biology, CIBIO, Universidade do Porto, Campus Agrário de Vairão, Rua Padre Armando Quintas, nr. 7, 4485-661, Vairão, Portugal
e-mail: jabrantes@cibio.up.pt
Contribution: chapter 9

Marielle Afanassieff
Univ Lyon, Université Claude Bernard Lyon 1, Inserm, INRAE, Stem Cell and Brain Research Institute U1208, USC1361, F-69500, Bron, France
e-mail: marielle.afanassieff@inserm.fr
Contribution: chapter 16

Daniel Allain
GenPhySE, INRAE, 31326 Castanet Tolosan, France
e-mail: daniel.allain@inrae.fr and d.allain@free.fr (Retired)
Contribution: chapter 7

André M. Almeida
Instituto de Tecnologia Química e Biológica António Xavier, Universidade Nova de Lisboa, 2780-157 Oeiras, Portugal
IBET – Instituto de Biologia Experimental e Tecnológica, Oeiras, Portugal and CIISA – Centro Interdisciplinar de Investigação em Sanidade Animal, Faculdade de Medicina Veterinária da Universidade de Lisboa, Lisboa, Portugal
LEAF, Instituto Superior de Agronomia, Universidade de Lisboa, Lisboa, Portugal
e-mail: aalmeida@isa.ulisboa.pt
Contribution: chapter 14

Susana S. Araújo
Instituto de Tecnologia Química e Biológica António Xavier, Universidade Nova de Lisboa, 2780-157 Oeiras, Portugal
e-mail: saraujo@itqb.unl.pt
Contribution: chapter 14

Manuel Baselga
Institute for Animal Science and Technology, Universitàt Politecnica de Valencia,Valencia, Spain
e-mail: mbaselga@dca.upv.es (Retired)
Contribution: chapter 13

Agustin Blasco
Institute for Animal Science and Technology, Universitat Politècnica de València, València, Spain
e-mail: ablasco@dca.upv.es
Contributions: chapter 10, 12

Paul Boettcher
Food and Agriculture Organization of the United Nations, 00153 Rome, Italy
e-mail: paul.boettcher@fao.org
Contribution: chapter 2

Cristiano Boiti
Department of Veterinary Medicine, University of Perugia, Perugia, Italy
e-mail: boiti.cristiano@gmail.com (Retired)
Contribution: chapter 11

Gerard Bolet
GenPhySE, INRAE, ENVT, Université de Toulouse, 31326 Castanet Tolosan, France
e-mail: gerard.bolet@inrae.fr (Retired)
Contribution: chapter 2

Darko Bosnakovski
Faculty of Medical Sciences, University Goce Delcev, Štip, Republic of North Macedonia
e-mail: darko.bosnakovski@ugd.edu.mk
Contribution: chapter 16

Zsuzsanna Bősze
National Agricultural Research and Innovation Centre, Agricultural Biotechnology Institute, Gödöllő, Hungary
e-mail: bosze.zsuzsanna@abc.naik.hu
Contributions: chapter 15, 17

Samuel Boucher
Labovet Conseil (Réseau Cristal), Bp 539, 85505 Les Herbiers, France
e-mail: s.boucher@labovet.fr
Contribution: chapter 2

Samuele Bovo
Department of Agricultural and Food Sciences, Division of Animal Sciences, University of Bologna, Viale Giuseppe Fanin 46, 40127 Bologna, Italy
e-mail: samuele.bovo@unibo.it
Contribution: chapter 3

Gudrun A. Brockmann
Albrecht Daniel Thaer – Institut für Agrar- und Gartenbauwissenschaften Humboldt Universität zu Berlin, Invalidenstr. 42, 10115 Berlin, Germany
e-mail: gudrun.brockmann@agrar.hu-berlin.de
Contribution: chapter 4

Luca Buttazzoni
Centro di ricerca di Zootecnia e Acquacoltura, Consiglio per la ricerca in agricoltura e l'analisi dell'economia agraria (CREA), Monterotondo (Roma), Italy
e-mail: luca.buttazzoni@crea.gov.it
Contribution: chapter 13

Cesare Castellini
Department of Agricultural, Food and Environmental Sciences, University of Perugia, Perugia, Italy
e-mail: cesare.castellini@unipg.it
Contributions: chapter 11, 18

Pascale Chavatte-Palmer
INRAE UMR 1198 Biologie du Développement et Reproduction, Physiologie Animale et Systèmes d'élevage, 78352, Jouy-en-Josas, France
e-mail: pascale.chavatte-palmer@inrae.fr
Contribution: chapter 11

Peter Chrenek
Department of Biochemistry and Biotechnology, Faculty of Biotechnology and Food Sciences, Slovak University of Agriculture in Nitra, Tr. A. Hlinku 2, 949 76 Nitra, Slovak Republic NPPC—Research Institute for Animal Production Nitra, Hlohovecká 2, 951 41 Lužianky, Slovak Republic
e-mail: chrenekp@cvzv.sk
Contributions: chapter 11, 15, 17

Veronique Duranthon
Université Paris-Saclay, INRAE, ENVA, UVSQ, BREED, 78350, Jouy-en-Josas, France
e-mail: veronique.duranthon@inrae.fr
Contribution: chapter 14

Pedro José Esteves
InBIO-Research Network in Biodiversity and Evolutionary Biology, CIBIO, Universidade do Porto, Campus Agrário de Vairão, Rua Padre Armando Quintas, nr. 7, 4485-661, Vairão, Portugal
Departamento de Biologia, Faculdade de Ciências, Universidade do Porto, 4169-007, Porto, Portugal
CITS-Centro de Investigação em Tecnologias de Saúde, CESPU, Gandra, Portugal
e-mail: pjesteves@cibio.up.pt
Contribution: chapter 9

Jianglin Fan
Department of Molecular Pathology, Interdisciplinary Graduate School of Medicine, University of Yamanashi, Yamanashi, Japan
e-mail: jianglin@yamanashi.ac.jp
Contribution: chapter 18

Bernd Fischer
Department of Anatomy and Cell Biology, Martin Luther University, Faculty of Medicine, Halle (Saale), Germany
e-mail: bernd.fischer@uk-halle.de
Contribution: chapter 11

Luca Fontanesi
Department of Agricultural and Food Sciences, Division of Animal Sciences, University of Bologna, Viale Giuseppe Fanin 46, 40127 Bologna, Italy
e-mail: luca.fontanesi@unibo.it
Contributions: chapter 1, 3, 4, 6, 8, 17

María-Luz García
Department of Agrofood Technology, Escuela Politécnica Superior de Orihuela, Universidad Miguel Hernández de Elche, Orihuela, Spain
e-mail: mariluz.garcia@umh.es
Contribution: chapter 13

Hervé Garreau
GenPhySE, INRAE, ENVT, Université de Toulouse, 31326 Castanet Tolosan, France
e-mail: herve.garreau@inrae.fr
Contributions: chapter 2, 9, 13

Elen Gócza
National Agricultural Research and Innovation Centre, Agricultural Biotechnology Institute, Animal Biotechnology Department, Gödöllő, Hungary
e-mail: gocza.elen@abc.naik.hu
Contribution: chapter 16

Mélanie Gunia
GenPhySE, INRAE, ENVT, Université de Toulouse, 31326 Castanet Tolosan, France
e-mail: melanie.gunia@inrae.fr
Contribution: chapter 9

Pilar Hernández
Institute for Animal Science and Technology, Universitat Politècnica de València, València, Spain
e-mail: phernan@dca.upv.es
Contribution: chapter 10

László Hiripi
National Agricultural Research and Innovation Centre, Agricultural Biotechnology Institute, Gödöllő, Hungary
e-mail: hiripi@abc.hu
Contribution: chapter 15

Arata Honda
Institute of Laboratory Animals, Kyoto University Graduate School of Medecine, Yoshidakonoe-cho, Sakyo-ku, Kyoto, Japan
e-mail: a-honda@med.miyazaki-u.ac.jp
Contribution: chapter 16

Adriana Kolesárová
Department of Animal Physiology, Faculty of Biotechnology and Food Sciences, Slovak University of Agriculture in Nitra, Nitra, Slovak Republic
e-mail: adriana.kolesarova@uniag.sk
Contribution: chapter 11

Ana Lemos de Matos
Biodesign Center for Immunotherapy, Vaccines, and Virotherapy (B-CIVV), The Biodesign Institute, Arizona State University, Tempe, AZ, 85287, USA
e-mail: anamatos@asu.edu
Contribution: chapter 9

Rose G. Mage
Laboratory of Immune System Biology, National Institute of Allergy and Infectious Diseases, National Institutes of Health, Bethesda, Maryland, USA
e-mail: rmage@niaid.nih.gov
Contribution: chapter 5

Manolis Matzapetakis
Instituto de Tecnologia Química e Biológica António Xavier, Universidade Nova de Lisboa, 2780-157 Oeiras, Portugal
e-mail: matzman@itqb.unl.pt
Contribution: chapter 14

István Nagy
Department of Animal Genetics and Biotechnology, Kaposvár University, Kaposvár, Hungary
e-mail: nagy.istvan@ke.hu
Contributions: chapter 10, 13

Clive P. Page
Sackler Institute of Pulmonary Pharmacology, Institute of Pharmaceutical Science, King's College London, London, United Kingdom
e-mail: clive.page@kcl.ac.uk
Contribution: chapter 18

Mariana Palma
Instituto de Tecnologia Química e Biológica António Xavier, Universidade Nova de Lisboa, 2780-157 Oeiras, Portugal
Centre for Funtional Ecology, Departmento de Ciências da Vida, Universidade de Coimbra, Coimbra, Portugal
e-mail: marianamrpalma@gmail.com
Contribution: chapter 14

Miriam Piles
Institute for Food and Agriculture Research and Technology, Caldes de Montbui, Barcelona, Spain
e-mail: miriam.piles@irta.cat
Contribution: chapter 12, 13

Andrea Rau
Université Paris-Saclay, INRAE, AgroParisTech, GABI, 78350 Jouy-en-Josas, France
e-mail: andrea.rau@inrae.fr
Contribution: chapter 14

Anisa Ribani
Department of Agricultural and Food Sciences, Division of Animal Sciences, University of Bologna, Viale Giuseppe Fanin 46, 40127 Bologna, Italy
e-mail: anisa.ribani2@unibo.it
Contribution: chapter 1

Andreas Ritsch
Department of Internal Medicine I, Innsbruck Medical University, Innsbruck, Austria
e-mail: andreas.ritsch@i-med.ac.at
Contribution: chapter 15

Claire Rogel-Gaillard
Université Paris-Saclay, INRAE, AgroParisTech, GABI, 78350, Jouy-en-Josas, France
e-mail: claire.rogel-gaillard@inrae.fr
Contribution: chapter 5

Jun Pablo Sanchez
Institute for Food and Agriculture Research and Technology, Caldes de Montbui, Barcelona, Spain
e-mail: JuanPablo.Sanchez@irta.cat
Contribution: chapter 12

Maria Antonia Santacreu
Institut for Animal Science and Technology, Universitàt Politecnica de Valencia, Valencia, Spain
e-mail: msantacr@dca.upv.es
Contribution: chapter 12

Pierre Savatier
Univ Lyon, Université Claude Bernard Lyon 1, Inserm, Inra, Stem Cell and Brain Research Institute U1208, USC1361, Bron, France
e-mail: pierre.savatier@inserm.fr
Contribution: chapter 16

Giuseppina Schiavo
Department of Agricultural and Food Sciences, Division of Animal Sciences, University of Bologna, Viale Giuseppe Fanin 46, 40127 Bologna, Italy
e-mail: giuseppina.schiavo2@unibo.it
Contribution: chapter 3

Domenico Spina
Sackler Institute of Pulmonary Pharmacology, Institute of Pharmaceutical Science, King's College London, London, United Kingdom
Passed away
Contribution: chapter 18

Ina Sternstein
Albrecht Daniel Thaer - Institut für Agrar- und Gartenbauwissenschaften Humboldt Universität zu Berlin, Invalidenstr. 42, 10115 Berlin, Germany
e-mail: ina.sternstein.1@rz.hu-berlin.de (Retired)
Contribution: chapter 4

Zsolt Szendrő
Department of Animal Genetics and Biotechnology, Kaposvár University, Kaposvár, Hungary
e-mail: szendro.zsolt@ke.hu
Contribution: chapter 13

Valerio Joe Utzeri
Department of Agricultural and Food Sciences, Division of Animal Sciences, University of Bologna, Viale Giuseppe Fanin 46, 40127 Bologna, Italy
e-mail: valeriojoe.utzeri2@unibo.it
Contribution: chapter 1

Preface

The European rabbit (*Oryctolagus cuniculus*), also known only as 'rabbit' (the simple name commonly used in this book) or, in some contexts referring to the domesticated animals, as 'domestic rabbit', is a multi-purpose species: it is considered a livestock (for meat and fur production), a fancy animal or a pet with a broad spectrum of different phenotypes, an animal model used to address many biological questions, a bioreactor for the production of antibodies and other biomolecules and a pest or a wild resource species in several regions.

World rabbit meat production reaches about 1.5 million tons and it is, on average, increasing by 3–4% each year. Rabbit fancy breeders maintain a large collection of breeds and lines (differing in size, morphological traits, coat colour, etc.) that constitute very valuable natural genetic resources. The rabbit is one of the most common animal models in biomedicine. It is used in all research areas, ranging from basic research to clinical disciplines. Its size and its specific anatomy and biology make the rabbit a more appropriate model than the rodents in several fields. Pharmaceutical companies have developed therapeutic rabbit antibodies and transgenic rabbits for production of humanized proteins used as therapeutic agents for several disease treatments. Polyclonal and monoclonal antibodies produced from rabbits are widely marketed commercially. Its worldwide annual market is estimated to be 3–4 billion dollars with an annual growth rate of ~10–15%.

The rabbit serves as a keystone species and an ecosystem engineer in its native range. The rabbit has been introduced in South America, Australia and many islands in which it is considered a pest. The use of viruses (e.g. myxoma virus) has been tested experimentally to control these feral populations.

Domestication of the rabbit has been a progressive and recent event (from 1200 to 400 years ago or less) as compared to other livestock species, making this unique lagomorph a very interesting model to evaluate domestication processes comparing domesticated stocks and wild populations that are present together in the domestication centre (Spain–France) as well as in most other European countries. By the way, the rabbit is considered the only animal species that has been domesticated only in western Europe.

Since the rediscovery of Mendel's laws at the beginning of the 1900s, the rabbit has been a cornerstone of the genetics of the mammals. The pioneering studies of William E. Castle, who published in 1930 the first genetic textbook dedicated to the rabbit (*The Genetics of the Domestic Rabbit*), together with the studies of several other founders of this emerging discipline, shed the first light on the genetics of the rabbit, starting from explaining the variety of coat colour diversity segregating in the species.

The genomics era has recently visited the rabbit. The rabbit genome has been sequenced, opening new perspectives in all fields mentioned above, eliminating disadvantages and limits of the rabbit compared to other species and creating additional opportunities in several rabbit applied biology areas and related fields. An 'omic' vision of rabbit biology is becoming reality by integrating genomic information with transcriptomics, proteomics, metabolomics and so on.

This book is addressed to a broad audience, including students, teachers, researchers, veterinarians and rabbit breeders. The purpose of the book is to present in one location a comprehensive overview of the progress of genetics in the rabbit, with a modern vision that integrates genomics to obtain a complete picture of the state of the art and of the applications in this species, defined according to the multiple uses and multi-faceted places that this species has in applied and fundamental biology. The 18 chapters cover several fields of genetics and genomics: Chapters 1 and 2 present the rabbit within the evolutionary framework, including the systematics, its domestication and an overview of the genetic resources (breeds and lines) that have been developed after domestication. Chapters 3–5 cover the rabbit genome, cytogenetics and genetic maps and immunogenetics in this species. Chapters 6–8 present the genetics and molecular genetics of coat colours, fibre traits and other morphological traits and defects. Chapters 9–13 cover the genetics of complex traits (disease resistance, growth and meat production traits, reproduction traits), reproduction technologies and genetic improvement in the meat rabbits. Chapters 14–18 present the omics vision, the biotech and biomodelling perspectives and applications of the rabbit. The continuous progress in genetics and genomics in this species made it impossible to cover all recent and relevant literature and studies, and some publications might not have been cited. We hope that any relevant omissions can be brought to our attention.

This book is the result of international efforts among scientists that started with the European Union-funded COST Action 'A Collaborative European Network on Rabbit Genome Biology – RGB-Net', which acquired the contributions of several other specialists. I am very grateful to all of them who, with patience, have waited for the time of publication of their work. I acknowledge the professional help of CABI in producing the book that I hope will serve as a useful reference for all who study or work with this fascinating species.

The book is particularly dedicated to my family (my wife Giovanna, my son Davide and my daughter Sara) who are supporting me in the daily work of a scientist, a job that is a sort of mission, to discover and spread knowledge, as reported in this book.

Luca Fontanesi
Bologna, 8th March 2021

1 The Evolution, Domestication and World Distribution of the European Rabbit (*Oryctolagus cuniculus*)

Luca Fontanesi[1]*, Valerio Joe Utzeri[1] and Anisa Ribani[1]
[1]*Department of Agricultural and Food Sciences, Division of Animal Sciences, University of Bologna, Italy*

1.1 The Order Lagomorpha

The European rabbit (*Oryctolagus cuniculus*, Linnaeus 1758) is a mammal belonging to the order Lagomorpha.

Lagomorphs are such a distinct group of mammalian herbivores that the very word 'lagomorph' is a circular reference meaning 'hare-shaped' (Chapman and Flux, 1990; Fontanesi *et al.*, 2016). A unique anatomical feature that characterizes lagomorphs is the presence of small peg-like teeth immediately behind the upper-front incisors. For this feature, lagomorphs are also known as Duplicidentata. Therefore, instead of four incisor teeth characteristic of rodents (also known as Simplicidentata), lagomorphs have six. The additional pair is reduced in size. Another anatomical characteristic of the animals of this order is the presence of an elongated rostrum of the skull, reinforced by a latticework of bone, which is a fenestration to reduce the weight (Chapman and Flux, 1990; Rose, 2006). The herbivorous diet and the terrestrial mode of life are the primitive conditions of this order (López-Martínez, 1985). All lagomorphs are also characterized by a specific feeding behaviour known as caecotrophy, that is the re-ingestion of soft cecal-derived faeces needed to assure essential vitamin uptake, the digestion of the vegetarian diet and water reintroduction (Hörnicke, 1981).

The order Lagomorpha was recognized as a distinct order within the class Mammalia in 1912, separated from the order Rodentia within which lagomorphs were originally placed (Gidely, 1912; Landry, 1999). Lagomorphs are, however, considered to be closely related to the rodents from which they diverged about 62–100 million years ago (Mya), and together they constitute the clade Glires (Chuan-Kuei *et al.*, 1987; Benton and Donoghue, 2007). Lagomorphs, rodents and primates are placed in the major mammalian clade of the Euarchontoglires (O'Leary *et al.*, 2013).

Modern lagomorphs might be evolved from the ancestral lineage from which derived the †Mimotonidae and †Eurymilydae sister taxa, following the Cretaceous-Paleogene (K-Pg) boundary around 65 Mya (Averianov, 1994; Meng *et al.*, 2003; Asher *et al.*, 2005; López-Martínez, 2008).

The systematics of the order are not completely clear yet and currently under revision by the International Union for Conservation of Nature and Natural Resources (IUCN), Species Survival Commission (SSC), Global Mammal Assessment, and Lagomorph Specialist Group (LSG). The order is divided into two families

Contact e-mail address: luca.fontanesi@unibo.it

DOI: 10.1079/9781780643342.0001

(Fig. 1.1; Hoffmann and Smith, 2005; Chapman and Flux, 2008; Fontanesi *et al.*, 2016): (i) Ochotonidae (the pikas), with 26 teeth, and (ii) Leporidae (the jackrabbits and hares, and the rabbits), with 28 teeth. According to the most accepted taxonomy, these two families account for a total of 92 living species (Chapman and Flux, 1990; Fontanesi *et al.*, 2016; Melo-Ferreira and Alves, 2018).

Fossil and molecular estimates are not completely in agreement in the identification of the evolutionary split between these two extant families. Fossil data seem consistent in indicating that the divergence period occurred before 42 Mya (Ruedas *et al.*, 2018) even if molecular estimates based on mitochondrial and nuclear genome sequences have placed this family split in a more recent period (about 30 Mya; Matthee *et al.*, 2004). Merging fossil data and molecular estimates, the differentiation time has been placed around 50 Mya (Meredith *et al.*, 2011).

The list of all recognized extant lagomorph species is reported in Table 1.1. The family Ochotonidae comprises 29 species of small (70–300 g) egg-shaped mammals with distinct rounded ears and no visible tail. All species have been assigned to the genus *Ochotona*. They have hind legs not much longer than the fore legs. Most of these species are vocal. There are two major groups of pikas (Smith, 1988; Smith *et al.*, 1990; Lissovsky, 2014): (i) those that live in rocks or talus have low reproductive rates and are generally long-lived; they are territorial either as individuals or pairs and have extremely low rates of social interaction; and (ii) those that live in meadow or steppe habitat and burrow; they have high reproductive rates and are generally short-lived; these species form extended families and are highly social. These groups have been divided into five subgenera (Alienauroa, Conothoa, Lagotona, Ochotona, and Pika), commonly recognized mainly considering their distribution (Hoffmann and Smith, 2005; Lissovsky, 2014). A recent phylogenomic analysis recognized four subgenera of extant pikas (Alienauroa, Conothoa, Ochotona and Pika), with the mountain group Conothoa being the sister group of all other pikas (Wang *et al.*, 2020).

The family Leporidae comprises 32 species of hares (genus *Lepus*) and 31 species of rabbits (Fig. 1.1; Table 1.1). The hares are the largest lagomorphs (2–5 kg), having long ears and hind legs. Of the group of rabbits, 18 species belong to the genus *Sylvilagus* whereas the remaining species are mainly included in monotypic genera (Fig. 1.1; Table 1.1), some of which are Evolutionarily Distinct and Globally Endangered (EDGE) species (Verde Arregoitia *et al.*, 2015). The rabbits have also somewhat long ears, but not as long as the hares, and present a more rounded body type. Other rabbits include a variety of unique forms found around the world: for example, the Riverine rabbit (*Bunolagus monticularis*) in South Africa; the

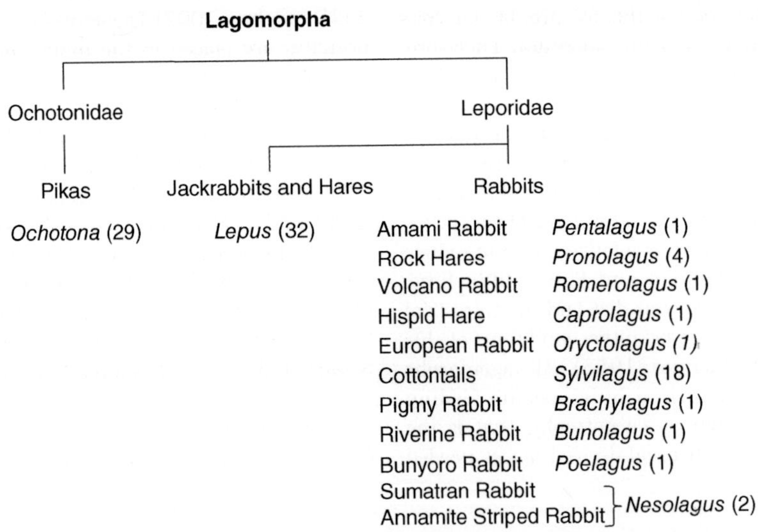

Fig. 1.1. Schematic representation of systematics of the order Lagomorpha (number in parenthesis indicates the number of species within the genera). (Modified from Fontanesi *et al.*, 2016)

Table 1.1. List of lagomorph species. (From Chapman and Flux, 1990; Fontanesi *et al.*, 2016; Melo-Ferreira and Alves, 2018)

Family/genus	Species	Common name
Ochotonidae		
Ochotona	*O. alpina*	Alpine Pika
	O. argentata	Helan Shan Pika
	O. cansus	Gansu Pika
	O. collaris	Collared Pika
	O. coreana	Korean Pika
	O. curzoniae	Plateau Pika
	O. dauurica	Daurian Pika
	O. erythrotis	Chinese Red Pika
	O. forresti	Forrest's Pika
	O. gloveri	Glover's Pika
	O. hoffmanni	Hoffmann's Pika
	O. hyperborea	Northern Pika
	O. iliensis	Ili Pika
	O. koslowi	Koslov's Pika
	O. ladacensis	Ladak Pika
	O. macrotis	Large-eared Pika
	O. matchurica	Mantchurian Pika
	O. nubrica	Nubra Pika
	O. opaca	Kazakh Pika
	O. pallasii	Pallas's Pika
	O. princeps	American Pika
	O. puslla	Steppe Pika
	O. roylei	Royle's Pika
	O. rufescens	Afghan Pika
	O. rutila	Turkestan Red Pika
	O. syrinx	Qinling Pika
	O. thibotana	Moupin Pika
	O. thomasi	Thomas's Pika
	O. tutuchanensis	Turuchan Pika
Leporidae		
(the rabbits)	*Brachylagus idahoensis*	Pygmy Rabbit
	Bunolagus monticularis	Riverine Rabbit
	Caprolagus hispidus	Hispid Hare
	Nesolagus netscheri	Sumatran Striped Rabbit
	Nesolagus timminsi	Annamite Striped Rabbit
	Oryctolagus cuniculus	European rabbit
	Pentalagus furnessi	Amami Rabbit
	Poelagus majorita	Bunyoro Rabbit
	Pronolagus crassicaudatus	Natal Red Rock Hare
	Pronolagus randensis	Jameson's Red Rock Hare
	Pronolagus rupestris	Smith's Red Rock Hare
	Pronolagus saundersiae	Hewitt's Red Rock Hare
	Romerolagus diazi	Volcano Rabbit
Sylvilagus	*S. aquaticus*	Swamp Rabbit
	S. audubonii	Desert Cottontail
	S. bachmani	Brush Rabbit
	S. brasiliensis (sensu stricto)	Tapetí, Andean Cottontail, Rio de Janeiro Dwarf Cottontail
	S. cognatus	Manzano Mountain Cottontail
	S. cunicularius	Mexican Cottontail
	S. dicei	Dice's Cottontail
	S. floridanus	Eastern Cottontail

Continued

Table 1.1. Continued.

Family/genus	Species	Common name
	S. gabbi	Gabb's Cottontail
	S. graysoni	Tres María Cottontail
	S. insonus	Omiltemi Rabbit
	S. mansuetus	San José Brush Rabbit
	S. nuttallii	Mountain Cottontail
	S. obscurus	Appalachian Cottontail
	S. palustris	Marsh Rabbit
	S. robustus	Davis Mountains Cottontail
	S. transitionalis	New England Cottontail
	S. varynaensis	Venezuelan Lowland Rabbit
Lepus	*L. alleni*	Antelope Jackrabbit
	L. americanus	Snowshoe Hare
	L. articus	Arctic Hare
	L. brachyurus	Japanese Hare
	L. californicus	Black-tailed Jackrabbit
	L. callotis	White-sided Jackrabbit
	L. capensis	Cape Hare
	L. castroviejoi	Broom Hare
	L. comus	Yunnan Hare
	L. coreanus	Korean Hare
	L. corsicanus	Corsican Hare
	L. europaeus	European Hare
	L. fagani	Ethiopian Hare
	L. flavigularis	Tehuantepec Jackrabbit
	L. granatensis	Iberian Hare
	L. habessinicus	Abyssinian Hare
	L. hainanus	Hainan Hare
	L. insularis	Black Jackrabbit
	L. mandshuricus	Manchurian Hare
	L. nigricollis	Indian Hare
	L. oiostolus	Woolly Hare
	L. othus	Alaskan Hare
	L. peguensis	Burmese hare
	L. saxatilis	Cape Scrub Hare
	L. sinensis	Chinese Hare
	L. starcki	Ethiopian Highland Hare
	L. tibetanus	Desert Hare
	L. timidus	Mountain Hare
	L. tolai	Tolai Hare
	L. townsendii	White-tailed Jackrabbit
	L. victoriae	African Savanna Hare
	L. yarkandensis	Yarkland Hare

Hispid hare (*Caprolagus hispidus*) of the Terai region of India and Nepal; the black Amami Island rabbit (*Pentalagus furnessi*) that occupies isolated islands in the far south of Japan; the Annamite striped rabbit (*Nesolagus timminsi*) of south-east Asia; the Volcano rabbit (*Romerolagus diazi*) that lives at high elevations on volcanoes surrounding Mexico City; and the European rabbit (*Oryctolagus cuniculus*) that originally occupied the Iberian Peninsula and that was subsequently spread in many other regions.

1.2 The European Rabbit (*Oryctolagus cuniculus*)

The European rabbit (*Oryctolagus cuniculus*) includes both wild and domesticated animals.

Oryctolagus cuniculus is also the only domesticated species of the order Lagomorpha and the only example of an animal domestication process that occurred exclusively in western Europe. Its distribution has been largely expanded and modified from its ancestral area (Flux, 1994), mainly recognized in the Iberian Peninsula, where the species might have been emerged in the mid-Pleistocene (López-Martínez, 2008). The subsequent geographical expansion of this species and the consequent successful colonization of a large variety of regions and ecosystems are mainly human-derived and occurred during historical times (Thompson and King, 1994). This species has also the most within-species phenotypic diversity due to human-driven artificial directional selection that produced many different breeds and lines, as a consequence of its domestication.

The wild original forms are mainly distributed in its native and contiguous regions (the Iberian Peninsula and the south of France). Feral forms, derived by repeated human-mediated dispersal of domesticated rabbits into the wild (and, in some cases, together with dispersal of wild forms), are present in many islands and continents, including Australia and New Zealand (Flux, 1994), often with devastating consequences for native flora and fauna and the agricultural system, leading some countries to fight the problem with drastic methods (Fenner and Ross, 1994; Williams *et al.*, 1995; Vere *et al.*, 2004; Cooke *et al.*, 2013; Pedler *et al.*, 2016).

Nowadays, the European rabbit is considered one of the most widespread species globally even if it was not able to colonize most of Africa and the north of America (Thompson and King, 1994). Despite attempts of introduction in these continents, probably the competition with other leporid species that exist in these territories and/or the potential natural control that might have been derived by the sensitivity of *O. cuniculus* to coexisting lagomorph pathogens could explain the incapacity of the European rabbit to spread into these areas (Cooke *et al.*, 2018).

The evolutionary history, domestication and expansion of the European rabbit can be inferred from fossil and archaeological records, molecular information and historical documents, which together can contribute to reconstructing remote and more recent events that shaped this species and contributed to its successful (fortunate or unfortunate) colonization of the globe and its shift to a domesticated species.

1.2.1 The evolutionary history of the European rabbit: fossil records

Fossil records indicate that the Iberian Peninsula is the most probable ancestral area of the European rabbit. The oldest fossils attributed to the modern *Oryctolagus cuniculus* species identified in this peninsula resulted from the Middle Pleistocene, around 0.6 Mya (Donard, 1982; López-Martínez 1989, 2008). This is in some ways in contrast to the ~2 Mya estimated by molecular data that traced back the divergence of two morphologically similar rabbit populations coexisting in different parts of the Iberian Peninsula (Branco *et al.*, 2000).

The first recorded *Oryctolagus* species (†*O. laynensis*) was from Middle Pliocene (about 3.5 Mya) from Spain and possibly from the South-east of France (López-Martínez, 2008). Despite the fact that the biogeographical origin of *Oryctolagus* seems confirmed by several evidences, the phylogenetic origin of the genus is not completely clear. Precursors of *Oryctolagus* in its ancestral area and its surroundings were the widespread Leporinae †*Alilepus* and after †*Trischizolagus*, the latter being proposed as the direct ancestor of the *Oryctolagus* and *Lepus* genera (López-Martínez, 2008). Phylogenetic analyses based on mitochondrial DNA sequences gave other contradictory results, linking *Oryctolagus* either to *Lepus*, *Bunolagus* or *Caprolagus* (Halanych and Robinson, 1999; Matthee *et al.*, 2004). From the fossil records, *Oryctolagus* and *Lepus* did not share a common ancestral area, although both might be related to their ancestor genera †*Alilepus* and †*Trischizolagus*. No transitional forms are known between †*O. laynensis* and the modern *O. cuniculus*. Other *Oryctolagus* species, †*O. lacosti* (recorded between 2.5 and 0.6 Mya in Spain, France, Italy and in some isolated sites in Hungary and Greece) and †*O. burgi* (described in Middle Pleistocene of Italy) have more differences than that between †*O. laynensis* and the modern *O. cuniculus* (López-Martínez, 2008). In the Middle Pleistocene of the central Iberian Peninsula and in France, two subspecies or separated groups, contemporaneous with the large-sized lineage †*O. lacosti*-†*O. burgi*, have been recorded: †*O. c. lunelensis* and †*O. c. grenalensis*. Only the modern rabbit survived in the late Pleistocene, spreading to northern Europe and peri-Mediterranean areas. Then, during the maximum glacial period and Early Holocene, the European rabbit returned to be confined to the Iberian Peninsula and southern France (López-Martínez, 2008).

1.2.2 The evolutionary history of the European rabbit: molecular data

The first studies that applied molecular data to obtain information useful to analyse the evolutionary dynamics of *O. cuniculus* were based on biochemical markers (Coggan *et al.*, 1974; Richardson *et al.*, 1980; Arana *et al.*, 1989; Ferrand and Rocha, 1992; Peterka and Hartl, 1992). These markers and the designs of the related studies were partially informative and could provide first-hand information on the variability present in the different rabbit populations that preliminarily suggested the partition of the existing diversity within the European rabbit species.

Subsequent studies that analysed the mitochondrial genome in wild European rabbits identified two mitochondrial DNA (mtDNA) lineages (clades A and B) with about 4.5% nucleotide divergence (Ennafaa *et al.*, 1987; Biju-Duval *et al.*, 1991; Monnerot *et al.*, 1994). These maternally derived genetic lineages were reported to overlap to two subspecies (*O. c. algirus* and *O. c. cuniculus*) distinguished by slight phenotypic differences in size and cranial morphology (Sharples *et al.*, 1996), also matching preliminary genetic partitioning determined by nuclear polymorphic loci (van der Loo *et al.*, 1991, 1999). These two subspecies are distributed parapatrically: *O. c. algirus*, originally localized in the south-west of the Iberian Peninsula; and *O. c. cuniculus*, originally localized in the north-east of the Iberian Peninsula and south of France (Biju-Duval *et al.*, 1991; Monnerot *et al.*, 1994; Branco *et al.*, 2000, 2002). The *O. c. cuniculus* population from the south of France shares the type B mitochondrial lineage with the population in north-east Iberia but with a reduced level of genetic diversity that is consistent with a bottleneck effect derived by the subsequent expansion of this Iberian subspecies in the close France region (Monnerot *et al.*, 1994; Branco *et al.*, 2000, 2002; Carneiro *et al.*, 2014a).

The two subspecies of the European rabbit (*Oryctolagus cuniculus*) provide a window into the early stages of speciation. This phylogeographical pattern suggests that two groups of European rabbits were isolated for a certain time, creating the conditions for independent evolutionary trajectories. The separation might be derived by the retreat into glacial refugia during Quaternary ice ages. Later, these isolated groups expanded and partially overlapped when barriers were eliminated over the post-glacial period that recreated the possibility for a recolonization of the Iberian Peninsula (Branco *et al.*, 2000). Therefore, throughout the Pleistocene, these two subspecies likely experienced multiple events of isolation and contact following climatic change dynamics. Mitochondrial DNA data suggested a history of large and stable populations associated with clade A, allowing the diversification and maintenance of many sub-haplotype variants. A history of small or fluctuating and possibly more fragmented populations is explained by the phylogeographical pattern observed for the sub-haplotypes of clade B (Branco *et al.*, 2000, 2002).

The *O. c. cuniculus* population from the South of France shares the type B mitochondrial lineage with the population in the north-east Iberia but with a reduced level of genetic diversity that is consistent with a bottleneck effect caused by the subsequent colonization of clade B Iberian subspecies of the close French region (Branco *et al.*, 2000; Carneiro *et al.*, 2014a). The lower levels of genetic diversity identified in the south of France compared to that of the other side of the Pyrenees (reported using both mtDNA and microsatellite markers) indicate that rabbit populations in that region could have experienced a few cycles of extinction and recolonization, probably derived by recurrent climatic modifications in this area (Hardy *et al.*, 1995; Queney *et al.*, 2001; Ferrand, 2008; Alves *et al.*, 2015). This colonization process led to about 12% reduction of genetic diversity estimated using microsatellite data (Alves *et al.*, 2015).

The current contact and natural overlapping zone between the two mtDNA lineages, further confirmed by analyses of nuclear genome variability, including Y and X chromosome markers (Geraldes *et al.*, 2006, 2008; Carneiro *et al.*, 2009, 2010, 2013), bisects the Iberian Peninsula along a diagonal that goes from north-west to south-east (Fig. 1.2). This 'hybrid zone' has been well characterized at the level of DNA variability, indicating some contrasting patterns of differentiation at multiple loci and according to the types of investigated markers (i.e. mtDNA, nuclear genome markers in sex chromosomes or close to centromeres; Branco *et al.*, 2000; Geraldes *et al.*, 2006, 2008; Carneiro *et al.*, 2009, 2010, 2013; Alda and Doadrio, 2014). This picture suggests that the differences between the two subspecies are maintained by a balance between dispersal and natural selection against hybrids (Carneiro *et al.*, 2013).

Carneiro *et al.* (2014b) analysed extensive nuclear genome variants distributed on all rabbit chromosomes (about 300 k polymorphic sites) and reported low to moderate overall levels of differentiation between the two subspecies. Only ~200 genomic regions, dispersed throughout the genome, showed high differentiation, consistent with a signature of reduced gene flow between the two subspecies (Carneiro *et al.*, 2014b). Differentiated chromosome regions had

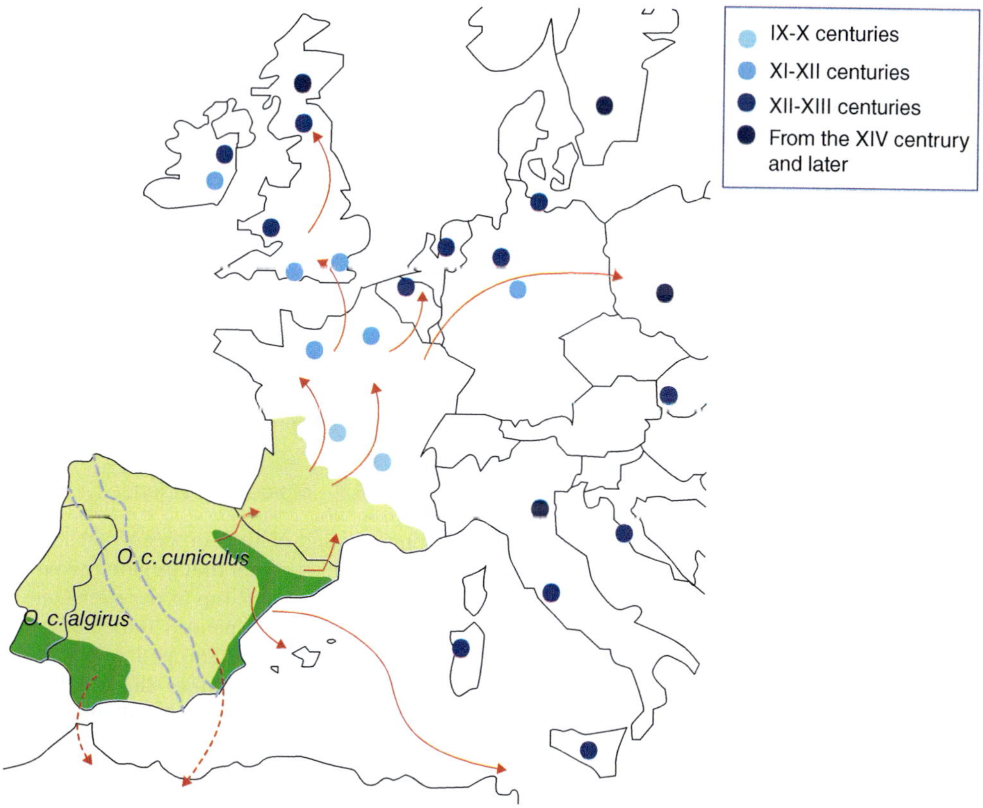

Fig. 1.2. Map of the original distribution of the rabbit and of the Medioeval dispersal in western Europe. The pale green area indicates the natural range of the rabbit derived by the constraints that occurred during the Last Glacial Maximum, with the approximate areas of the two subspecies indicated (*O. c. algirus* and *O. c. cuniculus*), divided by the two dashed lines, which border the overlapping region. The dark green areas indicate the putative refugial areas for the two main populations. The arrows indicate both natural and human-mediated diffusion. The coloured dots summarize the information on the historically and archaeologically documented appearance and transfer of the western Europe. (Adapted from information reported in Ferrand and Branco, 2007; Ferrand, 2008; Irving-Pease *et al.*, 2018a, and other literatures cited in the text)

small size and usually were smaller than 200 kb and contained very few genes. Regions of high differentiation were enriched on the X-chromosome and near centromeres. This picture was completed by a subsequent study that analysed whole genome-resequencing data of the two subspecies that display partial reproductive isolation (Rafati *et al.*, 2018). Geographic cline analysis was able to identify about 250 genomic regions characterized by steep changes in allele frequency across the natural geographic region of contact of the two subspecies (Rafati *et al.*, 2018). These chromosome portions included genes that might cause some reproductive dysfunctions in the hybrids raising the hypothesis that incomplete reproductive barriers are determined by the effects of many loci and that regulatory variants (and not large chromosomal rearrangements) are likely the primary factors that determine reproductive incompatibilities (Rafati *et al.*, 2018).

This original geographic distribution in the Iberian Peninsula of the two subspecies has been in part altered as a consequence of very recent human activities (in the 1980s and 1990s) carried out by hunters and conservationists that translocated rabbits of lineage A in several localities within the distribution area of lineage B, and vice versa (Delibes-Mateos *et al.*, 2008).

1.2.3 The domestication process of the European rabbit

All definitions of animal domestication indicate a relationship between humans and target animal populations (Zeder, 2006). Animal domestication can be considered a long-term and multi-stage process that gradually has led to morphological, biological and behavioural changes in the animal populations by means of directional selection, introgression and admixture and by starting from the wild ancestral counterparts (Larson and Burger, 2013). Not all animal species followed the same trajectory in this process. Three domestication pathways have been described (Zeder, 2012): the commensal pathway, when wild animals habituated to humans after being attracted by their waste; the prey pathway, when animals were initially hunted and then managed by humans; and the directed pathway that does not involve the preliminary steps of habituation or management and begins with the capture of wild animals with the aim to control their breeding and reproduction. This pathway took place over much shorter timeframes and was accompanied by a bottleneck. The domestication process of the European rabbit probably followed the directed pathway (Larson and Burger, 2013; Irving-Pease *et al.*, 2018a). The European rabbit is a burrowing species and this hiding characteristic probably facilitated the possibility to tame some of these animals while a superficially similar species like a hare, which depends on speed for escape, could not be easily tamed, and this aspect could explain why it was not targeted for domestication despite its similarities with the rabbit (Clutton-Brock, 2012).

The domestication process of the European rabbit can be described at the molecular level. The case of the European rabbit is unique among all domestic animal species as the ancestral wild population from which the domestication process derived is still alive and it can be compared in parallel to the domesticated populations. As the process occurred in historical time, zooarchaeological discoveries and historical sources can complement molecular information. The first waves of spread in Europe of the European rabbit can also be considered, at least in part, to be elements of this process.

1.2.3.1 Molecular evidences

The domestication of the European rabbit was based on the genetic pool that originally colonized the south of France (up to the river Loire) and that subsequently expanded in the north of France and the north of Europe by means of human translocation activities that occurred mainly during the Middle Age (Callou, 1995, 2003). The domestication process led to a second subsequent reduction of genetic diversity, after that which occurred when the south of France wild populations were derived from the Iberian populations.

All rabbits of domestic breeds belong to the B mitochondrial lineage (Biju-Duval *et al.*, 1991; Monnerot *et al.*, 1994; Queney *et al.*, 2002). Mitotypes identified in the domestic rabbits are a subset of the B haplotypes identified in the north-east Iberian wild rabbits and then identified in the south-west of France supporting that the domestication occurred from the genetic material that colonized France (Monnerot *et al.*, 1994; Hardy *et al.*, 1995; Queney *et al.*, 2002). The most frequent mitotype in the current European domestic populations is B1, followed by a few other haplotypes (Queney *et al.*, 2002). Similar patterns have also been observed in breeds and domestic populations of China, Egypt and Kenya, confirming that they share the same root of European domestication and subsequent exportation from Europe (Long *et al.*, 2003; Emam *et al.*, 2016; Owuor *et al.*, 2019).

The reduction of genetic diversity observed at the mitochondrial DNA level was also evidenced using protein markers and other nuclear genome markers, i.e. microsatellites and single nucleotide polymorphisms (Queney *et al.*, 2002; Ferrand and Branco, 2007; Carneiro *et al.*, 2011, 2014a; Alves *et al.*, 2015). Microsatellite data estimated that the initial domestication process accounted for losses of about 20% of the pre-existing levels of genetic diversity in the French wild population (Alves *et al.*, 2015) whereas sequencing data estimated a much larger reduction that was about the double of that estimated with microsatellite markers (Carneiro *et al.*, 2011, 2014a). This discrepancy might be due to different statistical properties of the two methods used to estimate genetic diversity, to the higher mutation rate of microsatellites in comparison with nucleotide substitutions that enabled a faster reconstitution of genetic diversity levels in domestic rabbits and to different numbers of domestic animals that were investigated in the two studies (that could or could not have captured a real picture of the genetic variability available in the domestic populations) (Alves *et al.*, 2015). The reduction of genetic diversity is compatible with the small effective population size that might be derived by the fact that the early rabbit domestication occurred in a geographically and temporally defined region and period (i.e. France in the Middle Age and close to the monasteries or castles) (Zeuner, 1963; Callou, 2003) and that subsequent recurrent backcrosses with the wild ancestors (described frequently in other domestic animals, i.e. pigs and dogs) might not have been very frequent in rabbits (Alves *et al.*, 2015). Recurrent backcrosses with wild rabbits would have limited the genetic distance between the wild and domestic populations. There is also no clear evidence for a highly unequal contribution of males and females to the domestic rabbit gene pool (Carneiro *et al.*, 2011) even if this matter has not been studied into detail.

However, estimating the exact time of domestication would require sampling the wild population from which domestic rabbits arose (which is not obvious in case of unclear population divergence during the domestication process) and the conversion of molecular time estimates into precise temporal periods would require robust mutational rates that are difficult to calculate. Therefore, molecular dating approaches to domestication should be critically evaluated considering other evidences and the domestication process as a gradual and continuous selective process (Irving-Pease *et al.*, 2018a).

Whole-genome resequencing analysis has shown that very few loci have gone to complete fixation in domestic rabbits compared to the wild ancestors and none of the fixed variants is in coding sites or at non-coding conserved sites (Carneiro *et al.*, 2014a). Allele frequency shifts in the domestic gene pools, however, were detected at many loci spread across the genome. Almost all domestic alleles (indicated in this way because of their higher frequency in the domestic rabbits) were also found in wild rabbits, implying that directional selection events associated with rabbit domestication are consistent with polygenic and soft-sweep modes of selection that primarily acted on standing genetic variation in regulatory regions of the genome (Carneiro *et al.*, 2014a). Many of the domestic alleles are in regulatory regions of genes affecting brain and neuronal development. This indication is consistent with the view that the most critical phenotypic changes during the initial steps of animal domestication probably involved behavioural traits that allowed animals to tolerate humans and the environment humans offered and involved adaption of the reproduction cycles to the new production systems (Carneiro *et al.*, 2014a; 2015). The paucity of specific fixed domestication genes in rabbits can be interpreted in the direction that no single genetic change was either necessary or sufficient for the domestication of this species. This is also in line with the current view that the tame behaviour has a complex genetic component and that the domestication of the rabbit occurred as a consequence of the effect of many mutations of small effects, rather than by changes at only a few loci with large effects. This conclusion might serve as model for the domestication of other species (Carneiro *et al.*, 2014a). Changes in brain architecture were probably the results of this genetic shift that occurred during the domestication process. High-resolution brain magnetic resonance imaging showed that domestication reduced amygdala volume and enlarged medial prefrontal cortex volume, supporting that areas driving fear have lost volume while areas modulating negative effect have gained volume during the domestication of the rabbit (Brusini *et al.*, 2018).

Another reduction of genetic diversity of about 20% was derived by the process of breed formation (this estimate is averaged across breeds; Alves *et al.*, 2015) after or concurrent to the domestication of the rabbit. This high level of reduction (similar to reduction of genetic diversity occurring during the first step of domestication) might be due to the fact that the constitution of breeds is usually an extreme process that includes strong founder and bottleneck effects, artificial selection and, again, reduction of effective population size (Alves *et al.*, 2015). Considering that the domestication of the rabbit occurred in historical time, combining historical records with molecular data it is tempting to consider that both domestication and breed formation might have been continuous and unseparated processes in the domestication history of this species. However, domestic rabbits exhibit a clear and detectable genetic substructure corresponding to the breeds which are genetically well differentiated mainly due to changes in allele frequencies (Alves *et al.*, 2015). This might be derived by the short generation interval and the high selection pressure that can be applied in this species and that can quickly modify allele frequency structures in rabbit populations. Similar levels of genetic diversity observed across breeds (Queney *et al.*, 2002) are consistent with historical records which indicate that the most modern rabbit breeds have also been derived by crossbreeding between pre-existing varieties or morphs and that outcrossing has been frequently used for introgression of desirable coat colour variants into other varieties (Whitmann, 2004; Alves *et al.*, 2015).

Important molecular signs of domestication can be considered to be the occurrence and spread of coat colour mutations that constitute the most important distinctive morphological trait (i.e. coat colour) of most rabbit breeds (see Chapter 6).

1.2.3.2 The first waves of rabbit distribution in Europe: archaeological and historical sources

As mentioned, the natural range of *O. cuniculus* derived by the constraints that occurred during the Last Glacial Maximum was the Iberian Peninsula and south-west France. This is well supported by fossil and zooarchaeological records (Donard, 1982; Callou, 1995; López-Martinez, 2008). This area is regarded as the only land occupied by the European rabbit until Classic Antiquity when, as a consequence of transportation by men, this species was first spread in the west and central Mediterranean basin, in a large part of Europe and in several other regions of the globe (Flux and Fullagar, 1992; Flux, 1994).

A few issues should be considered for the interpretation of zooarchaeological records that identify European rabbit specimens: in most cases they are incomplete and should be regarded with caution, particularly if they would be the only elements useful to construct historical trajectories; the burrowing behaviour of this species could complicate the identification based only on stratigraphy records; European rabbit bones could be easily misclassified as derived from hares, complicating the interpretation of the reports of several sites; and excavation strategies could systematically miss the presence of rabbits if sieving for small bones is not applied (Callou, 2003; Irving-Pease *et al.*, 2018b). Moreover, few ancient DNA studies have been carried out to link zooarchaeological information with molecular data, leaving some doubts and unclarified questions on the presence of European rabbit bones in explored sites (Hardy *et al.*, 1994a, 1994b, 1995; Monnerot *et al.*, 1994).

Analysis of bone remains from excavations throughout Iberia showed that this lagomorph was a crucial part of the diet of Anatomically Modern Humans but was relatively underutilized during the Mousterian, when Neanderthals were present. Game biomass from this small mammal that was abundant in this region contributed to feed hunters in a period of dramatic loss of large mammalian fauna (Fa *et al.*,

2013). Remains from the oldest Epipalaeolithic sites (11500–9300 cal BC) of the Iberian Peninsula demonstrated that hunting activity was based specifically on capturing European rabbits that were represented by more than 90% of the total number of identified specimens in 74% of the explored sites (17 out of 23) (Saña, 2013). It seems that in this period there could be some significant regional differences related to hunting strategies in Iberia and that rabbit hunting was more frequent in the eastern peninsula. Analysis of Mesolithic remains (9300–5700 cal BC) showed a significant drop in the number of sites specialized in rabbit hunting in this peninsula and only five out of 19 (26%) could be classified according to this specialized hunting strategy (Saña, 2013). The trend of decreasing exploitation of rabbits from the Epipalaeolithic to the Mesolithic period has also been documented in the south-west and south-east of France (Cochard and Brugal, 2004). This decrease also continues in all the Iberian Neolithic sites (5700–2500 cal BC) indicating that in this period there was a consolidation of the shift towards a greater exploitation of animal domestic resources, i.e. on pigs and cattle (Saña, 2013).

Excavations undertaken in France did not show any rabbits in all deposits of the Holocene (or earlier) in the regions north of the Loire, whereas rabbit bones were abundant in Charente Maritime, Gironde, Haute Garonne, Hérault, Bouches-du-Rhône, Ain, Dordogne, Corrèze and Ardèche (Donard, 1982; Rogers *et al.*, 1994) some of which were used for ancient DNA analyses (Hardy *et al.*, 1994a, 1994b; 1995).

It is not completely clear what roles Phoenicians, and later the Romans, played in the early distribution of the European rabbit in the Mediterranean area even if several scholars reported that these peoples might have largely contributed to the first waves of colonization over the first millennium BC and a few centuries later. The earliest reported introduction of European rabbits in a Mediterranean island was estimated to occur around the 14th–13th century BC as suggested by bones identified in the Menorca island (Balearic Archipelago) (Reumer and Sanders, 1984). This introduction might be derived by ancient settlers from the Iberian mainland, at the time of the Talayotic culture, the settlers that lived on the island before the Romans (Reumer and Sanders, 1984; Sanders and Reumer, 1984). The Romans conquered the archipelago around 1100 BC and might have contributed to the introductions in these Mediterranean islands. The introduced rabbits were suggested to be of the *O. c. algirus* subspecies (morphologically classified in another subspecies, i.e. *O. c. huxleyi*, that is not distinct at the molecular level and thus not recognized), also indicated to be widely distributed by the Phoenicians in the Mediterranean basin (Zeuner, 1963; Robinson, 1984; Gibb, 1990). Analysis of mtDNA of modern European rabbits of another Balearic island (Mallorca) indicted that they belonged to the mitochondrial clade B, therefore the *O. c. algirus* subspecies is not currently present in this island (Seixas *et al.*, 2014). Nine haplotypes were found among the modern Mallorcan rabbits (already identified in wild rabbits from Spain and in France) suggesting a complex dynamic of introductions that may reflect recurrent waves of faunal replacements induced from repeated man-mediated translocations (Seixas *et al.*, 2014).

Archaeological reports in these islands showed that rabbits were present much earlier than the date indicated by historical sources that reported the presence of rabbits in the same islands. Strabo, a Greek geographer and historian (64–63 BC – 20–25 AD), in his work *Geography* (Γεωγραφικά – 14–23 AD), reported the introduction in a Balearic Island of a pair of rabbits from the opposite continent (i.e. Spain). According to what he reported (Volume III, Chapter V), the rabbit population that derived from this introduction became so numerous that it impacted negatively on the vegetation and stability of the houses by burrowing beneath and the inhabitants were forced to request the support of the Romans. A similar tale was also reported in the same volume (Chapter II) indicating the damage that rabbits infesting the whole of Iberia, reaching Marseilles and several islands. In this text it was indicated that 'formerly the inhabitants of the Gymnesian islands [Mallorca and Menorca] sent a deputation to the Romans soliciting that a new land might be given them, as they were quite driven out of their country by these animals, being no longer able to stand against their vast multitudes'.

This story is also reported by the Roman author Gaius Plinius Secundus, known as Pliny the Elder (23–79 AD), in his encyclopedic *Naturalis*

Historia [Volume VIII, Chapter 81 (55)]. He wrote about a species of hare, in Spain, which is called coney (rabbit, cuniculus) 'that is extremely prolific and produced famine in the Balearic Islands, by destroying the harvests'. The inhabitants of these islands begged the Emperor Augustus Caeser the aid of Roman soldiers to counter the too-rapid increase of these animals. In his text, he mentioned the use of ferrets to catch the rabbits in their burrows and also the *laurices*, considered a delicate food based on the young rabbits ('either when cut from out of the body of the mother, or taken from the breast, without having the entrails removed'). Evidence of the strong association between Spain and rabbits during the Roman period is also deduced from the Roman coins under the Emperor Hadrian (117–138 AD) in which Hispania (the Latin name of Spain) is written in the face together with a rabbit and a woman holding an olive branch, as symbols of fertility and of this country.

In the central Mediterranean basin, zooarchaeological remains date the presence of the European rabbit in a period that ranged from the Bronze Age to the 2nd–3rd century AD (Massetti and De Marinis, 2008). Remains were identified to be related to this time window in a few islands: the site of Mursia in the Pantelleria island (Sicilian Channel) dated back to the late Bronze Age (Wilkens, 1987); the sanctuary of Juno at Tas Silg on Malta (1st century BC– 1st century AD); the archaeological sites of the islands of Nisida and Capri in the Gulf of Naples and of the island Zembra in Tunisia (Barrett-Hamilton, 1912; Vigne, 1988; Albarella, 1992; Massetti and De Marinis, 2008).

Ancient DNA analysis of European rabbit remains excavated in Zembra (200–600 cal AD) and mtDNA analysis of modern European rabbits sampled on the same island agreed on the presence of only B haplotypes of the *O. c. cuniculus* subspecies in the two periods, i.e. late Roman Empire time and present time (Ennafaa *et al.*, 1987; Hardy *et al.*, 1994a, 1994b, 1995; Monnerot *et al.*, 1994), matching what was reported for the modern samples of the Mallorca island. Therefore, multiple evidences agreed to support that the spread of European rabbits in the Mediterranean area was based on *O. c. cuniculus* subspecies, despite what was previously supposed (Zeuner, 1963; Robinson, 1984; Gibb, 1990; Flux, 1994). More recent introduction or re-introduction events could also have happened and/or escapes from domestic stocks might also have occurred together with recorded events of extinctions in the Mediterranean regions (Flux and Fullagar, 1992; Massetti and De Marinis, 2008).

An interesting match between zooarchaeological records and a classical historical source is also available for the presence of the rabbits in the island of Nisida (mentioned above). Athenaeus, a Greek author who lived at the end of the 2nd and the beginning of the 3rd century AD, in his *Deipnosophistae* (Δειπνοσοφισταί; IX, 63) mentioned that voyagers 'have seen a great many [κούνικλος, rabbit] in ...[a] voyage from Dicæarchia to Naples...there is an island not far from the mainland, opposite the lower side of Dicæarchia [Nisida], inhabited by only a very scanty population, but having a great number of rabbits'.

The European rabbit was also imported in other eastern Mediterranean islands in the last period of the Roman Empire and at the beginning of the Middle Ages (Massetti and De Marinis, 2008). The fact that Classical Greek authors like Xenophon (*c.*430–354 BC) and Aristotle (384–322 BC) did not mention rabbits in their writings has been interpreted as an indirect demonstration of the fact that these animals were not known in this region at that time and support a later introduction (Zeuner, 1963).

Other Classical sources mentioning the presence of the European rabbit in the Mediterranean basin are from: (i) a doubtful citation of Polybius (*c.*200–118 BC) who in his *Histories* (Ἱστορίαι, vol. XII, 3.8–4.6) reported in Corsica the presence of a type of hare (*kyniclos*, translated as rabbit) that could more plausibly be identified as the extinct *Prolagus sardus*, which was probably the only lagomorph present in this island and in Sardinia at that time (Massetti and De Marinis, 2008); (ii) the report of Marcus Terentius Varro (a Roman scholar; 116–27 BC) who, in his history of agriculture *De Re Rustica* (Vol. III), wrote instructions on how to keep rabbits (conies) in the *leporaria*, that are considered the precursors of the medieval warrens (Zeuner, 1963); (iii) the poet Gaius Valerius Catullus (86–40 BC) who linked rabbits to the Iberian country and people (Poem XXXIX); (iv) the *De Re Coquinaria*, a compiled collection (dated from the

3rd to the 4th century AD) of Roman cookery recipes attributed to Marcus Gavius Apicius (a wealthy Roman gourmet – who lived in a period across the 1st century BC and the 1st century AD), that includes dishes made of rabbit.

Subsequent historical sources are from the Medieval period from which a few anecdotes were then reported with misinterpretation of their general meaning (Nachtsheim, 1949; Zeuner, 1963; Rogers *et al.*, 1994). One of them is related to the consumption of *laurices* (mentioned in a writing, dated *c.*584 AD, attributed to St Gregory of Tours, 538–594 AD) that would have been admitted in the Lent period because it was not considered as meat. No other reports related to this potential use appeared before or later than this writing, confirming that this fact cannot be interpreted as the event that would have initiated the domestication of the European rabbit by French monks. There is no historical document supporting the reason for which monks started to breed rabbits in this period and would have been to obtain animal proteins without infringing religious rules. However, it seems plausible to suppose that practices to keep rabbits in warrens might also have been continued in the high Medieval period in France where wild rabbits were naturally present and that French monasteries between the 6th and 10th centuries AD could have attempted to breed these animals to secure meat (Zeuner, 1963). However, no historical documents are available in this period to testify this practice. The exploitation of the Euro pean rabbit as meat source in the high Middle Age in Spain and probably in France might have been mainly based on hunting as also deduced from the encyclopaedia of the Spanish archbishop, theologian and encyclopaedist Saint Isidore of Seville (*c.*560–636 AD), who explained the etymology of the name *cuniculus* (rabbit) as being derived from *caniculus* because dogs (*canis*) were used for hunting these animals from their holes (Etymologiae, Vol. XII).

The first archaeological evidence of the presence of European rabbits in the north of France is dated to the 9th century AD after which archaeological remains appear more frequently (Callou, 2003). This suggests that a second wave of spread towards the north of Europe started in this period probably driven by contacts between monasteries as deduced from a letter of 1149 AD from the Abbot Wibald of Corvey (a Benedictine monastery on the Wesser, Germany) to Abbot Gerald of Solignac (France) who asked for two pairs of rabbits and archaeological records from monastic sites with rabbit remains dated from the 11th–12th (site of Charité-sur-Loire in France; Audoin-Rouzeau, 1984) to the 12th–13th centuries (Belgian sites of Ename Abbey and Dune Abbey; Gautier, 1984; Ervynck *et al.*, 1999). It is not clear how the frequency and intensity of this distribution in Europe was, as just a few documents from this period are available. Starting from monastic centres, European rabbits were subsequently and more frequently associated with the secular elite (Sykes and Curl, 2010) and garennae, the corresponding of warrens (probably not only linked to rabbits in this period but also associated with a landscape for hunting a variety of small game animals), became important parts of the seigneurial culture in the North of France starting from the 11th century (Gautier, 2007; Sykes and Curl, 2010). It has been suggested that the way of keeping rabbits in warrens might not have facilitated selection for tameness and for this reason this practice cannot be considered as a first step towards domestication (Zeuner, 1963).

Subsequently, European rabbits were reported to arrive in the island of Amrum in the North Sea around the year 1230, transported by King Valdemar I of Denmark (Nachtsheim, 1949). European rabbits were mentioned to be usually sold in markets or were part of intense commercialization as reported in the book of the 12th-century *Conejeria de Toledo* (the Rabbitries of Toledo) and the despatch of 6000 pelts from Castile to Devon in 1221 (Delort, 1984). In the 13th–14th centuries, rabbits were then found in The Netherlands, in Belgium and in Germany (Thomson, 1951; Van Damme and Ervynck, 1988; Lauwerier and Zeiler, 2001).

The earliest records on the arrival of the European rabbit in England described the establishments of warrens (or introductions) on several islands, including Drake's Island, Devon, in 1135, Isles of Scilly, Cornwall, in 1176, Lundy Island in 1274 and Stockholm Island in 1324 (Thompson, 1994), and in the mainland with the Dartmoor warren (known from a deed dated to between 1135 and 1272), the cunicularium mentioned in the grant of land by Simon le Bret to the canons of Waltham in Essex (1187–1194) and the reference in the Close Rolls (1235) which reports the donation of King Henry III of 10 live rabbits obtained from his park at Guildford (Veale, 1957; Henderson, 1997; Bartlett, 2000; Sykes and Curl, 2010). Archaeological excavations of the Royal Palace at Guildford confirmed that in this period (*c.*1230–1268; Sykes *et al.*, 2005) rabbits increased their presence and several other evidences indicated that in the first half of this century (from 1230 to 1250) warrens quickly spread in mainland Britain and arrived by 1264 in Scotland (Veale, 1957; Gilbert, 1979). Despite the rise in the number of warrens that mainly started in this period, it should be considered that the establishment of rabbit colonies was quite difficult and expensive and that at the beginning it was mainly driven by the elite that wanted to secure the source of luxury meat. Therefore, rabbits and warrens were usually associated with castles and monasteries and became symbols of lordship (Bailey, 1988; Sykes and Curl, 2010). The relevance and value of rabbit meat increased during the 14th century and became part of the menus of great banquets like that of the coronation of King Henry IV (1399). Hunting of rabbits in warrens was not considered as true hunting and it could be an acceptable activity for men of the cloth (for whom hunting was not allowed) and for ladies. This is evident in the iconography of that time in which women are frequently depicted in the acts of catching rabbits. The most famous example is found in a few scenes of the *Queen Mary's Psalter* (*c.*1315) where two ladies are depicted rabbiting with the help of ferrets, cages and clubs. For seven centuries from their first introduction to Britain, rabbits were constantly bred in warrens for meat production and fur (Thompson, 1994). The economic relevance of the warren-based production system lasted till the second half of the 19th century when the import of carcases and fur from Australia and the development of more efficient agricultural systems contributed to the abandonment of the wild rabbit production (Thompson, 1994). Molecular evidence from modern wild rabbits indicated that the colonization of Britain derived from *O. c. cuniculus* having the B mitotype (Monnerot *et al.*, 1994).

Rabbit hunting was probably common in France in the 13th–14th centuries as evidenced by the iconography and by several documents, including the famous medieval book on hunting, *Livre de Chasse* (1387–1389) written by Gaston III, Count of Foix (known as Phoebus or Fébus) with a scene of rabbit hunting depicted with rabbits of pale and brown coat colour varieties. Warrens became quite frequent also in France in that period. The monarchy attempted to restrict the rights on the creation of new warrens and on the enlargements or re-establishment of old ones with the ordinance of King John II (1336) and of King Charles VI (1413). In the 17th century the prime minister of King Louis XIV ordered the destruction of rabbits in the royal forests as these animals became numerous and damaging. This ordinance was then cancelled during the French Revolution that annulled also the privilege of the gentry to control the warrens even if rabbits continued to be enclosed and controlled in France till the Empire of Napoleon III (Rogers *et al.*, 1994).

In the 13th–14th centuries, wild rabbits were quite common in central and south Italy and were usually hunted, as evidenced by the numerous archaeological sites in which rabbit bones were identified (Callou, 2003; De Venuto, 2009; Rizzo *et al.*, 2012). The use of wild rabbits was very frequent in Sicily where many remains were identified in the Medieval site of Brucato (Bresc, 1980). Mitochondrial haplotypes of the B lineage have been the only mitotypes identified in modern wild rabbits sampled in Sicily (Valvo *et al.*, 2017) in line with previous evidences of the spread of *O. c. cuniculus* in Europe. Despite the reported presence of hunted rabbits in Italian Medieval archaeological sites, rabbit meat was absent in Italian Middle Age recipes, suggesting that this species was not a common component of the diet in most parts of Italy in that period, probably due to the incomplete distribution in this Peninsula (Piccinni, 1982). However, its use for fur production increased. The painting of the Madonna of the Rabbit (*c.*1530) by Titian (Tiziano Vecellio) produced in Italy and now at the Louvre Museum is the first document on the appearance of the white coat colour variant in this species.

Other coat colour varieties were reported by the French-Dutch philologist and historian Joseph Justus Scaliger (1540–1609) who mentioned black, yellow, blue, piebald and, again, white rabbits. Olivier de Serres, a French agricultural scientist, in his *Le Théâtre d'Agriculture* (1606) distinguished rabbits into a few races according to the type of coat: *le lapin commun* (normal grey), *le lapin riche* (with grey-black-silver coat colour) and *le lapin Angora*. He also distinguished rabbits according to their origin or raising methods that give different flavour to the meat: warren rabbits (from the garenne) and domestic rabbits; hutched rabbits and wild rabbits. Darwin (1868) cited Gervaise Markham who in 1631 described coat and fur features in the rabbit suggesting that selection and breeding for fur production was already established at that time in England. In the same text, Darwin cited Aldrovandi who in 1637 described larger races. Van Leeuwenhoek, the inventor of the microscope, reported on the practice of crossbreeding white varieties with wild rabbits to obtain coloured furs (more requested at that time) and arrived to demonstrate dominance in rabbits by using coat colours much earlier than Mendel's law was developed (Sirks, 1959; Zeuner, 1963). This period might be considered the beginning of the formation of rabbit breeds and the consolidation of the domestication of the European rabbit.

1.2.4 World distribution of the European rabbit and genetic perspectives

The world distribution of the European rabbit depended largely on human activities (Flux, 1994). Both domestic and wild-type European rabbits have been transported in many parts of the world (including more than 800 islands) for many purposes, depending on the historical period in which the transfer occurred and on who carried out these actions: for sport, to farm for meat or fur production, as food for other animals or bait for lobster pots, to control vegetation, amuse tourists, and even to conserve representative populations from myxomatosis (Flux and Fullagar, 1992; Flux, 1994).

Following the early introductions in Britain and the development of the warren system, wild rabbits slowly proceeded north to Scotland where they arrived in 1793 (Barrett-Hamilton, 1912). Many other successful introductions contributed to the spread of European rabbits in northern, central and eastern European countries, including Germany, a few islands of Norway, the south part of Sweden, Poland, a spot in Lithuania and areas in the Czech Republic, Slovakia, Hungary, Romania, Italy and Ukraine and in a few Croatian islands (Flux, 1994). Other attempts, however, failed for several reasons, some of which are unknown.

Most of the introductions in other continents were from the *O. c. cuniculus* subspecies, as genetic stocks from this lineage were used to colonize Europe, which constituted the genetic reservoirs (of wild, feral or domestic origin) used for the colonization of many regions of the world, including Australia, New Zealand and some regions of South America. However, direct evidence of that is not always available as molecular genetic studies have been carried out in a few populations (Zenger, 1996; Long *et al.*, 2003; Zenger *et al.*, 2003; Emam *et al.*, 2016; Brajkovic *et al.*, 2017; Valvo *et al.*, 2017; Owuor *et al.*, 2019).

An interesting exception in the history of rabbit introduction is the colonization of the Portuguese Atlantic islands of the Madeira, including the Porto Santo island, mentioned by Darwin (1868), Azorean and Canary archipelagos. In these islands, wild rabbits were from the *O. c. algirus* subspecies as first deduced from morphological and then from molecular data, using polymorphisms at antibody loci, microsatellite and mitochondrial DNA markers (Franca, 1913; Gibb, 1990; Esteves, 2003; Fonseca, 2005; Ferrand and Branco, 2007). The introduction of wild rabbits in the Porto Santo island is dated back to the year 1418 by Bartolomeo Perestrello, a Portuguese navigator and explorer of Italian origin (considered the first to discoverer this island), even if earlier introductions probably occurred in the 14th century (Trouessart, 1917; Flux and Fullagar, 1992). The story reports that a single pregnant doe was released in the island and from it in a few years this land was populated by a large number of rabbits which completely devastated the vegetation and led to the collapse of the Portuguese settlement. Darwin (1868) erroneously described the Porto Santo rabbit population as an example of speciation that occurred since the first introduction in historical time (derived by the isolation and by the local selection pressure). His error was due to the fact that he compared the Porto Santo animals with the English wild rabbits, which actually derive from the *O. c. cuniculus* subspecies, and not with wild rabbits from the south of Iberia.

The most explosive and damaging expansion of the European rabbit occurred in Australia.

The first five domestic rabbits, probably silver-grey, were released in 1788 around Sydney. After, a few other importations were based on domestic varieties. In 1859, the release of 20 wild rabbits imported from England at Geelong (Victoria) was much more effective with an explosion of the number of rabbits in the subsequent period and the beginning of the geographic expansion that, after other releases that continued around 1870, reached an estimated rate of 125 km/year in New South Wales and 300 km/year along the drainage channels in the Simpson Desert (Flux, 1994; Myers *et al.*, 1994; Williams *et al.*, 1995). The expansion of the European rabbit is well documented in Australia and followed different rates, according to the environmental conditions that this pest encountered. The economic damage that the European rabbit has been causing to Australia is difficult to estimate as this pest impacts on the endemic wildlife flora and fauna and on the agricultural economy at different levels but it could be around several hundreds of millions of dollars per year (Williams *et al.*, 1995). Therefore, different counter actions have been experimented. The introduction of myxomatosis in 1950 as a weapon for biological control drastically reduced the Australian rabbit population that re-emerged after a few years from the virus outbreaks (Fenner and Ross, 1994; Williams *et al.*, 1995). The subsequent introduction of rabbit haemorrhagic disease virus (RHDV) in 1995 caused a rapid decrease in the Australian rabbit population that then recovered (Mutze *et al.*, 1998).

The effects of the potential bottlenecks and the results of the rabbit population dynamics in Australia have been genetically investigated by a few authors with contrasting interpretations that might be derived by the complexity and the overlapping of events and conditions on the diffusion of the European rabbit in this country (Fuller *et al.*, 1996, 1997; Zenger *et al.*, 2003; Schwensow *et al.*, 2017a; Iannella *et al.*, 2019). The absence of reduced genetic diversity observed with microsatellite data analysed in Australian populations was indicated to be derived by a rapid population expansion at the time of the establishment in Australia (Zenger *et al.*, 2003). Genome-wide population analyses supported a history of multiple independent introductions across the continent, as also demonstrated by historical records, followed by regional dispersal. The resulting genetic structure and sub-structure, with evidence for RHDV-driven selection, may contribute to variation across the country in rabbit resistance to the viral biocontrols (Schwensow *et al.*, 2017a, b; Iannella *et al.*, 2019). The natural experimental designs created by the occurrence of RHDV and myxomatosis made it possible to compare genome features of wild rabbits that survived the infection waves against that of wild rabbits that died and it was possible to obtain information of the genetic mechanisms determining virus disease resistance (Schwensow *et al.*, 2017a; Alves *et al.*, 2019).

Population genetic structures of wild and/or feral European rabbits distributed in many parts of the world (Ziege *et al.*, 2020) and the possibility to track back the origin of the European rabbit introductions by using DNA information can open interesting avenues for future investigations that could see this generalist colonizer as a model in a variety of natural environmental and genetic experimentations (Thulin *et al.*, 2017).

References

Albarella, U. (1992) La fauna. In: Arthur, P. (ed.) *L'Isola e il Santo. La chiesa di San Costanzo alla Marina Grande di Capri, scavi 1990*. Editoriale Scientifica, Napoli, pp. 53–58.

Alda, F. and Doadrio, I. (2014) Spatial genetic structure across a hybrid zone between European rabbit subspecies. *PeerJ* 2, e582.

Alves, J.M., Carneiro, M., Afonso, S., Lopes, S., Garreau, H. *et al.* (2015) Levels and patterns of genetic diversity and population structure in domestic rabbits. *PLOS ONE* 10, e0144687.

Alves, J.M., Carneiro, M., Cheng, J.Y., Lemos de Matos, A., Rahman, M.M. *et al.* (2019) Parallel adaptation of rabbit populations to myxoma virus. *Science* 363, 1319–1326.

Arana, A., Zaragoza, P., Rodellar, C. and Amorena, B. (1989) Blood biochemical polymorphisms as markers for genetic characteristics of wild Spanish and domestic rabbits. *Genetica* 79, 1–9.

Asher, R.J., Meng, J., Wible, J.R., McKenna, M.C., Rougier, G.W. *et al.* (2005) Stem Lagomorpha and the antiquity of Glires. *Science* 307, 1091–1094.

Audoin-Rouzeau, F. (1984) *Ossements animaux du Moyen Age au monastère de La Charité-sur-Loire.* Publications de la Sorbonne, Paris.

Averianov, A.O. (1994) Early Eocene mimotonids of Kyrgyzstan and the problem of Mixodontia. *Acta Palaeontologica Polonica* 39, 393–411.

Bailey, M. (1988) The rabbit and the medieval East Anglian economy. *Agricultural History Review* 36, 1–20.

Barrett-Hamilton, G.E. (1912) *A History of British Mammals*. Vol. 2. Gurney & Jackson, London.

Bartlett, R. (2000) *England under the Norman and Angevin Kings*. Clarendon Press, Oxford.

Benton, M.J. and Donoghue, P.C. (2007) Paleontological evidence to date the tree of life. *Molecular Biology and Evolution* 24, 26–53.

Biju-Duval, C., Ennafaa, H., Dennebouy, N., Monnerot, M. and Mignotte, F. (1991) Mitochondrial DNA evolution in lagomorphs: origin of systematic heteroplasmy and organization of diversity in European rabbits. *Journal of Molecular Evolution* 33, 92–102.

Brajkovic, V., Beljan, S., Kaštelanac, I., Oršanić, M. and Ugarković, D. (2017) DNA sequence variation in the mitochondrial control region of *Oryctolagus cuniculus* from Croatia. *Agriculturae Conspectus Scientificus* 82, 117–121.

Branco, M., Ferrand, N. and Monnerot, M. (2000) Phylogeography of the European rabbit (*Oryctolagus cuniculus*) on the Iberian Peninsula inferred from RFLP analysis of the cytochrome b gene. *Heredity* 85, 307–317.

Branco, M., Monnerot, M., Ferrand, N. and Templeton, A.R. (2002) Postglacial dispersal of the European rabbit (*Oryctolagus cuniculus*) on the Iberian Peninsula reconstructed from nested clade and mismatch analyses of mitochondrial DNA genetic variation. *Evolution* 56, 792–803.

Bresc, H. (1980) La chasse en Sicilie (XIIe–XVe siecles). In: *La chasse au Moyen Age*. Coloque du Centre d'Études Médievales de Nice. Nizza 22–24 June 1979. pp. 201–217.

Brusini, I., Carneiro, M., Wang, C., Rubin, C.J., Ring, H. *et al.* (2018) Changes in brain architecture are consistent with altered fear processing in domestic rabbits. *Proceedings of the National Academy of Sciences of the United States of America* 115, 7380–7385.

Callou, C. (1995) Modifications de l'aire de répartition du lapin (*Oryctolagus cuniculus*) en France et en Espagne, du Pléistocène à l'époque actuelle. Etat de la question. *Anthropozoologica* 21, 95–114.

Callou, C. (2003) *De la garenne au clapier. Étude archéozoologique du Lapin en Europe occidentale.* Mémoires do Muséum National d'Histoire Naturelle, 189. Paris.

Carneiro, M., Ferrand, N. and Nachman, M.W. (2009) Recombination and speciation: loci near centromeres are more differentiated than loci near telomeres between subspecies of the European rabbit (*Oryctolagus cuniculus*). *Genetics* 181, 593–606.

Carneiro, M., Blanco-Aguiar, J.A., Villafuerte, R, Ferrand, N. and Nachman, M.W. (2010) The X-chromosome and speciation: patterns of differentiation at multiple loci on the X-chromosome and autosomes of the European rabbit (*Oryctolagus cuniculus*). *Evolution* 64, 3443–3460.

Carneiro, M., Afonso, S., Geraldes, A., Garreau, H., Bolet, G. *et al.* (2011) The genetic structure of domestic rabbits. *Molecular Biology and Evolution* 28, 1801–1816.

Carneiro, M., Baird, S.J.E., Afonso, S., Ramirez, E., Tarroso, P. *et al.* (2013) Steep clines within a highly permeable genome across a hybrid zone between two subspecies of the European rabbit. *Molecular Ecology* 22, 2511–2525.

Carneiro, M., Rubin, C.J., Di Palma, F., Albert, F.W., Alföldi, J. *et al.* (2014a) Rabbit genome analysis reveals a polygenic basis for phenotypic change during domestication. *Science* 345, 1074–1079.

Carneiro, M., Albert, F.W., Afonso, S., Pereira, R.J. and Burbano, H. (2014b) The genomic architecture of population divergence between subspecies of the European rabbit. *PLOS Genetics* 10, e1003519.

Carneiro, M., Piorno, V., Rubin, C.J., Alves, J.M., Ferrand, N. *et al.* (2015) Candidate genes underlying heritable differences in reproductive seasonality between wild and domestic rabbits. *Animal Genetics* 46, 418–425.

Chapman, J.A. and Flux, J.E.C. (1990) *Rabbits, Hares and Pikas: Status Survey and Conservation Action Plan*. International Union for Conservation of Nature and Natural Resources (IUCN), Information Press, Oxford.

Chapman, J.A. and Flux, J.E.C. (2008) Introduction to the Lagomorpha. In: Alves, P.C., Ferrand, N. and Hackländer, K. (eds) *Lagomorph Biology*. Springer, Berlin, Heidelberg, Germany, pp. 1–9.

Chuan-Kuei, L., Wilson, R.W., Dawson, M.R. and Krishtalka, L. (1987) The origin of rodents and lagomorphs. In: Genoways, H.H. (ed.) *Current Mammalogy*. Springer, Boston, Massachusetts.

Clutton-Brock, J. (2012) *Animals as Domesticates: A World View through History*. Michigan State University Press, East Lansing, Michigan.

Cochard, D. and Brugal, J.-P. (2004) Importance des functions de site dans les accumulations paléolithiques de léporidés. In: Brugal, J.-P. and Desse, J. (eds) *Petits animaux et sociétés humaines: du complément alimentaire aux ressources utilitaires: actes des XXIVe Rencontres internationales d'archéologie et d'histoire: d'Antibes, 23–25 octobre 2003*. Association pour la promotion et la diffusion des connaissances archéologiques, Antibes, pp. 283–296.

Coggan, M., Baldwin, J. and Richardson, B.J. (1974) Ecological genetics of the wild rabbit in Australia. I. Geographical distribution and biochemical characterization of phosphogluconate dehydrogenase variants. *Australian Journal of Biological Sciences* 27, 671–676.

Cooke, B.D., Chudleigh, P., Simpson, S. and Saunders, G. (2013) Economic benefits of the biological control of rabbits in Australia. *Australian Economic History Review* 53, 91–107.

Cooke, B.D., Flux, J.E.C. and Bonino, N. (2018) Introduced lagomorphs. In: Smith, A.T., Johnston, C.H., Alves, P.C. and Hackländer K. (eds) *Lagomorphs: Pikas, Rabbits and Hares of the World*. Johns Hopkins University Press, Baltimore, Maryland, pp. 13–17.

Darwin, C. (1868) *The Variation of Animals and Plants*. Orange Judd & Company, New York.

Delibes-Mateos, M., Ramírez, E., Ferreras, P. and Villafuerte, R. (2008) Translocations as a risk for the conservation of European wild rabbit *Oryctolagus cuniculus* lineages. *Oryx* 42, 259–264.

Delort, R. (1984) *Les animaux ont une historie*. Editions du Seuil, Paris.

De Venuto, G. (2009) Analisi archeozoologica comparata di campioni faunistici provenienti dalle stratigrafie di abbandono di fosse granarie in disuso della Puglia Meridionale. Contributo alla ricostruzione delle dinamiche di produzione e consumo della risorsa animale in Italia meridionale tra XIII e XIV secolo. *Atti del V Congresso Nazionale di Archeologia Medievale*, 30 Settembre–3 Ottobre 2009, Foggia-Manfredonia, pp. 712–716.

Donard, E. (1982) Recherches sur les léporinés Quaternaires (Pléistocéne moyen et superiéur, Holocène). These doct. Université de Bordeaux I, 2 vols.

Emam, A.M., Afonso, S., Azoz, A.A.A., González-Redondo, P., Mehaisen, G.M.K. *et al.* (2016) Origin of Egyptian and Spanish common rabbits: evidence from mitochondrial DNA cytochrome b sequence analysis. *Proceedings of the 11th World Rabbit Congress*, pp. 15–18.

Ennafaa, H., Monnerot, M., El Gaaïed, A. and Mounolou, J.C. (1987) Rabbit mitochondrial DNA: preliminary comparison between some domestic and wild animals. *Genetique, Selection, Evolution* 19, 279–288.

Ervynck, A., Van Neer, W. and Lentacker, A. (1999) Introduction and extinction of wild animal species in historical times: the evidence from Belgium. In: Benecke, N. (ed.) *The Holocene History of the European Vertebrate Fauna: Modern Aspects of Research*. Workshop, 6–9 April 1998, Berlin. Verlag Marie Leidorf GmbH, Rahden, Germany, pp. 399–407.

Esteves, P.J. (2003) Molecular and population genetic analysis of polymorphism at the antibody loci IgGCH2 and IgVH in lagomorphs. PhD thesis, University of Porto, Portugal.

Fa, J.E., Stewart, J.R., Lloveras, L. and Vargas, J.M. (2013) Rabbits and hominin survival in Iberia. *Journal of Human Evolution* 64, 233–241.

Fenner, F. and Ross, J. (1994) Myxomatosis. In: Thompson, H.V. and King, C.M. (eds) *The European Rabbit. The History and Biology of a Successful Colonizer*. Oxford Science Publications, Oxford, pp. 205–239.

Ferrand, N. (2008) Inferring the evolutionary history of the European rabbit (*Oryctolagus cuniculus*) from molecular markers. In: *Lagomorph Biology*. Springer, Berlin, Heidelberg, pp. 47–63.

Ferrand, N. and Branco, M. (2007) The evolutionary history of the European rabbit (*Oryctolagus cuniculus*): major patterns of population differentiation and geographic expansion inferred from protein polymorphism. In: Weiss, S. and Ferrand, N. (eds) *Phylogeography of Southern European Refugia*. Springer, Dordrecht, The Netherlands, pp. 207–235.

Ferrand, N. and Rocha, J. (1992) Demonstration of serum Albumin (ALB) polymorphism in wild rabbits, *Oryctolagus cuniculus*, by means of isoelectric focusing. *Animal Genetics* 23, 275–278.

Flux, J.E.C. (1994) World distribution. In: Thompson, H.V. and King, C.M. (eds) *The European Rabbit: The History and Biology of a Successful Colonizer*. Oxford University Press, Oxford, pp. 8–21.

Flux, J.E.C. and Fullagar, P.J. (1992) World distribution of the rabbit *Oryctolagus cuniculus* on islands. *Mammal Review* 22, 151–205.

Fonseca, A. (2005) Análise da origem e dispers~ao das populações de elho-bravo (Oryctolagus cuniculus) dos arquipélagos dos Açores, Madeira e Canárias através da utilização de marcadores genéticos nucleares e de DNA-mitocondrial. PhD thesis, University of the Azores, Portugal.

Fontanesi, L., Di Palma F., Flicek P., Smith A.T., Thulin C.G. *et al.* (2016) LaGomiCs – Lagomorph Genomics Consortium: an international collaborative effort for sequencing the genomes of an entire mammalian order. *Journal of Heredity* 107, 295–308.

Franca, C. (1913) Contribution á l'étude du lapin de Porto Santo (*Oryctolagus cuniculus Huxleyi* Haeckel). *Bulletin de la Société Portugaise des Sciences Naturelles* 6, I–14.

Fuller, S.J., Mather, P.B. and Wilson, J.C. (1996) Limited genetic differentiation among wild *Oryctolagus cuniculus* L. (rabbit) populations in arid eastern Australia. *Heredity* 77, 138–145.

Fuller, S.J., Wilson, J.C. and Mather, P.B. (1997) Patterns of differentiation among wild rabbit populations *Oryctolagus cuniculus* L. in arid and semiarid ecosystems of north-eastern Australia. *Molecular Ecology* 6, 145–153.

Gautier, A. (1984) Enkele dierresten uit de abdji Ter Duinen te Koksijde (*c*.1175–*c*.1250, *c*.1250–1318, 1593–1601 AD). *De Duinen* 13–14, 61–63.

Gautier, A. (2007) Deer parks in Sussex and the Godwinesons. *Anglo-Norman Studies* 29, 51–64.

Geraldes, A., Ferrand, N. and Nachman, M.W. (2006) Contrasting patterns of introgression at X-linked loci across the hybrid zone between subspecies of the European rabbit (*Oryctolagus cuniculus*). *Genetics* 173, 919–933.

Geraldes, A., Carneiro, M., Delibes-Mateos, M., Villafuerte, R., Nachman, M.W. and Ferrand, N. (2008) Reduced introgression of the Y chromosome between subspecies of the European rabbit (*Oryctolagus cuniculus*) in the Iberian Peninsula. *Molecular Ecology* 17, 4489–4499.

Gibb, J.A. (1990) The European rabbit *Oryctolagus cuniculus*. In: Chapman, J.A. and Flux, J. (eds) *Rabbits, Hares and Pikas: Status Survey and Conservation Action Plan.* International Union for Conservation of Nature and Natural Resources, Gland, Switzerland, pp. 116–120.

Gidely, J.W. (1912) The lagomorphs are an independent order. *Science* 36, 285–286.

Gilbert, J.M. (1979) *Hunting and Hunting Reserves in Medieval Scotland*. Donald, Edinburgh, UK.

Halanych, K.M. and Robinson, T.J. (1999) Multiple substitutions affect the phylogenetic utility of Citochrome b and 12S rDNA data: examining a rapid radiation in Leporid (Lagomorpha) evolution. *Journal of Molecular Evolution* 48, 369–379.

Hardy, C., Casane, D., Vigne, J.D., Callou, C., Dennebouy, N. *et al.* (1994a) Ancient DNA from Bronze Age bones of European rabbit (*Oryctolagus cuniculus*). *Experientia* 50, 564–570.

Hardy, C., Vigne, J.-D., Casane, D., Dennebouy, N., Mounolou, J.-C. and Monnerot, M. (1994b) Origin of European rabbit (*Oryctolagus cuniculus*) in a Mediterranean island: zooarchaeology and ancient DNA examination. *Journal of Evolutionary Biology* 7, 217–226.

Hardy, C., Callou, C., Vigne, J.-D., Casane, D., Dennebouy, N. *et al.* (1995) Rabbit mitochondrial DNA diversity from prehistoric to modern times. *Journal of Molecular Evolution* 40, 227–237.

Henderson, A. (1997) From coney to rabbit: the story of a managed colonizer. *The Naturalist* 122, 101–121.

Hoffmann, R.S. and Smith, A.T. (2005) Lagomorphs. In: Wilson, D.E. and Reeder, D.M. (eds) *Mammal Species of the World*, 3rd edn. Johns Hopkins University Press, Baltimore, Maryland, pp. 185–211.

Hörnicke, H. (1981) Utilization of caecal digesta by caecotrophy (soft faeces ingestion) in the rabbit. *Livestock Production Science* 8, 361–366.

Iannella, A., Peacock, D., Cassey, P. and Schwensow, N. (2019) Genetic perspectives on the historical introduction of the European rabbit (*Oryctolagus cuniculus*) to Australia. *Biological Invasions* 21, 603–614.

Irving-Pease, E.K., Ryan, H., Jamieson, A., Dimopoulos, E.A., Larson, G. and Frantz, L.A. (2018a) Paleogenomics of animal domestication. In: Lindqvist, C. and Rajora, O.P. (eds) *Paleogenomics: Genome-Scale Analysis of Ancient DNA*. Springer International Publishing AG, Springer Nature, Cham, Switzerland, pp. 225–272.

Irving-Pease, E.K., Frantz, L.A., Sykes, N., Callou, C. and Larson, G. (2018b) Rabbits and the specious origins of domestication. *Trends in Ecology and Evolution* 33, 149–152.

Landry, S.O. (1999) A proposal for a new classification and nomenclature for the Glires (Lagomorpha and Rodentia). *Zoosystematics and Evolution* 75, 283–316.

Larson, G. and Burger, J. (2013) A population genetics view of animal domestication. *Trends in Genetics* 29, 197–205.

Lauwerier, R.C.G.M. and Zeiler, J.T. (2001) Wishful thinking and the introduction of the rabbit to the Low Countries. *Environmental Archaeology* 6, 87–90.

Lissovsky, A.A. (2014) Taxonomic revision of pikas Ochotona (Lagomorpha, Mammalia) at the species level. *Mammalia* 78, 199–216.

Long, J.-R., Qiu, X.-P., Zeng, F.-T., Tang, L.-M. and Zhang, Y.-P. (2003) Origin of rabbit (*Oryctolagus cuniculus*) in China: evidence from mitochondrial DNA control region sequence analysis. *Animal Genetics* 34, 82–87.

López-Martínez, N. (1985) Reconstruction of ancestral cranioskeletal features in the order Lagomorpha. In: Luckett, W.P. and Hartenberger, J.L. (eds) *Evolutionary Relationships among Rodents*. NATO Advanced Science Institutes (ASI) Series (Series A: Life Sciences), vol. 92. Springer, Boston, Massachusetts.

López-Martínez, N. (1989) Revisión sistemática y bioestratigráfica de los Lagomorpha (Mammalia) del Terciario y Cuaternario inferior de Espana. PhD thesis, Univiversity of Madrid, 1977. Mem. Museo Paleont, Univiversity of Zaragoza, Madrid.

López-Martínez, N. (2008) The lagomorph fossil record and the origin of the European rabbit. In: Alves, P.C., Ferrand, N. and Hackländer, K. (eds) *Lagomorph Biology*. Springer, Berlin, Heidelberg, Germany, pp. 27–46.

Massetti, M. and De Marinis, A.M. (2008) Prehistoric and historic artificial dispersal of lagomorphs on the Mediterranean islands. In: Alves, P.C., Ferrand, N. and Hackländer, K. (eds) *Lagomorph Biology: Evolution, Ecology, and Conservation*. Springer-Verlag, Berlin, pp. 13–25.

Matthee, C.A., Van Vuuren, B.J., Bell, D. and Robinson, T.J. (2004) A molecular supermatrix of the rabbits and hares (Leporidae) allows for the identification of five intercontinental exchanges during the Miocene. *Systematic Biology* 53, 433–447.

Melo-Ferreira J. and Alves P.C. (2018) Systematics of Lagomorphs. In: Smith A.T., Johnston, C.H., Alves, P.C. and Hackländer, K. (eds) *Lagomorphs: Pikas, Rabbits and Hares of the World*. Johns Hopkins University Press, Baltimore, Maryland, pp. 9–12.

Meng, J., Hu, Y.M. and Li, C.K. (2003) The osteology of Rhombomylus (Mammalia, Glires): implications for phylogeny and evolution of glires. *Bulletin of the American Museum of Natural History* 275, 1–247.

Meredith, R.W., Janečka, J.E., Gatesy, J., Ryder, O.A. and Fisher, C.A. (2011) Impacts of the Cretaceous Terrestrial Revolution and KPg extinction on mammal diversification. *Science* 334, 521–524.

Monnerot, M., Vigne, J.D., Biju-Duval, C., Casane, D., Callou, C. *et al.* (1994) Rabbit and man: genetic and historic approach. *Genetics Selection Evolution* 26 (Suppl. 1), 167S–182S.

Mutze, G., Cooke, B. and Alexander, P. (1998) The initial impact of rabbit hemorrhagic disease on European rabbit populations in South Australia. *Journal of Wildlife Diseases* 34, 221–227.

Myers, K., Parer, I., Wood, D. and Cooke, B.D. (1994) The rabbit in Australia. In: Thompon, H.V. and King, C.M. (eds) *The European Rabbit: The History and Biology of a Successful Colonizer*. Oxford University Press, Oxford, pp. 108–157.

Nachtsheim, H. (1949) *Vom Wildtier zum Haustier*, Paul Parey, Hamburg and Berlin, Germany.

O'Leary, M.A., Bloch, J.I., Flynn, J.J., Gaudin, T.J. and Giallombardo, A. (2013) The placental mammal ancestor and the post-K-Pg radiation of placentals. *Science* 339, 662–667.

Owuor, S.A., Mamati, E.G. and Kasili, R.W. (2019) Origin, genetic diversity, and population structure of rabbits (*Oryctolagus cuniculus*) in Kenya. *BioMed Research International* 2019, 7056940.

Pedler, R.D., Brandle, R., Read, J.L., Southgate, R., Bird, P. and Moseby, K.E. (2016) Rabbit biocontrol and landscape-scale recovery of threatened desert mammals. *Conservation Biology* 30, 774–782.

Peterka, M. and Hartl, G.B. (1992) Biochemical-genetic variation and differentiation in wild and domestic rabbits: on the significance of genetic distances, dendrograms and the estimation of divergence times in domestication studies. *Journal of Zoological Systematics and Evolutionary Research* 30, 129–141.

Piccinni, G. (1982) Note sull'alimentazione medievale. *Studi Storici* 23, 603–615.

Queney, G., Ferrand, N., Weiss, S., Mougel, F. and Monnerot, M. (2001) Stationary distributions of microsatellite loci between divergent population groups of the European rabbit (*Oryctolagus cuniculus*). *Molecular Biology and Evolution* 18, 2169–2178.

Queney, G., Vachot, A.M., Brun, J.M., Dennebouy, N., Mulsant, P. and Monnerot, M. (2002) Different levels of human intervention in domestic rabbits: effects on genetic diversity. *Journal of Heredity* 93, 205–209.

Rafati, N., Blanco-Aguiar, J.A., Rubin, C.J., Sayyab, S., Sabatino, S.J. *et al.* (2018) A genomic map of clinal variation across the European rabbit hybrid zone. *Molecular Ecology* 27, 1457–1478.

Reumer, J.W.F. and Sanders, E.A.C. (1984) Changes in the vertebrate fauna of Menorca in prehistoric and classical times. *Zeitschrift für Säugetierkunde* 49, 321–325.

Richardson, B.J., Rogers, P.M. and Hewitt, G.M. (1980) Ecological genetics of the wild rabbit in Australia. II. Protein variation in British, French and Australian rabbits and the geographical distribution of the variation in Australia. *Australian Journal of Biological Sciences* 33, 371–384.

Rizzo, M.S., Danile, L., Romano, D., Scibona, M. and Zambito, L. (2012) Il villaggio di Colmitella (Racalmuto, Ag): primi dati dallo scavo archeologico di un insediamento rurale di età altomedievale e medievale. In: Redi, F. and Forgione, A. (eds) *VI Congresso Nazionale di Archeologia Medievale*, pp. 419–424.
Robinson, R. (1984) Rabbit. In: Mason, I.L. (ed.) *Evolution of Domesticated Animals*. Longman Group Limited, New York, pp. 239–246.
Rogers, P.M., Arthur, C.P. and Soriguer, R.C. (1994) The rabbit in continental Europe. In: Thompson, H.V. and King, C.M. (eds) *The European Rabbit: The History and Biology of a Successful Colonizer*. Oxford University Press, Oxford, pp. 22–63.
Rose, K.D. (2006) *The Beginning of the Age of Mammals*. Johns Hopkins University Press, Baltimore, Maryland.
Ruedas, L.A., Mora, J.M. and Lanier H.C. (2018) Evolution of Lagomorphs. In: Smith, A.T., Johnston, C.H., Alves, P.C. and Hackländer K. (eds) *Lagomorphs: Pikas, Rabbits and Hares of the World*. Johns Hopkins University Press, Baltimore, Maryland, pp. 4–8.
Saña, M. (2013) Domestication of animals in the Iberian Peninsula. In: College, S., Conolly, J., Dobney, K., Manning, K. and Shennan, S. (eds) *Origins and Spread of Domestic Animals in Southwest Asia and Europe*. Publications of the Institute of Archaeology, University College, London, vol. 59. Routledge, Taylor & Francis Group, London and New York, pp. 195–220.
Sanders, E.A.C. and Reumer, J.W. (1984) The influence of prehistoric and Roman migrations on the vertebrate fauna of Menorca (Spain). *BAR International Series* 229, 119–144.
Schwensow, N.I., Mazzoni, C.J., Marmesat, E., Fickel, J. and Peacock, D. (2017a) High adaptive variability and virus-driven selection on major histocompatibility complex (MHC) genes in invasive wild rabbits in Australia. *Biological Invasions* 19, 1255–1271.
Schwensow, N.I., Detering, H., Pederson, S., Mazzoni, C. and Sinclair, R. (2017b) Resistance to RHD virus in wild Australian rabbits: comparison of susceptible and resistant individuals using a genome-wide approach. *Molecular Ecology* 26, 4551–4561.
Seixas, F.A., Juste, J., Campos, P.F., Carneiro, M., Ferrand, N. *et al.* (2014) Colonization history of Mallorca Island by the European rabbit, *Oryctolagus cuniculus*, and the Iberian hare, *Lepus granatensis* (Lagomorpha: Leporidae). *Biological Journal of the Linnean Society* 111, 748–760.
Sharples, C.M., Fa, J.E. and Bell, D.J. (1996) Geographical variation in size in the European rabbit *Oryctolagus cuniculus* (Lagomorpha: Leporidae) in western Europe and North Africa. *Zoological Journal of the Linnean Society* 117, 141–158.
Sirks, M.J. (1959) Leeuwenhoek on dominance in rabbits (1683). *Genetica* 30, 391–392.
Smith, A.T. (1988) Patterns of pika (Genus *Ochotona*) life history variation. In: Boyce, M.S. (ed.) *Evolution of Life Histories: Theory and Patterns from Mammals*. Yale University Press, New Haven, Connecticut, pp. 233–256.
Smith, A.T., Formozov, A.N., Hoffmann, R.S., Zheng, C., Erbajeva, M.A. (1990) The pikas. In: Chapman, J.A. and Flux, J.E.C. (eds) *Rabbits, Hares and Pikas: Status Survey and Conservation Action Plan*. International Union for Conservation of Nature and Natural Resources (IUCN), Information Press, Oxford, pp. 14–60.
Sykes, N.J. and Curl, J. (2010) The rabbit. In: O'Connor, T. and Sykes, N. (eds) *Extinctions and Invasions: A Social History of British Fauna*. Windgather Press, Oxford, pp. 116–126.
Sykes, N.J., Ingrem, C. and Ayres, K. (2005) The bones and shell. In: Poulton, R. (ed.) *A Medieval Royal Complex at Guildford: Excavations at the Castle and Palace*. Surrey Archaeological Society, Guildford, UK, pp. 116–132.
Thompson, H.V. (1994) The rabbit in Britain. In: Thompson, H.V. and King, C.M. (eds) *The European Rabbit: The History and Biology of a Successful Colonizer*. Oxford University Press, Oxford, pp. 64–107.
Thompson, H.V. and King, C.M. (1994) *The European Rabbit: The History and Biology of a Successful Colonizer*. Oxford Science Publications, Oxford.
Thomson, A.P.D. (1951) A history of the ferret. *Journal of History of Medicine and Allied Sciences* 6, 471–480.
Thulin, C.G., Alves, P.C., Djan, M., Fontanesi, L. and Peacock, D. (2017) Wild opportunities with dedomestication genetics of rabbits. *Restoration Ecology* 25, 330–332.
Trouessart, E.L. (1917) Le lapin de Porto Santo et le lapin nègre de la Camargue. *Bullelin du Muséum d'Histoire Naturelle* 1917, 366–373.
Valvo, M.L., Russo, R., Mancuso, F.P. and Palla, F. (2017) mtDNA diversity in a rabbit population from Sicily (Italy). *Turkish Journal of Zoology* 41, 645–653.

Van Damme, D. and Ervynck, A. (1988) Medieval ferrets and rabbits in the castle of Laarne (East Flanders, Belgium): a contribution to the history of a predator and its prey. *Helinium* 28, 278–284.

van der Loo, W., Ferrand, N. and Soriguer, R. (1991) Estimation of gene diversity at the b locus of the constant region of the immunoglobulin light chain in natural populations of European rabbit (*Oryctolagus cuniculus*) in Portugal, Andalusia and on the Azorean islands. *Genetics* 127, 789–799.

van der Loo, W., Mougel, F., Sánchez, M.S., Bouton, C., Castien, E. *et al.* (1999) Cytonuclear disequilibria in wild populations of rabbit (*Oryctolagus cuniculus* L.) suggest unequal allele turnover rates at the b locus (IGKC1). *Immunogenetics* 49, 629–643.

Veale, E.M. (1957) The rabbit in England. *English History Review* 5, 85–90.

Verde Arregoitia, L.D., Leach, K., Reid, N., and Fisher, D.O. (2015) Diversity, extinction, and threat status in Lagomorphs. *Ecography* 38, 1155–1165.

Vere, D.T., Jones, R.E. and Saunders, G. (2004) The economic benefits of rabbit control in Australian temperate pastures by the introduction of rabbit haemorrhagic disease. *Agricultural Economics* 30, 143–155.

Vigne, J.D. (1988) Données préliminaires sur l'histoire du peuplement mammalien de l'îlot de Zembra (Tunisie). *Mammalia* 52, 567–574.

Wang, X., Liang, D., Jin, W., Tang, M., Liu, S. and Zhang, P. (2020) Out of Tibet: genomic perspectives on the evolutionary history of extant pikas. *Molecular Biology and Evolution* 37, 1577–1592.

Whitman, B.D. (2004) *Domestic Rabbits & Their Histories: Breeds of the World*. Leathers Publishing, Overland Park, Kansas.

Wilkens, B. (1987) La fauna dell'età del Bronzo di Mursia. Nota preliminare. Atti della Società Toscana di Scienze Naturali, *Serie A*, 94, 215–224.

Williams, C.K., Parer, I., Coman, B.J., Burley, J. and Braysher, M.L. (1995) *Managing Vertebrate Pests: Rabbits*. Bureau of Resource Sciences/CSIRO Division of Wildlife and Ecology, Australian Government Publishing Service, Canberra.

Zeder, M.A. (2006) Central questions in the domestication of plants and animals. *Evolutionary Anthropology* 15, 105–117.

Zeder, M.A. (2012) The domestication of animals. *Journal of Anthropological Research* 68, 161–190.

Zenger, K.R. (1996) The genetic variation and evolution of the mitochondrial DNA non-coding region in Australian wild rabbit populations *(Oryctolagus Cuniculus (L))*. MSc thesis, University of Western Sydney – Hawkesbury, Richmond, Australia.

Zenger, K.R., Richardson, B.J. and Vachot-Griffin, A.M. (2003) A rapid population expansion retains genetic diversity within European rabbits in Australia. *Molecular Ecology* 12, 789–794.

Zeuner, F.E. (1963) *A History of Domesticated Animals*. Harper & Row Publishers, Evanston, UK.

Ziege, M., Theodorou, P., Jüngling, H., Merker, S., Plath, M. *et al.* (2020) Population genetics of the European rabbit along a rural-to-urban gradient. *Scientific Reports* 10, 1–12.

2 Rabbit Breeds and Lines and Genetic Resources

Samuel Boucher[1], Hervé Garreau[2]*, Paul Boettcher[3] and Gerard Bolet[2]
[1]*Labovet Conseil (Réseau Cristal), Les Herbiers, France;* [2]*GenPhySE, INRAE, Université de Toulouse, Castanet-Tolosan, France;* [3]*Food and Agriculture Organization of the United Nations, Rome, Italy*

2.1 What Does the Term 'Race' or Breed, as Applied to Rabbits, Mean Today?

The term 'race' probably comes from the 14th-century Italian *razza*, used in the sense of species. According to Quittet (1965), 'A race or breed is the normal, constant culmination of efforts to improve a population'. This idea echoes the patient and meticulous observations of Charles Darwin, who, in 1859 in *The Origin of Species*, wrote that 'nature gives successive variations; man adds them up in certain directions useful to him'. These concepts highlight the role ascribed to humans in forming breeds. Quittet continued, in the *Revue de l'Elevage*, giving the following definition: 'Within a species, a breed is a collection of individuals having a certain number of morphological and physiological characteristics in common which they perpetuate when they breed amongst themselves'. Moreover, the FAO (1999) defines a breed as being formed by 'either a sub-specific group of domestic livestock with definable and identifiable external characteristics that enable it to be separated by visual appraisal from other similarly defined groups within the same species or a group for which geographical and/or cultural separation from phenotypically similar groups has led to acceptance of its separate identity'. Article 2 section 1 of the French decree n. 69667, dated 14 June 1969, applying the Law on Breeding, states that 'to be able to be recognized, a breed must cover a group of animals of the same species having an adequate number of common characteristics: the breed model is defined by enumerating these hereditary characteristics, giving an indication of the mean intensity of their expression in the group under consideration' (Arnold, 2005). A breed is not therefore a group of clones – the notion of breed involves a certain inevitable variability when working with live organisms (Denis, 1982).

However, 'the breed as the geneticist endeavours to define it is only a moment in the biological history of the group,' explained Jacques Rufié (1982) in his *Traité du vivant*. In these few words, he straightaway posed the question of time: a breed is ephemeral and changing; it evolves. We have chosen to summarize the various definitions and to consider a breed to be 'a group of individuals which evolve within the same domesticated species, which have the same principal, particular genes, and which resemble each other and their progenitors but not always their descendants' (Boucher, 1988).

Contact e-mail address: herve.garreau@inrae.fr

DOI: 10.1079/9781780643342.0002

This definition applies very well to rabbit breeds, which are often picked out by their coat, among other morphological characteristics. However, there are breeds where their representatives are heterozygous for characteristics related to fur colour. This is the case for several domestic breeds. A peculiar example is constituted by the spotted markings of rabbits of the papillon breeds (Giant Papillon or Checkered Giant, English Spot, Rhinelander, Miniature Papillon, Dwarf Papillon, Dalmatian, Rex, etc.), which have the heterozygous genotype *En/en* at the *English spotting* locus (see Chapter 6) and, consequently, do not have homogeneous offspring: crossing two papillon rabbits only 50% of the kits are papillon, the others being either completely coloured (genotype *en/en*) or almost white (genotype *En/En*). This is also the case for rabbits with coloured extremities such as Marten Sable where crossing produces some individuals close to Himalayan rabbits. The individuals which do not resemble their progenitors are not considered as belonging to the breed by amateur breeders and their standard committees (Searle, 1968; Arnold, 1986; Boucher, 1988).

2.2 The Origin of Breeds

The wild species *Oryctolagus cuniculus*, from which all domestic rabbits are descended, has originated in Spain where the oldest fossil was discovered (Arnold, 1986; see Chapter 1). The rabbit has been known to humans for a very long time and was very accurately described from 1765 in the tenth volume of the *Encyclopédie des Sciences*. This very old description could be considered as the first standard.

The method of rearing, or rather, we should say, of keeping in captivity, means that it could not be possible to talk of domestication yet. In the early rearing systems selection was natural. We have to wait until the 16th century for rabbits to be kept other than in monasteries. Enclosed warrens were then developed (closed areas where the animals were concentrated). In this context, natural selection was still strong.

However, grouping the animals together, their being relatively protected from natural predators and the unavoidable increase in consanguinity increased the appearance of peculiar characteristics (size, weight, colour, fur structure, ears and other traits). At the same time, the rate of occurrence of genetic problems or diseases increased and individuals showing them being no longer naturally eliminated (Boucher and Nouaille, 2013). The gustatory quality of these, doubtless, quite 'degenerate' animals was mediocre because of the lack of real natural or artificial selection, and rabbits from natural warrens were still preferred.

In the 19th century, with the abolition of the hunting right privileges of the nobility, rabbit hutches appeared enabling separation by age and sex. The rabbit attained a certain degree of importance and finally found its place among domesticated animals. Multiplication of animals that were markedly different from wild rabbits could now occur: conscious selection could now take place. Interest centred essentially on the animal's fur but also on its meat. Ten or so breeds were listed in the middle of the 19th century and particularly the *Riche* or *Argenté* rabbit, the Lop (with drooping ears), the Angora, the Himalayan (also called Russian or China White), the Nicard (the possible ancestor of dwarf rabbits) and the Giant (Knight, 1881; Boucher, 1993; Arnold, 2005; Le Gal, personal communication). All these names should, however, be considered with caution, since different names are found in the literature for a single population, depending on the countries and period.

In the second half of the 19th century, in Britain, where the fanciers multiplied the pet lines, and in France, the number of new breeds continued to increase, starting from the few breeds mentioned previously and crossing between them and with farm populations (Melay, 1900, 1908; Boucher, 1993; Le Gal, personal communication).

2.3 Creation of Breeds

As Arnold (2005) wrote, the stages of obtaining 'racial populations (of rabbits)' can be summarized as follows: (i) appearance, through successive mating within a group of animals or by inter-group crossing of a type as close as possible to a predefined set of characteristics; (ii) multiplication between them of individuals showing the type more or less markedly, within a previously isolated group, until a certain state of apparent homogeneity appears in descendants (achieving success in this second stage may be very slow); (iii) final perfection solely by breeders selecting the type sought, with the help of all the refined and proven techniques of genetic improvement. All the recognized breeds of rabbits have been created more or less in this way. Some, the primary breeds, arose from geographical isolation (e.g. Normand). Others arose from geographical isolation and were then artificially selected (e.g. Fauve de Bourgogne or Burgundy Fawn, New Zealand Red, Giant Himalayan). In still others, the so-called Mendelian breeds, an 'abnormality' appearing spontaneously in a mutation, has been reproduced (e.g. Rex, Angora, Himalayan, Thuringer or Thuringian, Havana, Marten Sable). Finally, inter-breed crosses have produced synthetic breeds such as the Bouscat Giant White from Argenté de Champagne (or Champagne d'Argent), Flemish Giant and Angora. These synthetic breeds are very close to the lines that constitute nowadays the meat rabbit lines (Fig. 2.1).

2.4 Recognition of an Isolated Population as a Breed or a Line

Primary descriptions are made by the breeders creating the breeds, but since 1962 there has been a real compendium of standards in France managed by a technical commission. This compendium is regularly updated and, after an observation period, includes the new breeds and varieties. Rabbit varieties very often depend on the fur. A French decree dated 14 June 1969 stated that 'variety must correspond to the number of animals of a breed which particular selection treatment has distinguished from other animals of the breed'. Depending on the species, a divergent population may also be qualified as a branch, type, strain or line. The line, or strain, corresponds to obtaining a relatively homogeneous group of animals which can be characterized by a certain level of performance, in a given environment. The line or strain derives from a closed stock of animals or from a group of animals of whatever origin, which are subjected to continuous selection in a predetermined direction (Minvielle, 1990).

In the majority of countries, all rabbit breeds are described in a compendium of national standards (Associazione Nazionale Coniglicoltori Italiani, Standaard van de in Nederland Erkende Konijnenrassen, Commission des Standards Belges, Landesverband Luxemburger Kleintierzuchtervereine, Fédération Suisse de Cuniculture, Fédération Française de Cuniculiculture, Zentralverband Deutscher Rasse-Kaninchenzüchter). In contrast, lines are not described by a national

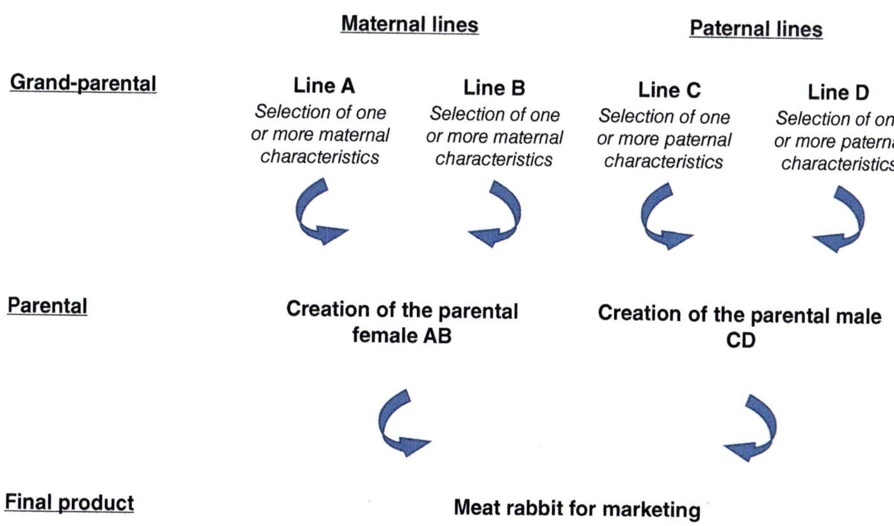

Fig. 2.1. Schematic representation of the production of a meat rabbit.

standard. The French compendium (Commission Technique et des Standards) is the worldwide reference for French breeds. Many European countries have published a compendium of standards for rabbit breeds. In Europe, the standards for breeds created in a given country but recognized in at least three countries are compiled into a compendium updated by the Standards Commission of the European Association of Poultry, Pigeon, Bird, Rabbit and Cavy Breeders.

A standard describes a breed, generally under six headings. The first three are identical for all the descriptions of breed and concern the general appearance (e.g. build, stance, dewlap), weight, size and fur. Then there are three headings which consider the characteristics particular to each race and which constitute its special features. These may be the colour, outline, shape and length of the ears, the head or any other particular characteristic. This is the description the judge will refer to when he or she is in action. The judge will compare the animal to be examined with a fictitious animal described in the standard. Nevertheless, this description is still deliberately somewhat vague to allow breeds to evolve. To freeze them and stop their improvement would indeed be an error as a breed still evolves over time. For example, in two studies, on the commercial line Orylag that originates from the Rex rabbit (Salvetti *et al.*, 2008) and on the Brun Marron de Lorraine breed (Boucher *et al.*, 2007), the phenotypic characteristics of animals produced by re-implanting embryos collected 14 years earlier were compared with contemporary animals of the line or breed, which showed that the animals had evolved. A line by definition evolves and a breed, defined by a deliberately open standard, shows phenotypic evolution over time.

2.5 Rabbit Breeds on an International Level

Although rabbits were domesticated in Europe, they have now spread throughout the world. The FAO maintains a global database of livestock breeds to store information provided by member countries. The database is publicly accessible through the Domestic Animal Diversity Information System (DAD-IS – http://www.fao.org/dad-is, accessed 22 December 2020). Member countries are expected to report in DAD-IS all of the livestock breeds present in their territory and regularly update information regarding population size. Data on population size are used in official United Nation processes to monitor animal genetic diversity, being used as Indicators for the Sustainable Development Goals and the Convention on Biological Diversity. Details about the physical and productive characteristics of each breed can also be inserted, as well as information about breed management, including the history of the breed's development, the primary ways in which the breed is utilized and relevant institutions such as breeders' associations and public or private conservation programmes.

According to DAD-IS, domestic rabbits are found in 70 different countries representing most continents. This is almost certainly an underestimate, due to incomplete recording. FAO members tend to report data more extensively for the major livestock species, such as cattle, chickens, goats, pigs and sheep. FAO uses a system for classifying breeds geographically as either 'local' or 'transboundary'. Local breeds are breeds reported to be present in only one country. Transboundary breeds are those reported in at least two countries. Transboundary breeds are further classified as either 'Regional' or 'International'. Regional transboundary breeds are found only in neighbouring countries within the same region, whereas international refers to breeds found in multiple regions.

In total, 672 different national rabbit breed populations have been reported in DAD-IS. These populations include 335 different breeds, as defined by individual countries. The difference between these two numbers represents populations of transboundary breeds in different countries. The 335 breeds include 276 local and 59 transboundary breeds. Among the transboundary breeds, 34 are regional and 25 are international.

Europe is, by a wide margin, the world leader in terms of rabbit breed diversity according to breed number reported in DAD-IS. This result is not surprising given the fact that Europe comprises the centre of rabbit domestication; in addition, European countries generally report more breeds than do other regions across livestock species (FAO, 2007). European countries report the presence of 230 local breeds; no other region claims as many as 20 (Asian countries report 18 local rabbit breeds). European and Asian countries have a greater tendency to develop more local breeds, rather than simply using transboundary breeds, than do Latin American and African countries. On average, 40–50% of the national rabbit breed populations in Europe and Asian countries are local. This proportion is less than 25% in Africa and Latin America.

In terms of countries, Germany is by far the leader in breed diversity, reporting the presence of 110 rabbit breeds. France, Luxemburg, The Netherlands, Slovenia and Slovakia each have about half this total, reporting 62, 59, 54, 43 and 40 breeds, respectively. Most of this difference is because of a much greater number of local breeds; Germany lists 78 local breeds, whereas the second-place, France, reports only 36. Outside of Europe, China reports 13 breeds, of which six are local, whereas Uruguay reports two local and ten transboundary breeds.

As was explained earlier in this chapter, the definition of breed is somewhat ambiguous and combines scientific, historical, genetic and social aspects. As a result, its definition can vary from circumstance to circumstance and from country to country. This fact may partially explain the large number of breeds reported in Germany. For example, Germany discriminates the Satin population type into 13 different breeds, distinguishable according to characteristics such as body size and coat colour. Most other countries list only one Satin breed, whereas both the American Rabbit Breeders' Association (http://www.arba.net, accessed 22 December 2020) and the British Rabbit Council (http://www.thebrc.org, accessed 22 December 2020) recognize only three distinct Satin breeds, with different colour varieties. Interpretation likely depends on the breeding common practices undertaken within each country, e.g. whether different colour strains are interbreeds or maintained as genetically isolated populations.

In terms of specific breeds, the Californian is the most popular breed on the global level, according to DAD-IS. Forty-one countries report the presence of this breed. Other popular breeds include the New Zealand White, reported by 29 countries, and the Chinchilla reported in 22 countries.

2.6 A Breed Has One or More Purposes

Analysing the history of rabbit breeds, it is possible to see that many of them were created for useful purposes (Boucher and Nouaille, 2013) (Table 2.1). Most special coats were selected, for example, in a period when fur was prized (Boucher, 1993). Castor Rex, where the fur has only downy hair, is the most typical example. The very recent creation of Orylag, one of the branches selected from Castor Rex, shows that breeds still have commercial niches to occupy.

Rabbit fur is also highly prized for the quality of its structure. There are two distinct classes: French Angora has hairs which are plucked and are long and fine, without too much duvet, and the hair of German strains is often shaved, coarser and rich in duvet. Within the same breed there can exist slightly different or very divergent strains (Rougeot and Thebault, 1984).

Finally, rabbits are bred for their meat. There are meat breeds with recognized gustatory qualities (e.g. Fauve de Bourgogne, New Zealand, Argenté de Champagne, Californian, among others). They all have a standard describing the ideal sought, because, as regards rabbit breeds, the phenotype is indeed important. A population very well selected for a given character but very heterogeneous in terms of coat, for example, would have little chance of being recognized as a breed. Because of this, we have, at the same time, lines which are well characterized but not recognized by a standard; they are bred with the aim of improving the animals to make them more prolific and develop better. These lines essentially meet the demands of professional breeders producing meat rabbits and allow the rabbit to form part of an economic system where a breeder can live from his production. Improved lines are essential for the rational production of meat rabbits and began to see the light of day in France in the 1970s, with the rationalization of production which was based on three main elements: animals imported from the USA (New Zealand White and Californian), complete granulated food, and breeding on a wire grid (Boucher, 2011).

Table 2.1. Rabbit breeds classified according to their main use, size, product quality or specialization (for meat, fur, hair and as a pet) and main susceptibilities or general features (1: Main quality; 2: Secondary quality).

Breeds	Meat	Fur	Hair	As a pet	Susceptibility, qualities or general features
Large breeds					**Susceptible to staphylococci**
Flemish Giant	1	–	–	–	Very rapid growth
Bouscat Giant White	2	1	–		Laboratory animals
French lop	1	–	–	2	Maternal qualities
French Giant papillon	2	1	–	2	Obligatory heterozygous *En/en*
Medium-sized breeds					
Alaska	2	1	–	–	Solid, strong bones
Argenté de Champagne	1	2	–	–	Maternal qualities
Saint Hubert Silver	1	2	–	–	Maternal qualities
English lop	–	–	–	1	Fragile, susceptible to otitis
Beveren	1	2	–	–	–
Hotot	1	2	–	2	–
Vendée White	1	2	–	–	Fine skeleton
Vienna White	1	2	–	2	Robust
Vienna Blue	1	2	–	–	Maternal qualities
Vienna Black	1	2	–	–	Maternal qualities
Vienna Grey	1	2	–	–	–
Vienna Blue Grey	1	2	–	–	–
Californian	1	–	–	–	Stands up well on grids
Thuringer (Thuringian)	1	2	–	2	Unusual colour genes, robust
Fauve de Bourgogne	1	2	–	–	Hardy
Giant Chinchilla	2	1	–	–	Maternal qualities
Giant Himalayan	1	–	–	2	Maternal qualities
Gris du Bourbonnais	1	–	–		Hardiness
Japanese	2	–	–	1	Susceptible, many malformations
Belgian Hare	2	–	–	1	Compact meat, lively
New Zealand	1	–	–	–	Stands up well on grids
Silver Fox	1	2	–	–	–
Normand	1	–	–	–	Hardiness, good general qualities
Rhinelander	2	1	–	2	–
					Fur breeds
Rex	2	1	–	2	Fine bones, susceptible to staphylococci
Satin	–	1	–	2	–
French Angora	–	–	1		Particular form of myxomatosis
Swiss fox	–	–	2	1	Particular form of myxomatosis
Small breeds					**generally hardy**
English silver	–	2	–	1	Fine bones
Brun marron de Lorraine	1	2	–	–	Very hardy
Chinchilla	2	1	–	–	Maternal qualities
Saxen gold	2	1	–	–	–
Marburg	2	1	–	–	Unusual coat
Tan	1	2	–	–	Hardy
Havana	2	1	–	–	Unique coloration genes
Dutch	2	–	–	1	Maternal qualities
Lux	2	1	–	–	Unknown coloration genes
English Spot	–	–	–	1	Unique spotting

Continued

Table 2.1. Continued.

Breeds	Meat	Fur	Hair	As a pet	Susceptibility, qualities or general features
Miniature papillon	–	–	–	1	–
Lop miniature	2	–	–	1	Maternal qualities
Perlfee	–	–	–	1	–
Himalayan	1	–	–	–	Maternal qualities
Sablé des Vosges	2	1	–	2	Exceptional fur
Marten sable	2	1	–	2	Unique coat
Dwarf breeds					**general respiratory sensitivity**
Netherland dwarf	2	–	–	1	Low productivity
Dwarf	2	–	–	1	Low productivity
Lop dwarf	2	–	–	1	Good productivity
Angora dwarf	–	–	2	1	Particular form of myxomatosis
Rex dwarf	–	–	–	1	Fragile, susceptible to staphylococci

2.7 Evolution of Breeds and Lines

One or several traits are selected within a line. Thus, several lines are selected to create an end-product intended for the rabbit meat market. For the maternal lines, the females are selected on various criteria such as the number of nipples, prolificacy, uniformity of birth weight, etc. For the paternal lines, the males are selected on characteristics such as carcass yield, and growth rate. In rabbit breeding schemes, these different lines, produced in closed populations, are crossed to give a parental hybrid female (prolific and raising her young well) mated with a pure or crossed male from synthetic paternal lines (producing well structured, quick-growing animals) to give a final product, well suited to the market targeted, benefitting greatly from heterosis or hybrid vigour (Fig. 2.1). New Zealand, Californian, Gris du Bourbonnais, Fauve de Bourgogne, Argenté de Champagne, Bouscat Giant White, Flemish Giant, Vienna Blue, Dutch and French Giant Papillon are the main breeds which are or have been used for producing these lines. There is an inventory of the main commercial lines used in Europe for the production of meat rabbits in Table 2.2. The breeds of domestic animals of a given species are thus a tremendous genetic reservoir (still not totally used) in which to delve to produce a line which, perhaps one day, will be approved as a breed (Rochambeau, 1980).

Besides these utilitarian criteria only concerning the agronomic aspect of the rabbit, humans have at the same time been able to develop natural behavioural habits allowing the animal a place in a human home. Thus, while seeking perfect rotundity (at least for individuals intended for 'breed' competitions) a dwarf rabbit was created with the appearance of a cuddly toy.

Because humans are not able to focus on all criteria at once, artificial selection leads to a degree of bias. By focusing all our attention on the fur, we forget other characteristics. For this reason, in its first hours of glory (in the 1920s), Rex had the bad reputation of being a very fragile animal, extremely susceptible to syphilis (Boucher, 2004). In the same way, selecting dwarf animals with rounder and rounder heads has increased respiratory problems (by shortening sinuses, which no longer filter the air breathed in well enough) and increased the rate of prognathism and poorly growing teeth. By selecting individuals such as Sable des Vosges or Rex with very fine fur, paw problems are encouraged if rabbits continue to be raised on grids. Conversely, the New Zealand, with its very dense pilosity under the paws, is particularly well suited to being raised in this way. Selection must be done at a given moment, in a given environment, and any undesirable appearance of a new characteristic must be ruthlessly tracked down.

Table 2.2. Main commercial lines used in Europe for the production of meat rabbits.

	France			Spain		Hungary	Italy	
	Hypharm Hyplus	Hyla	Hycole	Polytechnic University of Valencia	IRTA	University of Kaposvár	Martini Group	ANCI
Maternal lines	GD14	GPD	GPC	A, V, H, LP	PRAT	Pannon White	A	Bianca Italiana X Pezzata Italiana
	GD25	GPC	GPD			Pannon Ka	B	
Paternal lines	PS40	Hylamax	Medium	R	Caldes	Pannon Terminal	C	Argentata Italiana[1]
	PS59		XXL				D	
	PS119		Coloured Medium					

[1]Italian Silver

This is the only way to eliminate abnormalities or genetic diseases from a given population.

There are many breeds of rabbits. Each decade, a few more are created. Their genetic pool is frequently kept as a secret. However, considering that nowadays rabbit breeding is mainly oriented towards productivity criteria, a lower number of breeds are created compared to the past.

To date, a few dozen qualities (or defects) have been identified and studied in rabbits, but many more are still unrecognized and consequently unexploited. Professional rabbit breeders only make use of a small percentage of the genetic potential of rabbit lines and breeds. Nevertheless, some characteristics, previously unexploited in a rational fashion, appear in so-called 'industrial' lines. For example, the requirement for rabbits with coloured eyes (a regular request from the consumer) has led to the introduced of previously neglected breeds to selection and insemination centres. Moreover, quite a large number of farms are also involved in conservation programmes of endangered breeds. The number of animals of some lines/breeds is, however, decreasing rapidly, which endangers the maintenance of certain original characteristics. For this reason, a programme of deep-freezing embryos of particular lines of threatened breeds has been undertaken involving INRA and the French rabbit breeder association (FFC) (Joly *et al.*, 1998). Similar programmes of *ex situ* conservation have been started in other countries.

2.8 Synthetic Lines in Developing Countries

Genetic improvement is mainly based on selection and/or crossbreeding of above-described breeds or strains. Selection allows the accumulation of additive genetic progress, while the main interest of the crossbreeding between breeds or strains is to profit from their complementarity and exploit heterosis (Bidanel, 1992). In Europe, specialized strains are selected to produce F1 females, which are crossed with a sire of another strain to produce terminal products. This solution has the advantage of exploiting at each generation the entirety of the effect of heterosis (direct and maternal), but it requires a complex scheme, based on the maintenance and selection of the pure stocks and the multiplication and diffusion of the crossbred females (Fig. 2.1). This is the reason why these turnkey programmes are rarely used in developing countries, except in some large-scale private units. They need an important investment and create a technical and economic dependence which does not usually fit the social and economic environment.

Another solution is to create a synthetic line by crossing females of a local population or breed, well adapted to the environment, with imported males or semen from a selected strain to produce an F1 population; it will be bred without selection during a few generations, avoiding inbreeding, and constitute a nucleus submitted to selection. This solution makes it possible to provide farmers with improved animals, while ensuring their independence. It offers the farmers the faculty to adapt their strategy of renewal of their herd to their possibilities: they can practise self-replacement without loss of the genetic level, buy males or the two sexes to the nucleus, permanently or punctually to profit from the genetic progress carried out. It does not exclude the terminal crossbreeding with a male. Table 2.3 summarizes the main characteristics of such programmes in developing countries.

2.8.1 Western Asia

In Saudi Arabia, a national project of rabbit production was established to detect the possibilities of producing meat rabbit under industrialized and hot conditions (Khalil *et al.*, 2002, 2005). For this reason, special emphasis was paid to construct a genetic improvement programme to develop new lines of meat rabbits convenient for this hot country. Accordingly, V-line rabbits were imported in 2000 from the University of Valencia (Spain) and were crossed with desert Saudi rabbits (Gabali). This programme was based on an evidence stating that V-line rabbits and their crosses could produce efficiently under hot climatic conditions (Al-Sobayil and Khalil, 2002; Khalil *et al.*, 2002). From this programme, two synthetic lines (Saudi-2 as a maternal line with the structure of ((¾V ¼S)2)2 and Saudi-3 as a paternal line with the structure of ((¾S ¼V)2)2 were developed from crossing Saudi Gabali with V-line rabbits, both selected for litter weight at weaning and individual weight at 84 days (Table 2.3). Details concerning the development of these new lines were presented by Khalil *et al.* (2002, 2005) and Al-Saef *et al.* (2008). The superiority of these synthetic lines over original breeds was evidenced by Khalil and Al-Saef (2012) and Khalil and El-Zarie (2012).

2.8.2 North Africa

In Algeria, an attempt to introduce selected strains and develop rabbit meat production (between 1985 and 1988) failed because of many factors, among which were the lack of knowledge of the animal, the absence of an adapted industrial feedstuff, and the absence of a prophylactic programme. Afterwards, the strategy of development of this species was based on the use and upgrading of local populations but Kadi *et al.* (2008) underlined the weakness of this industry. Thus, since 1990, the Institut Technique de l'Elevage (ITELV) and some universities, especially that of Tizi Ouzou, set up programmes of characterization of these populations and control of their productive performances. They highlighted the defects of these populations, namely their too-weak prolificacy and their low adult weight, but also their qualities, namely a good adaptation to the local climatic conditions, without any loss of productivity in summer (Lakabi *et al.*, 2004; Zerrouki *et al.*, 2005a, 2005b, 2007). To provide farmers with more productive animals, the ITELV, in collaboration with the INRA, chose to create a synthetic line obtained by a crossing between a local population and the INRA 2666 strain (Gacem and Bolet, 2005). After four generations of homogenization, this synthetic line was compared during 18 months to two local populations in the same conditions. The synthetic line's does were heavier (+200 to 420 g) and the observed litter size showed its superiority (+1.9 to 2.5 born alive) There were no genotype x season interactions which changed the genotypes ranking. It means that the synthetic line is as well adapted as local populations to local climatic conditions and much more productive. So, this comparison confirms the interest of this synthetic line to develop rabbit production in Algeria. It is now selected for litter size and weight at slaughter age with a BLUP index including direct and maternal effects and disseminated by ITELV to co-operatives and multiplication farms (Gacem *et al.*, 2008, 2009; Lebas *et al.*, 2010).

In Egypt, great efforts have been made since 1998 to select for one exotic maternal line under local conditions and to develop and select lines based partially on local breeds. An Egyptian-Spanish programme was established involving Alexandria University, Animal Production Research Institute (APRI, Cairo) and Benha University. V-line rabbits were imported in 1998 from Spain and various selection experiments were practised. The first line was developed from crossing Baladi Red with V-line and this maternal line named APRI was selected for litter weight at weaning (Youssef *et al.*, 2008; Abou Khadiga

Table 2.3. Programmes used for creating and selecting synthetic lines in developing countries.

Synthetic line and origin	Founder breeds	Selection criteria	Selection methodology	Number of generations	Selection response per generation
Maternal lines:					
Saudi-2, Saudi Arabia, Khalil *et al.* (2005)	Line V and Saudi Gabali	LWW + W12	BLUP animal-repeatability model	10	LSB= 0.18 kit/litter LSW= 0.16 kit/litter LWW= 62 g/litter WW= 8.6 g/kit
APRI, Egypt, Youssef *et al.* (2008)	Line V, Baladi Red	LWW	BLUP animal-repeatability model	5	
Paternal lines:					
Alexandria, Egypt, El-Raffa, (2007)	Line V, Baladi Black	ADG (28-63 d)	Individual selection using BLUP	5	
Saudi-3, Saudi Arabia, Khalil *et al.* (2002, 2005)	Line V and Saudi Gabali	LWW + W12	Individual selection using BLUP	8	W12= 38 g; ADG= 0.6 g; LSB= 0.14 kit/litter; LSW= 0.12 kit/litter; LWW= 35 g/litter
Multi-purpose lines:					
ITELV, Algeria Gacem *et al.* (2008)	INRA2666 and local populations	LSB and 63 days weight (direct and maternal effects)	BLUP selection	4 (homogenization) +2 (selection)	
Botucatu, Brazil, Moura *et al.* (2001)	Norfolk English line	Weaning litter + ADG (28-70 d)	Selection index		
Moshtohor, Egypt, Iraqi *et al.* (2008)	Sinai Gabali, line V	LWW+ 56-d weight	Two-stage selection using BLUP		

ADG: average daily gain; LSB = litter size at birth; LSW = litter size at weaning; LWW = litter weight at weaning; WW = weaning weight; W12: weight at 12 weeks.

et al., 2012). A synthetic paternal line named Alexandria was originated in Alexandria University from crossing V-line with Baladi Black and selection was performed for daily weight gain during 28–63 days of age (El-Raffa, 2007). In 2003, a selection programme was started to produce a synthetic multi-purpose line named Moshtohor resulting from crossing Sinai Gabali with V-line and selection was performed for litter weight at weaning and live weight at 56 days (Iraqi *et al.*, 2007, 2008).

In Tunisia, local populations of the southwest (Tozeur and Gafsa oases) have been characterized (Ben Larbi *et al.*, 2008) and very poor performances have been evidenced. On the other hand, European strains have been imported about 20 years ago and are maintained in some farms with a limited selection (Lebas and Bolet, 2008). There is a strong political will to develop rabbit meat production in medium-size units. The ministry of agriculture and an interprofessional structure, the GIPAC, in collaboration with INRA, will develop in these farms a programme of data recording to characterize these strains; afterwards, according to the results, a programme of selection and diffusion will be implemented (Bouslama *et al.*, unpublished data).

2.8.3 Sub-Saharan Africa

In Benin, a nucleus of rabbits from local populations of the south has been implemented in the CECURI (University of Abomey-Calavi) (Kpodekon and Coudert, 1993) and, after many years of efforts focusing on sanitary aspects, a selection programme has been recently initiated (Akpo *et al.*, 2008) with the help of INRA. However, the project of crossing this local population with an INRA strain to create a synthetic line is in progress.

In Nigeria, Oseni (2008a) emphasized the lack of applied research for rabbit production; Abu *et al.* (2008) and Oseni *et al.* (2008) characterized the weakness of traditional systems. Oseni (2008b) recently proposed a genetic improvement programme; its strategy is to exploit locally available heterogeneous populations to create a closed nucleus, define breeding goals and selection criteria and provide backyard small units with improved animals.

2.8.4 Latin America

In Brazil, a multi-purpose selection programme was initiated in 1992 to develop a multi-purpose line named Botucatu, using a selection index including litter size and weight at weaning and post-weaning growth traits (Moura *et al.*, 2001).

In Mexico, a strategy of providing 'family packages' to promote the production and consumption of rabbit meat began many years ago and seems to continue (Mendoza *et al.*, 2008), but no information is available about the origin and selection of bucks and does.

Examples of the Development of a few Rabbit Breeds (Boucher, 1993)

1. An example of a primary breed arising from geographical isolation: the *silver* rabbit

We can perhaps trace the history of the contemporary breeds farthest back with *silver* rabbits. This form of colouration was already being described in the 17th century. As reported by Arnold (2005), in 1631 Gervaise Markham wrote: 'the skin, that is accounted the richest which hath the equallest mixture of black and white hair together, yet the black rather shadowing the white.....when another skin is worth two or three pence, they are worth two shillings'. Darwin (1868) made the following comment on this text: 'From this full description we see that silver-grey rabbits existed in England at this period; and, what is far more important, we see that the breeding or selection of rabbits was then carefully attended to'. In 1854, Mariot-Didieux described four varieties of rabbits including one, the *Riche* rabbit, which was 'a more or less dark silver-grey'. He explained that light greys were most appreciated by furriers and that their meat was good. It is true that one of the breeds descended from these *Riche* rabbits, the *Argenté de Champagne*, is today a breed for the table, widely used in the creation of black-eyed commercial lines. In England, Knight (1881) and others, from the end of the 19th century, spoke of different shades of silvering (today the English silver is bred in three shades: dark, medium or light) and indicated that the Cream silver (a fawn rabbit with genes allowing the expression of silvering) was common around Paris, including in the *Jardin d'Acclimatation*.

Continued

Continued

Silvered rabbits have thus produced several breeds (Argenté de Champagne, English silver, etc.) and varieties. Through the natural geographical isolation at first of the wild then the domestic populations, several breeds were able to be obtained, from a single colouration model – a sort of selection marker – which have been exploited for their meat and/or their fur, depending on the period or location.

2. An example of artificially selected geographical isolation: the Fauve de Bourgogne rabbit

Albert Renard, a rabbit breeder at Celle Saint Cyr in the French department of Yonne, found a population of rabbits of average size (3.5 kg) with a wild grey coat which was more phaeomelanic than usual. 'Its geographical area is quite restricted,' he wrote in the journal *L'Acclimatation* of 23 November 1919, although today it is spreading a little everywhere. It is found frequently in Burgundy, from Dijon, in the Côte d'Or department. It is seen in the Nièvre department as far as Clamecy and then enters the Yonne where it is more common than anywhere else, but hardly extends beyond a line between Sens and Montargis, in the Gâtinais region. It does not cross the boundary between the departments of Yonne and Aube.

According to his words it was 'an ordinary domestic rabbit bred haphazardly like so many others by small Burgundian farmers'. Renard related the history of the *Fauve de Bourgogne*, and explained that it was not 'a new breed obtained by crossing different breeds' but was a population 'the existence of which, according to the information collected from good sources, dates from at least a century ago, if not longer'. Still in this article, Renard explained that the breeders of these rabbits 'were not concerned with the principles of hygiene, selection or consanguinity' and this was one of the essential points of the selection for him, because, as he wrote, 'it is undeniable that with such methods, mortality must have been high and natural selection occurred eliminating the least robust'.

From the start, Renard wanted to improve this geographically isolated population to create a uniformly fawn rabbit with an intense colour. One of his good ideas was the creation of the Fauve de Bourgogne breeders association (AELFB) which, still today, is carrying on his selection work, and has made a slightly yellow common domestic rabbit one of our most successful French breeds, found nowadays throughout the world.

3. An example of a synthetic breed: the Bouscat Giant White

The Bouscat Giant White was obtained by M. and Mme Dulong, breeders in Bouscat in the Gironde region, from the French white Angora, the Argenté de Champagne and probably the wild grey Flemish Giant. From 1910, their aim was to obtain 'a short-haired rabbit, which had fur as white, and with the same silver highlights and the same silky filaments, as the Angora'. The first crosses between the Angora and the Argenté de Champagne gave a 'pretty, small, white rabbit, pretty, but small', according to Mme Dulong. It was then, quite logically, crossed with what was the heaviest breed of the period (and still is today): the Flemish Giant. The result was a large rabbit with short, white fur and which has what today is said to be 'frosted' fur, i.e. directional hair sufficiently long to extend beyond the undercoat hair giving a particularly shiny effect to the fur. Once again, the formation of a club allowed the breed to be developed in France.

References

Abou Khadiga, G., Yousef, Y.M.K. and Baselga, M. (2012) Characterization of reproductive performance of the APRI line of rabbits. In: *Proceedings of the 10th World Rabbit Congress*, Sharm El- Sheikh, Egypt, WRSA, ISSN 2308-1910, pp. 743–747.

Abu, O.A., Onifade, A.A., Abanikannda, O.T.F. and Obiyan, R.I. (2008) Status and promotional strategies for rabbit production in Nigeria. In: *Proceedings of the 9th World Rabbit Congress*, Verona, Italy, WRSA, ISSN 2308-1910, pp. 1499–1503.

Akpo, Y., Kpodekon, T.M., Tanimomo, E., Djago, A.Y., Youssao, A.K.I. and Coudert, P. (2008) Evaluation of the reproductive performance of a local population of rabbits in south Benin. In: *Proceedings of the 9th World Rabbit Congress*, Verona, Italy, WRSA, ISSN 2308-1910, pp. 29–33.

Al-Saef, A.M., Khalil M.H., Al-Homidann A.H., Al-Dobaibn S.N., Al-Sobayiln K.A. *et al.* (2008) Crossbreeding effects for litter and lactation traits in a Saudi project to develop new lines of rabbits suitable for hot climates. *Livestock Science* 118, 238–246.

Al-Sobayil, K. and Khalil, M. H. (2002) Semen characteristics of bucks in crossbreeding project involving Saudi Gabali with V line rabbits in Saudi Arabia. In: *Proceedings of the 3rd Scientific Conference on Rabbit Production in Hot Climates*, pp. 151–163 (abstract in *World Rabbit Science* (2004) 12, 191–192).

Arnold, J. (1986) Le lapin ses modèles de coloration. PhD thesis, Université Paris XI, France.

Arnold, J. (2005) *Parcours animalier, escapades zootechniques, cheminement cuniculicoles*. Ici, Paris.

Ben Larbi, M., Haddad, B. and Allalout, S. (2008) Characterization of traditional rabbit breeding system used in the South of Tunisia. In: *Proceedings of the 9th World Rabbit Congress*, Verona, Italy, WRSA, ISSN 2308-1910, pp. 1505–1508.

Bidanel, J.P. (1992) Comment exploiter la variabilité génétique entre races: du croisement simple à la souche synthétique. *INRA Productions Animales, Hors série 'Eléments de génétique quantitative et application aux populations animales'*, pp. 249–254.

Boucher, S. (1988) Qu'est-ce qu'une race aujourd'hui? Essai de définition à partir des populations raciales européennes de lapins domestiques issus d'Oryctolagus cuniculus. MSc thesis, Université Paris VII, France.

Boucher, S. (1993) Les robes du lapin (Oryctolagus cuniculus). Etude génétique des patrons 'blanc aux yeux bleus', 'agouti gris' et 'chamois'. PhD thesis, Faculté de Médecine de Nantes, France.

Boucher, S. (2004) L'apparition des principales maladies spécifiques du lapin et l'évolution de la médecine vétérinaire destinée à l'espèce en France. *Ethnozootechnie* 75, 73–80.

Boucher, S. (2011) La production cunicole française aujourd'hui. Une filière réactive. *Le Nouveau Praticien Vétérinaire élevage et santé* 4(16), 365–371.

Boucher, S. and Nouaille, L. (2013) *Maladies des lapins*. France Agricole, Paris, pp. 356–366.

Boucher, S., Menigoz, J.J., Salvetti, P., Joly, T., Arnold, J. *et al.* (2007) Comparaison phénotypique de lapins de race Brun Marron de Lorraine issus d'embryons congelés en 1992 avec des lapins de la même population vivant en 2007. In: *Proceedings of 12e journées de la recherche cunicole*, le Mans, France, pp. 137–140.

Darwin, C. (1859) *The Origin of Species*. John Murray, London.

Darwin, C. (1868) Lapins domestiques. In: *De la variation des animaux et des plantes*. C. Reinwald et Cie, Paris, pp. 113–142.

Denis, B. (1982) A propos de la notion de race: points de vue d'un zootechnicien. *Ethnozootechnie* 29, 61–67.

El-Raffa, A.M. (2007) Formation of a rabbit synthetic line (Alexandria line) and primary analysis of its productive and reproductive performance. *Egyptian Poultry Science* 27(2), 321–334

FAO (1999) *The Global Strategy for the Management of Farm Animal Genetic Resources. Executive Brief*. FAO, Rome.

FAO (2007) *The State of the World's Animal Genetic Resources for Food and Agriculture*. (Barbara Rischkowsky & Dafydd Pilling (eds)). FAO, Rome.

Gacem, M. and Bolet, G. (2005). Création d'une lignée issue du croisement entre une population locale et une souche européenne pour améliorer la production cunicole en Algérie. In: Bolet, G. (ed.) *Proceedings of 11èmes Journées de la Recherche Cunicole.* ITAVI, Paris, pp. 15–18.

Gacem, M., Zerrouki, N., Lebas, F. and Bolet, G. (2008) Strategy for developing rabbit meat production in Algeria: creation and selection of a synthetic strain In: *Proceedings of the 9th World Rabbit Congress*, Verona, Italy, WRSA, ISSN 2308-1910, pp. 85–89.

Gacem, M., Zerrouki, N., Lebas, F. and Bolet, G. (2009) Comparaison des performances de production d'une souche synthétique de lapins avec deux populations locales disponibles en Algérie. In: Bolet, G. (ed.) *Proceedings of 13èmes Journées de la Recherche Cunicole.* ITAVI, Paris, pp. 149–152.

Iraqi, M.M., Shenana, M.E. and Baselga, M. (2007) Some factors affecting production and milk composition characters in crossbreeding experiment involving Gabali and V-line rabbits in Egypt. *World Rabbit Science* 15, 151–159.

Iraqi, M.M., Afifi, E.A., Baselga, M., Khalil, M.H. and García, M.L. (2008) Additive and heterotic components for post-weaning growth traits in a crossing project of V-line with Gabali rabbits in Egypt. In: *Proceedings of the 9th World Rabbit Congress*, Verona, Italy, WRSA, ISSN 2308-1910, pp. 131–135.

Joly, T., De Rochambeau, H. and Renard, J.P. (1998) Cryobanking of embryos to preserve *ex situ* rabbit genetic resources: practical aspects. *Genetic Selection Evolution* 30 (Suppl), S259–S269.

Kadi, S.A., Djellal, F. and Berchiche, M. (2008) Commercialisation of rabbit's meat in Tizi-Ouzou area, Algeria. In: *Proceedings of the 9th World Rabbit Congress*, Verona, Italy, WRSA, ISSN 2308-1910, pp. 1559–1563.

Khalil, M.H. and Al-Saef, A.M. (2012) Genetic groups comparisons for litter and lactational traits and feeding parameters in program of synthesizing new lines of rabbits. In: *Proceedings of the 10th World Rabbit Congress*, Sharm El- Sheikh, Egypt, WRSA, ISSN 2308-1910, pp. 235–239.

Khalil, M.H. and El-Zarie, M.F. (2012) Genetic groups comparisons for growth, carcass, meat quality and blood parameters in program of synthesizing new lines of rabbits (Egypt/Saudi Arabia). In: *Proceedings of the 10th World Rabbit Congress*, Sharm El- Sheikh, Egypt, WRSA, ISSN 2308-1910, pp. 241–245.

Khalil, M.H., Al-Sobayel K., Hermes, I.H. and Al-Homidan, A.H. (2002) Crossbreeding effects for post-weaning growth, rectal and ear temperature and respiration rates in crossing Saudi Gabali with Spanish V line rabbits. In: *Proceedings of the 7th World Congress on Genetics Applied to Livestock Production*, Montpellier, France, Communication N° 04-12.

Khalil, M.H., García, M.L., Al-Dobaib, S.N., AL-Homidan, A.H. and Baselga, M. (2005) Genetic evaluation of crossbreeding project involving Saudi and Spanish V-Line rabbits to synthesize new maternal lines in Saudi Arabia: I. Pre-weaning litter, lactation traits and feeding parameters. In: *Proceedings of the 4th International Conference of Rabbit Production in Hot Climate*, Sharm El-Sheikh, Egypt, Egyptian Rabbit Science Association, pp. 89–99.

Knight, K.W. (1881) *The Book of the Rabbit*. Upcott Gill, London.

Kpodekon, T.M. and Coudert, P. (1993) Impact d'un centre cunicole de recherche et d'information sur la recherche et le developpment de la cuniculture au Benin. *World Rabbit Science* 1, 25–30.

Lakabi, D., Zerrouki, N., Lebas, F. and Berchiche, M. (2004) Growth performances and slaughter traits of a local Kabylian population of rabbits reared in Algeria: effects of sex and rearing season. In: *Proceedings of the 8th World Rabbit Congress*, Puebla, Mexico, ISSN 2308-1910, pp.1396–1402.

Lebas, F. and Bolet, G. (2008) Impressions sur l'élevage du lapin en Tunisie. *Cuniculture Magazine* 35, 68–76. Available at: http://www.cuniculture.info/Docs/Magazine/Magazine2008/mag35-068.htm#03 (accessed 23 December 2020).

Lebas, F., Gacem, M., Meftah, I., Zerrouki, N. and Bolet, G. (2010) Comparison of reproduction performances of a rabbit synthetic line and of rabbits of local populations in Algeria, in 2 breeding locations – first results. In: *Proceedings of the 6th International Congress on Rabbit Production in Hot Climates*, Assiut University, Egypt.

Mariot-Didieux, D. (1854) *Guide pratique de l'éducateur de lapins*, 5th edn. Hetzel, Paris.

Melay, E. (1900) *Les races de lapins*. Imprimerie Frécourt, Paris.

Melay, E. (1908) *Caractère sportif des principales races de lapins.* Charles Amat, Paris.

Mendoza, B.J., Diaz, Z.S., Velazquez, O.V., Alonso, F.M.U. and Ortega, B.E. (2008) Social and economic contribution of rabbit production in the state of Mexico. In: *Proceedings of the 9th World Rabbit Congress*, Verona, Italy, WRSA, ISSN 2308-1910, pp. 1575–1578.

Minvielle, F. (1990) *Principes d'amélioration génétique des animaux domestiques*. INRA, Paris. Les Presses de l'Université de Laval, Québec.

Moura, A.S.A.M.T., Costa, A.R.C. and Polastre, R. (2001) Litter and growth traits through a multi-purpose index. *World Rabbit Science* 9, 77–86.

Oseni, S.O. (2008a) Research output in rabbit science and production emanating from Sub-Saharan Africa – implications for future research and development efforts. In: *Proceedings of the 9th World Rabbit Congress*, Verona, Italy, WRSA, ISSN 2308-1910, pp. 1591–1596.

Oseni, S.O. (2008b) A proposal for the genetic improvement of rabbits for smallholder units in Nigeria. In: *Proceedings of the 9th World Rabbit Congress*, Verona, Italy, WRSA, ISSN 2308-1910, pp. 1585–1590.

Oseni, S.O., Ajayi, B.A., Komolafe, S.O., Siyanbola, O., Ishola, M. and Madamidola, G. (2008) Smallholder rabbit production in Soutwestern Nigeria: current status, emerging issues and ways forward. In: *Proceedings of the 9th World Rabbit Congress*, Verona, Italy, WRSA, ISSN 2308-1910, pp. 1597–1602.

Quittet, E. (1965) *Races ovines françaises*. La maison Rustique, Paris, pp. 26–30.

Rochambeau, H. (1980) Elevage du lapin et races pures. *Ethnozootechnie* 27, 55–59.

Rougeot, J. and Thebault, G. (1984) *Le lapin angora, sa toison son élevage*. Editions du Point Vétérinaire, Maisons Alfort, France.

Rufié, J. (1982) *Traité du vivant*. Fayard, Paris, pp. 66–67; 394–401.

Salvetti, P., Allain, D., Bolet, G., Hurtaud, J., Boucher, S. *et al.* (2008) La Cryobanque Nationale Lapin: un outil de gestion dynamique des ressources génétiques cunicoles. *Les Actes du BRG* 7, 391–404.

Searle, A.G. (1968) *Comparative Genetics of Coat Color in Mammals*. Logo Press Limited, Academic Press Inc., London.

Youssef, Y.K., Iraqi, M.M., El-Raffa, A.M., Afifi, E.A., Khalil, M.H. *et al.* (2008) A joint project to synthesize new lines of rabbits in Egypt and Saudi Arabia: emphasis for results and prospects. In: *Proceedings of the 9th World Rabbit Congress*, Verona, Italy, WRSA, ISSN 2308-1910, pp.1637–1641.

Zerrouki, N., Bolet, G., Berchiche, M. and Lebas, F. (2005a) Evaluation of breeding performance of a local Algerian rabbit population raised in the Tizi-Ouzou area (Kabylia). *World Rabbit Science* 13, 29–37.

Zerrouki, N., Kadi, S.A., Berchiche, M. and Bolet, G. (2005b) Evaluation de la productivité des lapines d'une population locale algérienne, en station expérimentale et dans des élevages. In: Bolet, G. (ed.) *Proceedings of 11èmes Journées de la Recherche Cunicole.* ITAVI, Paris, pp. 11–14.

Zerrouki, N., Kadi, S.A., Lebas, F. and Bolet, G. (2007) Characterisation of a Kabylian population of rabbits in Algeria: birth to weaning growth performance. *World Rabbit Science* 15, 111–114.

3 The Genome of the European Rabbit and Genomic Tools

Luca Fontanesi[1]*, Samuele Bovo[1] and Giuseppina Schiavo[1]
[1]*Department of Agricultural and Food Sciences, Division of Animal Sciences, University of Bologna, Italy*

3.1 Introduction

Genomics, which has intrinsically revolutionized most of the genetic approaches in all livestock species and in many other animals, mainly relies on DNA sequencing and high-throughput genotyping platforms coupled with data analyses. Particularly, next-generation sequencing and related technologies introduced in this field have been largely contributing to the exponential production of genomic data also in rabbits for several applications that span from the study of genetic variability and the characterization of the transcriptome and epigenome. Transcriptomic data have been used for the functional annotation of the rabbit genome and for gene expression analyses in many different tissues and conditions. Sequence data generated from the rabbit are expected to increase exponentially. The interpretation and use of these data will benefit the understanding of the biology of this species.

The most important resource that derived by the application of genomic approaches in the rabbit during the last years is the complete sequence of its genome (Carneiro *et al.*, 2014). This fundamental resource has opened new opportunities to study this multi-purpose species for its different applications and will contribute to sustain applications in the future.

3.2 The Genome of *Oryctolagus cuniculus*

The genome of the European rabbit is divided into 21 autosomes plus the sexual chromosomes X and Y. Therefore, the 2n is equal to 44. Autosomes and the X chromosome have been sequenced and assembled producing a low coverage and first genome draft (oryCun1.0), generated within the Mammalian Genome Project by the Broad Institute (Lindblad-Toh *et al.*, 2011). This preliminary version, that was used for an evolutionary analysis across mammals, has been improved and a second reference genome version has been subsequently assembled (oryCun2.0; Carneiro *et al.*, 2014). oryCun2.0 was sequenced to a ~7.5X and made available in Ensembl and NCBI databases (http://www.ensembl.org/Oryctolagus_cuniculus/Info; https://www.ncbi.nlm.nih.gov/assembly/GCF_000003625.3, accessed 23 December 2020). The genome assembly is on chromosome level and accounts for a total of about 2.74 Gbp, 82% of which has been anchored to chromosomes. Fig. 3.1A reports the size in Mbp of the assembled chromosomes. In more details, the genome consists of 84,024 contigs assembled into 3318 scaffolds from which 22 chromosomes have been built (i.e. 21 autosomes and the X chromosome). This genome version also includes 3218 unplaced (Un)

Contact e-mail address: luca.fontanesi@unibo.it

DOI: 10.1079/9781780643342.0003

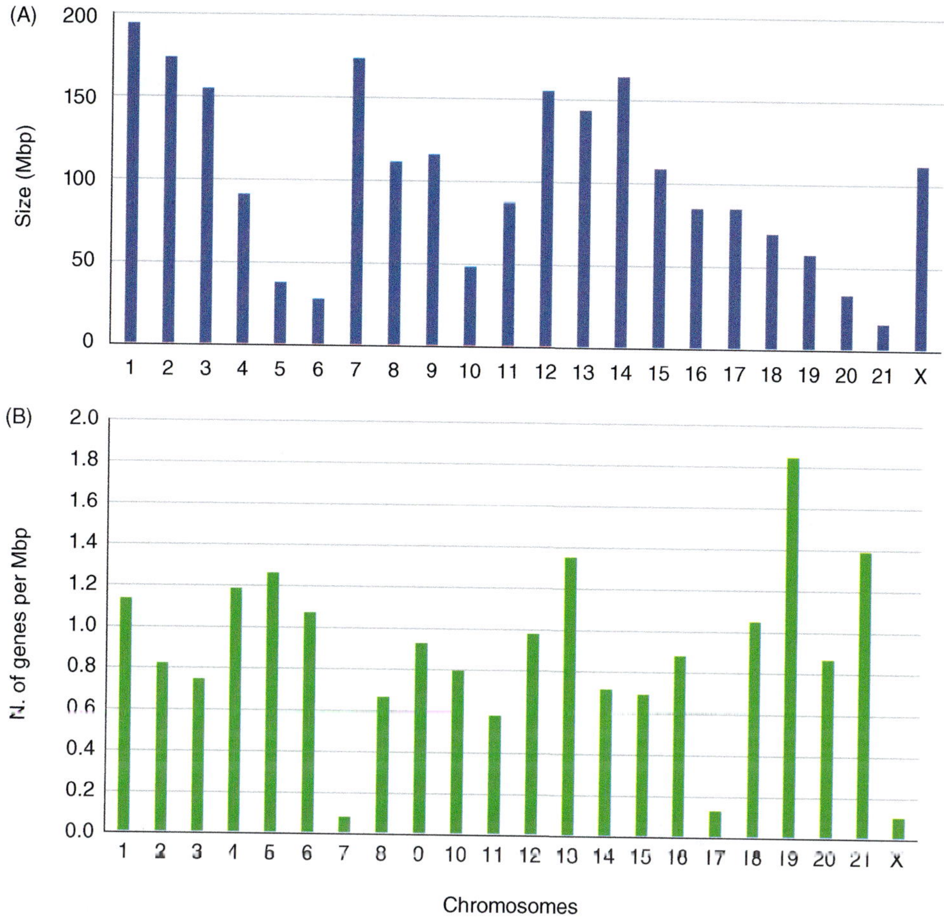

Fig. 3.1. Numbers from the rabbit genome. (A) Size in Mb of the 21 autosomes and of the X chromosome; (B) Gene density (defined as the number of annotated genes per Mb) for the different chromosomes.

sequences (not anchored to any chromosome) of total length 489,693,461 bases with names of chrUn0001 thru chrUn3219. In addition to the nuclear genome, oryCun2.0 contains a non-nuclear assembly of the mitochondrial genome of 17,245 nucleotides.

To obtain an approximate evaluation of the quality of the assembly of a genome, the used parameter is the N50 size. N50 size is the length such that 50% of the assembled genome lies in blocks of the N50 size or longer. The oryCun2.0 N50 length for scaffolds is 35,972,871 kb whereas the N50 size for contigs is 64.648 kb. To have a first evaluation of the quality of this assembly we could compare it to the N50 size of the latest cattle genome assembly (ARS-UCD1.2 submitted by the USDA in 2018), that is one of the most precise genome assemblies among all livestock species. For the *Bos taurus* genome, the N50 size for contigs is 25,896.116 kb and for the scaffolds is 103,308.737 kb. That means that the quality of the cattle genome is much better than that of the oryCun2.0 genome version. In the near future, the complementary use of long-read sequencing technologies (e.g. PacBio, Nanopore) will make it possible to further improve the genome assembly of all species, including that of the rabbit.

The annotation process of the rabbit genome available in the Ensembl database has followed a defined annotation pipeline that included different steps to identify an initial set of genomic features, including repeat features (using RepeatMasker), species specific cDNA and proteins, projected protein-coding annotation from the human assembly GRCh38.p12, a set of proteins retrieved from UniProt to cover the rabbit proteome and RNA-seq datasets. Then a filtering phase selected the protein-coding transcript models to construct the final protein-coding gene set. Other pipelines annotated Immunoglobulin and T-cell receptor genes, non-coding RNAs and pseudogenes.

The current annotation of the oryCun2.0 genome version (Ensembl release 98) lists a total of 20,612 coding genes, 8319 non-coding genes (2520 small non-coding genes, 5736 long non-coding genes and 63 other non-coding genes), 656 pseudogenes and a total of 51,853 gene transcripts. Fig. 3.1B shows the gene density (defined as the number of annotated genes per Mb) on the different assembled chromosomes. Chromosomes 19 and 21 have the highest gene density whereas chromosomes 7, X and 17 have the lowest number of annotated genes per Mb. Fig. 3.2 reports the distribution of the number of repeated features identified in the rabbit genome.

Genetic variations are not integrated yet in the Ensembl browsers (Ensembl release 98), even if extensive studies have been carried out on this species at the genome-wise level. The first study that reported massive data of single nucleotide polymorphisms (SNPs) over the rabbit genome has been obtained using the Ion Torrent Personal Genome Machine to sequence reduced representation libraries in a DNA pool-seq approach (Bertolini *et al.*, 2014). This study identified about 62.5 k SNPs (479 of which were missense mutations, and 16 were stop-gain mutations) by sequencing sampled fractions of the rabbit genome that covered about 0.1 Gb with a detection rate of one SNP per about 1.7 kb of sampled genome. Then, Carneiro *et al.* (2014) described the sequencing and assembly of oryCun2.0 identifying about 50 million high-quality SNPs and 5.6 million insertions/deletions (indels) by whole-genome sequencing of several DNA-pools from different domestic breeds and wild strains.

The genome of a total of 30 rabbits belonging to three different rabbit breeds or lines (New Zealand White, Japanese White and Watanabe heritable hyperlipidemic line) was sequenced at 12–13X producing about 29.8 million of SNPs and 1.6 million of indels (Wang *et al.*, 2016).

Another study applied restriction-site-associated DNA sequencing (RAD-seq) to discover genome-wide SNPs in about 100 rabbits from four Chinese breeds (Ren *et al.*, 2019). A total of about 7 million SNPs were identified but only ~114 k SNPs were considered for the analysis of population genetic structure of the investigated

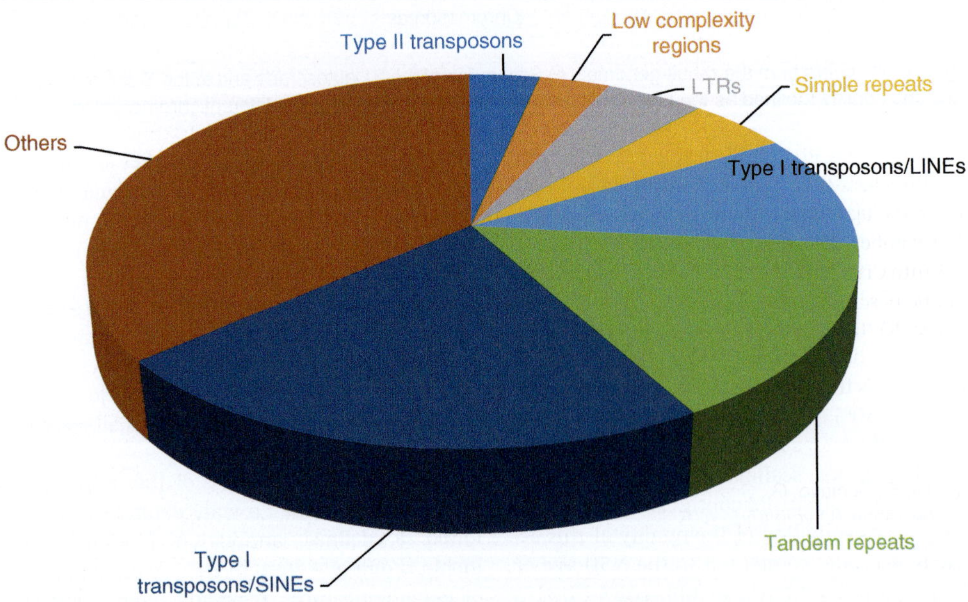

Fig. 3.2. Proportion of the number of repeat features in the rabbit genome.

breeds. Variability in the rabbit genome has been investigated using a genotyping-by-sequencing approach designed to identify signature of selection derived by the adaptation of wild (feral) Australian rabbits to a viral disease (Schwensow *et al.*, 2020).

An additional level of variability in the rabbit genome is constituted by copy number variations (CNVs). Copy number variations are defined as interspecific gains or losses of ≥1 kb of genomic DNA. They are very frequent in all mammalian genomes which represent the most important source of variability in terms of nucleotides involved (covering, on the whole, ~0.4–25% of a genome; Redon *et al.*, 2006; Conrad *et al.*, 2010), with potential functional relevance. One study investigated CNVs in the rabbit genome using the array comparative genome hybridization (aCGH) technique based on high-density probes spread all over the oryCun2.0 genome (Fontanesi *et al.*, 2012). A total of 155 copy number variation regions (CNVRs), identified by overlapping or partially overlapping CNV events in the four analysed rabbits, covered about 6.62 Mb (~0.3% of the rabbit genome). These 155 CNVRs included 95 gains, 59 losses and one with both gain and loss, localized on all chromosomes except on chromosome 20. As more extensive analyses of CNVs are not available in the rabbit, thus far these variants are not annotated in the oryCun2.0 genome version.

3.3 Genomic Tools and Applications

The databases already mentioned above and their browsers (Ensembl: http://www.ensembl.org/Oryctolagus_cuniculus/Info/Index, accessed 23 December 2020) and University of California Santa Cruz (UCSC) Genome Browser: http://genome.ucsc.edu/index.html (accessed 23 December 2020) are the most important bioinformatic tools publicly available for the rabbit genome. Another database (RabGTD) compiling both genome and transcriptome data produced by next-generation sequencing (retrieved from the NCBI Sequence Read Archive, SRA; https://www.ncbi.nlm.nih.gov/sra, accessed 23 December 2020) has been constructed (Zhou *et al.*, 2018). RabGTD gathered genomic variations coming from 79 rabbit genomes and gene expression profiles of 86 tissue samples of domestic rabbits and included tools to query and visualize polymorphisms (https://www.picb.ac.cn/RabGTD/, accessed 23 December 2020).

A high-throughput genotyping tool has also recently been developed for the rabbit. The genotyping array (OrcunSNP array) is based on the Axiom genotyping platform of Affymetrix (now Thermo Fisher Scientific) and includes 199,692 SNPs (200 k) selected among polymorphisms identified by Carneiro *et al.* (2014) and Bertolini *et al.* (2014) with a few other functional markers. This tool has been used in a few genome-wide association studies that detected genomic regions affecting meat quality traits and reproduction traits in meat rabbit lines (Casto-Rebollo *et al.*, 2020; Sosa-Madrid *et al.*, 2020a, 2020b, 2020c; Bovo *et al.*, 2021). It is expected that this tool will become more widely used in many other population genomic studies in this species. A high-throughput genotyping array can be used for cost-effective analyses of the genome to identify and screen simple and complex genetic traits, identify population structures and relationships among animals, families, lines and breeds, identify signatures of selections or infer population histories. Other applications could be for parentage analysis, traceability of rabbit products and animals, for the construction of high-resolution genetic maps and for the design of genomic selection programmes in this species.

References

Bertolini, F., Schiavo, G., Scotti, E., Ribani, A., Martelli P.L. *et al.* (2014) High throughput SNP discovery in the rabbit (*Oryctolagus cuniculus*) genome by next generation semiconductor-based sequencing. *Animal Genetics* 45, 304–307.

Bovo, S., Schiavo, G., Utzeri, V.J., Ribani, A., Schiavitto, M. *et al.* (2021) A genome-wide association study for the number of teats in European rabbits (*Oryctolagus cuniculus*) identifies several candidate genes affecting this trait. *Animal Genetics* 52, 237–243.

Carneiro, M., Rubin, C.J., Di Palma, F., Albert, F.W., Alföldi, J. *et al.* (2014) Rabbit genome analysis reveals a polygenic basis for phenotypic change during domestication. *Science* 345, 1074–1079.

Casto-Rebollo, C., Argente, M.J., García, M.L., Pena, R. and Ibáñez-Escriche, N. (2020) Identification of functional mutations associated with environmental variance of litter size in rabbits. *Genetics Selection Evolution* 52, 22.

Conrad, D.F., Pinto, D., Redon, R., Feuk, L. and Gokcumen, O. (2010) Origins and functional impact of copy number variation in the human genome. *Nature* 464, 704–712.

Fontanesi, L., Martelli, P.L., Scotti, E., Russo, V., Rogel-Gaillard, C. *et al.* (2012) Exploring copy number variation in the rabbit (*Oryctolagus cuniculus*) genome by array comparative genome hybridization. *Genomics* 100, 245–251.

Lindblad-Toh, K., Garber, M., Zuk, O., Lin, M.F. and Parker, B.J. (2011) A high-resolution map of human evolutionary constraint using 29 mammals. *Nature* 478, 476–482.

Redon, R., Ishikawa, S., Fitch, K.R., Feuk, L. and Perry, G.H. (2006) Global variation in copy number in the human genome. *Nature* 444, 444–454.

Ren, A., Du, K., Jia, X., Yang, R., Wang, J. *et al.* (2019) Genetic diversity and population structure of four Chinese rabbit breeds. *PLOS ONE* 14, e0222503.

Schwensow, N., Pederson, S., Peacock, D., Cooke, B. and Cassey, P. (2020) Adaptive changes in the genomes of wild rabbits after 16 years of viral epidemics. *Molecular Ecology* 29, 3777–3794.

Sosa-Madrid, B.S., Hernández, P., Blasco, A., Haley, C.S., Fontanesi, L. *et al.* (2020a) Genomic regions influencing intramuscular fat in divergently selected rabbit lines. *Animal Genetics* 51, 58–69.

Sosa-Madrid, B.S., Santacreu, M.A., Blasco, A., Fontanesi, L., Pena, R.N. and Ibáñez-Escriche, N. (2020b) A genome-wide association study in divergently selected lines in rabbits reveals novel genomic regions associated with litter size traits. *Journal of Animal Breeding and Genetics* 137, 123–138.

Sosa-Madrid, B.S., Varona, L., Blasco, A., Hernández, P., Casto-Rebollo, C. and Ibáñez-Escriche, N. (2020c) The effect of divergent selection for intramuscular fat on the domestic rabbit genome. *Animal* 14, 2225–2235.

Wang, Z., Zhang, J., Li, H., Li, J. and Niimi, M. (2016) Hyperlipidemia-associated gene variations and expression patterns revealed by whole-genome and transcriptome sequencing of rabbit models. *Scientific Reports* 6, 26942.

Zhou, L., Xiao, Q., Bi, J., Wang, Z. and Li, Y. (2018) RabGTD: a comprehensive database of rabbit genome and transcriptome. *Database (Oxford)* 2018. DOI: 10.1093/database/bay075

4 Cytogenetics, Physical and Genetic Maps and QTL Mapping in the European Rabbit

Ina Sternstein[1], Gudrun A. Brockmann[1] and Luca Fontanesi[2*]
[1]Albrecht Daniel Thaer – Institut für Agrar- und Gartenbauwissenschaften Humboldt Universität zu Berlin, Germany; [2]Department of Agricultural and Food Sciences, Division of Animal Sciences, University of Bologna, Italy

4.1 The Karyotype of *Oryctolagus cuniculus*

Painter (1926) first established the correct diploid karyotype of *Oryctolagus cuniculus*, which comprises 21 pairs of autosomes and the X and Y chromosomes (n = 22; 2n = 44). Since this first description, followed then by the work of Melander (1956), who described the position of the centromere in the European rabbit chromosomes, several banded karyotypes have been published (Echard, 1973; Ducayen *et al.*, 1974; Stock, 1976; Chan *et al.*, 1977; Schroder *et al.*, 1978; Hageltorn and Gustavsson, 1979) by using different staining techniques. To avoid confusions, regarding the numbering of the chromosomes, a standardized system was required. The first standardized karyotype for the rabbit was established at the First International Conference for the Standardization of Banded Karyotypes of Domestic Animals in 1976 (Ford *et al.*, 1980). The Committee for Standardized Karyotype of *Oryctolagus cuniculus* modified this karyotype in 1981 by switching the chromosome numbers 16 and 17 (Committee for Standardized Karyotype of *Oryctolagus cuniculus*, 1981). This standardized karyotype provides chromosome and band numbering based on G-band patterns (Fig. 4.1). To define a correlation between G- and R-banded rabbit chromosomes, Hayes *et al.* (2002) constructed new GTG- (G-bands by trypsin using Giemsa) and RBG- (R-bands by bromodeoxyuridine using Giemsa) banded mid-metaphase rabbit karyotypes and performed classical karyotyping and fluorescent *in situ* hybridization (FISH) localization of chromosome-specific markers on both G- and R-banded chromosomes.

4.2 Gene Mapping in the Domestic Rabbit: Physical Gene Maps

The physical location of genes provided the possibility to assign linkage groups to chromosomes. This information, in turn, was important to link cytogenetic-derived chromosome nomenclature to the sequenced chromosome scaffold derived from the sequencing of the European rabbit genome (Carneiro *et al.*, 2014). The complete sequence of the nuclear genome of *Oryctolagus cuniculus* constitutes the definitive physical map that can be improved based on subsequent versions.

Since the Mammalian Genome Project founded by the National Institutes of Health was started in 2004, the *Oryctolagus cuniculus*

Contact e-mail address: luca.fontanesi@unibo.it

DOI: 10.1079/9781780643342.0004

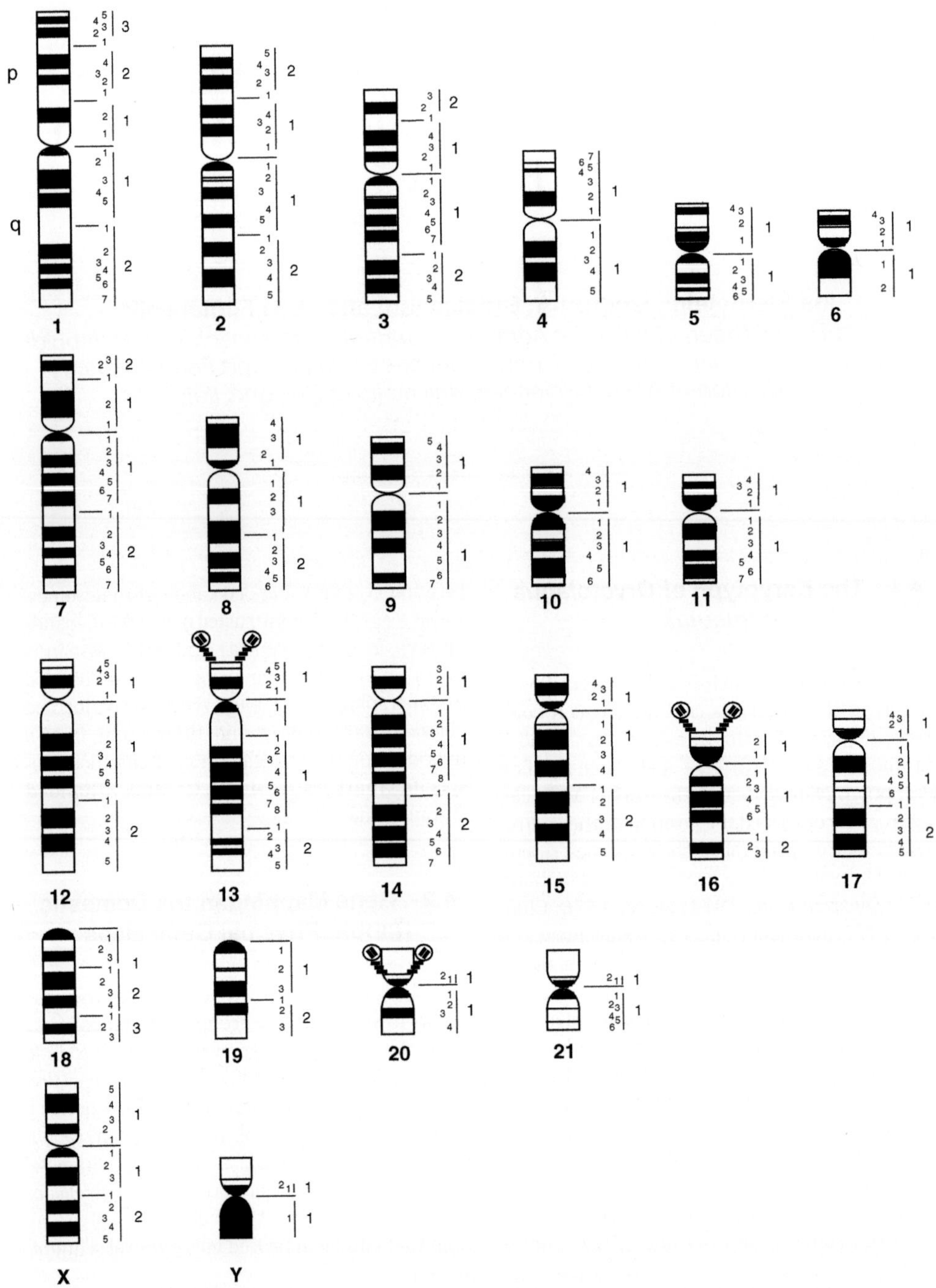

Fig. 4.1. Standardized karyotype of the European rabbit defined by the Committee for Standardized Karyotype of *Oryctolagus cuniculus* (1981). Ideograms illustrating the G-banding patterns of the chromosomes have been redesigned following the standard karyotype patterns.

was one of 24 mammals that was sequenced. The details of the first twofold assembly were published in 2005 (see Chapter 3). Today, the genome is sequenced with a sevenfold coverage. Approximately 82% of the 2.74 Gb sequences have been anchored to chromosomes, comprising all autosomes, the X chromosome and the mitochondrial genome. The remaining unanchored sequence contigs were assigned to virtual chromosomes. The current status on the physically mapped genes and the estimated length of the chromosomes is described in Chapter 3.

4.2.1 Mapping using somatic cell hybrids

A major advancement in assigning genes to rabbit chromosomes was achieved by the generation and use of somatic cell hybrids in the early 80s. A hamster-rabbit somatic cell hybrid panel allowed the mapping of several genes to specific chromosomes (Echard *et al.*, 1981; Soulie and de Grouchy, 1982). The first genes that were mapped were hypoxanthine-guanine phosphoribosyltransferase (*HPRT*), glucose-6-phosphate dehydrogenase (*G6PD*), phosphoglycerate kinase 1 (*PGK*) and galactosidase, alpha (*GLA*), which were all assigned to the X chromosome (Echard *et al.*, 1981). In 1982, synteny between the rabbit gene loci lactate dehydrogenase B (*LDHB*) and triosephosphate isomerase (*TPI*) as well as between lactate dehydrogenase A (*LDHA*) and tissue acid phosphatase 2 (*ACP2*) was observed (Echard *et al.*, 1982). Furthermore, the rabbit gene locus uteroglobin (*UGL*) was syntenic with *LDHA* and *ACP2* (Gellin *et al.*, 1983). These two loci and 11 other genes were assigned to specific chromosomes (Soulie and de Grouchy, 1982, 1983). *LDHA* and galactose-1-phosphate uridyltransferase (*GALT*) were mapped to chromosome 1, *LDHB* and glyceraldehyde-3-phosphate dehydrogenase (*GADPH*) to chromosome 4, glutathione peroxidase (*GPX*) and aminoacylase (*ACY*) to chromosome 9, phosphoglucomutase 1 (*PGM1*) to chromosome 13, guanylate kinase (*GUK*) and malate dehydrogenases 2 (*MDH2*) to chromosome 15, and glutathione reductase (*GSR*) to chromosome 19. Peptidase B (*PEPB*), nucleoside phosphorylase (*NP*) and inosine triphosphatase (*ITP*) were assigned to chromosome 17 [in their original paper they were assigned to chromosome 16 using the karyotype according to the Reading Conference in 1976 (Ford *et al.*, 1980)]. Medrano and Dutrillaux (1984) were able to map immunoglobulin kappa (*IGK*) to chromosome 3 and immunoglobulin heavy (*IGH*) to chromosome 17 (originally chromosome 16) using mouse–rabbit cell hybrids.

4.2.2 Mapping by *in situ* hybridization

In 1969, *in situ* hybridization was introduced into rabbit cytogenetics by Pardue and Gall (1969). At that time, radioactive gene probes were used. Using ^{125}I-labeled ribosomal RNA (*rRNA*), genes were localized on chromosomes 13, 16, 20, and on the telomeric region of the long arm on chromosome 21 (Martin-DeLeon, 1980). The first genes assigned to a band were casein alpha (*CASA*) and casein beta (*CASB*), which were assigned to band q24 of chromosome 12 (Gellin *et al.*, 1985). Later, Kirsten rat sarcoma viral oncogene homolog (*KRAS2*) was assigned to 16p11-q11 (Martin-DeLeon and Picciano, 1988), followed by the assignments of the creatine kinase B (*CKBB*) gene to 20q13-ter (Mahoney *et al.*, 1988), creatine kinase M (*CKMM*) gene to 19q11-q12 (Martin-DeLeon *et al.*, 1989), beta-like globin gene cluster (*HBBC*), parathyroid hormone (*PTH*) gene and Harvey-ras 1 proto-oncogene (*HRAS1*) to 1q14-q21, alpha-like globin gene cluster (*HBAC*) to 6p12 (Xu and Hardison, 1989, 1991), and phosphorylase kinase, alpha (*PKHA*) and beta (*PHKB*) genes to Xq12 and the proximal end of 5q, respectively (Debecker and Martin-DeLeon, 1992).

While the mapping resolution was rather poor with the use of radioactive probes, the physical mapping of genes became more accurate with the introduction of fluorescent *in situ* hybridization (FISH). Using FISH, additional genes could be integrated into the cytogenetic map of the rabbit, such as: retinoblastoma (*RB1*) gene to 8q21, thymidylate synthetase (*TYMS*) gene to 9q12-13, superoxide dismutase 1 (*SOD1*) gene to 6p12, superoxide dismutase 2 (*SOD2*) gene to 12q14-16 and thymidine kinase (*TK1*) gene to 19q22-23 (Lemieux and Dutrillaux, 1992), the Na-phosphate cotransporter 1 (*NPT1*) and 2 (*NPT2*) genes localized to 12q11 and 3p11, respectively (Kos *et al.*, 1996), calcium-sensing receptor (*CASR*) gene to 14q11 (Martin-DeLeon *et al.*, 1999), the whey acidic protein (*WAP*) gene to 10q16 and the three major histocompatibility complex (MHC) related genes *DQA*, *TNFA* and *R19* to 12q11 (Rogel-Gaillard *et al.*, 2000, 2001).

Using FISH, the genes encoding prolactin receptor (*PRLR*) and growth hormone receptor (*GHR*) were mapped to q12 of chromosome 11 (Zijlstra *et al.*, 2002). These new positions were not in agreement with the before-published comparative mapping data (Korstanje *et al.*, 1999; see this chapter, synteny mapping), which suggested the position of these genes on chromosome 10 due to homology between human chromosome 5 and rabbit chromosome 10. The discrepancy could result from an ambiguity error between chromosome 10 and chromosome 11 (Zijlstra *et al.*, 2002). However, with today's knowledge, rearrangements between different rabbit breeds cannot be excluded. Additional FISH mapping led to the localization of the following genes: peripheral myelin protein 2 (*PMP2*) to 3q14, metallothionein 1 (*MT1*) to 5p12, T-cell receptor beta locus (*TRB*) to 7p23, transferrin (*TF*) to14p11-q11, hexokinase 1 (*HK1*) to 18q21, cytochrome P450, subfamily IIC, polypeptide 4 (*CYP2C4*) to 18q24, arachidonate 15-lipoxygenase (*ALOX15*) to 19q12, angiotensin I converting enzyme (*ACE*) to 19q22 (Zijlstra *et al.*, 2002), insulin-like growth factor 2 (*IGF2*) to 1q27, transforming growth factor, alpha (*TGFA*) to 2q21, insulin-like growth factor 1 (*IGF1*) to 3q15, deoxyribonuclease I (*DNASE1*) to 6p14, coagulation factor VII (*F7*) to 8q26, cytochrome b-5 (*CYB5*) 9q17, calreticulin (*CALR*) to 10p12, solute carrier family 2, member 1 (*SLC2A1*) to 13q31-32, alpha-2-HS-glycoprotein (*AHSG*) to 14q17-18, polymeric immunoglobulin receptor (*PIGR*) to 16q21, lipase, hepatic (*LIPC*) to 17q11, transforming growth factor, beta 3 (*TGFB3*) to 20q12, nitric oxide synthase 1 (*NOS1*) to 21q14, sex determining region Y (*SRY*) to Yp12 (Hayes *et al.*, 2002), *PMP2* to 3q11, *MT1* to 5q11, *TCRB* to 7pter, *ALOX* to 19q11 (Korstanje *et al.*, 2003), and the microsatellites Sat13 to 1p12, Sol33 to 3q11 and D1Utr6 to 1q24-q26 (Zijlstra *et al.*, 2002). Based on the FISH mapping result of the gene *DNASE1* on the p arm of OCU6 and not on the q arm as expected from the cross-species comparison with the human chromosomal band (HSA16p13.3), the authors concluded that the rabbit chromosome 6q (OCU6p) corresponds to HSA16 and OCU6q to HSA7 (Hayes *et al.*, 2002).

A further improvement of the cytogenetic map of the rabbit was obtained by FISH mapping of 235 genes (Chantry-Darmon *et al.*, 2003, 2005a). The mapping of these genes contributed to the discovery of a conserved segment between rabbit chromosomal band 4q15.3 and part of human chromosome 22. Together with these genes, 305 rabbit microsatellite sequences were described and mapped, 177 of them were assigned to 139 defined cytogenetic positions distributed across all rabbit chromosomes except OCU21 (Chantry-Darmon *et al.*, 2005b).

4.3 Synteny Mapping

Using the human genome information, the gene map of the rabbit could be further improved. This might still be useful even though the rabbit genome sequence is now available, considering that a lot of scaffolds still need to be assigned to a chromosome. The work of Dutrillaux *et al.* (1980) compared the R-banded karyotypes of rabbit and human and found about 50 structural rearrangements that differentiate the rabbit from the human karyotype. However, in using the R-banding comparison, some rabbit chromosome arms still remained unassigned or could not be clearly identified.

The technique of reciprocal chromosome painting between mammalian species (Wienberg and Stanyon, 1995) and the massive mapping of rabbit genes by FISH have facilitated the construction of a first homology map between the human and rabbit genome (Korstanje *et al.*, 1999). The results revealed extensive genome conservation between the two species. The whole rabbit chromosomes 12, 19 and X were homologous to human chromosomes 6, 17 and X, respectively. The other human chromosomes were homologous to two or sometimes three rabbit chromosomes. Some hybridization patterns found are remarkably conserved throughout the mammalian orders. For example, the chromosome segments homologous to human chromosomes 3 and 21, 4 and 8p, 7 and 16, 12 and 22,

14 and 15, 16 and 19, which reside on rabbit chromosomes 14, 2, 6, 21, 17 and 5, respectively, are also found in other mammals like the cat, the mink, the mouse, the pig and the cow. The chromosome segments homologous to human chromosomes 1 and 10, which are in synteny with rabbit chromosome 16, are also found in the mouse and the grey squirrel. It seems that these regions with conserved synteny are ancestral to all mammalian species (Ferguson-Smith and Trifonov, 2007).

The syntenic regions between *Homo sapiens* (GRCh38 genome version) and European rabbit (oryCun2.0 genome version) extracted from their pairwise alignment in Ensembl release 76 (run in February 2020) identified 242 synteny blocks. This synteny map covers about 73% of the human genome and 79% of the rabbit genome. The graphical representation of the human–rabbit synteny map for the rabbit autosomes (OCU) versus the human chromosomes (HSA) is shown in Fig. 4.2.

4.4 Genetic Linkage Maps

With respect to inheritance patterns, genetic markers provide a powerful tool to scrutinize the transfer of maternal and paternal alleles to offspring. By observing parental and offspring marker alleles, markers can be identified which are inherited together, more often than expected, by chance, an observation that is named linkage disequilibrium. The analysis of linkage between markers can be used not only to identify markers which are inherited together but also to determine the genetic distance between markers and their order on a chromosome. Linkage mapping is based on the recognition of recombination events through successive generations as a result of crossing over of chromatids of homologous pairs of chromosomes between polymorphic markers during meiosis. The recombination rate is used as a measure for the genetic distance between the markers. Since linkage mapping is dependent on the occurrence of polymorphic markers, reference pedigrees for linkage mapping are usually generated by crossing genetically different breeds. In rabbits, initially, coat colour (Castle, 1924a), blood protein groups (Sawin *et al.*, 1944), immune proteins (Tissot and Cohen, 1974) and enzyme isoforms (van Zutphen, 1974) were used for mapping studies. Blood protein groups and biochemical enzymes contributed substantially to early linkage mapping efforts in this species as in most other mammalian species. Good markers for genome mapping are microsatellites and SNP because they are easy to type, very abundant and distributed across the whole genome. In particular, SNP emerged as the most widely used markers as they can by genotyped at high-density time and cost-effectiveness, on SNP chips and by next-generation sequencing shots (Bertolini *et al.*, 2014; Sosa-Madrid *et al.*, 2020).

The construction of genetic linkage maps is based on examination of co-segregation of marker alleles through a pedigree. Order and distances of linked markers along a chromosome are defined by the position and frequency of recombination between markers during meiosis. In part, the frequency of meiotic crossover events depends on the length of a chromosome, the distance from the centromere, the specific pedigree, and the sex. Large chromosomes recombine more often through meiosis than smaller ones. An ideal mapping pedigree is produced by crossing genetically divergent breeds and consists of parents and two subsequent generations of full sibs. The meiosis of F_1 parents to produce the F_2 generation is particularly informative according to the Mendelian Laws of segregation of alleles, if fixation of alternative alleles in the two parental breeds is assumed. The rabbit is an ideal animal to produce a large number of informative meiosis seen in offspring of F_2 intercross or in a backcross pedigree.

4.4.1 History of the rabbit linkage map

First linkage in rabbits was reported in 1919 between *Dutch* (*du*) and *English spotting* (*En*) by Castle (1919). Later on, linkage was found between the *Angora* locus (*L*) and *du* and *En* (Castle, 1924a). Only a short time later, linkage between the *Albino* (*c*) and the *Brown* (*b*) pigmentation loci was reported by Castle (1924b). Linkage between the *Yellow fat* (*y*) and *c* loci was described by Pease (1928). The distance between *du* and *En* was determined to be 1.2 cM (Castle and Sawin, 1941).

In the 1960s, the rabbit was used for immunological studies, thus, various immunological

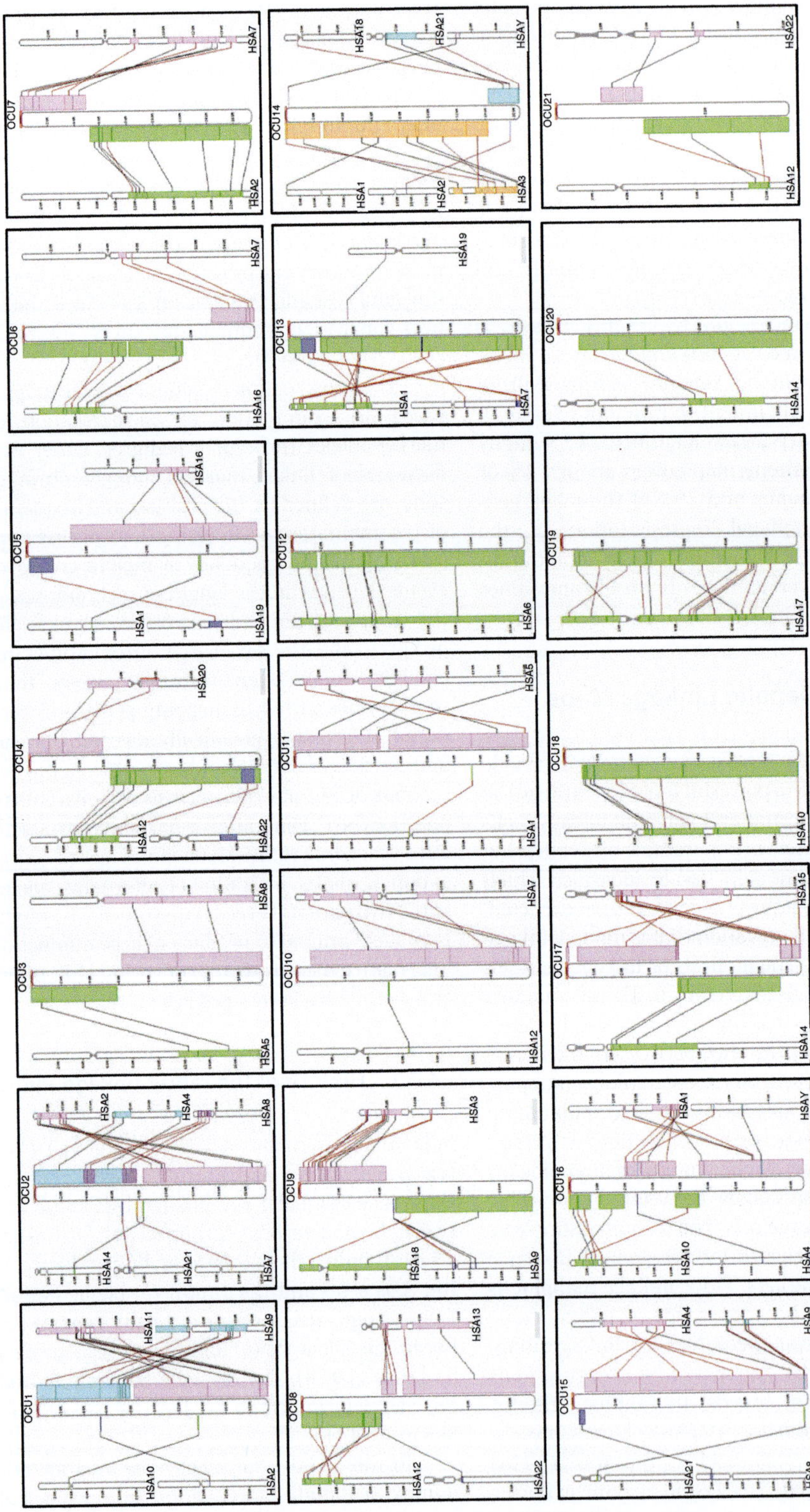

Fig. 4.2. Synteny maps between the human and the European rabbit chromosomes obtained using the Ensembl Synteny analysis tool (Cunningham *et al.*, 2019). Chromosome X which is in almost complete synteny was not reported.

markers were added to the genetic map (Hagen *et al.*, 1978). In the 1980s a few biochemical markers were added to the linkage map. Summarizing these results, Fox (1994) presented the first comprehensive linkage map, comprising 10 autosomal rabbit linkage groups (LG), 11 chromosomes and five X-linked loci. The 10 autosomal linkage groups were named LG1 to LGX. The whole map included a total of 69 assigned loci.

Subsequently, three F_2 reference populations were generated to improve the genetic map of the rabbit. Korstanje *et al.* (2001a, 2001b, 2003) at Utrecht University (The Netherlands) generated a F_2 intercross population between the rabbit inbred strains AX/JU and IIIVO/JU. Chantry-Darmon *et al.* (2006) at INRA (France) used a cross between three rabbit inbred strains (INRA 2066, Castor Orylag and Laghmere), and Sternstein *et al.* (2015) at Humboldt-Universität Berlin (Germany) used an intercross between Giant Grey and New Zealand White. In addition to the F_2 intercross a backcross population was generated from an initial cross between the inbred strains AX/JU and IIIVO/JU. This cross was used to generate the first male linkage map by using 103 amplified fragment length polymorphism (AFLP) markers with a map length of 583 cM (Van Haeringen *et al.*, 2002). Furthermore, two backcross populations with the partially inbred strains OS/J, WH/J and X/J were used for linkage analysis (Korstanje *et al.*, 2001a). For these backcross populations two linkage groups could be assigned to OCU1.

The mapping of 36 microsatellites in the intercross AX/JU and IIIVO/JU led to the identification of new linkage groups to chromosomes 1, 3, 5, 6, 7, 12 and 19 (Korstanje *et al.*, 2001a, 2001b, 2003). The length of the linkage groups was approximately 470 cM. In the cross of the three rabbit inbred strains INRA 2066, Castor Orylag and Laghmere, for the first time an integrated genetic and cytogenetic map was constructed. In this cross, 109 microsatellites and two phenotypic markers were mapped (Chantry-Darmon *et al.*, 2006). The linkage analysis could integrate 90 markers into 20 linkage groups on 20 chromosomes. The other 21 microsatellites could be mapped to separate linkage groups. While 19 out of the 21 markers could be mapped precisely to a cytogenetic position, the two other microsatellite markers could only be assigned to a chromosome. No marker could be mapped to chromosomes 20, 21 and X. The constructed genetic map spans 2766.6 cM averaged across the male and female maps. The male and female maps were 2942 and 2930 cM long, respectively. In the cross of Giant Grey x New Zealand White 186 microsatellite markers and three SNPs could be integrated in to 21 linkage groups residing on 19 autosomes and the X chromosome (Sternstein *et al.*, 2015). The total length of all autosomes reached 1419 cM with an average marker distance of 7.8 cM. In contrast to the findings of Chantry-Darmon *et al.* (2006), the total map length calculated from maternal meioses was 1.4 times longer than from paternal meioses. These sex-specific results are consistent with findings in several other species. For example, sex-specific differences in recombination rates were also reported in pigs (Archibald *et al.*, 1995), cattle (Kappes *et al.*, 1997) and mice (Cox *et al.*, 2009). However, in distinct regions on OCU4 (LG4b), OCU9 and OCU16 maternal maps were shorter than paternal maps, a result which is also consistent with other species, e.g. pigs (Geldermann *et al.*, 2003), and mice (Cox *et al.*, 2009).

Other developments in the European rabbit linkage map, including links to the reference genome, are expected with the use of the commercial SNP genotyping tool which can analyse about 200,000 biallelic markers.

4.4.2 Comparison of linkage maps in European rabbit populations

In all three reference populations, the linkage groups on chromosomes 1, 3, 5, 7, 8, 11, 14, 16, and 19 have the same marker order, although differences with regard to marker distances do occur. Differences in the marker order were identified for the linkage groups on chromosomes 4, 6, 9, 13, 15, and 18. The other remaining chromosomes (2, 10, 12, 17, 20, 21, X and Y) cannot be compared, because different markers were used in the different populations (Table 4.1). Information presented in this table is important to link data of classical genetic maps to genome coordinates for further exploitation and integration with SNP genotyping data that will be produced in the future.

Table 4.1. Summary of the data associated with microsatellites integrated into genetic maps.

Marker symbol (associated gene)	Accession number	Localization (Literature)	Physical position Chr.: Mb (start)[1]	Linkage group	Berlin Map Position (cM)	Linkage group	INRA Map Position (cM)	Utrecht Map Position (cM)
Chromosome 1								
INRACCDDV0130	AJ874471	1p33[k]	1:3.036859			LG1a	0.0	
D1Utr1	AF389372	1[c]	1:4.365497					0.0
D1L1B10	AF398352	nd[2]	1:11.669725	LG1a	0.0			
Sol51	X94685	nd[2]	1:18.934007	LG1a	5.8			
D1Utr2	AF389367	1[c]	1:23.922373	LG1a	7.0			12.9
D1L2B4	AF389358	nd[2]	1:31.864989	LG1a	14.4			
D1Utr7	AF389355	1[c]	1:46.450792	LG1a	21.4			
INRACCDDV0236 (*TJP2*)	AJ874569	1p21.3-p21.1[h]	1:55.571101	LG1a	27.1			
INRACCDDV0269 (*TJP2*)	AJ874595	1p21.3-p21.1[h]	1:55.598578	LG1a	27.1	LG1a	43.4	
Sat13	X99892	1p12[b,f]	1:63.440888	LG1a	31.7			27.2
INRACCDDV0345 (*PSAT1*)	AJ874661	1p12[j]	1:64.769638	LG1a	32.7			
INRACCDDV0240 (*DAPK1*)	AJ874573	1p11dist[h]	1:73.481778	LG1a	36.8			
INRACCDDV0272 (*PTCH*)	AJ874598	1p11[h]	1:74.514704			LG1a	63.5	
INRACCDDV0299	AJ874621	nd[2]	1:75.221585	LG1a	37.4			
D1Utr3	AF389359	1[c]	UN0044:236839	LG1a	39.9			35.0
Est-5	-	1[c]	-					40.5
INRACCDDV0204	AJ874541	nd[2]	1:77.245221			LG1a	75.3	
D1L7C11	AF389369	nd[2]	1:101.865544	LG1a	53.1			
INRACCDDV0271 (*HTR3B*)	AJ874597	1q14[h]	1:102.427689	LG1a	53.1	LG1a	99.0	
INRACCDDV0252 (*HTR3B*)	AJ874583	1q14[h]	1:102.503351	LG1a	53.1			
INRACCDDV0320 (*SLN*)	AJ874640	1q14[j]	1:108.750414	LG1a	56.8	LG1a	111.2	
D1L8C9	AF389374	nd[2]	1:109.935529	LG1a	57.4			
OCPRG5 (*PRG5*)	M14547.1	nd[2]	1:115.601359	LG1a	61.1			
Tyrosinase (*c-locus*)	AF210660	1[c]	1:127.563000			LG1a	136.3	60.9
INRACCDDV0136	AJ874476	nd[2]	1:129.254095	LG1a	70.5			
D1Utr4	AF389353	1[c]	1:140.830836	LG1a	74.4	LG1a	152.4	67.3
INRACCDDV0302	AJ874624	nd[2]	UN0035:2.586312	LG1a	88.8			
INRACCDDV0169	AJ874508	1q21.5[k]	1:158.578780	LG1a	90.4	LG1a	176.5	
D1Utr5	AF389357	1[c]	1:167.752362	LG1a	96.0			86.9
D1Utr6	AF389354	1q24-q26[c,f]	1:185.813547	LG1a	110.6			97
INRACCDDV0298	AJ874620	1q27dist[k]	UN0303:119096	LG1a	126.8		nl[3]	

Chromosome 2								
INRACCDDV0192	AJ874530	2p21.3dist[k]	2:29.010004	LG2a	0.0		nl[3]	
INRACCDDV0096	AJ874439	2q14dist[k]	2:97.960977			LG2a	0.0	
INRACCDDV0173	AJ874511	2q14-q25[k]	2:103.450832	LG2a	41.4		nl[3]	
INRACCDDV0070 (*TGFA*)	AJ874414	2q21[e]	2:115.725659			LG2a	8.8	
INRACCDDV0077 (*FSHR*)	AJ874421	2q22[h]	2:137.458194	LG2a	72.7			
Chromosome 3								
D6L3H10	AF421925	nd[2]	UN0969:35347	LG3a	0.0			
INRACCDDV0036 (*CD14*)	AJ874398	3p21prox[h]	3:22.734617	LG3a	26.4	LG3a	0.0	
INRACCDDV0159	AJ874499	nd[2]	3:31.608853	LG3a	34.9			
D3Utr2	AF421903	3[i]	3:36.821929			LG3a	12.6	0.0
D3Utr1	Z54345.1	3[i]	3:41.628443					5.2
INRACCDDV0110	AJ874452	nd[2]	3:51.793955	LG3a	49.5			
INRACCDDV0111	AJ874453	nd[2]	3:51.794186	LG3a	50.1			
Sat5	X99887	nd[2]	3:52.120053	LG3a	50.7			
Sol33	X94683	3q11[i]	3:67.506294	LG3a	54.6			16.0
INRACCDDV0158	AJ874498	nd[2]	3:67.808093	LG3a	54.6			
INRACCDDV0225 (*ASPH*)	AJ874558	3q16[h]	3:89.390826	LG3a	59.5			
INRACCDDV0129	AJ874470	nd[2]	3:97.054015	LG3a	72.0			
Sat3 (PLP2/PMP2)	J03744	3q14-3q15[i,f]	3:97.128709	LG3a	72.0			24.6
D3Utr3	AJ874396	3[i]	3:130.580965					43.3
INRACCDDV0203	AJ874540	3q22-q23[k]	3:132.701292	LG3a	90.5		nl[3]	
Chromosome 4								
INRACCDDV0340 (*NCOA6*)	AJ874657	4p13[k]	4:4.997753	LG4a	0.0			
INRACCDDV0314 (*PRNP*)	AJ874635	4p13[h]	4:11.430617	LG4a	16.6			
INRACCDDV0333 (*PRNP*)	AJ874650	4p13[h]	4:11.438746	LG4a	19.7			
INRACCDDV0022 (*ERBB3*)	AJ874385	4q11[h]	4:39.642327	LG4a	30.8	LG4b	0.0	
INRACCDDV0040 (*ERBB3*)	AJ874400	4q11[h]	4:39.692417	LG4a	31.3			
D6L2H3	AF421920	nd[2]	UN0016:1.423853	LG4a	32.2			
INRACCDDV0100	AJ874442	4[l]	4:40.214873	LG4a	33.9		nl[3]	
INRACCDDV0182	AJ874520	4q13[k]	4:47.789199	LG4a	36.6	LG4a	0.0	
INRACCDDV0090	AJ874433	nd[2]	4:53.765969			LG4b	26.6	
INRACCDDV0248 (*PMCH*)	AJ874579	4q15.1-q15.2[h]	4:81.470379	LG4b	0.0	LG4b	68.4	
INRACCDDV0228	AJ874561	4[l]	UN0068:787523	LG4b	52.5	LG4a	0.0	
INRACCDDV0194	AJ874532	nd[2]	UN0112:177215	LG4b	61.5			

Continued

Table 4.1. Continued.

Marker symbol (associated gene)	Accession number	Localization (Literature)	Physical position Chr.: Mb (start)[1]	Linkage group	Berlin Map Position (cM)	Linkage group	INRA Map Position (cM)	Utrecht Map Position (cM)
Chromosome 5								
INRACCDDV0282	AJ874606	5[l]	UN0660:37685	LG5a	0.0	LG5a	0.0	
D5Utr4	AF421907	5[i]	UN0091:788280	LG5a	11.3			0.0
D5L1C3	AF421908	nd[2]	UN0024:2.971584	LG5a	21.6			
D5Utr3	AF421905	5[i]	5:2.065870			LG5a	43.7	1.5
INRACCDDV0142	AJ874482	5[l]	5:3.956473	LG5a	22.2	LG5a	32.7	
D5Utr2	AF421913	5[i]	UN0024:544852	LG5a	26.0			6.7
Es-1	-	5[i]	nd					18.9
D5Utr1 (*MT1*)	X07791	5[i]	5:12.933984					19.6
Est-2	-	5[i]	nd					26.6
Est4, Est6	-	5[i]	nd					28.5
EstX	-	5[i]	nd					33.0
HP	-	5[i]	nd					37.0
INRACCDDV0039 (LCAT)	AJ874399	5q14[h]	5:23.468490	LG5a	54.2			
INRACCDDV0211 (HAS3)	AJ874545	5q14[h]	5:24.513058	LG5a	64.2	LG5a	77.7	
Chromosome 6								
D6Utr1 (*HBA*)	M74142	6[i]	UN0452:119139					81.4
INRACCDDV0290	AJ874613	6p14prox[k]	6:4.700318	LG6a	0.0	LG6a	0.0	
INRACCDDV0287	AJ874610	nd[2]	6:7.089687	LG6a	4.7			
D6Utr2	AF421922	6[i]	6:9.765056	LG6a	14.9			42.6
D6L2F1	AF421919	nd[2]	6:16.072792	LG6a	23.4			
INRACCDDV0214 (*EIF3S8*)	AJ874548	6p12prox[h]	6:18.674086	LG6a	27.8	LG6a	34.2	
D6L2B5	AF421918	nd[2]	UN0054:856279	LG6a	29.1			
INRACCDDV0120	AJ874462	6q12prox[k]	6:26.244933	LG6a	30.5			
D6Utr3	AF421915	6[i]	UN0163:156403					29.1
D6Utr4	AF421916	6[i]	6:25.038137	LG6a	31.6	LG6a	26.4	16.6
D6Utr5	AF421923	6[i]	UN0062:567					0.0
INRACCDDV0292	AJ874615	6[l]	UN0150:40539			LG6a	41.8	
INRACCDDV0187	AJ874525	6[l]	X:33.398793			LG6b	0.0	
INRACCDDV0213 (*PRKCB1*)	AJ874547	6p12-p13[h]	X:9.618638			LG6b	7.7	
INRACCDDV0127	AJ874468	6[l]	X:24.702536			LG6b	45.9	

Chromosome 7								
D7Utr1 (*TCRB*)	M26312	7[i]	UN0060:1.646812					0.0
D7Utr6 (*PODXL*)	NM001082766	7[i]	7:10.098532	LG7a	0.0			23.9
INRACCDDV0311 (*CALU*)	AJ874633	7p21prox[k]	7:15.768489	LG7a	2.9			
INRACCDDV0231 (*GPR37*)	AJ874564	7p21-p12[h]	7:19.767546	LG7a	5.2			
INRACCDDV0221 (*GPR37*)	AJ874555	7p21-p12[h]	7:19.825055	LG7a	8.9		nl[3]	
Sat12	X99891	nd[2]	7:30.647593	LG7a	17.7			
INRACCDDV0093	AJ874436	nd[2]	7:41.036817	LG7a	22.7			
D7L2F2	AF421933	nd[2]	7:60.680119	LG7a	29.6			
INRACCDDV0323 (*PROC*)	AJ874642	7q14[k]	7:59.306464	LG7a	30.5			
INRACCDDV0336 (*PROC*)	AJ874653	7q14[k]	7:59.289400	LG7a	33.7			
D7Utr2	AF421934	7[i]	7:80.549370					67.7
INRACCDDV0163	AJ874502	7q21dist[k]	7:117.531438	LG7a	68.0		nl[3]	
INRACCDDV0164	AJ874503	nd[2]	7:117.531629	LG7a	69.3			
D7Utr3	AF421935	7[i]	7:118.659538	LG7a	70.9			87.9
MSTN	A7LH84	nd[2]	7:130.429151	LG7a	75.9			
D7Utr4	AF421932	7[i]	7:151.015154	LG7a	88.1			102.2
D7L1B10	AF421926	nd[2]	7:157.320400	LG7a	90.8			
INRACCDDV0092	AJ874435	7q25[k]	7:157.491661	LG7a	92.4			
D7Utr5	AF421930	7[i]	7:161.912736	LG7a	96.2			108.6
D12L1H3	AF421943	nd[2]	7:167.212085	LG7a	98.1			
Chromosome 8								
INRACCDDV0276	AJ874602	8[l]	UN1449:7894			LG8a	0.0	
INRACCDDV0074	AJ874418	8[l]	8:37.469977	LG8a	0.0	LG8a	20.1	
INRACCDDV0080 (*SLC6A12*)	AJ874423	8p12[h]	8:37.478525	LG8a	0.0			
INRACCDDV0087 (*SLC6A12*)	AJ874430	8p12[h]	8:37.475734	LG8a	0.7	LG8a	52.1	
INRACCDDV0165	AJ874504	nd[2]	UN0007:2.539579	LG8a	6.1			
INRACCDDV0341 (*TPT1*)	AJ874658	8q13.3-q21[k]	8:56.020048	LG8a	19.0	LG8a	80.1	
INRACCDDV0021 (*SLC15A1*)	AJ874384	8q24[h]	8:100.476949	LG8a	49.6	LG8a	119.7	
INRACCDDV0157	AJ874497	8p11[k]	8:45.058626				nl[3]	

Continued

Table 4.1. Continued.

Marker symbol (associated gene)	Accession number	Localization (Literature)	Physical position Chr.: Mb (start)[1]	Linkage group	Berlin Map Position (cM)	Linkage group	INRA Map Position (cM)	Utrecht Map Position (cM)
Chromosome 9								
INRACCDDV0184	AJ874522	9[l]	UN0005:3.059813	LG9a	0.0	LG9a	0.0	
D0Utr16/ D12L1C2	AF421939	nd[2]	UN0005:220525	LG9a	10.0			
INRACCDDV0005 (*GPX1*)	AJ874371	9p13[h]	9:16.925406	LG9a	33.3			
INRACCDDV0274 (*ITIH3*)	AJ874600	9p13prox[h]	9:19.829246	LG9a	35.5	LG9a	32.8	
INRACCDDV0296 (*NPC1*)	AJ874618	nd[2]	9:29.041225	LG9a	44.6			
INRACCDDV0016 (*NPC1*)	AJ874380	9q13[h]	9:64.721594	LG9a	60.2			
INRACCDDV0010	AJ874375	9q13[h]	9:64.784532	LG9a	60.2	LG9a	87	
INRACCDDV0146	AJ874486	9[l]	9:66.056608	LG9a	61.5	LG9a	49.6	
INRACCDDV0344 (*MAPRE2*)	AJ874660	9q14.2[k]	9:76.418691	LG9a	73.1			
INRACCDDV0155	AJ874495	9q15.1[k]	9:87.333291	LG9a	87.9	LG9a	119.5	
INRACCDDV0200	AJ874537	9q15.3[k]	9:96.072343			LG9a	152.5	
INRACCDDV0017 (*CYB5*)	AJ874381	9q17[e]	9:114.418655	LG9a	102.4	LG9a	194.7	
Chromosome 10								
INRACCDDV0138	AJ874478	10q12[k]	10:3.394692				nl[3]	
INRACCDDV0004 (*CALR*)	AJ874370	10p12[e]	UN0287:20813	LG10a	0.0			
INRACCDDV0006 (*ICAM5*)	AJ874372	10p12[h]	UN0135:129426	LG10a	13.5			
INRACCDDV0145	AJ874485	nd[2]	10:13.560088	LG10a	31.4			
Sat7	X99888	nd[2]	10:21.599399	LG10a	41.7			
INRACCDDV0025 (*STK17A*)	AJ874388	10q15[h]	10:26.306649	LG10a	44.6			
INRACCDDV0076 (*DLX5*)	AJ874420	10q15[h]	10:32.224844	LG10a	50.1			
D10Utr1 (*WAP*)	NM_001082390	10[a]	10:45.322775	LG10a	65.6			
INRACCDDV0304 (*EGFR*)	AJ874626	10q16ter[k]	10:45.727001	LG10a	74.6		nl[3]	
Chromosome 11								
INRACCDDV0183	AJ874521	11p11.1-p11.2[k]	11:27.133319	LG11a	0.0	LG11a	0.0	
INRACCDDV0108	AJ874450	11q13prox[k]	11:52.391639	LG11a	10.2		nl[3]	
INRACCDDV0086 (*GHR*)	AJ874429	11q13[e]	11:63.633971			LG11a	31.9	
INRACCDDV0237 (*NNT*)	AJ874570	11q13-q14[h]	11:64.364077	LG11a	19.6	LG11a	66.0	

Chromosome 11								
INRACCDDV0072 (*DSP*)	AJ874416	12p15prox[h]	12:1.341023				nl[3]	
D12L4A1	AF421945	nd[2]	UN0049:773653	LG12a	0.0			
D12L1E11	AF421941	nd[2]	UN0017:1.450742	LG12a	9.9			
INRACCDDV0191	AJ874529	nd[2]	12:12.395105	LG12a	25.9			
D12Utr1 (*RLADPA1*)	M22640	12q11[d]	12:23.542148	LG12a	29.6			
OCRLADF4 (*RLADF4*)	X60986	nd[2]	UN0072:11751	LG12a	32.0			0.0
INRACCDDV0260 (*RNGTT*)	AJ874590	12q15[h]	12:75.751148			LG12a	4.5	
INRACCDDV0201	AJ874538	12q16[k]	12:77.126750	LG12a	64.6	LG12a	0.0	
D12Utr2	AY095444	12[i]	12:90.608707			LG12a	4.5	31.5
INRACCDDV0176	AJ874514	12q23dist[k]	12:128.433337	LG12a	94.5		nl[3]	
Chromosome 13								
INRACCDDV0198	AJ874536	nd[2]	13:4.371301	LG13a	0.0			
INRACCDDV0230	AJ874563	14p11dist[k]	13:10.557452	LG13a	8.7			
INRACCDDV0219 (*MYH11*)	AJ874553	6p12-p13[h]	13:10.596261	LG13a	8.7			
INRACCDDV0106	AJ874448	13[l]	13:12.692600	LG13a	12.4	LG13a	0.0	
OCELAMB (*ELAMB*)	M91004	nd[2]	13:21.851839	LG13a	15.0			
INRACCDDV0297	AJ874619	13q21prox[k]	13:26.717719	LG13a	16.3	LG13a	48.6	
OCCRP (*CRP*)	M14538.1	nd[2]	13:33.404947	LG13a	23.4			
INRACCDDV0293	AJ874616	nd[2]	13:33.420599	LG13a	26.6			
INRACCDDV0027 (*CD1B*)	AJ874390	13q21[h]	13:34.788232	LG13a	27.9	LG13a	62.5	
INRACCDDV0153	AJ874493	nd[2]	13:42.013148	LG13a	35.3			
INRACCDDV0139	AJ874479	13[l]	13:47.564061	LG13a	38.3	LG13a	84.6	
INRACCDDV0270 (*BCAS2*)	AJ874596	13q22-q23[h]	13:49.682570	LG13a	40.0			
INRACCDDV0177	AJ874515	13[l]	13:57.547003	LG13a	47.9	LG13a	102.8	
INRACCDDV0137	AJ874477	13[l]	13:63.908.025			LG13a	109.4	
INRACCDDV0342	AJ874659	13q26-q27[k]	13:95.525.834			LG13a	150.7	
INRACCDDV0289	AJ874612	13[l]	13:111.882614	LG13a	76.4	LG13a	214.3	
INRACCDDV0014 (*SLC2A1*)	AJ874378	13q31-q32[e]	13:123.935975	LG13a	84.9	LG13a	181.5	
INRACCDDV0151	AJ874491	13[l]	UN0006:4.131875	LG13a	91.7	LG13a	180.5	
INRACCDDV0291	AJ874614	13[l]	UN0006:5.102365	LG13a	95.0	LG13a	243.3	
INRACCDDV0310 (*PTAFR*)	AJ874632	13q33[k]	13:137.381150	LG13a	99.2			

Continued

Table 4.1. Continued.

Marker symbol (associated gene)	Accession number	Localization (Literature)	Physical position Chr.: Mb (start)[1]	Linkage group	Berlin Map Position (cM)	Linkage group	INRA Map Position (cM)	Utrecht Map Position (cM)
Chromosome 14								
INRACCDDV0112	AJ874454	14[l]	14:14.651472			LG14a	0.0	
INRACCDDV0337 (*SIAH2*)	AJ874654	14q13prox[j]	14:44.599073	LG14a	0.0	LG14a	34.9	
INRACCDDV0166	AJ874505	14q13prox[k]	14:47.396502			LG14a	37.2	
INRACCDDV0313 (*HES1*)	AJ874634	14q21prox[h]	14:89.981493	LG14a	37.4		nl[3]	
INRACCDDV0140	AJ874480	14q23[k]	14:121.357640	LG14a	50.8	LG14a	70.3	
INRACCDDV0241 (*TIAM1*)	AJ874574	14q25[h]	14:161.237466	LG14a	79.6	LG14a	92.4	
INRACCDDV0162	AJ874501	14[l]	UN0209:452477	LG14a	86.6	LG14a	108.2	
Chromosome 15								
INRACCDDV0101	AJ874443	15[l]	11:35.511688			LG15a	0.0	
INRACCDDV0103	AJ874445	nd[2]	UN0003:7.861545	LG15a	0.0			
INRACCDDV0286	AJ874609	15q11.3[l]	15:18.057583	LG15a	14.9	LG15a	47.6	
INRACCDDV0125	AJ874466	15[l]	15:20.804619	LG15a	17.3	LG15a	24.9	
INRACCDDV0143	AJ874483	15q12[k]	15:108.340756	LG15a	23.3		nl[3]	
INRACCDDV0091	AJ874434	nd[2]	15:35.956631	LG15a	37.7			
INRACCDDV0288	AJ874611	15[l]	15:53.373658	LG15a	46.9	LG15a	64.7	
INRAL (*Angora*)		15[l]	nd			LG15a	69.4	
INRACCDDV0115	AJ874457	nd[2]	15:60.785483	LG15a	51.9			
INRACCDDV0044 (*ALB*)	AJ874402	15q23[h]	15:76.663725	LG15a	62.2			
INRACCDDV0306 (*GC*)	AJ874628	15q23dist[k]	15:78.231304	LG15a	62.2			
INRACCDDV0303 (*SLC4A4*)	AJ874625	15q23dist[k]	15:78.678247	LG15a	63.4	LG15a	136.1	
INRACCDDV0035 (*CSN3*)	AJ874397	15q23dist[h]	15:79.715078			LG15a	115.2	
Sat2 (*CSN1S1*)	M77195	nd[2]	15:79.905947	LG15a	63.4			
OCAS1CG (*CSN1S1*)	AY284844.1	nd[2]	15:79.905947	LG15a	63.4			
INRACCDDV0018 (*CSN1S1*)	AJ874382	15q23[e]	15:79.939447	LG15a	64.3			
INRACCDDV0294	AJ874617	15[l]	UN0015:2.339043	LG15a	78.7	LG15a	90.0	

Chromosome 16								
INRACCDDV0148	AJ874488	16p12.1[k]	16:2.262135	LG16a	0.0	LG16a	0.0	
INRACCDDV0279	AJ874604	nd[2]	16:23.048782	LG16a	29.4			
INRACCDDV0105	AJ874447	16[l]	16:54.615243			LG16a	32	
INRACCDDV0185	AJ874523	16c23[k]	16:82.956921	LG16a	91.2	LG16a	74.4	
Chromosome 17								
INRACCDDV0031 (*CA12*)	AJ874394	17c12[h]	17:8.874666			LG17a	0.0	
INRACCDDV0172	AJ874510	17c21dist[k]	17:55.188981	LG17a	0.0	LG17a	33.2	
INRACCDDV0217 (*GMFB*)	AJ874551	17q23prox[h]	17:73.200829	LG17a	10.6			
Sat8	X99889	nd[2]	17:78.000869	LG17a	15.1			
Chromosome 18								
INRACCDDV0280	AJ874605	18q12prox[k]	18:3.817803	LG18a	0.0		nl[3]	
INRACCDDV0218 (*ARFGEF1*)	AJ874552	3q14[h]	18:68.515123	LG18a	8.7			
INRACCDDV0119	AJ874461	18q21.1[k]	18:8.195053			LG18b	0.0	
INRACCDDV0123	AJ874464	18[l]	18:9.868534	LG18a	19.0	LG18b	8.1	
INRACCDDV0188	AJ874526	18[l]	18:23.594801	LG18a	36.7	LG18a	0.0	
INRACCDDV0190	AJ874528	18q22-q23[k]	18:32.532845			LG18a	0.0	
INRACCDDV0258 (*MINPP1*)	AJ874588	18q23[h]	18:35.388785	LG18a	43.3			
INRACCDDV0029 (*CYP2C18*)	AJ874392	18q31[h]	18:42.614826	LG18a	48.5			
INRACCDDV0023 (*CYP2C18*)	AJ874386	18q31[h]	18:42.669764	LG18a	48.5	LG18b	53.9	
INRACCDDV0063 (*CYP2C18*)	AJ874409	18q31[h]	18:42.699522	LG18a	48.5			
D0Utr10 (*CYP2C4*)	M74203	18q24[f]	18:43.242464	LG18a	48.6			
INRACCDDV0256 (*MSN*)	AJ874586	Xq12prox[h]	18:52.168480	LG18a	53.4			
INRACCDDV0104	AJ874446	18[l]	18:52.711932			LG18b	64.4	
INRACCDDV0168	AJ874507	18[l]	18:59.695227	LG18a	57.7	LG18b	47.0	
Chromosome 19								
D19Utr2	AF421952	19[i]	19:4.820808					0.0
INRACCDDV0234 (*NDEL1*)	AJ874567	19q12.3[h]	19:10.884688	LG19a	0.0	LG19a	0.0	
D19Utr1 (*ALOX*)	M33291	19[i]	19:12.174077					14.0
D19L1E12	AF421950	nd[2]	19:23.208880	LG19a	19.7			
INRACCDDV0102	AJ874444	19[l]	19:24.412844	LG19a	21.8	LG19a	25.2	
INRACCDDV0094	AJ874437	19q21prox[k]	19:36.155753	LG19a	30.9	LG19a	36.3	
D19Utr3	AF421951	19[i]	19:37.973556	LG19a	34.6			42.0
INRACCDDV0071 (*KRT12*)	AJ874415	19q21[h]	19:41.957794	LG19a	36.9	LG19a	45.9	

Continued

Table 4.1. Continued.

Marker symbol (associated gene)	Accession number	Localization (Literature)	Physical position Chr.: Mb (start)[1]	Linkage group	Berlin Map Position (cM)	Linkage group	INRA Map Position (cM)	Utrecht Map Position (cM)
INRACCDDV0033 (*ITGB3*)	AJ874396	19q21[h]	19:46.768982			LG19a	52.5	
INRACCDDV0193	AJ874531	19[l]	19:55.797718	LG19a	54.2	LG19a	60.4	
Es-3		19[i]	nd					77.5
D19Utr4	AF421949	19[i]	nd	LG19a	67.8			84.5
Chromosome X								
INRACCDDV0213 (*PRKCB1*)	AJ874547	6p12prox[h]	X:9.618638	LGX	0.0			
INRACCDDV0127	AJ874468	6[l]	X:24.702536	LGX	18.1			
INRACCDDV0126	AJ874467	nd[2]	X:24.702290	LGX	19.6			
DXUtr1	AF389361	X[c]	UN0085:798734	LGX	119.6			
INRACCDDV0084 (*TGFB3*)	AJ874427	20q12[e]	X:108.830383	LGX	139.7			

[1] (http://www.ensembl.org/Oryctolagus_cuniculus/, accessed 30 December 2020), Ensembl 73, oryCun2.0);
[2] nd = not determined;
[3] nl = not linked;
[a] Rogel-Gaillard *et al.* (2000);
[b] Korstanje *et al.* (2001a);
[c] Korstanje *et al.* (2001b);
[d] Rogel-Gaillard *et al.* (2001);
[e] Hayes *et al.* (2002);
[f] Zijlstra *et al.* (2002);
[h] Chantry-Darmon *et al.* (2003);
[i] Korstanje *et al.* (2003):
[j] Chantry-Darmon *et al.* (2005a);
[k] Chantry-Darmon *et al.* (2005b);
[l] Chantry-Darmon *et al.* (2006).

Some discrepancies among the published linkage maps and the assignment of the markers to the chromosomes can be deduced from the reported data. Fig. 4.3 illustrates an example of integrated linkage and physical maps for chromosome 1 (OCU1).

4.4.3 Physical versus genetic maps

In the age of whole-genome sequences, physical distances measured in base pairs of nucleotides provide the standard coordinates for navigating the positions and structures of the genome. Although genetic and physical maps are colinear, well-characterized differences and heterogeneities in the average frequency of meiotic recombination events that occur along chromosomes were detected within a genome and also between populations (Cox *et al.*, 2009).

For the rabbit genome, the relationship between genetic and physical maps is shown in Fig. 4.4. Although the average translation of the genetic to the physical distance in the rabbit is about 1 megabase to 1 centiMorgan (Table 4.2), there are remarkable differences in the recombination landscape between the three most widely used mapping populations (INRA, Berlin, Utrecht). The INRA population, consisting of a three-way cross, harbours more recombinations than the two other populations which are classical intercross populations between two breeds. This is best documented for chromosomes 1 and 13, where many markers were mapped in all mapping populations. The translation between genetic and physical maps is very similar in the two other crosses. For chromosomes, where only a few markers were mapped and these markers differed between the different populations, the map conversion is less well documented. Single markers which are highly deviant from the linear relationship, which can be found on several chromosomes (e.g. chromosomes 5, 6, 11, 15, 18) are most likely mapped to the wrong genetic position. The knowledge of the relationship between genetic and physical maps helps to correctly reconstruct the whole genome physical map.

4.5 QTL Studies in Rabbits

The described genetic maps constituted by DNA markers were used to design experimental studies that could identify chromosome regions harbouring variants affecting phenotypic traits (QTL). These traits might have a direct economic relevance (carcass composition and meat quality) or could be relevant in biomedical studies. Despite the relevance of this species in several fields, only a few QTL studies have been produced so far. However, classical QTL studies in the European rabbit based on reference populations are expected to be substituted by genome-wide association studies due to the availability of a commercial high-throughput SNP genotyping platform (e.g. Sosa-Madrid *et al.*, 2020). Other studies aimed to identify markers associated with several traits were based on variability in candidate genes. These studies are presented in the chapters dealing with different traits.

4.5.1 QTL studies for carcass composition and meat-quality traits

Up to now, only one QTL mapping study for carcass composition and meat-quality traits was performed in rabbits in a crossbred population of Giant Grey and New Zealand White (Sternstein *et al.*, 2015). On the basis of the generated genetic map (as described above), a highly significant QTL for different carcass weights was identified on chromosome 7, a significant QTL for bone mass was found on chromosome 9 and another one for drip loss on chromosome 12. Additional suggestive QTL were found on almost all chromosomes. The identified QTL explained 2.5% to 14.6% of the phenotypic variance in the F_2 population. The results provide starting points for fine mapping and candidate gene search for carcass composition and meat-quality traits in the rabbit.

4.5.2 QTL studies for atherosclerosis and hypercholesterolemia

Beside the agricultural use, the rabbit often is used as a model animal for human research. For this purpose, different rabbit inbred lines have been developed. To analyse the genetic determinants for dietary cholesterol susceptibility, two rabbit inbred strains (AX/JU and IIIVO/JU) with different dietary cholesterol reactivity were used to generate an F_2 intercross population (Korstanje *et al.*, 2001b) and a backcross population

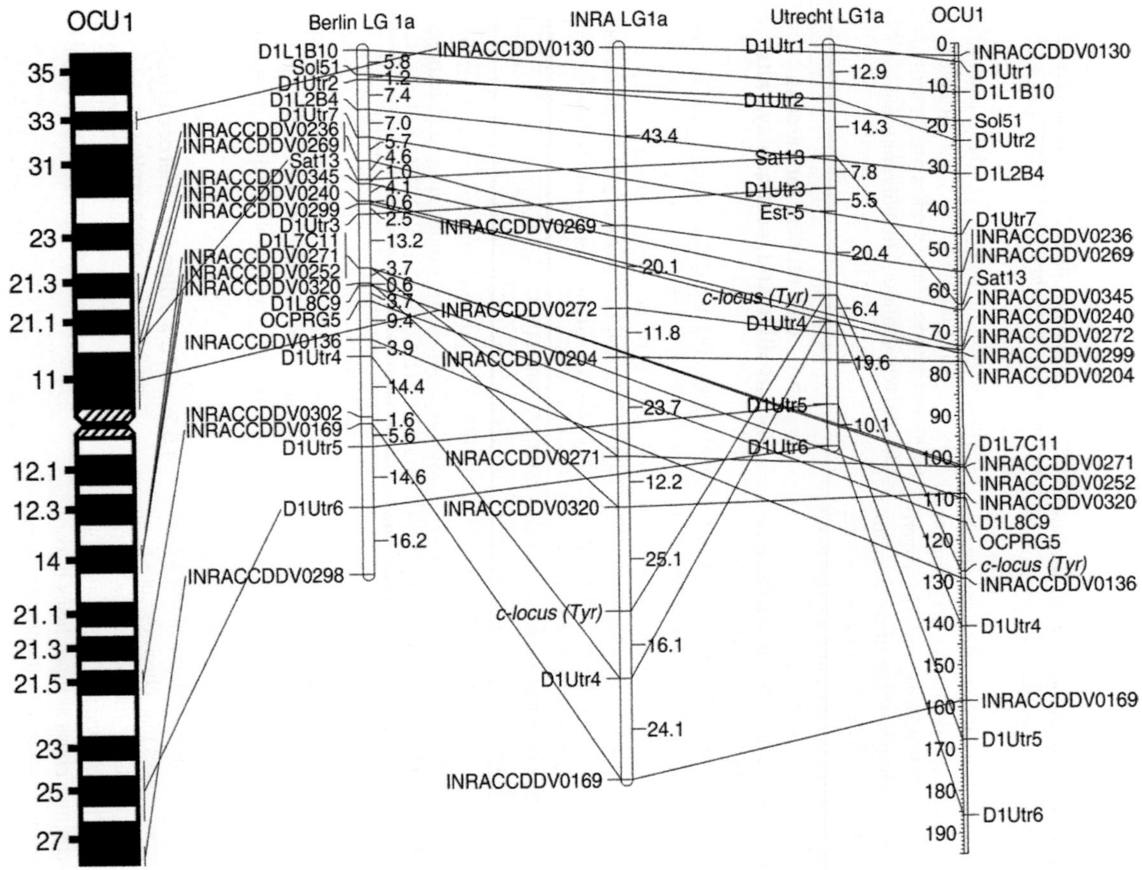

Fig. 4.3. Linkage groups of the rabbit assigned to chromosome 1 (OCU1) according to the linkage mapping results in the pedigrees Giant Grey x New Zealand White (Berlin-Map, Sternstein *et al.*, 2015), INRA2066, Castor Orylag and Laghmere (INRA-Map, Chantry-Darmon *et al.*, 2006) and AX/JU and IIIVO/JU (Utrecht-Map, Korstanje *et al.*, 2001a, 2003). From left to right, the maps are the cytogenetic map with band numbers, Berlin-Map, INRA-Map, Utrecht-Map with distances between markers in centiMorgan (cM), physical map with megabase pairs (Mb) scale, Ensembl version 74, oryCun2.0. Genetic markers that were used in two or more maps are connected by lines.

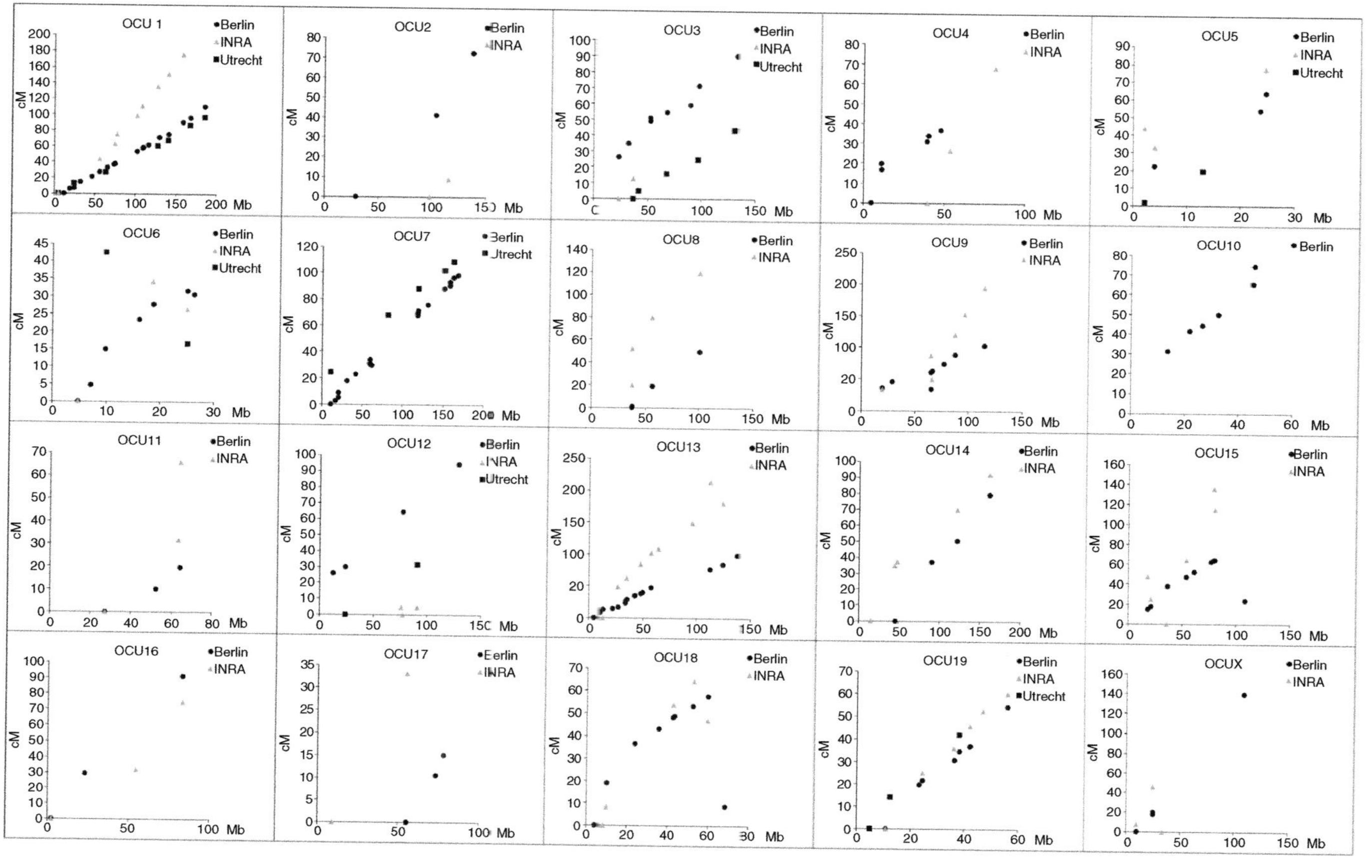

Fig. 4.4. Relationship between physical and genetic maps of all chromosomes. The genetic map (*x*-axis) is given in centiMorgans (cM), the physical map is given in megabases (Mb). Every dot, triangle or square represents a marker mapped in the Giant Grey x New Zealand White (Berlin) population, the INRA2066, Castor Orylag and Laghmere (INRA) population and AX/JU x IIIVO/JU (Utrecht) population. Information was reported only for the chromosomes where the number of genetically mapped markers could allow to define a meaningful relationship.

Table 4.2. Chromosomal recombination rate (cM/Mb). Information was reported only for the chromosomes where the number of genetically mapped markers could allow to estimate this ratio.

LG/ OCU	INRA-Map	Berlin-Map	Utrecht-Map
1	1.13	0.63	0.53
2	0.50	0.67	–
3	0.89	0.58	0.46
4	1.64	0.86	–
5	2.19	2.04	1.67
6	1.30	1.55	1.70
7	–	0.62	0.56
8	1.58	0.79	–
9	1.71	0.71	–
10	–	1.34	–
11	1.77	0.53	–
12	0.33/ 3.27	0.59	0.47
13	1.63	0.75	–
14	0.63	0.68	–
15	2.61	0.80	–
16	0.92	1.13	–
17	0.72	0.66	–
18	0.91	1.03	–
19	1.34	1.21	1.27
X	2.53	1.41	–
Average	1.45	0.93	0.95

(Van Haeringen *et al.*, 2001). Using microsatellites and biochemical markers a sex-specific QTL for the degree of aorta atherosclerosis was observed on LGVI (OCU5) in the F_2 intercross population (Korstanje *et al.*, 2001b). The finding was further supported by the identification of a QTL for basal serum HDL cholesterol on the same linkage group, using the AFLP technology in the backcross population (Van Haeringen *et al.*, 2001). In addition to this, a male genetic map was constructed, consisting of 12 linkage groups and 103 AFLP markers in a backcross population of the AX/JU and IIIVO/JU inbred strains (Van Haeringen *et al.*, 2002). Linkage analysis between the cholesterol-related traits and marker loci revealed a significant QTL for the relative weight of adrenal glands in males and suggestive linkages were found for basal serum total cholesterol levels in females, for serum total cholesterol response in males, and for hematocrit in males (Van Haeringen *et al.*, 2002).

4.6 Concluding Remarks

The generation of the physical and genetic maps provides accurate information about the consecutive order of markers and genes along the chromosomes in the genome of different species. The genetic maps of the European rabbit, compared to the genetic maps available in other species, are still quite primitive. Comparative mapping procedures between humans and rabbits might still be useful, considering that the rabbit genome version needs to be refined by assigning and placing scaffolds in established chromosome contigs. The integration of the rabbit genetic maps with SNPs data from the commercial high-throughput genotyping platform is also needed. Genetic maps have been successfully used to identify genomic regions affecting growth, carcass and fertility traits and other biomedically relevant traits. At present there is no QTL database for this species to gather all related information. This is mainly due to the limited number of studies that have been carried out on this matter in this species. More studies are expected in the future and it would also be useful to develop for the European rabbit a QTL database integrating results from QTL studies, genome-wide association studies and candidate gene analyses.

References

Archibald, A.L., Haley, C.S., Brown, J.F., Couperwhite, S., Mcqueen, H.A. *et al.* (1995) The PiGMaP consortium linkage map of the pig (*Sus scrofa*). *Mammalian Genome* 6, 157–175.

Bertolini, F., Schiavo, G., Scotti, E., Ribani, A., Martelli P.L. *et al.* (2014) High throughput SNP discovery in the rabbit (*Oryctolagus cuniculus*) genome by next generation semiconductor-based sequencing. *Animal Genetics* 45, 304–307.

Carneiro, M., Rubin, C.J., Di Palma, F., Albert, F.W., Alföldi, J. *et al.* (2014) Rabbit genome analysis reveals a polygenic basis for phenotypic change during domestication. *Science* 345, 1074–1079.

Castle, W.E. (1919) *Studies in Heredity in Rabbits, Rats, and Mice.* Carnegie Institution of Washington, Washington, DC.

Castle, W.E. (1924a) Linkage of Dutch, English, and Angora in rabbits. *Proceedings of the National Academy of Sciences of the USA* 10, 107–108.

Castle, W.E. (1924b) On the occurrence in rabbits of linkage in inheritance between albinism and brown pigmentation. *Proceedings of the National Academy of Sciences of the USA* 10, 486–488.

Castle, W.E. and Sawin, P.B. (1941) Genetic linkage in the rabbit. *Proceedings of the National Academy of Sciences of the USA* 27, 519–523.

Chan, F.P.H., Sergovich, F.R. and Shaver, E.L. (1977) Banding patterns in mitotic chromosomes of the rabbit (*Oryctolagus Cuniculus*). *Canadian Journal of Genetics and Cytology* 19, 625–632.

Chantry-Darmon, C., Rogel-Gaillard, C., Bertaud, M., Urien, C., Perrocheau, M. *et al.* (2003) 133 new gene localizations on the rabbit cytogenetic map. *Cytogenetic and Genome Research* 103, 192–201.

Chantry-Darmon, C., Bertaud, M., Urien, C., Chadi-Taourit, S., Perrocheau, M. *et al.* (2005a) Expanded comparative mapping between man and rabbit and detection of a new conserved segment between HSA22 and OCU4. *Cytogenetic and Genome Research* 111, 134–139.

Chantry-Darmon, C., Urien, C., Hayes, H., Bertaud, M., Chadi-Taourit, S. *et al.* (2005b) Construction of a cytogenetically anchored microsatellite map in rabbit. *Mammalian Genome* 16, 442–459.

Chantry-Darmon, C., Urien, C., De Rochambeau, H., Allain, D., Pena, B. *et al.* (2006) A first-generation microsatellite-based integrated genetic and cytogenetic map for the European rabbit (*Oryctolagus cuniculus*) and localization of angora and albino. *Animal Genetics* 37, 335–341.

Committee for Standardized Karyotype of *Oryctolagus cuniculus* (1981) Standard karyotype of the laboratory rabbit, *Oryctolagus cuniculus*. *Cytogenetics and Cell Genetics* 31, 240–248.

Cox, A., Ackert-Bicknell, C.L., Dumont, B.L., Ding, Y., Bell, J.T. *et al.* (2009) A new standard genetic map for the laboratory mouse. *Genetics* 182, 1335–1344.

Cunningham, F., Achuthan, P., Akanni, W., Allen, J., Amode, M.R *et al.* (2019) Ensembl 2019. *Nucleic Acids Research* 47, D745–D751.

Debeccker, A. and Martin-Deleon, P.A. (1992) Assignment of the rabbit genes for alpha (PHKA) and beta (PHKB) phosphorylase kinase subunits. *Cytogenetics and Cell Genetics* 61, 208–210.

Ducayen, M.B., Jagiello, G.M., Allderdice, P.W., Miller, W.A., Grey, R. and Fang, J.S. (1974) A study of the chromosomes of the normal male rabbit and two intersex rabbits with quinacrine fluorescence and Giemsa banding. *Cytologia (Tokyo)* 39, 839–845.

Dutrillaux, B., Viegas-Pequignot, E. and Couturier, J. (1980) Great homology of chromosome banding of the rabbit (*Oryctolagus cuniculus*) and primates, including man (author's transl.). *Annales de Génétique* 23, 22–25.

Echard, G. (1973) Bandes Chromosomiques de type G chez le lapin domestique (*Oryctolagus Cuniculus*). *Annales de Génétique et de Sélection Animale* 5, 425–434.

Echard, G., Gellin, J., Benne, F. and Gillois, M. (1981) The gene map of the rabbit (*Oryctolagus cuniculus* L.) I. Synteny between the rabbit gene loci coding for HPRT, PGK, G6PD, and GLA: their localization on the X chromosome. *Cytogenetics and Cell Genetics* 29, 176–183.

Echard, G., Gellin, J., Benne, F. and Gillois, M. (1982) The gene map of the rabbit, *Oryctolagus cuniculus* I. II. Analysis of the segregation of 11 enzymes in rabbit x hamster somatic cell hybrids: two syntenic groups, LDHB-TPI and LDHA-ACP2. *Cytogenetics and Cell Genetics* 34, 289–295.

Ferguson-Smith, M.A. and Trifonov, V. (2007) Mammalian karyotype evolution. *Nature Reviews Genetics* 8, 950–962.

Ford, C.E., Pollock, D.L. and Gustavsson, I. (1980) *Proceedings of the First International Conference for the Standardisation of Banded Karyotypes of Domestic Animals.* University of Reading, England. 2-6 August 1976. *Hereditas* 92, 145–162.

Fox, R.R. (1994) Taxonomy and genetics. In: Manning, P.J., Ringler, D.H. and Newcomer, C.E. (eds) *The Biology of the Laboratory Rabbit*, 2nd edn. Academic Press, San Diego, California.

Geldermann, H., Müller, E., Moser, G., Reiner, G., Bartenschlager, H. *et al.* (2003) Genome-wide linkage and QTL mapping in porcine F2 families generated from Pietrain, Meishan and Wild Boar crosses. *Journal of Animal Breeding and Genetics* 120, 363–393.

Gellin, J., Dalens, M., Echard, G. and Hatey, F. (1983) Carte genique du lapin (*Oryctolagus cuniculus* L.): syntenie entre les genes uteroglobine, lactate deshydrogenase A et phosphatase acide 2. *Génétique Sélection Évolution* 15, 489–494.

Gellin, J., Echard, G., Yerle, M., Dalens, M., Chevalet, C. and Gillois, M. (1985) Localization of the alpha and beta casein genes to the q24 region of chromosome 12 in the rabbit (*Oryctolagus cuniculus* L.) by *in situ* hybridization. *Cytogenetics and Cell Genetics* 39, 220–223.

Hageltorn, M. and Gustavsson, I. (1979) Identification by banding techniques of the chromosomes of the domestic rabbit (*Oryctolagus cuniculus* L.). *Hereditas* 90, 269–279.

Hagen, K.L., Suzuki, Y., Tissot, R. and Cohen, C. (1978) The hemopexin locus: its assignment to linkage group I in the laboratory rabbit (*Oryctolagus cuniculus*) and evidence for a fourth allele. *Animal Blood Groups and Biochemical Genetics* 9, 151–159.

Hayes, H., Rogel-Gaillard, C., Zijlstra, C., De Haan, N.A., Urien, C. *et al.* (2002) Establishment of an R-banded rabbit karyotype nomenclature by FISH localization of 23 chromosome-specific genes on both G- and R-banded chromosomes. *Cytogenetic and Genome Research* 98, 199–205.

Kappes, S.M., Keele, J.W., Stone, R.T., Mcgraw, R.A., Sonstegard, T.S. *et al.* (1997) A second-generation linkage map of the bovine genome. *Genome Research* 7, 235–249.

Korstanje, R., O'Brien, P.C., Yang, F., Rens, W., Bosma, A.A. *et al.* (1999) Complete homology maps of the rabbit (*Oryctolagus cuniculus*) and human by reciprocal chromosome painting. *Cytogenetics and Cell Genetics* 86, 317–322.

Korstanje, R., Gillissen, G.F., Den Bieman, M.G., Versteeg, S.A., Van Oost, B. *et al.* (2001a) Mapping of rabbit chromosome 1 markers generated from a microsatellite-enriched chromosome-specific library. *Animal Genetics* 32, 308–312.

Korstanje, R., Gillissen, G.F., Kodde, L.P., Den Bieman, M., Lankhorst, A. *et al.* (2001b) Mapping of microsatellite loci and association of aorta atherosclerosis with LG VI markers in the rabbit. *Physiological Genomics* 6, 11–18.

Korstanje, R., Gillissen, G.F., Versteeg, S.A., Van Oost, B.A., Bosma, A.A. *et al.* (2003) Mapping of rabbit microsatellite markers using chromosome-specific libraries. *Journal of Heredity* 94, 161–169.

Kos, C.H., Tihy, F., Murer, H., Lemieux, N. and Tenenhouse, H.S. (1996) Comparative mapping of Na+-phosphate cotransporter genes, NPT1 and NPT2, in human and rabbit. *Cytogenetics and Cell Genetics* 75, 22–24.

Lemieux, N. and Dutrillaux, B. (1992) New gene assignments to rabbit chromosomes; implications for chromosome evolution. *Cytogenetics and Cell Genetics* 61, 132–134.

Mahoney, C.E., Picciano, S.R., Burton, K.M. and Martin-Deleon, P.A. (1988) Regional mapping of the creatine kinase b (CKBB) gene in rabbit (*Oryctolagus cuniculus*) and man using a rat cDNA probe. *Cytogenetics and Cell Genetics* 48, 160–163.

Martin-Deleon, P.A. (1980) Location of the 18S and 28S rRNA cistrons in the genome of the domestic rabbit (*Oryctolagus cuniculus* L). *Cytogenetics and Cell Genetics* 28, 34–40.

Martin-Deleon, P.A. and Picciano, S.R. (1988) Localization of the KRAS2 oncogene in the domestic rabbit (*Oryctolagus cuniculus*). *Cytogenetics and Cell Genetics* 48, 201–204.

Martin-Deleon, P.A., Mclaughlin, J., Mitzel, E. and Tailor, S. (1989) Mapping of the creatine kinase M gene to 19q11-q12 in the rabbit genome. *Cytogenetics and Cell Genetics* 50, 165–167.

Martin-Deleon, P.A., Canaff, L., Korstanje, R., Bhide, V., Selkirk, M. and Hendy, G.N. (1999) Rabbit calcium-sensing receptor (CASR) gene: chromosome location and evidence for related genes. *Cytogenetics and Cell Genetics* 86, 252–258.

Medrano, L. and Dutrillaux, B. (1984) Chromosomal location of immunoglobulin genes: partial mapping of these genes in the rabbit and comparison with Ig genes carrying chromosomes of man and mouse. *Advances in Cancer Research* 41, 323–367.

Melander, Y. (1956) The chromosome complement of the rabbit. *Hereditas* 42, 432–435.

Painter, T.S. (1926) Studies in mammalian spermatogenesis VI. The chromosomes of the rabbit. *Journal of Morphology* 43, 1–43.

Pardue, M.L. and Gall, J.G. (1969) Molecular hybridization of radioactive DNA to the DNA of cytological preparations. *Proceedings of the National Academy of Sciences of the USA* 64, 600–604.

Pease, H. (1928) Yellow Fat in Rabbits a linked character? *Verh. V. Intern. Kongr. Vererbungswissenschaft* 2, 1153–1156.

Rogel-Gaillard, C., Zijlstra, C., Bosma, A.A., Thepot, D., Fontaine, M.L. *et al.* (2000) Assignment of the rabbit whey acidic protein gene (WAP) to rabbit chromosome 10 by *in situ* hybridization and description of a large region surrounding this gene. *Cytogenetics and Cell Genetics* 89, 107–109.

Rogel-Gaillard, C., Piumi, F., Billault, A., Bourgeaux, N., Save, J.C. *et al.* (2001) Construction of a rabbit bacterial artificial chromosome (BAC) library: application to the mapping of the major histocompatibility complex to position 12q.1.1. *Mammalian Genome* 12, 253–255.

Sawin, P.B., Griffith, M.A. and Stuart, C.A. (1944) Genetic linkage of blood types in the rabbit. *Proceedings of the National Academy of Sciences of the USA* 30, 217–221.

Schroder, J., Suomalainen, H., Van Der Loo, W. and Schroder, E. (1978) Karyotypes in lymphocytes of two strains of rabbit, and two species of hare. *Hereditas* 88, 183–188.

Sosa-Madrid, B.S., Hernández, P., Blasco, A., Haley, C.S., Fontanesi, L. *et al.* (2020) Genomic regions influencing intramuscular fat in divergently selected rabbit lines. *Animal Genetics* 51, 58–69.

Soulie, J. and De Grouchy, J. (1982) Of rabbit and man: comparative gene mapping. *Human Genetics* 60, 172–175.

Soulie, J. and De Grouchy, J. (1983) New gene assignments in the rabbit (*Oryctolagus cuniculus*). Comparison with other species. *Human Genetics* 63, 48–52.

Sternstein, I., Reissmann, M., Maj, D., Bieniek, J. and Brockmann, G.A. (2015) A comprehensive linkage map and QTL map for carcass traits in a cross between Giant Grey and New Zealand White rabbits. *BMC Genetics* 16, 16.

Stock, A.D. (1976) Chromosome banding pattern relationships of hares, rabbits, and pikas (order Lagomorpha). A phyletic interpretation. *Cytogenetics and Cell Genetics* 17, 78–88.

Tissot, R.G. and Cohen, C. (1974) Histocompatibility in the rabbit. Linkage between RL-A, MLC, and the He blood group loci. *Transplantation* 18, 142–149.

Van Haeringen, W.A., Den Bieman, M., Gillissen, G.F., Lankhorst, A.E., Kuiper, M. T.R., Van Zutphen, L.F.M. and Van Lith, H.A. (2001) Mapping of a QTL for serum HDL cholesterol in the rabbit using AFLP technology. *Journal of Heredity* 92, 322–326.

Van Haeringen, W.A., Den Bieman, M.G., Lankhorst, A.E., Van Lith, H.A. and Van Zutphen, L.F. (2002) Application of AFLP markers for QTL mapping in the rabbit. *Genome* 45, 914–921.

Van Zutphen, L.F. (1974) Serum esterase genetics in rabbits. II. Genetic analysis of the prealbumin esterase system, including atropinesterase and cocainesterase polymorphism. *Biochemical Genetics* 12, 327–340.

Wienberg, J. and Stanyon, R. (1995) Chromosome painting in mammals as an approach to comparative genomics. *Current Opinions on Genetics and Development* 5, 792–797.

Xu, J. and Hardison, R.C. (1989) Localization of the beta-like globin gene cluster and the genes for parathyroid hormone and c-Harvey-ras 1 to region q14-q21 of rabbit chromosome 1 by *in situ* hybridization. *Cytogenetics and Cell Genetics* 52, 157–161.

Xu, J. and Hardison, R.C. (1991) Localization of the alpha-like globin gene cluster to region q12 of rabbit chromosome 6 by *in situ* hybridization. *Genomics* 9, 362–365.

Zijlstra, C., De Haan, N.A., Korstanje, R., Rogel-Gaillard, C., Piumi, F. *et al.* (2002) Fourteen chromosomal localizations and an update of the cytogenetic map of the rabbit. *Cytogenetic and Genome Research* 97, 191–199.

5 Immunogenetics in the Rabbit

Rose G. Mage[1]* and Claire Rogel-Gaillard[2]

[1]*Laboratory of Immune System Biology, National Institute of Allergy and Infectious Diseases, National Institutes of Health, Bethesda, Maryland, USA;* [2]*Université Paris-Saclay, INRAE, AgroParisTech, GABI, Jouy-en-Josas, France*

5.1 Introduction

It is likely that more than 1% of the protein-coding genes in the rabbit genome participate in some aspects of the immune system. This chapter on immunogenetics in the rabbit cannot include descriptions of all such genes but will focus on some genes with genetic and genomic sequence information including those encoding: soluble circulating immunoglobulin molecules (Igs) and their surface-bound forms on B lymphocytes (BCRs); T-cell receptors on T lymphocyte surfaces, (TCRs); the rabbit Leukocyte Antigen (RLA) complex (proteins on cells that function to present antigen fragments to TCRs); and some cytokine genes that encode key regulators of T- and B-cell responses. A recent description of Lagomorph immunology can be found in Pinheiro *et al.* (2016).

Immunogenetic studies add to understanding of diseases that result from abnormalities of the immune system as well as the roles that genes play in responses to infectious diseases. Specific immune responses include those against foreign antigens present on or in viruses, bacteria, fungi, and parasites that cause infections, as well as responses to vaccines developed to protect against some of the infectious diseases they cause. Allergic diseases also reflect specific immune responses to a wide array of environmental allergens that lead to allergic reactions such as asthma, hives and other forms of dermatitis. Silverstein (2001) provides a valuable historical perspective on early concepts in immunology. Paul Ehrlich used the term 'horror autotoxicus' at the beginning of the 20th century to describe the observation that destructive antibodies are not normally produced against our own cells and tissues. However, we now know that T-cell and B-cell tolerance must be established and maintained to avoid development of serious autoimmune diseases when immune responses are directed against self-antigens on or in an individual's own cells and tissues.

The rabbit has long been used as an animal model for infectious, cardiovascular and autoimmune diseases (Rogel-Gaillard *et al.*, 2009). Unusual or unique features of rabbit B-lymphocyte development and sequence diversification during immune responses contribute to their ability to produce protective antibodies. Rabbit antibodies to a large number of target antigens are commercially available because they are very stable reagents with high specificity and affinity. Polyclonal products are developed for basic research, as well as by biotech companies that provide reagents for basic and clinical tests, and by pharmaceutical companies for therapeutics; e.g. Thymoglobulin (reviewed in Deeks and Keating, 2009). These efforts will benefit

Contact e-mail address: rmage@niaid.nih.gov

DOI: 10.1079/9781780643342.0005

from the draft assembly of the *Oryctolagus cuniculus* whole genome sequence (oryCun2.0; GCA 000003625.1. Nov/ 2009) and first whole genome analyses (Carneiro *et al.*, 2014). Rabbit monoclonal antibodies (mAbs) are reviewed by a few authors (Rader, 2009; and Weber *et al.*, 2017).

A multi-authored review (Esteves *et al.*, 2018) describes some of the many diseases for which rabbits are among the best or only animal model of human diseases. Rabbits are also used for investigations of autoimmune diseases such as Systemic Lupus Erythematosus (Mage and Rai, 2012).

Even greater benefit will come from improvements of the current assembly and annotation of gene sequences in this and new sequences from rabbits of different breeds. OryCun2.0 is a high-quality draft assembly of Chromosomes 1-22 and X, but with only 6.51x coverage and 3218 unplaced scaffolds (chrUn), the assemblies of scaffolds and chromosomes have numerous small and some very large gaps in regions with Ig and other immune system-related genes. OryCun2.0 DNA was from a single female rabbit of a partially inbred strain that is now extinct. These Thorbecke rabbits had developmental anomalies and were more susceptible to experimental infection with *Mycobacterium tuberculosis* than normal rabbits such as New Zealand White (NZW) (Dorman *et al.*, 2004; Mendez *et al.*, 2008). Fortunately, an important cluster of cytokine genes was included in the 1% of human genome sequence targeted for comparative sequencing as part of a pilot phase of the ENCODE project (Margulies *et al.*, 2007). The additional assembled genomic sequence used DNA from an NZW rabbit (Accession: PRJNA13686 ID: 13686). This allowed comparisons of independent assemblies of the region from two rabbits (Gertz *et al.*, 2011). Unfortunately, the ENCODE project did not include Ig, TCR and MHC genes.

5.2 Immunoglobulins (Igs)

5.2.1 Genes encoding Igs

A detailed review of information available in 2006 was published (Mage *et al.*, 2006).

With the completion of the oryCun2.0 sequence and assembly, additional information is now available (Gertz *et al.*, 2013). At the time of this writing, the last updated new information about Ig genes in oryCun2.0 was added to the International ImMunoGeneTics Information System site (IMGT) on 12 March 2013 at http://imgt.org/IMGTveterinary/#Rabbit (accessed 28 February 2021).

5.2.1.1 Definitions

Table 5.1 lists some terms used to describe the genes that rearrange to encode mRNA that is spliced to then encode the final structures of Ig molecules.

A series of gene rearrangements occur before mRNA can be transcribed, processed, and proteins expressed. For heavy chains, one of the diversity (*DH*) genes must rearrange to one of the joining (*JH*) genes. In rabbits the *VH1a* allotype-encoding gene closest to the *DH* genes (Becker and Knight, 1990) usually rearranges to form *VHDHJH*. The joins must be in frame, i.e. the codon at the join must be in the same translational frame across the join. Once a functional rearrangement has occurred, rearrangement on the second chromosome is inhibited. Individual Ig-producing cells thus express the functional products of the rearrangement on only one chromosome (allelic exclusion). This also applies to kappa or lambda light chains where a *V* gene must rearrange to a *J* gene and produce an in-frame product. Rabbit Ig proteins are produced as both cell surface receptors (BCRs) on B cells and as molecules secreted from mature B cells and plasma cells. BCRs have transmembrane and cytoplasmic portions of the heavy-chain constant regions that are encoded by additional exons. Production of surface or secreted protein depends on alternative splicing of mRNA. Complexes of specialized proteins including VDJ recombination-activating proteins RAG 1 and RAG 2 are required for gene rearrangements to form both BCRs and TCRs. At some stages of B-cell and T-cell development the enzyme terminal deoxynucleotidyl transferase (TdT) adds non-templated bases not present in the genome (N regions) during gene rearrangements at the junctions of *V* to *D* and *D* to *J* and/or *V* to *J*. Further diversification of the rearranged sequences occurs in B cells by somatic hypermutation and gene conversion, but this does not occur in T-cells. Activation Induced Deaminase (AID) is required for both types of sequence diversification as well as for

Table 5.1. Definitions of basic terms and genes that rearrange and assemble to encode rabbit Ig molecules.

Genes	Components of Ig molecules encoded (estimated number)
Heavy Chains	
IGHV	Variable-regions (>200)
IGHD	Diversity-regions (14-20)
IGHJ	Joining-regions (six - three functional))
IGHM	IgM one copy, four constant domains CH1, CH2, hinge region, CH3, CH4
IGHG	IgG one copy, three constant domains CH1, hinge region, CH2, CH3
IGHE	IgE one copy, four constant domains CH1, CH2, hinge region, CH3, CH4
IGHA	IgA 15 copies (isotypes) per haploid genome; CH1+hinge, CH2, CH3
Light Chains	
Kappa	
IGKV	Variable-regions (~100-150)
IGKJ1-5	Joining-regions IGKC1 (five - one functional)
IGKJ1-3	Joining-regions IGKC2 (three - functional associated)
IGKC1	Duplicated constant-region IGKC1
IGKC2	Duplicated constant-region IGKC2
Lambda	
IGLV	Variable-regions (~43)
IGLJ5	Joining-region
IGLC5	Constant-region
IGLJ6	Joining-region
IGLC6	Constant-region

class switching that results in the rearranged *VDJ* relocating into a new genomic location to produce a different class of Ig.

The Ig classes have constant regions with different functional roles such as binding complement and Fc Receptors (IgM and some IgGs), or to receptors on mast cells or basophils (IgE). Rabbit Igs are composed of symmetrical heavy- and light-chain pairs usually connected by interchain disulfide bonds. IgG in serum and on the cell surface consists of two such pairs. IgM in secreted form has five pairs and includes a joining chain (*J*). Serum IgA is generally dimeric and also contains the J chain. An additional secretory component is associated with secretory IgA found in the gut. In rabbits, normal development of gut-associated lymphoid tissues (GALT) and production of IgA requires gut flora. Studies in man and mice emphasize that IgA plays a key role in maintaining gut homeostasis, symbiotic relationships with commensal gut microbes, and protection against pathogens (Mantis *et al.*, 2011; Brandtzaeg, 2013). The 15 different isotypes of IgA in rabbits may have different functional roles in different parts of the body (Spieker-Polet *et al.*, 1993). Fig. 5.1 shows simplified diagrams of the general organization of genes encoding Ig heavy, kappa and lambda chains in rabbits.

5.2.2 Heavy-chain genes

The heavy-chain-encoding genes were not mapped to a chromosome in oryCun2.0. Gertz *et al.* (2013) identified the location of *IGH* near the q telomere of Chromosome 20 by fluorescence *in situ* hybridization (FISH) analyses on RBP-banded chromosomes. Synteny analysis with human chromosome 14 predicted that orthologous genes in approximately 3 Mb at the q telomeric end of rabbit chromosome 20 are absent from the assembly. Some, but not all, missing genes were found on unplaced scaffolds including at least 79 containing approximately 372 *IGHV* genes. The donor of DNA used for the oryCun2.0 sequence was heterozygous at the heavy-chain locus. Thus, the number of *IGHV* genes per haploid genome is uncertain. The sequence of *VH1a1* is probably incorrectly placed in the oryCun2.0 assembly of scaffold (chrUn0742) because it is surrounded by other *VH* genes. The *VH1a2* gene that encodes the VHa2 allotype, known to be closest to *DH*, *JH* and *CH* genes, is found near one end of another unplaced scaffold chrUn0439 that contains the genes extending through *IGHE* that encodes IgE (Fig. 5.1a and Fig. 1 in Gertz *et al.*, 2013).

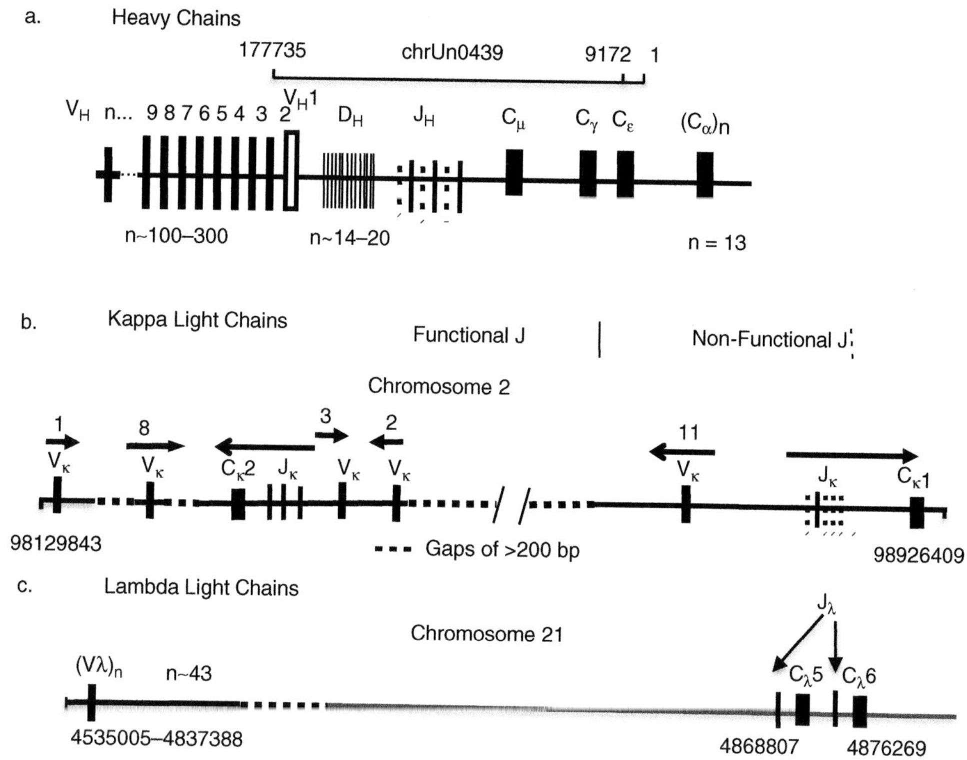

Fig. 5.1. Diagrammatic summary of genomic organization of rabbit Ig genes.

a. Positions shown are on the assembly of chrUn0439 that contains *VH1a2* through the region encoding IgE.

b. The duplicated kappa light-chain locus as assembled on Chromosome 2 in oryCun2.0 with positions of *IGKV*, *IGKJ*, and *IGKC* genes. The assembly has nine gaps of more than 200 bp (dotted lines). Arrows indicate transcriptional orientations. In the central region with dotted lines, the orientation of transcription of a few Vk is on the forward strand and the remainder on the reverse strand.

c. The lambda light-chain locus on Chromosome 21. There are 3 Mb missing from the p telomeric end of the chromosome in the assembly. Numbers in the current IMGT reference sequences are 3 Mb lower because assembled scaffold NW_003159316.1 was used for the IMGT annotations.

Ros *et al.* (2004) prepared a BAC library from a rabbit of the VH1a2 allotype. They reported the sequences of BAC clones (GenBank accession numbers AY386694, AY386695, AY386696 and AY386697) that encompass regions containing 34 *IGHV, IGHD, IGHJ* and the constant regions of IgM, IgG, IgE, and four of the known IgA isotypes. Partial sequencing of a Fosmid clone (AY386698) identified an additional *IGHA* (see Fig. 1 in Gertz *et al.*, 2013). Scaffold chrUn0439 does not include the expected 15 copies of *IGHA* genes encoding IgA isotypes. Partial matches to some but not all *IGHA* genes were found on four different unplaced scaffolds.

5.2.3 Light chain genes – Kappa and Lambda variable and constant regions

The oryCun2.0 donor rabbit was homozygous for the *IGKC1*-encoded allotype b5 and carried the two known functional lambda light-chain isotypes *IGLC5* and *IGLC6* that encode serologically detectable variants c21 and c7 (Gertz *et al.*, 2013). In the oryCun2.0 assembly, *IGKC1, IGKC2, IGKJ* (Jκ), and associated enhancers are on chromosome 2. Fig. 5.1b illustrates the unusual duplication of the kappa light-chain constant regions. The *IGKC2* and *IGKC1* on chromosome 2 in the assembly are in opposite transcriptional orientations about 0.8 Mb apart but this may be an underestimate of the distance because of numerous gaps in the assembly. Arrows indicate the transcriptional orientations of some of the 94 *IGKV* (Vκ) genes, pseudogenes and gene fragments found on chromosome 2. Dotted lines indicate gaps of more than 200 bp. A full tabulation of Vκ including the additional 21 Vκ genes in three unplaced scaffolds is given in online resource Table 5 of Gertz *et al.* (2013). Lambda light-chain genes are assembled on chromosome 21 in a region syntenic to the human lambda light-chain-encoding region on human chromosome 22. Fig. 5.1c summarizes the locations of the known functional *IGLJ5*, *IGLC5*, *IGLJ6*, *IGCL6* and ~43 Vλ genes or pseudogenes identified on chromosome 21. Some *IGLC* genes are missing, possibly due to gaps in the assembly. *IGLC* gene sequences are also found in three unplaced scaffolds. Although the sequences of half or more of the heavy chain *VH* and light-chain *Vκ* and *Vλ* genes may be listed as 'non-functional' pseudogenes, these are potentially functional as sources of donor sequences in gene conversions that diversify rearranged Ig genes during B-cell development and specific immune responses discussed in section 3.

5.3 Ig Protein Structure

Table 5.2 lists terms used to describe the structural basis for recognition of antigens by antibodies and some key proteins that contribute to antibody diversification and Ig functions. Complexities, extensions and exceptions to some basic definitions in the table are reviewed in Sela-Culang *et al.* (2013).

5.3.1 Genetic variants detectable with alloantisera (allotypes)

Alloantisera used to detect genetically controlled differences in protein products were developed by immunization of one or more individual rabbits with purified immunoglobulin of other rabbit(s). Rabbit genetic variants and allotypes provided markers used for documenting allelic exclusion and showing that genes encoding the V and C regions on the same chromosome are usually rearranged intrachromosomally and expressed together (cis expression during rearrangement and class switching). Somatic and germline recombination between *VH* and *CH* genes will be discussed further in section 5.3.3.

Table 5.2. Structures of antigens and Igs that contribute to functional interactions.

Term	
Variable region (V)	Amino terminal first domain of antibody heavy (H) and light (L) chains
CDRs	Complementarity determining regions (hypervariable loops)
FR	Framework regions that maintain the folded V region structures
Paratope	Antibody combining site formed by six hypervariable loops three from VH and three from light-chain V domain
Epitope	A structure recognized by antibody*
Fab, F(ab')$_2$	Amino terminal domains VH and CH1 of heavy and light chains**
Fc	The CH2 and CH3 constant domains of rabbit IgG ***
Hinge region	Sequence usually encoded by a separate exon that provides flexibility of Y-shaped N-terminal domains and encodes Cys that connects the heavy chains by one disulfide bond in rabbit IgG.
Fv	Fragment containing only variable regions of H or L chains
Single-chain Fv	H and L chain Fv engineered with a synthetic peptide to reconstitute a combining site

*Antigens may be proteins, carbohydrates, some lipids, and peptides or other small molecules (haptens) if coupled to a larger protein or carrier molecule.

**Fab or F(ab')$_2$ derive from protease digestion of purified rabbit IgG with Papain or Pepsin.

***Papain digestion also yields crystalline Fc. The CH2 domain has carbohydrate associated with it. Fc regions may bind and activate complement as well as a variety of activating or inhibitory receptors on other cells of the immune system.

Lists of the genetic variants found in common laboratory rabbit strains can be found in Mage *et al.* (2006). Additional variants have been identified in wild rabbits. Fig. 5.2 shows the localization of common allotypes and variants found on IgG molecules detectable with antisera. Allotypes of IgG heavy chains include allelic VH1a1, a2, a3, other VH collectively termed VHn, CH hinge region d11, d12 and CH2 e14 and e15, and on kappa chains, allotypes b4, b5, b6, b9 of kappa 1 and bas1, bas 2 of kappa 2. Alloantisera also detect variants of the two functional isotypes of lambda light-chain constant region, c7 and c21. Rabbits of the now extinct inbred ACEP/J strain at the Jackson Laboratory that lacked c7 made antibodies to lambda light chains that we now know are encoded by *IGLC6*. Similarly, some rabbits lacking expressed *IGLC5* produced antibodies to the c21 type. The schematic diagrams of rabbit IgG molecules in Fig. 5.2 also show Fab, hinge and Fc regions and the disulfide bonds within and between the heavy and light chains. Conventional structures associated with IgG bearing lambda or kappa light chains associated with Cκ2 are shown in Fig. 5.2a. These do not have the unusual interdomain disulfide bond between the Vκ and Cκ regions shown in Fig. 5.2b. This extra disulfide bond connects a Cys at variable region position 80 in

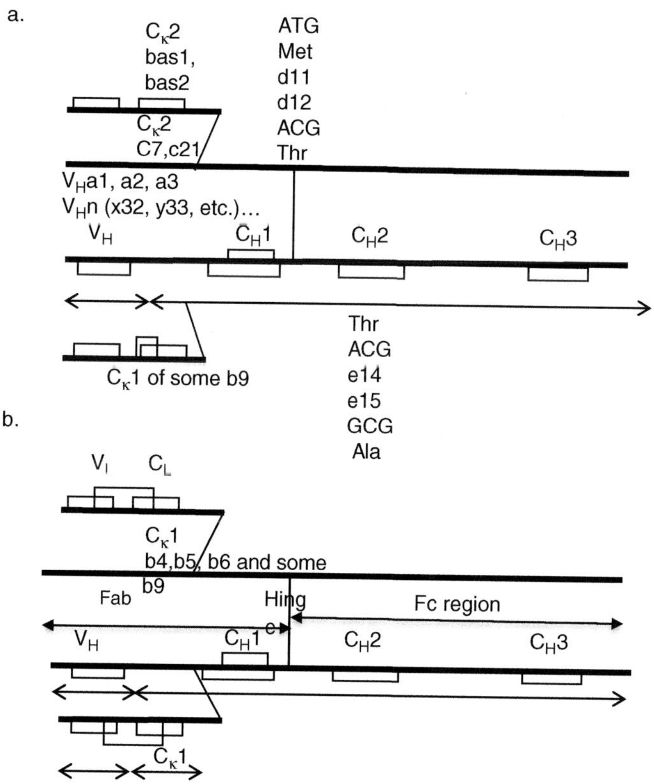

Fig. 5.2. Locations of variants or allotypes of *IGHV, IGHC, IGKC1, IGKC2* and *IGLC* genes found on rabbit IgG.

a. Light chains of CK2 and some CK1 of b9 type have disulfide bonds within the variable and constant domains that are typical of most species. The bas1 and bas 2 allotypes were discovered in mutant Basilea rabbits that fail to express parental b9 allotype because of a defect in the acceptor site for splicing Jκ to Cκ b9 (Lamoyi and Mage, 1985).

b. Light chains of CK1 and some of b9 type have an unusual additional disulfide bond between the variable and constant domains of the light chains.

Vκ genes to a constant region Cys at position 171 of the Cκ1 light chains of rabbits with allotypes b4, b5, b6, and some of the Vκ genes from b9 rabbits. In the OryCun2.0 assembly of chromosome 2, most Vκ genes or pseudogenes encoded the Cys at position 80. Among those surrounding *IGKC2* and associated *IGKJ1*, *IGKJ2-4* and *IGKJ3*, only two were found to encode Cys 80; the others encoded either Ala or Pro. The great stability of rabbit antibodies may in part result from stabilization of kappa 1 light-chain structures by the unusual inter-domain disulfide bond illustrated in Fig. 5.2b. Compared to b4 rabbits, those of the rare b9 and mutant bas types express a higher proportion of Vκ that lack the Cys 80. Yields of distinct and specific high-affinity Fab increased when rabbits of these types were immunized and recombinant rabbit–human Fab without the extra disulfide bond were generated by phage display (Popkov *et al.*, 2003). The Cys at position 80 in most rabbit Vκ genes leads to an unpaired thiol group that interferes with folding and expression of recombinant proteins because the method is designed to generate recombinant chimeric rabbit Fab molecules with human Cκ. This method is still widely used to generate selected highly specific high-affinity rabbit mAbs (reviewed in Rader, 2009).

5.3.2 Haplotypes and recombinants

Evidence for somatic and germline recombination between *VH* and *CH* (or between *IGHM* and the remaining *CH* genes within the heavy-chain locus was reviewed by Kelus and Steinberg (1991). Table 5.3 lists some of the haplotypes found in laboratory strains including additional allotypic markers on IgM and IgA and recombinant haplotypes observed during breeding in different laboratories (first reviewed in Mage *et al.*, 1982). Additional recombinants were reported in Kelus and Steinberg (1991) and Mage *et al.* (1992). The regions encompassing recombination sites were determined by restriction enzyme mapping of DNA from four recombinants. Two were found 5' (R3 and R4) and two 3' (R1 and R2) of the stretch of sequence containing the *DH* genes (Newman *et al.*, 1991; Mage *et al.*, 1992). Kelus and Steinberg (1991) summarized previous observations of recombinations and an additional large-scale experiment in their breeding colony to find recombinants between genes of the *IGH* locus that led to four more observations and an estimated recombination frequency of 0.1%. They discovered that all reported recombinations occurred in the male parent and suggested that they occurred during spermatogenesis, possibly during the multiple mitotic divisions of spermatogonia.

5.4 B Cell Development and Diversification of Sequences Encoding Igs

5.4.1 Early development of the B-cell repertoire

Protective circulating antibodies in newborn pups are acquired from the dam via hemochorial transplacental transfer and further transfer from colostrum, and milk. The development of the rabbit's own B lymphocytes that initiates in locations including fetal liver and omentum, continues in bone marrow during the perinatal period. Ig allotypes were used to measure decreases in maternally acquired antibodies. As the protection from maternal antibodies wanes at about 6 weeks of age, pups must produce their own protective antibodies and are particularly vulnerable to infections.

Rabbits and chickens both produce pre-immune and specific immune repertoires. A T-independent process in chickens initiates in the Bursa of Fabricius (bursa) prior to hatching. The crucial bursa-like role of appendix and other GALT in rabbit initiates after birth. Surgically or hormonally bursectomized chickens produce Igs but are incapable of producing specific antibodies. Neonatal appendectomy and removal of the developing sacculus rotundus and Peyer's patches similarly impairs the ability of rabbits to make specific protective immune responses. Early B-cell development in rabbit GALT also appears to be T-cell independent (Yermalli and Knight, 2013). In contrast to the specific T-dependent immune response discussed below in section 5.5, primary pre-immune repertoire development and expansion of appendix B cells requires endogenous and

Table 5.3. Some haplotypes of the complex heavy-chain locus found in laboratory strains and recombinant haplotypes observed during breeding.

Designation	VH	μ	γ	α
	a x y	**ms n**	**d e**	**f g**
1A	1	16(81)	12,15	73,74
1B	1 33,30	17(80)	12,15	71,75
1C	1 33,30	17(80)	11,15	72,74
1I	1 33,30	17(80)	12,14	69,77
1J	1	16(81)	12,15	70,76
1R2[1I-(F-C)]*	1 33,30	17(80)	11,15	72,74
1R6[1F-(F-I)]*	1	17(80)	12,14	69,77
1F	1	17(80)	12,15	71,75
2R3 F-C*	2 32, 33	17(80)	11,15	72,74
2E	2 32, 33	17(80)	12,15	71,75
2F	2 32, 33	17(80)	12,15	69,77
2R1 F-I*	2 32, 33	17(80)	12,14	69,77
2ali F-I**	(2) 32, 33	17(80)	12,14	69,77
2M	2 32, 33	16(81)	12,15	73,74
3G	3 32	17(80)	12,15	71,75
3H	3 32	16(81)	11,15	72,74
3R4 H-I*	3 32	17(80)	12,14	69,77
3R7 [H(ali F-I)]*	3 32	16(81)	12,14	69,77
3R8 [3F-(F-I)]*	3	17(80)	12,14	69,77
3F	3	17(80)	12,15	71,75

Asterisks (*) in Table 5.3 identify haplotypes derived from recombinations that occurred during breeding of laboratory rabbits. The parental haplotypes from which the recombinants were derived are shown in brackets when one of the parental types was derived from a previous recombinant haplotype. **The ali F-I was derived from the F-I recombinant (Kelus and Weiss, 1986). It has an approximately 10 kb deletion in the region containing *VH1a2* (Allegrucci *et al.*, 1990; Knight and Becker, 1990; Ros *et al.*, 2004).

microbial superantigens as stimuli. Gut-associated lymphoid tissue (GALT) including the developing appendix, sacculus rotundus and Peyer's patches are important sites for rabbit B-lymphocyte development (reviewed in Severson and Knight, 2014). Dasso *et al.* (2000) found that the morphology of bursal follicles is similar to those in young human and rabbit appendix. Quantitation of T-cell content indicated that if the human appendix also functions as a primary lymphoid organ, it might occur during the time when few T-cells are present. In contrast with bursal involution, both rabbit and human appendix follicles do not involute but become less elongated after the newborn period and then resemble Peyer's patches. More recently, Weller *et al.* (2008) discovered a potential human pre-immune repertoire in a population of circulating IgM-positive B cells from children aged 2 or younger with somatic diversification patterns that did not resemble those undergoing clonal selection.

5.4.2 Diversification by gene conversion and somatic hypermutation of rearranged rabbit heavy- and light-chain gene sequences

Rabbit B-cell development resembles chicken in another important way. Rearranged rabbit heavy- and light-chain sequences derived from the genome (germline) undergo sequence diversification by a somatic gene conversion-like mechanism in addition to the somatic hypermutation pathway typically used to introduce point mutations in human and mouse B cells. A recent report of gene conversion-like events in man also contains an excellent overview of this subject (Duvvuri and Wu, 2012).

5.4.2.1 Gene conversion

Gene conversion was first documented as a mechanism to diversify sequences in chickens where only the first *VH* gene in the locus is functional but upstream 'pseudogenes' serve as donors. Rabbit is the most well-studied of mammals although other species such as sheep and cattle probably also utilize gene conversion-like alterations of rearranged sequences. Rabbit *V* genes have similar framework region sequences. Thus, gene conversion blocks often appear to alter only the CDRs. Studies of clonally related sequences from single cells in splenic germinal centres found point mutations beyond gene-converted stretches and also superimposed upon sequences already introduced as blocks by gene conversion. Sequences of both rearranged heavy- and light-chain genes are altered by introduction of small or large blocks of sequences from other *V* genes (Sehgal *et al.*, 2000; and references therein). This mechanism provides a highly efficient means of generating new combining sites to produce a pre-immune repertoire in GALT of young rabbits and fine-tune the affinity of antibodies during immune responses in germinal centres of peripheral secondary lymphoid tissues such as spleen, lymph nodes, Peyer's patches and appendix in adult rabbits.

5.4.2.2 *Clonally related sequence diversification patterns in developing appendix differ from patterns during specific immune responses in secondary lymphoid organs*

Gene conversion and somatic hypermutation of rearranged heavy- and light-chain gene sequences initiates during the first few weeks of life. Between 3 and 9 weeks of age, when rabbit pre-immune B-cell populations expand in appendix and few, if any, T cells are present, Ig-producing cells display clonal sequence diversification patterns that differ from the patterns of diversification in spleen (Sehgal *et al.*, 2002). A quantitative algorithm that generated mutational lineage trees from sites of primary and secondary Ig gene diversification in rabbits confirmed the earlier conclusions. The trees derived from data during primary repertoire development in appendix differ significantly from the type of trees seen when antigen-specific selection occurs in splenic germinal centres (Mehr *et al.*, 2004). Affinity maturation occurs in splenic germinal centres in response to immunization. A review by Goodnow *et al.* (2010) describes the complex regulatory pathways of positive and negative selection of B-lymphocytes as B- and T-lymphocytes encounter antigens and enter developing germinal centres in peripheral lymphoid organs such as spleen and lymph nodes. Selection for survival and expansion occurs while at the same time eliminating potential autoantibody-producing cells. B cells require continuous survival signals. Competition between B cells with receptors specific for very low amounts of immunizing antigen bound to specialized follicular dendritic cells in germinal centers leads to survival of those with high affinity that out-compete other specific B cells and receive survival signals from specialized T-follicular helper cells.

5.5 T-Cell Receptors

As in man and mouse, rabbit T cells bear T-cell receptors (TCR) that are either composed of alpha and beta chains or of gamma and delta chains (α, β or γ, δ). Gene rearrangements similar to those described above for Ig genes utilize similar factors including RAG1, RAG2 and TdT for important additions of non-templated N regions.

5.5.1 TCR Gamma (TRG)

In the murine and human immune systems, α, β T cells are the more abundant type (Isono *et al.*, 1995; Massari *et al.*, 2012). Cho *et al.* (2005) suggested that in species with highly organized GALT including rabbit, chicken, cattle and sheep, higher proportions of γ, δ T cells could reflect some extra-thymic development and expansion of cells in the gut. In contrast to α, β T cells that must recognize peptides in association with MHC molecules and undergo selection and maturation in the thymus, γ, δ T cells recognize unique structures on pathogens and tend to localize at barrier sites including skin, lungs and gut epithelium. Cho *et al.* (2005) isolated, sequenced and characterized one *TRGC*, 22 distinct *TRGV*, and two *TRGJ* genes by probing and sequencing rabbit germ-line encoded *TRGV* in cosmid and lambda phage libraries spanning 200 kb. However, although also finding two *TRGJ* and one *TRGC*, Massari *et al.* (2012), only found a 70 kb stretch with 10 *TRGV* on chromosome 10 in the oryCun2.0 assembly. They stated, 'we assume that the rabbit locus is the smallest among the mammalian loci identified to date'. Both groups note the resemblance of the rabbit and human *TRG* loci. Whether the size discrepancy is due to gaps in oryCun2.0 assembly, differences between rabbits or other reasons remains to be determined. The regions of rabbit chromosome 10 and human chromosome 7 where the *TRG* genes reside are syntenic.

5.5.2 TCR beta (TRB)

The general organization of the *TRBV, TRBJ and TRBC* genes resembles those of man and mouse. In addition to *TRBC1* and *TRBC2* some rabbits have a third chimeric copy of a *TRBC* gene that appears to have resulted from unequal crossing over leading to segments from both *TRBC1* and *TRBC2*. Mage (1998) contains tabulations and references to available information. The locus is not mapped to a chromosome in oryCun2.0 but is assembled between positions 120,000 and 166,000 in unplaced scaffold chrUn0060 (NW_003159384.1).

5.5.3 TCR alpha and delta

The organization of the complex locus encoding components of the rearranging genes that form alpha or delta chains in the rabbit also appears to be similar to the human locus.

Rabbit and human *TRAV, TRAJ, TRAC, TRDV, TRDD* and *TRDC* genes are present on syntenic regions of chromosomes 17 and 14 respectively.

5.6 The Rabbit Leukocyte Antigen (RLA) Complex

Historically, the MHC has been extensively studied in many animal species as a key player in adaptive immunity by encoding the highly polymorphic class I and II molecules involved in peptide presentation to TCRs and responsible for allograft rejection. In addition, the MHC has been found in all jawed vertebrate species examined including cartilaginous fishes, and stands as an outstanding model region for genome evolution studies (Danchin and Pontarotti, 2004; Kelley *et al.*, 2005). The Rabbit MHC is referred to as the Rabbit Leukocyte Antigen complex, following a species-specific nomenclature (e.g. HLA for Human Leukocyte Antigen, SLA for Swine Leukocyte Antigen, BoLA for Bovine Leucocyte Antigen).

5.6.1 Mapping and organization

The MHC is one of the regions with highest gene density in mammalian genomes and spans several megabases. The MHC locus contains the class I and II histocompatibility genes but also a wide range of other genes with immunity-related, non-immunity-related as well as unknown functions. It has been estimated that about 40–60% of the expressed genes have immune system function (MHC Sequencing Consortium, 1999). The MHC locus has been divided into three subregions (class I, II and III) according to the gene content. The class I and II subregions contain MHC class I and II genes, respectively. The class III subregion located at the junction between class I and II does not contain histocompatibility genes but key genes such as the complement genes, tenascin, heat shock protein genes, and tumour necrosis factor (MHC Sequencing Consortium, 1999). In rabbit, the class III region was initially reported to be located outside the class I and class II adjacent regions (Chouchane and Kindt, 1992), in contrast to other mammalian species. More recently, it was shown that the RLA complex maps to a unique locus on chromosome 12 at position q1.1 (Rogel-Gaillard *et al.*, 2001). This RLA map location is orthologous to the human HLA locus at position HSA 6p21, as expected from reciprocal comparative chromosomal painting (Korstanje *et al.*, 1999). This location is confirmed in the oryCun2.0 assembly.

In this chapter, we have considered that the MHC boundaries are provided by the genes *MOG* (myelin oligodendrocyte glycoprotein) and *KIFC1* (kinesin family member C1), at the end of class I and II subregions, respectively (Fig. 5.3). By referring to this genomic segment *MOG-KIFC1*, the HLA complex spans 3.6 megabases and the first extensive map identified 224 gene loci among which 128 were predicted to be expressed (MHC Sequencing Consortium, 1999). For rabbit, in the oryCun2.0 assembly *MOG* maps at chromosome 12:12:21984941-21999226 and *KIFC1*

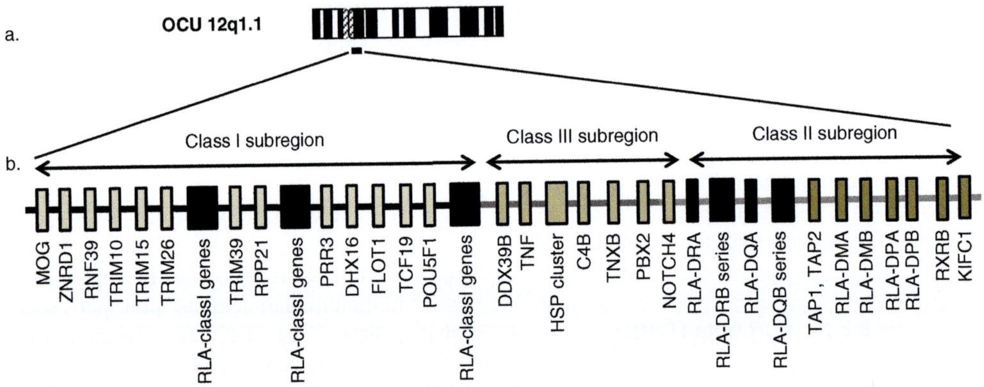

Fig. 5.3. Rabbit MHC (RLA complex) genomic mapping.
a. Mapping of the RLA complex on chromosome 12 at position q1.1.
b. Genomic organization of the RLA complex established from BAC contigs using the rabbit BAC library reported by Rogel-Gaillard *et al.* (2001). The genomic sequence from ZNRD1 to VARSL is available with the BAC sequence accession numbers AC234889, AC234890, AC234891, AC234997, KJ810984, KJ810977, KJ810983, KJ810978, AC234892, AC235084, AC236481, AC234893. The genomic sequence from CDSN to NOTCH4 is available with the BAC sequence accession numbers AC235550, KJ810979, KJ810980, AC234894, AC234895, AC235568, AC234896, AC235095, AC234897. The genomic sequence from RLA-DRA to RLA-DQB is available with the BAC sequence accession numbers KJ810981, AC234898, AC234899, KJ810982, AC235476. The genomic sequence from RLA-DMA to KIFC1 is available with the BAC accession numbers AC234900, AC234901, AC234902.

at chromosome 12:23777969-23793012:1 suggesting that the RLA region spans 1.8 megabases. A partial BAC-based contig of the RLA complex has been built by screening a BAC library derived from a NZW rabbit homozygous for MHC (Rogel-Gaillard *et al.*, 2001). All BACs have been sequenced and are provided EMBL/NCBI/DDBJ accession numbers (see legend of Fig. 5.3). This BAC contig does not cover the whole RLA region and a few gaps still have to be filled. However, the existing non-overlapping sub-contigs span a total length of 2 megabases, meaning that the current status of the RLA region provided by the oryCun2.0 assembly is probably not complete. A general map of the RLA complex is presented in Fig. 5.3.

5.6.2 The MHC class I genes

The MHC class I genes are a series of duplicated genes and pseudogenes that include classical (class Ia) and non-classical genes (class Ib). The class I genes encode the heavy chains of the class I molecules.

In humans, the functional MHC class Ia genes are HLA-A, -B and -C, and the class Ib genes HLA-E, -F and -G. The MHC class Ia genes are highly polymorphic and expressed in all nucleated cells. Conversely, the MHC class Ib genes present limited polymorphism, tissue-specific expression and alternate transcripts that encode various protein isoforms. In rabbit, the number of class I genes and pseudogenes is not precisely known but has been estimated between 8 and 13 (Rebière *et al.*, 1987). The genes are found in two main clusters. One cluster is in the vicinity of TRIM26, TRIM39 and PRR3 and the other is at the telomeric end of the class I subregion after POU5F1 (Fig. 5.3). By contrast to the HLA complex, but similar to the SLA complex (Renard *et al.*, 2006), no class I gene was found in the vicinity of *ZNRD1*. This finding confirms that the class I genes evolve in a species-specific manner and that no easy gene orthology can be drawn between the series of MHC class I genes in different species. The MHC class Ia molecules are transmembrane glycoproteins of 44 kilodaltons (kDa) that belong to the immunoglobulin superfamily. MHC class Ia molecules present endogenous and viral peptides to cytotoxic CD8+ lymphocytes. They have also been shown to regulate innate immunity as ligands for killer inhibitory receptors (KIRs) on NK cells in humans. They bind to inhibitory receptors on NK cells that include KIR (killer cell Ig-like receptors) in man, C-type lectin-like Ly49 molecules in mouse, and CD94/NKG2A heterodimers in human and mouse (Anfossi *et al.*, 2006). The MHC class Ia molecules are composed of an alpha chain non-covalently associated with a β chain encoded by the β2-microglobulin (B2M) gene that maps outside the MHC region. In rabbit, B2M maps to chromosome 17 (Ensembl ID ENSOCUG00000017117, position 17:27607330-27614076 on the reverse strand in oryCun2.0 assembly). The MHC class Ia genes have a common genomic organization with eight exons. Exon 1 corresponds to a leader peptide. Exons 2 and 3 encode the polymorphic α1 and α2 domains, respectively. The α1 and α2 domains fold together to create a groove, which binds peptides. Exon 4 corresponds to the α3 domain, an immunoglobulin-like region that binds to the CD8 co-receptor on CD8 positive T cells. Exon 5 encodes the transmembrane domain and the remaining exons 6 through 8 encode the cytoplasmic tail. The RLA class Ia genes have been reported to have a similar organization (Marche *et al.*, 1985).

Several transcripts have been reported and their features suggest that two clones referred to as pR11 (Accession numbers: K02441, K02442) and pR19 (Accession number: K02819) could correspond to RLA-class Ia genes whereas the clone pR27 (Accession numbers: M22384, M22385) could correspond to an RLA-class Ib gene due to a tissue-specific expression (Rebière *et al.*, 1987).

5.6.3 The MHC class II genes

The MHC class II region includes a series of alpha and beta genes that encode heterodimeric class II molecules. The class II molecules are expressed on professional antigen presenting cells (such as dendritic cells, macrophages and B cells) and present peptides derived from exogenous antigens that have been processed intracellularly and displayed on the cell surface to interact with the TCR and CD4 co-receptor on CD4+ T helper lymphocytes. Seven class II beta genes have been reported thus far in rabbits, including: one DQ beta (*DQB*), one DP beta (*DPB*) and five DR beta (*DRB*) genes (Sittisombut *et al.*, 1989). Characterization and expression of the rabbit *DQ*, *DR* and *DP* RLA class II gene series have been reported (LeGuern *et al.*, 1985; Sittisombut and Knight, 1986; Sittisombut 1988; Sittisombut *et al.*, 1988; Spieker-Polet *et al.*, 1990, 1993). In addition, *DM* genes (*DMA* and *DMB*) have been characterized and reported to be co-expressed with class II genes in lymphoid tissues, as are the *DM* genes of other mammals (Hermel *et al.*, 1999). The class II genes annotated in the oryCun2.0 assembly are summarized in Table 5.4.

5.6.4 Genetic variability of the RLA complex

One remarkable characteristic of MHC class Ia and II genes is the extremely high polymorphism carried by exons 2 and 3 that encode the α1 and α2 domains responsible for the peptide binding groove. This feature is not only a selective advantage for an individual to express molecules that bind different repertoires of peptides, but also for a population to have many variants segregating among its members. An underlying concept is that the higher the number of allelic variants, the higher putative protection against pathogens at the population level.

The high number of MHC polymorphic genes has led to the construction of haplotypes and for transplantation studies; a prospective diagnosis of isogenicity between donor and recipient is a prerequisite before any surgical operation. No large-scale design of serological testing has ever been developed in rabbit due to the very small number of available allele-specific anti-rabbit MHC monoclonal antibodies (Boyer *et al.*, 1995). Typing methods have been based on restriction fragment length polymorphism (RFLP) markers in the RLA class I and class II regions in inbred rabbit lines (Marche *et al.*, 1989) as well as outbred lines (Boyer *et al.*, 1995). Haplotypes were numbered by class I and class II designations and the composites are given a numerical identification. A total of 13 distinct RLA haplotypes have been reported using this method (Marche *et al.*, 1989). Boyer *et al.* (1995) refined this method for RFLP-based haplotyping by using similar class I probes and a DQ alpha probe

Table 5.4. Mapping of class I and II MHC genes in the oryCun2.0 assembly.

Classes	Rabbit genes	Ensembl ID	Mapping in the oryCun2.0 assembly[2]
Class I		ENSOCUG00000011016	12:20242252-20300679:1[3]
Class I		ENSOCUG00000010972	12:20208925-20285057:-1[3]
Class II	RLA-DR-ALPHA	ENSOCUG00000009103	12:23070319-23074942:1
Class II	RLA-DRB1	ENSOCUG00000029731	12:23116765-23198692:-1
Class II	RLA-DQA[1]	ENSOCUG00000002480	12:23254403-23259308:1
Class II	RLA-DQB1[1]	ENSOCUG00000002485	12:23143857-23287878:-1
Class II	RLA-DOB[1]	ENSOCUG00000029437	12:23329185-23335061:-1
Class II	TAP1	ENSOCUG00000002135	12:23361423-23370950:-1
Class II	TAP2	ENSOCUG00000002489	12:23346548-23354611:-1
Class II	RLA-DMA[1]	ENSOCUG00000022573	12:23467875-23469906:-1
Class II	RLA-DMB	ENSOCUG00000026567	12:23449981-23456715:-1
Class II	RLA-DOA[1]	ENSOCUG00000003319	12:23521562-23550151:-1
Class II	RLA-DPB1[1]	ENSOCUG00000008721	12:23554108-23560778:1

[1]The prefix HLA or SLA used in the oryCun2.0 assembly has been replaced by the prefix RLA in the table.
[2]The transcriptional orientation of the genes is indicated by 1 (forward) or -1 (reverse). Chromosomal positions as available in December 2019.
[3]The position maps outside the expected boundaries of the RLA complex, likely suggesting assembling issues in the oryCun2.0 assembly.

for the class II region. A total of 34 RFLP-based class I haplotypes and 16 RFLP-based class II haplotypes have been defined in two commercially available populations of NZW rabbits using this method. Allelic analyses are now based on resequencing of the variable exons, as recently reported for association studies with DRB (Oppelt *et al.*, 2010).

In rabbit, a nomenclature for the MHC class I and II genes has to be established, together with a nomenclature for the allelic variants. The Immuno Polymorphism Database (https://www.ebi.ac.uk/ipd/mhc/, accessed 31 December 2020) offers a repository for MHC allelic variability for a wide range of species but the rabbit is not yet included.

5.6.5 Insights provided by the current assembly of the rabbit genome

The oryCun2.0 assembly is very useful for searching gene by gene but, likely, does not yet provide an extensive description of the RLA region as a whole. In addition, an extensive annotation of all functional MHC genes is still missing. As presented in Table 5.4, putative RLA class I genes are found outside the predicted boundaries of the MHC region, suggesting either misassemblies in oryCun2.0 or RLA features not yet identified. Deep analyses are required to accurately relate transcript data with genomic annotation in the oryCun2.0 assembly (Table 5.4), and to reliably classify functional genes and pseudogenes. The availability of a reference sequenced haplotype is still pending and should pave the way towards the extensive characterization of other haplotypes by resequencing.

5.7 Cytokine Genes

Perkins *et al.* (2000) published sequences of rabbit cDNAs encoding IL2, IL4, IL6 and IL10. The cluster of genes encoding the coordinately regulated Th2 cytokines IL5, IL4 and IL13 that is present on human chromosome 5 is found on a syntenic region of rabbit chromosome 3 in the oryCun2.0 genome assembly. These cytokines are beneficial in combating helminth infections, but play negative roles in allergy and asthma, and also contribute to fibrosis of the lung and liver. Gertz *et al.* (2011) compared sequences of promoter regions, DNase hypersensitive sites, transcription factor binding sites and coding regions in the oryCun2.0 genome from a tuberculosis-susceptible strain with the syntenic regions in other species as well as with the corresponding regions in ENCODE ENm002 from a normal NZW rabbit. Fig. 5.4 summarizes the region

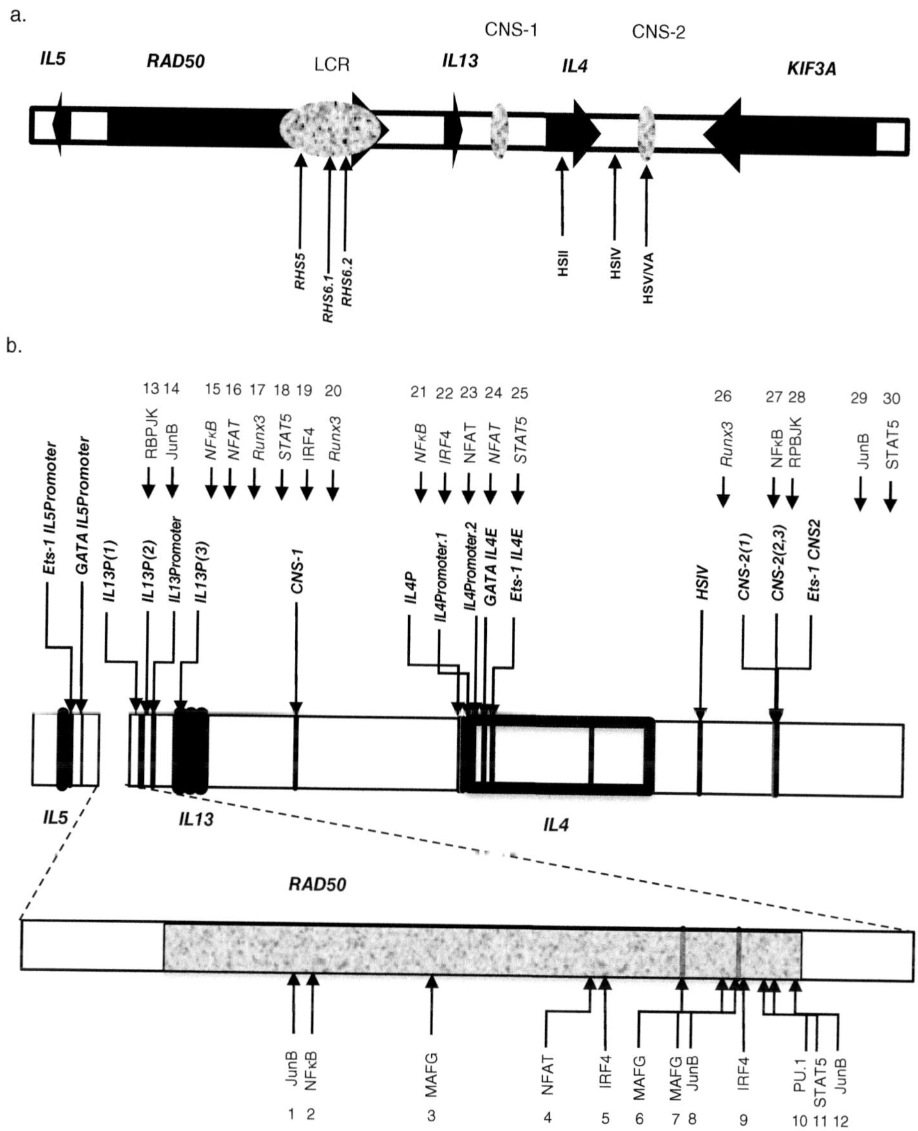

Fig. 5.4. Region encoding Th2 cytokines on rabbit chromosome 3. (Adapted from Gertz *et al.*, 2011)

a. Phylogenetically conserved overall structure of the Th2 cytokine region (not shown to scale) that contains *IL5*, the DNA repair gene *RAD50* with a locus control region (LCR), *IL13, IL4* with conserved non-coding sequences (CNS) 1 and 2 and *KIF3A* (Strempel *et al.*, 2010). The directions of transcription are shown with arrows. DNase hypersensitive sites include enhancers and a silencer (HSIV) that suppresses expression of IL4 in Th1 and immature T cells.

b. Transcription factor binding sites. The expanded view of the large *RAD50* gene is to ~30% of true relative length. Transcription factor binding sites are conserved across rabbit, mouse and man in *RAD50* (1-12) and in the regions containing *IL13* and *IL4* (13-30). Also shown are conserved Ets-1, GATA-binding and promoter sequences. *IL13P(3)* has probably been lost in rabbit as both the oryCun2.0 and ENCODE NZW rabbit sequences have gaps when this region aligns with the nine other species. Full details are presented in Gertz *et al.* (2011).

containing rabbit genes encoding Th2 cytokines, *IL5*, the DNA repair gene *RAD50 IL13*, *IL4* and *KIF3A* (Gertz *et al.*, 2011). The *KIF3A* and *RAD50* genes do not encode cytokines, yet their placement at this locus has been conserved throughout mammals (Strempel *et al.*, 2010). *RAD50* contains a locus control region and activated Th2 cells produce SATB1 that binds to CNS1 and 2 (Fig. 5.4b) and nine other sites from *IL5* past *KIF3A*. As discussed in Gertz *et al.* (2011), Th2 cell activation involves chromatin remodeling to bring together distant sites within the locus.

Although the donor of DNA for oryCun2.0 was more inbred than outbred NZW rabbits, difficulties in assembly of multiple copies of highly similar genes were compounded by potential heterozygosity at the complex loci described in this chapter. A second genomic sequence of the Th2 cytokine region permitted more complete evaluation of areas where gaps occurred in one of the two assemblies.

Acknowledgements

This work was supported by the Intramural Research Program of the NIH, NIAID and the Animal Genetics and Animal Physiology Divisions of INRAE.

References

Allegrucci, M., Newman, B.A., Young-Cooper, G.O., Alexander, C.B., Meier, D. *et al.* (1990) Altered phenotypic expression of immunoglobulin heavy-chain variable-region (VH) genes in Alicia rabbits probably reflects a small deletion in the VH genes closest to the joining region. *Proceedings of the National Academy of Sciences of the USA* 87, 5444–5448.

Anfossi, N., André, P., Guia, S., Falk, C.S., Roetynck, S. *et al.* (2006) Human NK cell education by inhibitory receptors for MHC class I. *Immunity* 25, 331–342.

Becker, R.S. and Knight, K.L. (1990) Somatic diversification of immunoglobulin heavy chain VDJ genes: evidence for somatic gene conversion in rabbits. *Cell* 63, 987–997.

Boyer, M.I., Bowen, C.V. and Danska, J.S. (1995) Restriction fragment-length polymorphism analysis of the major histocompatibility complex in New Zealand white rabbits. *Transplantation* 59, 1043–1046.

Brandtzaeg, P. (2013) Secretory IgA: designed for antimicrobial defense. *Frontiers in Immunology* 4, 222.

Carneiro, M., Rubin, C.J., Di Palma, F., Albert, F.W., Alföldi, J. *et al.* (2014) Rabbit genome analysis reveals a polygenic basis for phenotypic change during domestication. *Science* 345, 1074–1079.

Cho, K.S., Zhai, S.K. and Esteves, P.J. (2005) Characterization of the T-cell receptor gamma locus and analysis of the variable gene segment expression in rabbit. *Immunogenetics* 57, 352–363.

Chouchane, L. and Kindt, T.J. (1992) Mapping of the rabbit MHC reveals that class I genes are adjacent to the DR subregion and defines an insertion/deletion-related polymorphism in the class II region. *Journal of Immunology* 149, 1216–1222.

Danchin, E.G. and Pontarotti, P. (2004) Towards the reconstruction of the bilaterian ancestral pre-MHC region. *Trends in Genetics* 20, 587–591.

Dasso, J.F., Obiakor, H., Bach, H., Anderson, A.O. and Mage, R.G. (2000) A morphological and immunohistological study of the human and rabbit appendix for comparison with the avian bursa. *Developmental and Comparative Immunology* 24, 797–814.

Deeks, E.D. and Keating, D.M. (2009) Rabbit antithymocyte globulin (thymoglobulin): a review of its use in the prevention and treatment of acute renal allograft rejection. *Drugs* 69, 1483–1512.

Dorman, S.E., Hatem, C.L., Tyagi, S., Aird, K., Lopez-Molina, J., *et al.* (2004) Susceptibility to tuberculosis: clues from studies with inbred and outbred New Zealand White rabbits. *Infection and Immunity* 72, 1700–1705.

Duvvuri, B. and Wu, G.E. (2012) Gene conversion-like events in the diversification of human rearranged IGHV3-23*01 gene sequences. *Frontiers in Immunology* 3, 158.

Esteves, P.J., Abrantes, J., Baldauf, H.M., BenMohamed, L., Chen, Y. *et al.* (2018) The wide utility of rabbits as models of human diseases. *Experimental and Molecular Medicine* 50, 66.

Gertz, E.M., Agarwala, R., Mage, R.G. and Schaffer, A.A. (2011) Comparative analysis of genome sequences of the Th2 cytokine region of rabbit (*Oryctolagus cuniculus*) with those of nine different species. *Immunology and Immunogenetics Insights* 3, 59–82.

Gertz, E.M., Schaffer, A.A., Agarwala, R., Bonnet-Garnier, A., Rogel-Gaillard, C. *et al.* (2013) Accuracy and coverage assessment of *Oryctolagus cuniculus* (Rabbit) genes encoding immunoglobulins in the whole genome sequence assembly (OryCun2.0) and localization of the IGH locus to chromosome 20. *Immunogenetics* 65, 749–762.

Goodnow, C.C., Vinuesa, C.G., Randall, K.L., Mackay, F. and Brink, R. (2010) Control systems and decision making for antibody production. *Nature Immunology* 11, 681–688.

Hermel, E., Han, M., Hague, B., Kindt, T. and Monaco, J.J. (1999) Isolation and mapping of the rabbit DM genes. *Immunogenetics* 49, 295–302.

Isono T., Kim C.J. and Seto A. (1995) Sequence and diversity of rabbit T-cell receptor gamma chain genes. *Immunogenetics* 41, 295–300.

Kelley J., Walter L. and Trowsdale J. (2005) Comparative genomics of major histocompatibility complexes. *Immunogenetics* 56, 683–695.

Kelus, A.S. and Steinberg, C.M. (1991) Is there a high rate of mitotic recombination between the loci encoding immunoglobulin VH and CH regions in gonial cells? *Immunogenetics* 33, 255–259.

Kelus, A.S. and Weiss, S. (1986) Mutation affecting the expression of immunoglobulin variable regions in the rabbit. *Proceedings of the National Academy of Sciences of the USA* 83, 4883–4886.

Knight, K.L. and Becker, R.S. (1990) Molecular basis of the allelic inheritance of rabbit immunoglobulin VH allotypes: implications for the generation of antibody diversity. *Cell* 60, 963–970.

Korstanje, R., O'Brien, P.C., Yang, F., Rens, W., Bosma, A.A. *et al.* (1999) Complete homology maps of the rabbit (*Oryctolagus cuniculus*) and human by reciprocal chromosome painting. *Cytogenetics and Cell Genetics* 86, 317–322.

Lamoyi, E. and Mage, R.G. (1985) The lack of K1b9 light chains in Basilea rabbits is probably due to a mutation in an acceptor site for mRNA splicing. *Journal of Experimental Medicine* 162, 1149–1160.

LeGuern, C., Marche, P.N. and Kindt T.J. (1985) Molecular evidence for five distinct MHC class II alpha genes in the rabbit. *Immunogenetics* 22, 141–148.

Mage, R.G. (1998) Immunology of lagomorphs. In: Pastoret, P.P., Bazin, H., Griebel, P. and Govaerts. H. (eds) *Handbook of Vertebrate Immunology*. Academic Press, London, pp. 223–260.

Mage, R.G. and Rai, G. (2012) A rabbit model of systemic lupus erythematosus, useful for studies of neuropsychiatric SLE. In: Almoallim, H. (ed.) *Systemic Lupus Erythematosus.* InTech –Open Access Publisher, Rijeka, Croatia, pp. 201–212.

Mage, R.G., Dray, S., Gilman-Sachs, A., Hamers-Casterman, C., Hamers, R. *et al.* (1982) Rabbit heavy chain haplotypes – Allotypic determinants expressed by V_H-C_H recombinants. *Immunogenetics* 15, 287–297.

Mage, R.G., Young-Cooper, G.O., Alexander, C.B., Hanly, W.C. and Newman, B.A. (1992) A new V_H-C_H recombinant in the rabbit. *Immunogenetics* 35,131–135.

Mage, R.G., Lanning, D. and Knight, K.L. (2006) B cell and antibody repertoire development in rabbits: the requirement of gut-associated lymphoid tissues. *Developmental and Comparative Immunology* 30,137–153.

Mantis, N.J., Rol, N. and Corthesy, B. (2011) Secretory IgA's complex roles in immunity and homeostasis in the gut mucosa. *Mucosal Immunology* 4, 603–611.

Marche, P.N., Tykocinski, M.L., Max, E.E. and Kindt, T.J. (1985) Structure of a functional rabbit class I MHC gene: similarity to human class I genes. *Immunogenetics* 21, 71–82.

Marche, P.N., Rebiére, M.C., Laverrière, A., English, D.W., LeGuern, C. and Kindt, T.J. (1989) Definition of rabbit class I and class II MHC gene haplotypes using molecular typing procedures. *Immunogenetics* 29, 273–276.

Margulies, E.H., Cooper, G.M., Asimenos, G., Thomas, D.J., Dewey, C.N. *et al.* (2007) Analyses of deep mammalian sequence alignments and constraint predictions for 1% of the human genome. *Genome Research* 17, 760–774.

Massari, S., Ciccarese, S. and Antonacci, R. (2012) Structural and comparative analysis of the T cell receptor gamma (TRG) locus in *Oryctolagus cuniculus*. *Immunogenetics* 64, 773–779.

Mehr, R., Edelman, H., Sehgal, D. and Mage, R. (2004) Analysis of mutational lineage trees from sites of primary and secondary Ig gene diversification in rabbits and chickens. *Journal of Immunology* 172, 4790–4796.

Mendez, S., Hatem, C.L., Kesavan, A.K., Lopez-Molina, J., Pitt, M.L.M. *et al.* (2008) Susceptibility to tuberculosis: composition of tuberculous granulomas in Thorbecke and outbred New Zealand White rabbits. *Veterinary Immunology and Immunopathology* 122, 167–174.

MHC Sequencing Consortium (1999) Complete sequence and gene map of a human major histocompatibility complex. *Nature* 401, 921–923.

Newman, B.A., Young-Cooper, G.O., Alexander, C.B., Becker, R.S., Knight, K.L. *et al.* (1991) Molecular analysis of recombination sites within the immunoglobulin heavy chain locus of the rabbit. *Immunogenetics* 34, 101–109.

Oppelt, C., Starkloff, A., Rausch, P., Von Holst, D. and Rödel, H.G. (2010) Major histocompatibility complex variation and age-specific endoparasite load in subadult European rabbits. *Molecular Ecology* 19, 4155–4167.

Perkins, H.D., van Leeuwen, B.H., Hardy, C.M. and Kerr, P.J. (2000) The complete cDNA sequences of IL-2, IL-4, IL-6 and IL-10 from the European rabbit (*Oryctolagus cuniculus*). *Cytokine* 12, 555–565.

Pinheiro, A., Neves, F., de Matos, A.L., Abrantes, J., van der Loo, W. *et al.* (2016) An overview of the lagomorph immune system and its genetic diversity. *Immunogenetics* 68, 83–107.

Popkov, M., Mage, R.G., Alexander, C.B., Thundivalappil, S., Barbas III, C.F. and and Rader, C. (2003) Rabbit immune repertoires as sources for therapeutic monoclonal antibodies: the impact of kappa allotype-correlated variation in cysteine content on antibody libraries selected by phage display. *Journal of Molecular Biology* 325, 325–335.

Rader, C. (2009) Generation and selection of rabbit antibody libraries by phage display. *Methods in Molecular Biology* 525, 101–128.

Rebière, M.C., Marche P.N. and Kindt, T.J. (1987) A rabbit class I major histocompatibility complex gene with a T cell-specific expression pattern. *Journal of Immunology* 139, 2066–2074.

Renard, C., Hart, E., Sehra, H., Beasley, H., Coggill, P. *et al.* (2006) The genomic sequence and analysis of the swine major histocompatibility complex. *Genomics* 88, 96–110.

Rogel-Gaillard, C., Piumi, F., Billault, A., Bourgeaux, N., Save, J.C. *et al.* (2001) Construction of a rabbit bacterial artificial chromosome (BAC) library: application to the mapping of the major histocompatibility complex to position 12q1.1. *Mammalian Genome* 12, 253–255.

Rogel-Gaillard, C., Ferrand, N. and Hayes, H. (2009) Genome mapping in the rabbit. In: Kole, C. and Cockett, N.E. (eds) *Genome Mapping in Animals*. Vol. II Wildlife & Companion Animals. Springer-Verlag, Berlin, Heidelberg, New York, Tokyo.

Ros, F., Puels, J., Reichenberger, N., van Schooten, W.C.A., Buelow, R. and Platzer, J. (2004) Sequence analysis of 0.5Mb of the rabbit germline immunoglobulin heavy chain locus. *Gene* 330, 49–59.

Sehgal, D., Schiaffella, E., Anderson, A.O. and Mage, R.G. (2000) Generation of heterogeneous rabbit anti-DNP antibodies by gene conversion and hypermutation of rearranged VL and VH genes during clonal expansion of B cells in splenic germinal centers. *European Journal of Immunology* 30, 3634–3644.

Sehgal, D., Obiakor, H. and Mage, R.G. (2002) Distinct clonal Ig diversification patterns in young appendix compared to antigen-specific splenic clones. *Journal of Immunology* 168, 5424–5433.

Sela-Culang, I., Kunik, V. and Ofran, Y. (2013) The structural basis of antibody-antigen recognition. *Frontiers in Immunology* 4, 302.

Severson, K.M. and Knight K.L. (2014) Generation of the antibody repertoire in rabbits: role of gut-associated lymphoid tissues. In: Kaushik, A.K. and Pasman, Y. (eds) *Comparative Immunoglobulin Genetics*. Apple Academic Press Inc., Waretown, New Jersey.

Silverstein, A.M. (2001) Autoimmunity versus *horror autotoxicus*: the struggle for recognition. *Nature Immunology* 2, 279–281.

Sittisombut, N. (1988) Two distinct nuclear factors bind the conserved regulatory sequences of a rabbit major histocompatibility complex class II gene. *Molecular and Cell Biology* 8, 2034–2041.

Sittisombut, N. and Knight, K.L. (1986) Rabbit major histocompatibility complex. I. Isolation and characterization of three subregions of class II genes. *Journal of Immunology*136, 1871–1875.

Sittisombut, N., Mordacq, J. and Knight K.L. (1988) Rabbit MHC. II. Sequence analysis of the R-DP alpha- and beta-genes. *Journal of Immunology* 140, 3237–3243.

Sittisombut, N., Tissot, R.G. and Knight, K.L. (1989) Rabbit major histocompatibility complex. III. Multiple class II DR beta genes and restriction fragment length polymorphism of the class II alpha and beta genes. *Journal of Immunogenetics*16, 63–75.

Spieker-Polet, H., Sittisombut, N., Yam, P.C. and Knight, K.L. (1990) Rabbit major histocompatibility complex. IV. Expression of major histocompatibility complex class II genes. *Journal of Immunogenetics* 17, 123–132.

Spieker-Polet, H., Yam, P.C. and Knight, K.L. (1993) Differential expression of 13 IgA-heavy chain genes in rabbit lymphoid tissues. *Journal of Immunology* 150, 5457–5465.

Strempel, J.M., Grenningloh, R., Ho, I-C. and Vercelli, D. (2010) Phylogenetic and functional analysis identifies Ets-1 as a novel regulator of the TH2 cytokine gene locus. *Journal of Immunology* 184, 1309–1318.

Weber, J., Peng, H. and Rader, C. (2017) From rabbit antibody repertoires to rabbit monoclonal antibodies. *Experimental and Molecular Medicine* 49, e305–e305.

Weller, S., Mamani-Matsuda, M., Picard, C., Cordier, C., Lecoeuche, D. *et al.* (2008) Somatic diversification in the absence of antigen-driven responses is the hallmark of the IgM+ IgD+ CD27+ B cell repertoire in infants. *Journal of Experimental Medicine* 205, 1331–1342.

Yermalli, V.A. and Knight, K.L. (2013) Development of $CD27^+$ marginal zone B cells requires GALT. *European Journal of Immunology* 43, 1484–1488.

6 Genetics and Molecular Genetics of Coat Colour in the European Rabbit

Luca Fontanesi
Department of Agricultural and Food Sciences, Division of Animal Sciences, University of Bologna, Italy

6.1 Introduction

After the rediscovery of Mendel's laws at the beginning of the 20th century, the European rabbit, with the large variety of domestic lines and strains that, at that time, were already available, became a model mammal, together with the rodents, to study the genetic factors determining coat-colour variability (Castle, 1905; Punnett, 1912). The relevance of this phenotypic trait in the domestic rabbit is also evident by the name of many breeds and strains that include a characterizing coat colour or colour pattern and by the description of breed standards that are defined considering, as primary elements, these features (see Chapter 2).

Pigmentation in all mammals (including the European rabbit) is determined by the accumulation (or absence) of melanin pigments in the hair and skin. Melanins are synthesized starting from the enzymatic oxidation of the tyrosine amino acid, which is not the limiting factor as it is a basic biochemical element of the proteins. From this biochemical step, two chemically distinct types of melanin are then produced: (i) eumelanin (black or brown melanin); and (ii) pheomelanin (yellow or red melanin). The appearance of a coat colour is subsequently determined by the relative quantity or distribution of these two pigments producing the two extreme colours (black/brown and yellow/red), with the only accumulation of one of these two forms, and all intermediate grades according to their different proportion. For the final production of the coat colour, in addition to these basic biochemical components, other biological processes are involved. A simplified summary of the different steps and related mechanisms is reported below, which introduces the complexity of coat-colour determination where a large number of genes and their derived protein products or regulatory elements are involved (Lamoreaux *et al.*, 2010).

The metabolic pathways leading to the production of these two melanin pigments are mostly known. Tyrosinase (TYR) is the key enzyme in the melanin synthesis. It catalyses the first two metabolic steps that first hydroxylate tyrosine to obtain dihydroxyphenylalanine (DOPA) and then oxidase this molecule to produce DOPAquinone. Eumelanins are subsequently produced from metabolites originated from DOPAchrome and pheomelanins are derived from 5-S-cysteinylDOPA metabolites (Prota, 1992). Melanocytes are specialized cells in which melanins are synthesized and accumulated. Melanocytes are positioned between the derm and the epiderm and together with keratinocytes constitute the epidermal melanin unit. Melanogenesis occurs in the melanosomes that are organelles

Contact e-mail address: luca.fontanesi@unibo.it

 The Genetics and Genomics of the Rabbit (ed Luca Fontanesi)

DOI: 10.1079/9781780643342.0006

present in the cytoplasm of these specialized cells. The melanosomes derive from the smooth endoplasmic reticulum and their biogenesis and translocation include several steps of maturation and transformation. Premelanosomes develop into mature melanosomes through four stages. When melanosomes reach stage 4, they are transported to the periphery of the cells in which they have been developed and then they are transferred to keratinocytes. An exocytosis process occurring at this level is responsible for the transfers of the melanosomes into the hairs over their growing phase, conferring on these structures the related pigmentation. Pigmentation can be modified depending on a more or less reduced melanocyte activity or modified process of melanin accumulation or transfer, with the involvement of melanosomal protein processing and transfer. The melanocytes originate from the melanoblasts (the primordial cells) that, starting from the neural crest during the embryonic development, colonize different regions of the animal's body. After transformation into melanocytes, only the regions in which they reside can show pigmentation, whereas the areas that lack melanocytes have white patterns, like white-spotted or white-belted colourations. The position of the body in which melanocytes are located regulates, in some cases, the activities of these specialized cells and the result is another level of pigmentation variability.

A few reviews, written over the last century (Wright, 1917; Castle, 1930, 1940; Sawin, 1955; Robinson, 1958; Searle, 1968; Fox, 1994), compiled the evidences on the genetic factors affecting coat colour in the European rabbit that were derived by classical genetic studies. Information mainly came from crossbreeding and segregation analyses which identified several allelic series, matching similar series reported in other mammals for the putative homologous major loci. The list in rabbit included the following allelic series: *A* (*Agouti*), *B* (*Brown*), *C* (*Albino*), *D* (*Dilute*), *E* (*Extension*), *Du* (*Dutch*), *En* (*English spotting*), *V* (*Viennese white*), *Si* (*Silver*), *Re* (*Red Eye*) and *W* (*Wide Band*). Some of these loci have been subsequently characterized at the molecular level starting from a candidate gene approach based on homologous genes responsible for similar coat-colour phenotypes in mice and other species. Studies in rabbits identified the causative mutations of different alleles or markers associated with specific coat-colour phenotypes, confirming for most loci the assumptions on the mutated alleles reported in the classical studies with some new evidences and, in a few cases, contrasting or unexpected results. The *C* (*Albino*) series has been the first coat-colour locus in the European rabbit for which several alleles have been described at the DNA level (Aigner *et al.*, 2000), followed by the *E* and *A* loci (Fontanesi *et al.*, 2006, 2010a) and others that were subsequently characterized, at least in part. Table 6.1 summarizes the alleles and the molecular information available for the major coat-colour loci described in the European rabbit. From this summary, it is evident that the molecular characterization is still incomplete for a few loci and for a few alleles hypothesized at some loci.

6.2 The *Albino* Locus: Characterization of the *Tyrosinase* Gene

The *Albino* (*C*) locus encodes the rate-limiting enzyme (tyrosinase or TYR) required for the production of melanin within the melanosome. TYR catalyses the formation of melanin pigments in the melanosome, and the result of a normally functioning gene is essential to have pigmented melanosomes (Beermann *et al.*, 2004). Therefore, the *Albino* locus determines whether or not, and how much, melanin is produced. Considering its role, it is epistatic over many other coat-colour loci. The human syndrome that results from the complete absence of TYR activity is referred to as Oculocutaneous albinism type IA (OCA1A), whereas the diminished melanogenesis resulted by a defective TYR enzyme is known as OCA1B (Grønskov *et al.*, 2007). The coding region of the mammalian *TYR* gene is about 1.6 kb long and is composed of five exons. TYR has two active sites, each binding a copper ion and known as CuA and CuB, encoded by the first and the third exon, respectively (Goldfeder *et al.*, 2014; Kanteev *et al.*, 2015).

The rabbit *TYR* gene is located on chromosome 1 (OCU1) and encodes for a protein of 530 amino acids, of which the first 18 residues form the signal peptide (Aigner *et al.*, 2000; Carneiro *et al.*, 2014). The assignment of this gene to

Table 6.1. Loci, genes, alleles and mutations affecting coat colour in the rabbit.

Locus	LG[1]	Gene symbol	Gene name	Chr.[2]	Alleles (effects)	Mutations[4]	References[3]
C (Albino)	I	*TYR*	*tyrosinase*	OCU1	*C* (fully coloured)	Several wild-type alleles	Aigner *et al.* (2000); Utzeri *et al.* (2021)
					c^{chd} (dark chinchilla)	p.E294G and p.T358I	Aigner *et al.* (2000); Utzeri *et al.* (2021)
					c^{chm} (medium chinchilla)	Unknown (not confirmed yet by molecular studies)	–
					c^{chl} (light chinchilla)	Unknown (not confirmed yet by molecular studies)	–
					c^{h} (Himalayan)	p.E294G	Aigner *et al.* (2000); Utzeri *et al.* (2021)
					c (albino, total lack of pigments)	p.T373K	Aigner *et al.* (2000); Utzeri *et al.* (2021)
E (Extension)	VI	*MC1R*	*melanocortin 1 receptor*	GL018965	E^{+} (wild type)	Two wild-type alleles differing by two SNPs: c.[333A>G;555T>C]	Fontanesi *et al.* (2006)
					E^{D} (dominant black)	6 bp-in-frame deletion: c.280_285del6	Fontanesi *et al.* (2006)
					E^{S} (steel)	Hypothesized to be due to the same mutation of E^{D}	Fontanesi *et al.* (2006)
					e (red, non-extension of black)	30 bp-in frame deletion: c.304_333del30	Fontanesi *et al.* (2006)
					e^{J} (Japanese brindling)	6 bp-in frame deletion flanked by a G>A transition in 5': c.[124G>A;125_130del6]	Fontanesi *et al.* (2010c)
A (Agouti)	IV	*ASIP*	*agouti signalling protein*	OCU4	*A* (light-bellied agouti; wild type)	Several wild-type alleles	Fontanesi *et al.* (2010b)
					a (recessive black non-agouti)	c.5_6insA	Fontanesi *et al.* (2010b)
					a^{t} (black and tan)	p.L55M and p.L89P / 11 kb deletion spanning the promoter and first exon	Fontanesi *et al.* (2010b); Letko *et al.* (2020)

Continued

D (Dilute)	–	*MLPH*	*melanophilin*	GL018840	*D* (wild type, intense black and red)	Several wild-type alleles	Fontanesi *et al.* (2014a)
					d (dilution of black to blue and red to yellow)	c.585delG (g.549853delG)	Fontanesi *et al.* (2014a); Demars *et al.* (2018)
					d (dilution of black to blue and red to yellow: not confirmed in all diluted rabbits)	Two exon skipping mutation: c.111-5C>A	Lehner *et al.* (2013); Demars *et al.* (2018)
B (Brown)	I	*TYRP1*	*tyrosinase-related protein 1*	OCU1	*B* (wild type)	Several wild-type alleles	Utzeri *et al.* (2014)
					b (brown)	p.W190ter (g.41360196G>A)	Utzeri *et al.* (2014)
English Spotting	II	*KIT*	*v-kit Hardy-Zuckerman 4 feline sarcoma viral oncogene homolog*	OCU15	*en* (solid coloured, wild type, recessive)	Wild-type sequences	Fontanesi *et al.* (2014b)
					En (English spotted; partially dominant)	g.93948587T>C (in complete linkage disequilibrium with the segregating alleles)	Fontanesi *et al.* (2014b)
Dutch	II	–	–	–	*Du* (wild type)	Unknown	–
					du (Dutch pattern)	Unknown	–
Viennese white	–	–	–	–	*V* (wild type)	Unknown	–
					v (completely white with blue eyes)	Unknown	–
Silver	–	–	–	–	*Si* (wild type)	Unknown	–
					si (progressive greying)	Unknown	–

Continued

Table 6.1. Continued.

Locus	LG[1]	Gene symbol	Gene name	Chr.[2]	Alleles (effects)	Mutations[4]	References[3]
Red Eye	–	–	–	–	*Re* (wild-type)	Unknown	–
					re (red eye and diluted coat colour)	Unknown	–
Wide Band	IV	–	–	–	*W* (wild-type)	Unknown	–
					w (wide yellow band)	Unknown	–

[1]Linkage Group including the indicated locus, as defined by classical genetic studies (summarized in Fox, 1994).
[2]Chromosome in which the reported gene is annotated. Some genes have been placed in unassembled scaffolds. Information is derived from the oryCun2.0 genome version. OCU, followed by a number, indicates the rabbit chromosome in which the gene is annotated.
[3]References of the molecular studies. Other references of the classical genetic studies are mentioned in the text.
[4]Unknown means that the allele has not been characterized yet by molecular studies.

OCU1 is in agreement with earlier studies that established the linkage between the *Albino* and *Brown* loci in rabbits (Castle, 1924a), which was the seed for the constitution of the first linkage group in this species (Linkage Group I or LG I), that was subsequently mapped to OCU1 (Hatey *et al.*, 1987; Xu and Hardison, 1989).

Classical genetic studies have reported six alleles at the rabbit *Albino* locus (Castle, 1905, 1921; Sawin, 1932a, b; Robinson, 1958; Searle, 1968; Table 6.1). The full development of both black and yellow pigments is dependent in the first instance on the presence of the *C* allele. The result is the production of full colour with dark eyes (very dark pupil and dark brown iris). All other mutant alleles are recessive to *C*. They lead to a progressive decline in quantity of pigment produced. The *dark chinchilla* allele (c^{chd}) is dominant over the other mutated alleles for coat colour alone. It causes a reduction of phaeomelanin whereas eumelanin remains unaffected. It produces very dark pupil with iris colour that might be able to distinguish the heterozygote state with other lower alleles. The *medium chinchilla* allele (c^{chm}) does not lead to any pheomelanin production, it slightly reduces eumelanin production determining slightly sepia or sepia in homozygous state or in combination with other lower alleles. The pupil is reddish-black (or red in combination with lower alleles) and the iris is brown. The *light chinchilla* allele (c^{chl}) does not produce any pheomelanin and causes a reduced eumelanin production determining sepia (if homozygous) or pale sepia in heterozygous state with other lower alleles. The pupil is red and the iris is brown. The *Himalayan* allele (c^{h}) has a temperature-dependent action. It does not lead to any pheomelanin production whereas it determines a reduced eumelanin production even if located only at the extremities where the body temperature is lower than in the rest of the body, which appears white. Nachtsheim (1959) suggested the presence of more than one *Himalayan* allele, having extreme effects from extended pigmented parts to reduced pigmented regions located on ears and tail only. The *albino* allele (*c*) causes the complete absence of melanin and determines pink pupil and white iris. Table 6.2 summarizes the dominance relationships and expression of pigments for the different combinations of alleles.

The five coding exons of the rabbit *TYR* gene have been sequenced in European rabbits carrying different alleles at the *Albino* locus (Aigner *et al.*, 2000; Utzeri *et al.*, 2021). These studies made it possible to characterize at least four of the six alleles of the *C* series reported by classical genetic studies (Table 6.1). The *albino* allele (*c*) is determined by the p.T373K amino acid substitution. This missense mutation (allele with the K amino acid), identified in the albino New Zealand White rabbits, substitutes a conserved residue in the CuB binding site. Sequencing the *TYR* gene in several New Zealand White rabbits, Utzeri *et al.* (2021) reported that this causative mutation is included in at least two different *TYR* haplotypes. The same substitution causes OCA1A in humans (Sanabria *et al.*, 2012). Genome editing in rabbits that inserted this mutation in the *TYR* gene, confirmed the disrupting effect of the p.T373K amino acid substitution in producing albinism (Song *et al.*, 2018).

Other two missense mutations (p.E294G, identified in Californian rabbits showing the Himalayan phenotype; and the p.E294G together with the p.T358I, identified in Chinchilla rabbits) determine the c^{h} allele and one of the three c^{ch} alleles, respectively. It is, however, not clear to which chinchilla allele the mutated haplotype determined by these two missense mutations corresponds. Two Himalayan haplotypes and two chinchilla haplotypes determined by other synonymous mutations were identified by Utzeri *et al.* (2021). The sequence of the *C* allele has been determined in several fully coloured rabbits and other missense and synonymous mutations have been reported (Aigner *et al.*, 2000; Utzeri *et al.*, 2021). That means that more than one wild-type allele, determining the same or similar full production of melanin, exists at this locus in *Oryctolagus cuniculus*. Fig. 6.1 reports a schematic representation of the rabbit TYR protein with indicated amino acid changes that determine the described *C* series alleles.

CRISPR/Cas9-mediated deletions in the 3'-untranslated region of the rabbit *TYR* gene that eliminated putative polyadenylation sites created rabbits with a greying phenotype, probably due to reduced stability of the *TYR* mRNA, resulting in lower expression of tyrosinase (Song *et al.*, 2017).

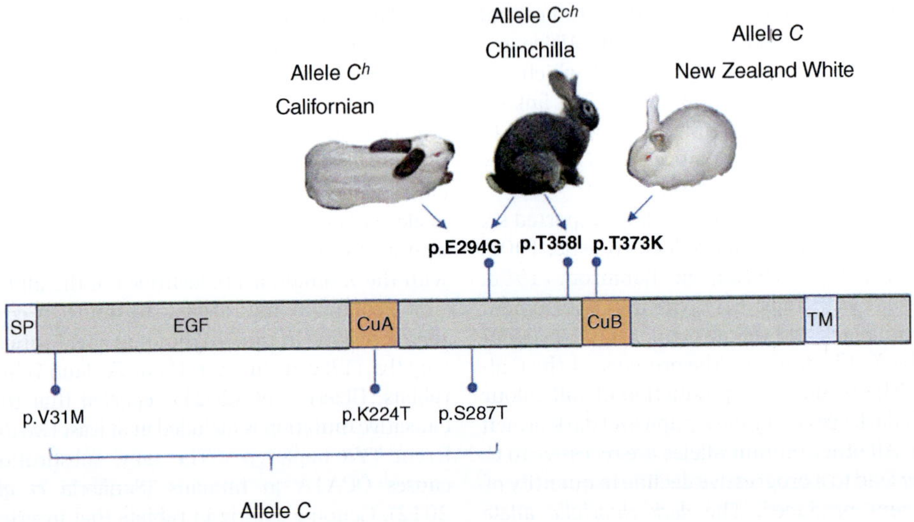

Fig. 6.1. A schematic representation of the rabbit tyrosinase protein with the indication of the different domains (SP: signal peptide; EGF: epidermal growth factor-like domain; CuA and CuB: copper binding; TM: transmembrane domain) and of the amino acid substitutions that determine the described alleles at the *Albino* locus. Several wild-type alleles (*C*) have been detected and the related variants have been reported (Aigner *et al.*, 2000; Utzeri *et al.*, 2021).

Table 6.2. Dominance relationships between alleles of the *C* series and expression of pigments for the different combinations of alleles. (Adapted from Robinson, 1958; Searle, 1968)

Genotype	Eumelanin level	Phaeomelanin level	Pupil	Iris colour
$C/-$	+ + + +	+ + +	black	brown
c^{chd}/c^{chd} c^{chd}/c^{h} c^{chd}/c	+ + + +	+ (dorsal stripe)	black	marbled blue
c^{chd}/c^{chm} c^{chd}/c^{chl}	+ + + +	+ (dorsal stripe)	black	brown
c^{chm}/c^{chm} c^{chm}/c^{chl}	+ + + (slightly sepia)	absent	Reddish-black	brown
c^{chm}/c^{h} c^{chm}/c c^{chl}/c^{chl}	+ + (sepia)	absent	red	brown
c^{chl}/c^{h} c^{chl}/c	+ + (pale sepia)	absent	red	brown
c^{h}/c^{h} c^{h}/c	+ + (extremities)	absent	pink	white
c/c	absent	absent	pink	white

6.3 The *Extension* Locus: Several Deletions in the *Melanocortin 1 Receptor* Gene

The *Extension* (*E*) locus encodes the melanocortin 1 receptor (MC1R) (Robbins *et al.*, 1993). MC1R is a seven transmembrane G protein coupled receptor that is primarily located on the surface of melanocytes. In normal conditions, this receptor alternatively binds the α melanocyte-stimulating hormone (αMSH) and the agouti signalling protein (ASIP), which induce eumelanin and phaeomelanin synthesis, respectively. MC1R controls which of these two types of melanin is produced. When it is activated, it triggers a chemical reaction cascade inside the melanocytes that leads to the production of eumelanin, whereas when it is blocked or not activated the melanocytes are stimulated to produce phaeomelanin. The rabbit *MC1R* gene is constituted by one single coding exon of about 920 bp, that in the oryCun2.0 genome version of the European rabbit genome is localized in an unassigned scaffold (GL018965 from position 152,309 to position 153,232). The *Extension* locus has been originally assigned to linkage group VI

(Fox, 1994), together with *Esterase* loci (*Es-1* and *Es-2*) and *Carboxylesterase* loci (*Est-1*, *Est-2*, *Est-4* and *Est-6*) described in Chapter 8.

Early classical genetic studies in the European rabbit have indicated five alleles at the *Extension* locus (Punnett, 1912, 1924, 1930; Castle, 1924b; Robinson, 1958; Searle, 1968; Table 6.1): E^D (dominant black), E^S (steel, weaker version of E^D), E (wild type, normal grey or normal extension of black), e^J (Japanese brindling, mosaic distribution of black and yellow) and e (non-extension of black, yellow/red with white belly). The order of dominance is the following (Punnett, 1930): $E^D > E^S > E > e^J > e$, with possible partial dominance of E^D over E, of E^S over E and of e^J over e. E^D and E^S tend to eliminate the agouti band of phaeomelanin and darken the belly. Therefore, homozygous E^D/E^D rabbits are uniformly black and non-agouti whereas E^D/E rabbits are agouti-black like E^S/E^S, E^S/e^J and E^S/e rabbits. E^S/E rabbits are steel-grey, E/e^J animals are agouti-steel mosaic and e^J/e rabbits are Japanese but with reduced black spots if compared with e^J/e^J rabbits (Fontanesi *et al.*, 2010b). Table 6.3 reports the dominance relationships and interactions of the alleles at this locus derived by Robinson (1958) and Searle (1968).

Sequencing the coding region of the rabbit *MC1R* gene in different breeds, Fontanesi *et al.* (2006, 2010c) identified three in-frame deletions. Fig. 6.2 indicates the positions of the resulting deleted amino acids in the 2D structure of the rabbit MC1R protein.

One deletion of 6 bp (c.280_285del6 or *Δ6* allele), which eliminates two amino acids in the second transmembrane domain, is associated with the dominant black coat colour and determines the E^D allele. As no other mutations associated to self-black or agouti-black were identified, it is not completely clear if this deletion could also be the causative mutation of the E^S allele. That means that E^D and E^S could be determined by the same mutation and the observed variation of intensities of black (steel) could be due to different states at this locus (i.e. homozygous vs heterozygous), due to the effect of modifier genes or due to epigenetic factors. These questions are still open and should be the matter of other specific studies. The homozygous genotype for this 6 bp deletion (*Δ6/Δ6*) was observed in all analysed Californian, Checkered Giant, Checkered Small, Dutch, Giant White (albino) and New Zealand White (albino) rabbits (Fontanesi *et al.*, 2006). As described above, the albinism of the completely white rabbits inhibits the production of all types of melanin and is epistatic over all other coat-colour loci. In all other mentioned breeds, the black of the coloured spots is due to the dominant black *Δ6* allele at the *MC1R* gene.

A larger in-frame deletion of 30 nucleotides (c.304_333del30; *Δ30* allele) determines the recessive *e* allele that in homozygous condition gives the red coat colour (Fontanesi *et al.*, 2006). The recessive red deletion eliminates 10 amino acids of the first extracellular loop (Fig. 6.2), disrupting the function of the MC1R protein. This allele was identified to be in homozygous state in all yellow/red rabbits belonging to the Burgundy Fawn, Gold Saxony, New Zealand Red and Thuringian breeds and in all yellow/red rabbits of other breeds (Fontanesi *et al.*, 2006).

Table 6.3. Dominance relationships and interactions of the alleles at the *Extension* locus (Robinson, 1958; Searle, 1968).

Genotype	Coat-colour appearance
E^D/E^D, E^D/E^S, E^D/e^J, E^D/e	Self-black
E^D/E, E^S/E^S, E^S/e^J, E^S/e	Agouti-black
E^S/E	Steel-grey
E/E, E/e	Agouti
E/e^J	Agouti-Steel mosaic
e^J/e^J	Japanese
e^J/e	Japanese with reduced black areas
e/e	Red/yellow

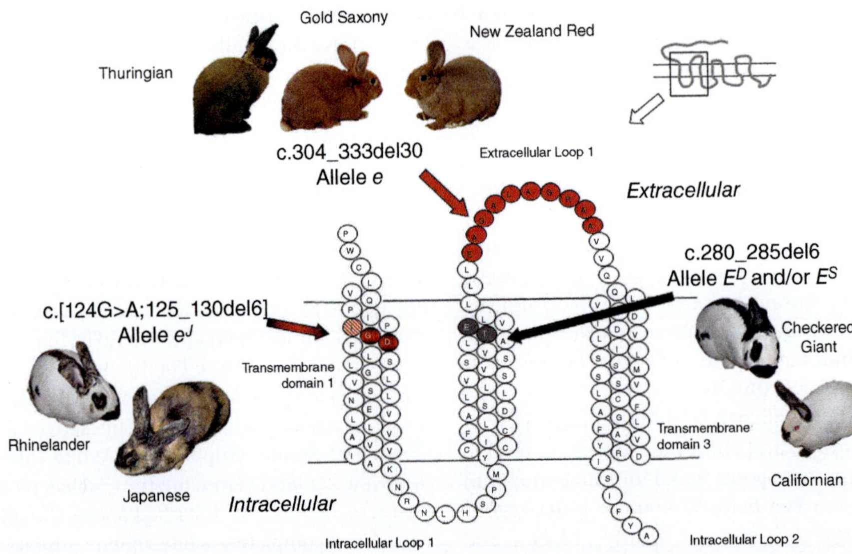

Fig. 6.2. The 2D structure of the rabbit MC1R protein with the position of the three in-frame deletions and some examples of breeds that are fixed for the identified mutations.

Another 6 bp deletion flanked by a G>A transition in 5' (c.[124G>A;125_130del6]) constitutes the e^J allele determining the so-called Japanese brindling coat-colour phenotype. The 6 bp deletion included 2 nucleotides of codon 42, three nucleotides of codon 43 and one nucleotide of codon 44. Together with the G>A transition, this 6 bp deletion eliminates two amino acids in the first transmembrane domain (D and G at positions 42 and 43 of the wild-type sequences) and, in addition, causes an amino acid substitution at position 44 (p.L44T). The e^J allele determining the Japanese brindling coat-colour variety probably appeared in France within the second half of the 19th century and subsequently was introduced in other countries (Castle, 1924b; Punnett, 1924). This coat pattern can be defined as a yellow coat mottled with black and seems determined by the presence of two different types of melanocytes in different skin areas, one producing eumelanin and another one producing pheomelanin (Fontanesi *et al.*, 2010b). The c.[124G>A;125_130del6] mutation is fixed in the Japanese and Rhinelander rabbit breeds. The former breed (named after its coat colour) has the classic mottled phenotype. The Rhinelander rabbit breed has a tricolour pattern with black, yellow and white areas. The white regions are caused by the absence of mature melanocytes due to a dominant *English spotting* locus allele (see below). The coloured patches are much larger than those of the tortoiseshell-like pattern probably because a reduced number of melanoblasts in the skin allows for a spatial expansion of the melanoblast-derived clones. RNA analysis of the skin regions with hairs of different colours (white, black and red) in Rhinelander rabbits reported that MC1R transcripts are present in the black regions whereas no transcripts of this gene are present in the red and white regions, even if the genomic DNA of the skin cells of these body parts has the homozygous genotype for the mutated e^J allele. This evidence excluded that somatic mutations could be involved in the pattern distribution of black and red regions. Therefore, it seems that peculiar regulatory mechanisms, driven by another mutation in linkage disequilibrium or by the same identified composite mutation, could determine the brindling phenotype. Epigenetic factors could be involved in determining this coat-colour phenotype (Fontanesi *et al.*, 2010b).

Gene editing based on CRISPR/Cas9 that targeted the *MC1R* gene in rabbits confirmed that disrupting mutations at this locus produces red/yellow coat colours (Xiao *et al.*, 2019).

6.4 The *Agouti* Locus: Mutations Affecting the *Agouti Signalling Protein* Gene

The *Agouti* locus encodes the agouti signalling protein (ASIP) which is a ~130 amino acid paracrine peptide that is involved in determining the switch from eumelanin to pheomelanin synthesis in melanocytes (Bultman *et al.*, 1992). The wild type (banded-hair) agouti coat colour is caused by the mutually exclusive binding of the MC1R by the ASIP protein or by the α-melanocyte-stimulating hormone (Lu *et al.*, 1994; Ollmann *et al.*, 1998). Recessive loss-of-function mutations of the *ASIP* gene described in many species, which impair either the protein function or reduce the level of its mRNA synthesis, result in lack of regulation activity of the MC1R protein which, in turn, leads only to eumelanin production and then dark coat colour (e.g. Bultman *et al.*, 1992; Kuramoto *et al.*, 2001; Miltenberger *et al.*, 2002; Kerns *et al.*, 2004; Norris and Whan, 2008; Fontanesi *et al.*, 2011).

The rabbit *Agouti* locus has been originally assigned to the Linkage Group (LG) IV, together with the *Dwarf* (*Dw*) locus and the *Wide Band* (*W*) locus (Sawin, 1934; Castle and Sawin, 1941; Fox, 1994). Three alleles have been detected at the *Agouti* locus by classical genetic analyses (Robinson, 1958; Searle, 1968; Table 6.1). The wild-type allele (*A*) resembles that of the mouse A^w (light-bellied agouti) allele described to be the true wild-type allele in murine species (Silvers, 1979). This allele (also indicated in rabbit as A^w; Searle, 1968) produces grey or greyish-brown dorsal fur but a much lighter, almost white, underside. The short guard-hairs and underfur all have a band of phaeomelanin whereas the long dorsal guard-hairs are unbanded and appear intensively dark at the tip but shading to blue towards the base due to the pigments that are more sparse at this level. Along the sides and flanks of the animals the fur changes to lighter tones. The belly fur is usually white or faintly flushed with cream and bluish undercolour (Robinson, 1958; Searle, 1968). The A^w allele is dominant over the other two alleles: a^t (black-and-tan pattern) and *a* (non-agouti or self). Rabbits homozygous for the a^t allele do not have banded hairs on the dorsal surface of the body where all hairs are non-agouti but still have the whitish belly colour as the agouti rabbits, with similar colour in other positions (i.e. undertail, line of jowl, eye-circles, feet pads and inside the ears; Robinson, 1958; Searle, 1968). The *tan* allele is dominant over the *self* allele. The non-agouti rabbits (i.e. homozygous for the *a* allele) are completely self-coloured over the whole body.

The rabbit *ASIP* gene has been placed on chromosome 4 (OCU4). On the oryCun2.0 genome version, it is annotated from nucleotide positions 5435027-5439803 of OCU4 (on the reverse strand). Fontanesi *et al.* (2010b) characterized by sequencing and gene expression analysis the rabbit A^w allele. The three coding exons, namely exons 2, 3 and 4, were separated by 1277 and 2911 bp intronic sequences. The predicted encoded rabbit ASIP protein is constituted by 133 amino acids. Other non-coding exons were identified in the 5' flanking region (Fig. 6.3). Transcription analysis in wild-type agouti rabbits revealed the presence of two major transcripts with different 5'-untranslated regions (1A and 1C, named according to the high level of identity of two corresponding regions already reported in mice; Vrieling *et al.*, 1994) having ventral (1A) or dorsal and ventral (1C) skin-specific expression. These transcripts were not expressed in other tissues. *ASIP* gene expression was, however, detected in several other tissues (brain, kidney, heart, fat, muscle, liver and spleen) distinguishing the rabbit expression pattern from what was observed in wild type mice, in which *ASIP* transcription is not ubiquitous (Bultman *et al.*, 1992; Vrieling *et al.*, 1994).

The rabbit *non-agouti* allele is determined by a frameshift mutation caused by a 1 nucleotide insertion in the coding region of exon 2 (c.5_6insA), determining the production of a predicted truncated protein of only 21 amino acids (Fontanesi *et al.*, 2010a). All rabbits of breeds with black coat colour or variation of black (for example, blue or silver), that also carried a wild-type *MC1R* allele (Fontanesi *et al.*, 2006), namely Alaska, Blanc de Hotot, Champagne d'Argent, English Spot, Havana, Mini Silver, Russian, Silver and Vienna Blue, were homozygous for the c.5_6insA mutation. For most of these breeds, classic genetic studies have indicated that their black coat colour was due to the homozygous *a*/*a* genotype at the *Agouti* locus (Castle, 1930; Robinson, 1958;

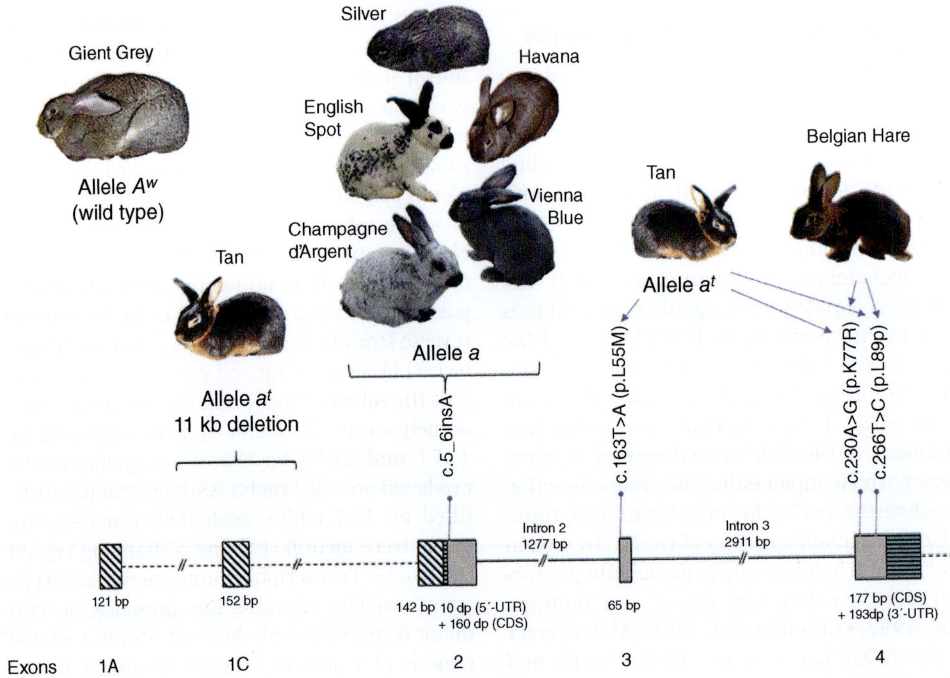

Fig. 6.3. A schematic representation of the genomic organization of the rabbit *ASIP* gene with the indication of the mutations determining the a^t and *a* alleles. The coding regions of exons 2, 3 and 4 are shown as solid boxes. Untranslated regions are shown as striped (horizontal or diagonal) boxes. Four *ASIP* transcripts with different 5'-untranslated have been identified. Only the two main untranslated exons (1A and 1C) that are conserved in rodents have been reported (Fontanesi *et al.*, 2010b). Representative breeds for the different alleles have been reported in the correspondence of the indicated mutations.

Fox, 1994). Californian, Checkered Giant, Checkered Small and Dutch breeds that have black coat-colour patterns but that carry (fixed or not fixed) the dominant allele at the *Extension* locus (*MC1R* c.280_285del6 allele; Fontanesi *et al.*, 2006), were also fixed or had high frequency of the *ASIP* c.5_6insA mutation. The *ASIP* c.5_6insA mutation was also detected in non-black rabbits of other breeds (i.e. New Zealand White, Thuringian and White Vienna) for which mutations in other loci (*Albino, Extension* and *Viennese White*, respectively) determine their coat colour (Castle, 1930; Robinson, 1958; Aigner *et al.*, 2000; Fontanesi *et al.*, 2006). Giant Grey rabbits were usually wild-type at the *ASIP* gene. A few animals of this breed were heterozygous for the c.5_6insA allele even if their coat colour was always of the typical wild-type grey, confirming that the A^w allele is dominant over the *non-agouti* allele (Fontanesi *et al.*, 2010a).

Three other missense mutations (p.L55M, pK77R and p.L89P) were reported by Fontanesi *et al.* (2010b) in the rabbit *ASIP* gene sequenced in several breeds. Even if *in silico* analysis of the predicted amino acid changes excluded that they could have any functional effects, it is worth mentioning that the SNP in exon 3 (c.163T>A) causing the p.L55M amino acid change (located in a conserved position of the basic amino-terminal domain of the ASIP protein) was identified only in Tan rabbits, which are expected to carry the a^t allele. The other two missense mutations were identified both in Tan and in Belgian Hare breeds. The presence of a large haplotype at the *Agouti* locus determining the a^t allele was also reported by Letko *et al.* (2020) who identified 75 a^t-associated variants including an 11 kb deletion. The deletion is located in the region of the hair cycle-specific *ASIP* promoter. This region is homologous to the site of the retroviral insertion causing the a^t allele in mice. Considering the perfect association between the genotypes at this deletion with the black and tan phenotype in rabbits, the 11 kb deletion is the most likely causative variant of the a^t allele in rabbits (Letko *et al.*, 2020). A schematic representation of the mutations identified in the rabbit *ASIP* gene with some representative breeds for these variants is reported in Fig. 6.3.

6.5 The *Dilute* Locus: Polymorphisms in the *Melanophilin* Gene

In rabbits, coat-colour dilution, which derives from an altered distribution of eumelanin and pheomelanin pigments in the hairs, is caused by the *Dilute* locus (Castle *et al.*, 1909; Castle, 1930; Robinson, 1958; Searle, 1968). Two alleles have been described at this locus (Table 6.1): the wild-type allele *D* and the mutated recessive allele *d*. As a result of the mutated allele, the distribution of the pigment granules in the hair is altered, which in turn determines a change in the gross appearance of the coat from black to blue, from grey to blue-grey, from chocolate to lilac and from yellow to beige or cream (Castle, 1930; Robinson, 1958; Searle, 1968). The first rabbit breed that was fixed for the non-agouti dilute or self blue coat-colour phenotype (having genotype *a/a* and *d/d* at the *Agouti* and *Dilute* loci) was the Vienna Blue, from which all other dilute coloured breeds might be derived (Castle, 1930; Fox, 1994).

Similar coat-colour dilutions are determined in mice by mutations in the myosin Va (*Myo5a*; the murine *Dilute* locus), *Rab27a* (*Ashen* locus) and melanophilin (*Mlph*; *Leaden* locus) genes (Mercer *et al.*, 1991; Wilson *et al.*, 2000; Matesic *et al.*, 2001). These genes encode proteins of the melanosome transport complex (Barral and Seabra, 2004). In these mutants, mature melanosomes cluster in the perinuclear area of the cell, rather than at the periphery, and pigment granules are released unevenly into the developing hair shaft causing a decrease in the amount of light absorbability and hence a lightened coat-colour phenotype (Marks and Seabra, 2001; Barral and Seabra, 2004).

A first molecular genetic study excluded the rabbit homologous gene of the mouse *Dilute* locus (i.e. *MYO5A*) as the determinant of the *Dilute* locus in rabbit (Fontanesi *et al.*, 2012). The subsequent family-based segregation analysis of rabbit melanophilin (*MLPH*) gene markers revealed that this is the gene responsible for the *Dilute* locus in *Oryctolagus cuniculus* (Fontanesi *et al.*, 2014a). Two potential candidate causative mutations determining the *d* allele have been reported (Lehner *et al.*, 2013; Fontanesi *et al.*, 2014a). The c.111-5C>A variant, located within intron 2 in an acceptor site for splicing, was reported as the most likely polymorphism leading to the skipping of exons 3 and 4 from the normal rabbit *MLPH* transcript, which was suggested to be the molecular event causing the diluted coat colour in rabbits (Lehner *et al.*, 2013). However, the identification of homozygous rabbits for this mutation having wild-type coat colour tended to exclude this variant as being the causative mutation of the *d* allele (Demars *et al.*, 2018). A second variant in the *MLPH* gene, c.585delG (indicated also g.549853delG, if referred to its position on the scaffold GL018840 of the oryCun2.0 genome version where the *MLPH* gene is located), corresponding to a 1-bp deletion in exon 6, leads to a frameshift and an altered amino acid sequence with a premature stop codon in a downstream exon (Fontanesi *et al.*, 2014a). All self blue or spotted blue rabbits (Vienna Blue, Castor Rex and Chinchilla diluted rabbits and Californian, Checkered Giant and English Spotted rabbits with blue spots) and light grey or cream rabbits (Fairy Marburg and Fairy Pearly rabbits) were homozygous for the c.585delG deletion (Fontanesi *et al.*, 2014a; Demars *et al.*, 2018). A few heterozygous rabbits for the g.549853delG mutation were identified in several breeds and rabbits without diluted coat colour, further confirming the recessive behaviour of this allele and the fact that *d* segregates in several rabbit populations (Fontanesi *et al.*, 2014a). The overall skin expression of *MLPH* transcripts, including several isoforms, was shown to be drastically decreased in animals carrying the c.585delG variant and with a coat-colour dilution, further supporting that this mutated variant is the most likely causative factor determining coat-colour dilution in many rabbit breeds (Demars *et al.*, 2018). The identification of the *MLPH* gene as the responsible gene for the *Dilute* coat-colour locus in rabbit provides a natural animal model for the human Griscelli syndrome type 3, which is an inherited condition determined by mutations in the same gene and characterized by unusually hypopigmented skin and light silvery-grey hair starting in infancy (Sanal *et al.*, 2002; Menasche *et al.*, 2003).

6.6 The *Brown* Locus: A Mutation in the *Tyrosinase-related Protein 1* Gene

The *Brown* locus gives the possibility to develop dark pigments in the fur, skin and eyes. Classical genetic studies in rabbits suggested the presence of two alleles at this locus (Table 6.1): a wild type *B* allele that can produce dense eumelanin throughout the coat and a recessive *b* allele that in homozygous condition (*b*/*b* genotype) produces brown pigmentation, as the animals are unable to produce black pigments. In rabbits with diluted coat colour, the effect of this genotype on the blue undercolour is a slight reduction of intensity to obtain a grey-blue colour (Castle, 1930; Robinson, 1958; Searle, 1968). The molecular characterization of the *Brown* locus, which has been described in several mammals (Searle, 1968), was first reported in mice when the tyrosinase-related protein 1 (*Tyrp1*) gene was assigned to the chromosome region in which this locus was mapped (Jackson, 1988). This gene encodes for a transmembrane melanosomal enzyme, which oxidases 5,6-dihydroxyindole-2-carboxylic acid (DHICA), stabilizes the TYR protein and contributes to the structure of the melanosome (Sarangarajan and Boissy, 2001). The effects of the mutated *b* allele in mice are: (i) the absence of the oxidase function which, in turn, determines an increased incorporation of DHICA and reduced inclusion of 5,6-dihydroxyindole (DHI) into the melanin (obtaining brown melanin); (ii) a modified interaction with TYR that influences the function of both enzymes; and (iii) modified shape and dimension of the melanosomes, potentially influencing the reflection of the light (Lamoreaux *et al.*, 2010).

As already mentioned, the *Brown* locus was included in the first linkage group established in rabbits that also comprised the *Albino* locus (Castle, 1924c), subsequently assigned to OCU1. The rabbit *TYRP1* gene based on seven coding exons is localized on this chromosome and is annotated on the reverse strand from nucleotide 41,345,981 to 41,362,120 (oryCun2.0 genome version). Utzeri *et al.* (2014) sequenced this gene in several rabbit breeds with different coat colours, including Havana rabbits, which have brown/chocolate coat colour. This study identified a mutation in exon 2 (g.41360196G>A) in Havana rabbits only. The g.41360196G>A polymorphism leads to a premature stop codon at position 190 of the deduced protein sequence (p.W190ter) that, in the wild-type form, contains 537 predicted amino acids. The predicted truncated TYRP1 protein of the brown/chocolate rabbits lacks almost completely the tyrosinase domain. Genotyping analysis of rabbits from 32 breeds and populations reported that this non-sense mutation was only present (and always in homozygous condition) in rabbits with brown/chocolate coat colour (Utzeri *et al.*, 2014). Therefore, this disrupting variant of the *TYRP1* gene can be considered the causative mutation of the *b* allele of the rabbit *Brown* locus even if functional analyses have not been carried out thus far.

6.7 The *English Spotting* Locus: Association with a *KIT* Gene Marker

The *English Spotting* locus is caused by a dominant allele indicated as *En* (determining a spotted phenotype, that can be very variable in extent) over the wild-type allele symbolized as *en* (determining a self-coloured phenotype in homozygous state; Castle, 1919; Table 6.1). The heterozygous *En*/*en* rabbits have far larger patches of coloured fur than the homozygous *En*/*En* rabbits (Robinson, 1958), which are called Chaplins (Fig. 6.4). The heterozygous state determines the characteristic spotted patterns of the English spot (or spotted) rabbits, Butterflies and Checkered Giants (Castle, 1930) that is requested by the standards of these breeds and that is desired by fancy breeders usually for show purposes. The coloured patches are presented in the colour that is determined by the allelic combination at other loci described above. Robinson (1958) reported a large degree of variation of expression of the English-type white spotting. His scale comprises 12 grades, from 1 (few white areas) to 12 (black marks only around the eyes). Grades from 1 to 8 include heterozygous *En*/*en* rabbits

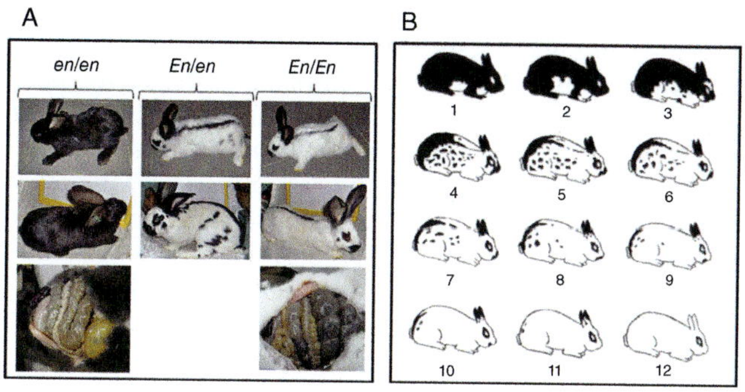

Fig. 6.4. Spotted phenotypes in rabbits. A) Examples of rabbits with the three genotypes at the *English Spotting* locus obtained by crossing Checkered Giant bucks and does. For the *en/en* and *En/En* genotypes, details of a normal intestine and of a megacolon defect are reported (Fontanesi *et al.*, 2014b). B) Variation of the grades of English-type white spotting as reported by Robinson (1958).

and grades from 8 to 11 or even 12 are indicators of the homozygous *En/En* state (Fig. 6.4). The degree of extension of the white might be due, in part, to modifying genes (Richardson, 1953). *En/En* rabbits are usually subvital compared to the heterozygous *En/en* and *en/en* rabbits because dominant homozygous animals are affected by an underlying megacolon (Nachtsheim, 1943; Richardson, 1953; Robinson, 1958; Wieberneit *et al.*, 1990; Böderek *et al.*, 1995; Wieberneit and Wegner, 1995; Fontanesi *et al.*, 2014b). The defect is, however, recessive since it is not observed in *En/en* rabbits. It has incomplete penetrance, probably depending on environmental conditions (e.g. diet, stressors and aging) and also on the contribution of other modifier genes (Fontanesi *et al.*, 2014b). The first studies that investigated the etiopathogenesis of this megacolon suggested the involvement of enteric nervous system abnormalities throughout the colon, but with incomplete characterization (Gerlitz *et al.*, 1993; Böderek *et al.*, 1995; Wieberneit and Wegner, 1995).

In rodents and other mammals, several genes that also affect coat colour have been implicated in the pathogenesis of megacolon. For example, inactivating mutations of the *endothelin receptor B* (*Ednrb*) gene in rodents cause aganglionic megacolon with spotted coat-colour patterns (Hosoda *et al.*, 1994; Gariepy *et al.*, 1996) and *EDNRB* mutations in humans are known to be associated with different forms of Hirschsprung disease (Puffenberger *et al.*, 1994; Attié *et al.*, 1995). However, a study in rabbit excluded *EDNRB* as the gene involved in determining the *English spotting* locus (Fontanesi *et al.*, 2010e).

A subsequent study targeted the v-kit Hardy-Zuckerman 4 feline sarcoma viral oncogene homolog (*KIT*) gene as candidate for this coat colour locus in rabbit (Fontanesi *et al.*, 2014b). Mutations in the *KIT* gene have been already described to affect pigmentation in many mammals (e.g. Chabot *et al.*, 1988; Fleischman *et al.*, 1991; Marklund *et al.*, 1998; Fontanesi *et al.*, 2010a, 2010d). This gene encodes the mast/stem cell growth factor receptor that is a large protein with an extracellular domain consisting of 5 Ig-like subdomains, a transmembrane region and a tyrosine kinase domain (Ray *et al.*, 2008). *KIT* is involved in regulating the migration of melanoblasts from the neural crest along the dorsolateral pathway to reach the final destinations in the skin and is also involved in the differentiation of these primordial cells (Besmer *et al.*, 1993; Thomas and Erickson, 2008). Its expression in the gut musculature is prominent only in interstitial cells of Cajal (ICC). These cells are known to play important roles in gut motility and their abnormalities are associated with several gastrointestinal motility disorders (Farrugia, 2008).

The *KIT* gene was sequenced in rabbits of different breeds to identify informative polymorphisms that were used in an association analysis based on an F1 population obtained by crossing Checkered Giant bucks and does with genotype *En/en* at the *English Spotting* locus (Fontanesi *et al.*, 2014b). Even if no obvious causative mutation was identified, a synonymous single nucleotide polymorphism identified in exon 5 (g.93948587T>C) perfectly cosegregated with the three coat-colour phenotypes (θ = 0.00 LOD = 75.56). All animals with the T/T genotype were *En/En* at the *English Spotting* locus and highly associated with the megacolon defect of these animals, whereas all rabbits with the C/C genotype were self-coloured (*en/en*) and all T/C heterozygous rabbits were *En/en*, like all Checkered Giant parents. The C/C and C/T rabbits did not show any megacolon (Fontanesi *et al.*, 2014b). These results were confirmed genotyping other Checkered Giants and rabbits of the Checkered pattern, indicating that undetected variants at the *KIT* gene are responsible for this specific *English Spotting* pattern. Other alleles at this locus might determine different spotted phenotypes as all English Spot rabbits analysed at the g.93948587T>C SNP had genotype C/C (Fontanesi *et al.*, 2014b). In addition to the genotyping results, it was interesting to note that *KIT* gene expression in cecum and colon specimens of *En/En* rabbits (obtained from the F1 families) was just a small fraction of that of *en/en* rabbits. *En/En* rabbits had also reduced and altered c-kit immunolabelled ICC compared to *en/en* rabbits. Neuronal and ICC abnormalities were evident in *En/En* cecum and colon tissues. Therefore, in addition to the effect of this locus on coat colour evidenced in Checkered Giant rabbits, the neuro-ICC changes in the *En/En* genotype that are reminiscent of the human non-aganglionic megacolon, could make Chaplin rabbits useful animal models for this defect (Fontanesi *et al.*, 2014b).

6.8 The Dutch Series

The characteristic form of *Dutch* type of white spotting consists of a belt of white encircling the front half of the body, including the front legs and some white on the nose (Searle, 1968). According to Robinson (1958), the *Dutch* pattern has, however, a great variation of extension of the white-spotting that ranges from an almost completely pigmented body to an almost completely unpigmented one, similarly as already described for the *English Spotting* locus. The genetic basis of the *Dutch* series has not been completely clarified yet and contrasting hypotheses have been proposed (Punnett and Pease, 1925; Castle, 1926). It seems that a primary recessive allele *du* is the major causative factor of this pattern even if a number of minor modifying genes might contribute to the expression of this phenotype (Robinson, 1958; Searle, 1968). The presence of more than one recessive allele at this series cannot be excluded (Castle, 1919, 1932). Castle (1924a) reported that the *Dutch*, the *English Spotting* and the *Angora* loci are on the same linkage group. As the *Dutch* and the *English Spotting* were reported to be very closely linked, considering that several complex mutations at the *KIT* gene in other species have been demonstrated to cause different spotted phenotypes, including belted patterns (e.g. Fontanesi and Russo, 2013), it is tempting to predict that variability of the *KIT* gene or mutations involving the surrounding chromosome region could be responsible for the spotted phenotypes that might range from Dutch types to English spotted types. The complete molecular characterization of the *Dutch* and *English Spotting* loci might clarify the complexity of the white spotting patterns that could not be well explained by classical genetic studies.

6.9 The *Viennese White* Locus

The colour factor of *Viennese White* (or *Vienna White*) was first interpreted as an allelomorph of albinism (Castle, 1922) and subsequently recognized as an independent locus. Utzeri *et al.* (2021) confirmed that variability in the *TYR* gene is not associated with the *Viennese White* phenotype. This locus might have some phenotypic effects similar to those of the *Dutch* type spotting. The mutated allele *v*, in homozygous *v/v* rabbits, determines a completely white coat but with blue eyes (Castle, 1922; Castle, 1930; Robinson, 1958; Searle, 1968). The wild-type allele is designated *V*. The heterozygous *V/v* rabbits usually have low grade of Dutch-like markings whereas *V/V* rabbits are usually self-coloured. Rabbits homozygous for the *v* allele can frequently develop epileptic seizures (Nachtsheim, 1939a, 1939b, 1941). The locus has not been characterized at the molecular level yet.

6.10 Silvering

Silvering or hereditary greying of hair has been mainly described in the Champagne d'Argent and Silver (also known as English Silver) breeds (Nachtsheim, 1929). The greying effect is supposed to be due to a recessive *si* allele which determines a progressive reduction in the number of pigmented hairs during subsequent hair growth cycles. It is still not completely clear if polygenic effects or modifier genes are involved (Robinson, 1958) or if the same genetic factors are responsible for the progressive greying in both Champagne d'Argent and Silver breeds. Rabbits of these two breeds are genetically black and their coat colour is due to the homozygous state for the non-agouti allele at the *Agouti* locus (Fontanesi *et al.*, 2010b). In these rabbits, the first hairs are normally pigmented, thus the coat colour is self-black. In Champagne d'Argent, the secondary guard hairs of the second hair cycle are generally unpigmented whereas the primary guard hairs and wool hairs, for the most part, are pigmented from tip to base. The general effect is that the coat appears grey with a bluish cast. In older Champagne d'Argent rabbits the coat is lightened further and many of the primary guard hairs and wool hairs are found to lack pigment. Similarly, in the Silver breed the silvering does not appear until after the first moult and consists in the loss of pigment in the secondary guard hair. Quevedo and Chase (1957) reported that the progressive silvering in the Champagne d'Argent rabbits involves a disappearance of melanocytes from the hair follicles and the consequent production of entirely unpigmented secondary guard hairs during the second hair generation and many of the primary guard hairs and wool hairs during later hair generations. Histologically, the unpigmented hair follicles are identical in appearance with the unpigmented follicles of white spotted rabbits (Quevedo and Chase, 1957).

6.11 The *Red Eye* Locus

A mutated allele (recognized with the symbol *re*) that mainly affects eye colour with a slight dilution of hair pigmentation has been described by Magnussen (1952, 1954, 1959a, 1959b). Homozygous *re*/*re* rabbits have pink albinotic eyes at birth and lack a cornea. Then, some pigments in the eye appear and the cornea starts to develop even if it remains underdeveloped (Robinson, 1958). The effect on coat colour appears limited to the neck region where agouti red-eyed rabbits are lighter than normal animals. The effect of the mutation might be modified by the genetic background at other coat-colour loci or modifier genes.

6.12 The *Wide Band* Locus

The mutation indicated with the symbol *w*, described by Sawin (1932a, 1934) to have a recessive mode of inheritance, doubles the normal length of the yellow pigmented band of the agouti hair. Homozygous *w*/*w* rabbits are consequently lighter than normal animals. The *Wide Band* (*W*) locus was reported to be linked to the *Agouti* locus (Sawin, 1934).

6.13 Conclusion

The molecular characterization of several coat-colour loci in rabbits has shown that variants determining the described alleles are distributed in different breeds where they contribute to establish the specific coat-colour phenotype that is one of the most important breed-characterizing feature in the domesticated rabbit populations. That means that variants, emerging in one population from one unique mutational event, were then transferred in other populations contributing to maintain within breeds and across breeds genetic variability. The constitution of the rabbit breeds could be considered the last step of the domestication process of this species and can be summarized as the activity that gathered together and fixed some alleles to originate the desired coat colour phenotypes. The molecular mechanisms determining a few loci (*Dutch*, *Viennese White*, *Silver*, *Red Eye* and *Wide Band*), some alleles for other partially characterized loci and other coat-colour variants that might be recently emerged in fancy breed stocks are still to be elucidated. Their characterization could probably be relevant to further add useful information on the biology of pigmentation in mammals and to establish new animal models.

References

Aigner, B., Besenfelder, U., Müller, M. and Brem G. (2000) Tyrosinase gene variants in different rabbit strains. *Mammalian Genome* 11, 700–702.

Attié, T., Till, M., Pelet, A., Amiel, J., Edery, P. *et al.* (1995) Mutation of the endothelin-receptor B gene in Waardenburg-Hirschsprung disease. *Human Molecular Genetics* 4, 2407–2409.

Barral, D.C. and Seabra, M.C. (2004) The melanosome as a model to study organelle motility in mammals. *Pigment & Cell Research* 17, 111–118.

Beermann, F., Orlow, S.J. and Lamoreux, M.L. (2004) The *Tyr* (*albino*) locus of the laboratory mouse. *Mammalian Genome* 15, 749–758.

Besmer, P., Manova, K., Duttlinger, R., Huang, E.J., Packer, A. *et al.* (1993) The kit-ligand (steel factor) and its receptor c-kit/W: pleiotropic roles in gametogenesis and melanogenesis. *Development (Suppl)* 1, 125–137.

Böderek, D., Türk, O., Lovén, E., Wieberneit, D. and Wegner, W. (1995) Pathophysiological and functional aspects of the Megacolon-Syndrome of homozygous Spotted rabbits. *Journal of Veterinary Medicine A* 42, 549–559.

Bultman, S.J., Michaud, E.J. and Woychik, R.P (1992) Molecular characterization of the mouse agouti locus. *Cell* 71, 1195–1204.

Carneiro, M., Rubin, C.J., Di Palma, F., Albert, F.W., Alföldi, J. *et al.* (2014) Rabbit genome analysis reveals a polygenic basis for phenotypic change during domestication. *Science* 345, 1074–1079.

Castle, W.E. (1905) Heredity of coat characters in guinea pigs and rabbits. *Carnegie Institution of Washington Publication no.* 23, 1–78.

Castle, W.E. (1919) Studies of heredity in rabbits, rats and mice. *Carnegie Institution of Washington Publication no.* 288, 4–28.

Castle, W.E. (1921) Genetics of the chinchilla rabbit. *Science* 53, 387–388.

Castle, W.E. (1922) Genetics of the Vienna White rabbit. *Science* 55, 269–270.

Castle, W.E. (1924a) Linkage of Dutch, English and Angora in rabbits. *Proceedings of the National Academy of Sciences of the USA* 10, 107–108.

Castle, W.E. (1924b) Genetics of the Japanese rabbit. *Journal of Genetics* 14, 225–229.

Castle, W.E. (1924c) On the occurrence in rabbits of linkage in inheritance between albinism and brown pigmentation. *Proceedings of the National Academy of Sciences of the USA* 10, 486–488.

Castle, W.E. (1926) On the pattern of the Dutch rabbit, a discussion of the results of Punnett and Pease. *Journal of Genetics* 16, 189–196.

Castle, W.E. (1930) *The Genetics of Domestic Rabbit*. Harvard University Press, London.

Castle, W.E. (1932) English and Dutch spotting and the genetics of the Hôtot rabbits. *Carnegie Institution of Washington Publication no.* 427, 3–13.

Castle, W.E. (1940) *Mammalian Genetics*. Harvard University Press, Cambridge, Massachusetts.

Castle, W.E. and Sawin, P.B. (1941) Genetic linkage in the rabbit. *Proceedings of the National Academy of Sciences of the USA* 27, 519–523.

Castle, W.E., Walter, H.E., Mullenix, R.C. and Cobb, S. (1909) Studies of inheritance in rabbits. *Carnegie Institution of Washington Publication* no. 114, 1–70.

Chabot, B., Stephenson, D.A., Chapman, V.M., Besmer, P. and Bernstein, A. (1988) The proto-oncogene c-kit encoding a transmembrane tyrosine kinase receptor maps to the mouse W locus. *Nature* 335, 88–89.

Demars, J., Iannuccelli, N., Utzeri, V.J., Auvinet, G., Riquet, J. *et al.* (2018) New insights into the melanophilin (*MLPH*) gene affecting coat color dilution in rabbits. *Genes* 9, 430.

Farrugia, G. (2008) Interstitial cells of Cajal in health and disease. *Neurogastroenterology & Motility* 20, 54–63.

Fleischman, R.A., Saltman, D.L., Stastny, V. and Zneimer S. (1991) Deletion of the c-kit protooncogene in the human developmental defect piebald trait. *Proceedings of the National Academy of Sciences of the USA* 88, 10885–10889.

Fontanesi, L. and Russo, V. (2013) Molecular genetics of coat colour in pigs. *Acta Agriculturae Slovenica* Suppl. 4, 15–20.

Fontanesi, L., Tazzoli, M., Beretti, F. and Russo V. (2006) Mutations in the melanocortin 1 receptor (*MC1R*) gene are associated with coat colours in the domestic rabbit (*Oryctolagus cuniculus*). *Animal Genetics* 37, 489–493.

Fontanesi, L., D'Alessandro, E., Scotti, E., Liotta, L., Crovetti, A. *et al.* (2010a) Genetic heterogeneity and selection signature at the KIT gene in pigs showing different coat colours and patterns. *Animal Genetics* 41, 478–492. Available at: https://doi.org/10.1111/j.1365-2052.2010.02054.x (accessed 9 January 2021).

Fontanesi, L., Forestier, L., Allain, D., Scotti, E., Beretti, F. *et al.* (2010b) Characterization of the rabbit agouti signaling protein (*ASIP*) gene: transcripts and phylogenetic analyses and identification of the causative mutation of the nonagouti black coat colour. *Genomics* 95, 166–175.

Fontanesi, L., Scotti, E., Colombo, M., Beretti, F., Forestier, L. *et al.* (2010c) A composite six bp in-frame deletion in the melanocortin 1 receptor (*MC1R*) gene is associated with the Japanese brindling coat colour in rabbits (*Oryctolagus cuniculus*). *BMC Genetics* 11, 59.

Fontanesi, L., Tazzoli, M., Russo, V. and Beever, J. (2010d) Genetic heterogeneity at the bovine *KIT* gene in cattle breeds carrying different putative alleles at the spotting locus. *Animal Genetics* 41, 295–303.

Fontanesi, L., Vargiolu, M., Scotti, E., Mazzoni, M., Clavenzani, P. *et al.* (2010e) Endothelin receptor B (*EDNRB*) is not the causative gene of the *English spotting* locus in the domestic rabbit (*Oryctolagus cuniculus*). *Animal Genetics* 41, 669–670.

Fontanesi, L., Dall'Olio, S., Beretti, F., Portolano, B. and Russo, V. (2011) Coat colours in the Massese sheep breed are associated with mutations in the agouti signalling protein (*ASIP*) and melanocortin 1 receptor (*MC1R*) genes. *Animal* 5, 8–17.

Fontanesi, L., Scotti, E., Dall'Olio, S., Oulmouden, A. and Russo, V. (2012) Identification and analysis of single nucleotide polymorphisms in the myosin VA (*MYO5A*) gene and its exclusion as the causative gene of the dilute coat colour locus in rabbit. *World Rabbit Science* 20, 35–41.

Fontanesi, L., Scotti, E., Allain, D. and Dall'Olio, S. (2014a) A frameshift mutation in the melanophilin (*MLPH*) gene causes the dilute coat colour in rabbit (*Oryctolagus cuniculus*) breeds. *Animal Genetics* 45, 248–255.

Fontanesi, L., Vargiolu, M., Scotti, E., Latorre, R., Faussone Pellegrini, M.S. *et al.* (2014b) The *KIT* gene is associated with the *English spotting* coat color locus and congenital megacolon in Checkered Giant rabbits (*Oryctolagus cuniculus*). *PLOS ONE* 9, e93750.

Fox, R.R. (1994) Taxonomy and genetics. In: Manning, P.J., Ringler, D.H. and Newcomer, C.E. (eds) *The Biology of the Laboratory Rabbit*, 2nd edn. Academic Press, San Diego, California, pp. 1–26.

Gariepy, C.E., Cass, D.T. and Yanagisawa, M. (1996) Null mutation of endothelin receptor type B gene in spotting lethal rats causes aganglionic megacolon and white coat color. *Proceedings of the National Academy of Sciences of the USA* 93, 867–872.

Gerlitz, S., Wessel, G., Wieberneit, D. and Wegner, W. (1993) Zur Problematik der Scheckenzucht bei Kaninchen. 3. Mitteilung: Variabilität des Pigmentierungsgrades, ganglionäre Darmwandversorgung, Beziehung zur Pathogenese – tierzüchterische und tierschützerische Aspekte. *Deutsche tierärztliche Wochenschrift* 100, 237–239.

Goldfeder, M., Kanteev, M., Isaschar-Ovdat, S., Adir, N. and Fishman, A. (2014) Determination of tyrosinase substrate-binding modes reveals mechanistic differences between type-3 copper proteins. *Nature Communication* 5, 1–5.

Grønskov, K., Ek, J. and Brondum-Nielsen, K. (2007) Oculocutaneous albinism. *Orphanet Journal of Rare Diseases* 2, 43.

Hatey, E., Bell, G., Echard, G., Gellin, J. and Langlois, C. (1987) Rabbit gene mapping: new syntenies. Ninth International Workshop on Human Gene Mapping. *Cytogenetics and Cell Genetics* 46, 627.

Hosoda, K., Hammer, R.E., Richardson, J.A., Baynash, A.G., Cheung, J.C. *et al.* (1994) Targeted and natural (piebald-lethal) mutations of endothelin-B receptor gene produce megacolon associated with spotted coat color in mice. *Cell* 79, 1267–1276.

Jackson, I.J. (1988) A cDNA encoding tyrosinase-related protein maps to the mouse brown locus. *Proceedings of the National Academy of Sciences of the USA* 85, 4391–4396.

Kanteev, M., Goldfeder, M. and Fishman, A. (2015) Structure–function correlations in tyrosinases. *Protein Science* 24, 1360–1369.

Kerns, J.A., Newton, J., Berryere, T.G., Rubin, E.M., Cheng, J.F. *et al.* (2004) Characterization of the dog Agouti gene and a nonagouti mutation in German Shepherd dogs. *Mammalian Genome* 15, 798–808.

Kuramoto, T., Nomato, T., Sugimura, T. and Ushijima, T. (2001) Cloning of the rat agouti gene and identification of the rat nonagouti mutation. *Mammalian Genome* 12, 469–471.

Lamoreaux, M.L., Delmas, V., Larue, L. and Bennett, D.C. (2010) *The Colors of Mice: A Model Genetic Network.* Wiley-Blackwell, Oxford.

Lehner, S., Gähle, M., Dierks, C., Stelter, R., Gerber, J. *et al.* (2013) Two-exon skipping within MLPH is associated with coat color dilution in rabbits. *PLOS ONE* 8, e84525.

Letko, A., Ammann, B., Jagannathan, V., Henkel, J., Leuthard, F. *et al.* (2020). A deletion spanning the promoter and first exon of the hair cycle-specific *ASIP* transcript isoform in black and tan rabbits. *Animal Genetics* 51, 137–140.

Lu, D., Willard, D., Patel, I.R., Kadwell, S., Overton, L. *et al.* (1994) Agouti protein is an antagonist of the melanocyte-stimulating-hormone receptor. *Nature* 371, 799–802.

Magnussen, K. (1952) Beitrag zur Genetik und Histologie eines isolierten Augenalbinismus beim Kaninchen. *Zeitschrift für Morphologie und Anthropologie* 44, 127–135.

Magnussen, K. (1954) Beitrag zur Genetik und Histologie eines isolierten Augen-Albinismus beim Kaninchen. *II. Zeitschrift für Morphologie und Anthropologie* 46, 24–29.

Magnussen, K. (1959a) Beitrag zur Genetik eines isolierten Augenalbinismus beim Kaninchen III. *Zeitschrift für Morphologie und Anthropologie* 49, 306–311.

Magnussen, K. (1959b) Beitrag zur Genetik und Histologie eines isolierten Augenalbinismus beim Kaninchen. *IV. Zeitschrift für Morphologie und Anthropologie* 50, 103–120.

Marklund, S., Kijas, J., Rodriguez-Martinez, H., Rönnstrand, L., Funa, K. *et al.* (1998) Molecular basis for the dominant white phenotype in the domestic pig. *Genome Research* 8, 826–833.

Marks, M.S. and Seabra, M.C. (2001) The melanosome: membrane dynamics in black and white. *Nature Reviews Molecular Cell Biology* 2, 738–748.

Matesic, L.E., Yip, R., Reuss, A.E., Swing, D.A., O'Sullivan, T.N. *et al.* (2001) Mutations in *Mlph*, encoding a member of the Rab effector family, cause the melanosome transport defects observed in leaden mice. *Proceedings of the National Academy of Sciences of the USA* 98, 10238–10243.

Menasche, G., Ho, C.H., Sanal, O., Feldmann, J., Tezcan, I. *et al.* (2003) Griscelli syndrome restricted to hypopigmentation results from a melanophilin defect (GS3) or a MYO5A F-exon deletion (GS1). *Journal of Clinical Investigation* 112, 450–456.

Mercer, J.A., Seperack, P.K., Strobel, M.C., Copeland, N.G. and Jenkins, N.A. (1991) Novel myosin heavy chain encoded by murine dilute coat colour locus. *Nature* 349, 709–713.

Miltenberger, R.J., Wakumatsu, K., Ito, S., Woychik, R.P., Russell, L.B. and Michaud, E.J. (2002) Molecular and phenotypic analysis of 25 recessive, homozygous-viable alleles at the mouse agouti locus. *Genetics* 160, 659–674.

Nachtsheim, H. (1929) Die Entstehung der Kaninchenrassen in Lichte ihrer Genetik. *Zeitschrift für Tierzüchtung und Züchtungsbiologie einschliesslich Tierernährung* 14, 53–109.

Nachtsheim, H. (1939a) Die Bedeutung des Cardiazolkrampfer für die Diagnose der erblichen epilepsie. Versuche an kampfbereiten und nicht krampfbereiten Kaninchen. *Deutsche medizinische Wochenschrift* 65, 168–171.

Nachtsheim, H. (1939b) Krampfbereitschaft und Genotypus. I. Die Epilipsie der Weissen Wiener Kaninchen. *Zeitschrift für menschliche Vererbungs- und Konstitutionslehre* 22, 781–810.

Nachtsheim, H. (1941) Krampfbereitschaft und Genotypus. II. Weitere untersuchungen zur epilepsie der Weissen Wiener Kaninchen. *Zeitschrift für menschliche Vererbungs- und Konstitutionslehre* 25, 229–244.

Nachtsheim, H. (1943) Ergebrisse und Probleme der Vergleichenden und Experimentellen Erbpathologie. *Jenaische Zeitschrift für Naturwissenschaft* 76, 81–108.

Nachtsheim, H. (1959) Problems of comparative genetics in mammals. *Proceedings of the 10th International Congress on Genetics* vol. 1, 187–198.

Norris, B.J. and Whan, V.A. (2008) A gene duplication affecting expression of the ovine *ASIP* gene is responsible for white and black sheep. *Genome Research* 18, 1282–1293.

Ollmann, M.M., Lamoreux, M.L., Wilson, B.D. and Barsh, G.S. (1998) Interaction of Agouti protein with the melanocortin 1 receptor in vitro and in vivo. *Genes & Development* 12, 316–330.

Prota G. (1992) *Melanins and Melanogenesis*. Academic Press, New York.

Puffenberger, E.G., Hosoda, K., Washington, S.S., Nakao, K., deWit, D. *et al.* (1994) A missense mutation of the endothelin-B receptor gene in multigenic Hirschsprung's disease. *Cell* 79, 1257–1266.

Punnett, R.C. (1912) Inheritance of coat colour in rabbits. *Journal of Genetics* 2, 221–238.

Punnett, R.C. (1924) On the "Japanese" rabbit. *Journal of Genetics* 14, 230–240.

Punnett, R.C. (1930) On the series of allelomorphs connected with the production of black pigment in rabbit. *Journal of Genetics* 23, 265–274.

Punnett, R.C. and Pease, M.S. (1925) On the pattern of the Dutch rabbit. *Journal of Genetics* 15, 375–412.

Quevedo, W.C. and Chase, H.B. (1957) Histological observations on the silvering process in the Champagne d'Argent rabbit. *Anatomical Record* 129, 87–95.

Ray, P., Krishnamoorthy, N. and Ray, A. (2008) Emerging functions of c-kit and its ligand stem cell factor in dendritic cells: regulators of T cell differentiation. *Cell Cycle* 7, 2826–2832.

Richardson, E.C. (1953) Inheritance of white in the "English" rabbit. *Heredity* 7, 150 (abstract).

Robbins, L.S., Nadeau, J.H., Johnson, K.R., Kelly, M.A., Roselli-Rehfuss, L. *et al.* (1993) Pigmentation phenotypes of variant extension locus alleles result from point mutations that alter MSH receptor function. *Cell* 72, 827–834.

Robinson, R. (1958) Genetic studies of the rabbit. *Bibliografia Genetica* 17, 229–558.

Sanabria, D., Groot, H., Guzmán, J. and Lattig, M.C. (2012) An overview of oculocutaneous albinism: TYR gene mutations in five Colombian individuals. *Biomedica* 32, 269–276.

Sanal, O., Ersoy, F., Tezcan, I., Metin, A., Yel, L. *et al.* (2002) Griscelli disease: genotype-phenotype correlation in an array of clinical heterogeneity. *Journal of Clinical Immunology* 22, 237–243.

Sarangarajan, R. and Boissy, R.E. (2001) Tyrp1 and oculocutaneous albinism type 3. *Pigment Cell Research* 14, 437–444.

Sawin, P.B. (1932a) Albino allelomorphs of the rabbit with special reference to blue-eyed chinchilla and its variations. *Carnegie Institution of Washington Publication no.* 427, 15–50.

Sawin, P.B. (1932b) Hereditary variation of the chinchilla rabbit in coat and eye color. *Journal of Heredity* 23, 39–46.

Sawin, P.B. (1934) Linkage of "wide band" and agouti genes. *Journal of Heredity* 25, 477–481.

Sawin, P.B. (1955) Recent genetics of the domestic rabbit. *Advances in Genetics* 7, 183–226.

Searle, A.G. (1968) *Comparative Genetics of Coat Colour in Mammals*. Logos Press, London.

Silvers, W.K. (1979) The agouti and extension series of alleles, umbrous, and sable, In: Silver W.K. (ed.) *The Coat Colors of Mice: A Model for Mammalian Gene Action and Interaction*. Springer-Verlag, New York, pp. 6–44.

Song, Y., Xu, Y., Deng, J., Chen, M., Lu, Y. *et al.* (2017) CRISPR/Cas9-mediated mutation of tyrosinase (Tyr) 3' UTR induce graying in rabbit. *Scientific Reports* 7, 1560.

Song, Y., Zhang, Y., Chen, M., Deng, J., Sui, T. *et al.* (2018) Functional validation of the albinism-associated tyrosinase T373K SNP by CRISPR/Cas9-mediated homology-directed repair (HDR) in rabbits. *EBioMedicine* 36, 517–525.

Thomas, A.J. and Erickson, C.A. (2008) The making of a melanocyte: the specification of melanoblasts from the neural crest. *Pigment Cell & Melanoma Research* 21, 598–610.

Utzeri, V.J., Ribani, A. and Fontanesi, L. (2014) A premature stop codon in the *TYRP1* gene is associated with brown coat colour in the European rabbit (*Oryctolagus cuniculus*). *Animal Genetics* 45, 600–603.

Utzeri, V.J., Ribani, A., Schiavo, G. and Fontanesi, L. (2021) Describing variability in the tyrosinase (*TYR*) gene, the *albino* coat colour locus, in domestic and wild European rabbits. *Italian Journal of Animal Science* 20, 181–187.

Vrieling, H., Duhl, D.M., Millar, S.E., Miller, K.A. and Barsh, G.S. (1994) Differences in dorsal and ventral pigmentation result from regional expression of the mouse agouti gene. *Proceedings of the National Academy of Sciences of the USA* 91, 5667–5671.

Wieberneit, D. and Wegner, W. (1995) Albino rabbits can suffer from Megacolon-Syndrome when they are homozygous for the "English-spot" gene (En En). *World Rabbit Science* 3, 19–26.

Wieberneit, D., Mahdi, N., Zacharias, K. and Wegner, W. (1990) Zur Problematik der Scheckenzucht bei Kaninchen. 1. Mitteilung: Mast- und Schlachtkörpereigenschaften, Organbefunde. *Deutsche tierärztliche Wochenschrift* 98, 352–354.

Wilson, S.M., Yip, R., Swing, D.A., O'Sullivan, T.N., Zhang, Y. *et al.* (2000) A mutation in Rab27a causes the vesicle transport defects observed in ashen mice. *Proceedings of the National Academy of Sciences of the USA* 97, 7933–7938.

Wright, S. (1917) Colour inheritance in mammals – IV. The rabbit. *Journal of Heredity* 8, 473–475.

Xiao, N., Li, H., Shafique, L., Zhao, S., Su, X. *et al.* (2019) A novel pale-yellow coat color of rabbits generated via *MC1R* mutation with CRISPR/Cas9 system. *Frontiers in Genetics* 10, 875.

Xu, J. and Hardison, R.C. (1989) Localization of the ß-like globin gene cluster and the genes for parathyroid hormone and c-Harvey-ras 1 to region ql4 —» q21 of rabbit chromosome 1 by *in situ* hybridization. *Cytogenetics and Cell Genetics* 52, 157–161.

7 Genetics of Fibre and Fur Production in Rabbits

Daniel Allain
GenPhySE, INRAE, Castanet Tolosan, France

7.1 Introduction

Variations in coat structure and coat colour of the rabbit were among the first characters of any animal to be studied at the beginning of the 20th century (Castle, 1905; Hurst, 1905; Castle *et al.*, 1909) because variations in the colour and hair structure are more conspicuous. The early investigations were undertaken from the various coat colour and hair structure present in fancy and specific rabbit breeds which afforded ready material for the confirmation of Mendelian inheritance.

Angora rabbit was probably the earliest commercial rabbit breed. It is established that the Romans kept and bred them and utilized Angora wool from at least 100 BC and the Angora breed has been developed in the Carpathian mountains where tribes maintained herds of captive Angoras, and bred them for their warmer and softer wool, as opposed to their prized goat wool.

Angora made their way to England where in the 1500s laws banned the export of prized 'English silk hares' as Angora rabbits were called there. Then, Angora were imported to the south-west of France in 1723 by English sailors who described the animals as originally coming from the Angora region of Turkey; and the Angora wool (also called 'the silk of Angora') industry started in France in 1765.

Properly prepared rabbit pelts from both wild and domestic rabbits were also valued during the Middle Ages. As far back as 1631, differential prices were noted between ordinary rabbit pelts and the pelts of quality riche rabbit in the Champagne region of France. The colour of the pelt originating from the development of new coat-coloured breeds was the main criterion determining the value of the pelts for a long time. Commercial breeds for fur production were then developed following the appearance of the Rex rabbit in France in 1919.

The emphasis on Angora and specific coloured-fur rabbit breeds drove development of fibre and fur production by rabbits until the 1930s when systematic studies appear to be made on the quantitative variation of coat qualities, fibre and pelt production. This chapter aims to present the genetics of fibre and fur production in the rabbit. It focuses first on general biology of hair covering and hair development of the rabbit. A clear understanding of the biological basis of the coat is essential for a better understanding of the genetic basis of fibre and fur production. It then examines the genetic basis of hair growth pattern and coat composition for fibre and fur production. Selection of rabbit on coat colour, a component of fur production, will be dealt with very briefly. Genetics and molecular genetics of coat colour is the subject of Chapter 6 of this book.

Contact e-mail address: daniel.allain@inrae.fr

 The Genetics and Genomics of the Rabbit (ed Luca Fontanesi)
DOI: 10.1079/9781780643342.0007

7.2 Biology of the Coat of the Rabbit

7.2.1 Structure and composition of the coat

The normal coat of the rabbit, which is its main protective organ, is made up of three kinds of hairs: (i) bristles or guard hair; (ii) awns or intermediate hair; and (iii) down or underfur fibres. The properties of the different hair types, the composition and the structure of the coat, all play a role in this protection. The bristles are long and coarse overhairs of the outer coat, which provide mechanical protection. The downs are short and fine underhairs of the inner coat, which provide the thermal protection. The awns having intermediate length and diameter are considered as overhairs.

The distribution of hair follicles within skin is not determined at random. They are assembled in homogeneous functional units known as follicle groups. Some hair follicles are specialized in the production of guard hairs, whereas others supply down hairs. The type of hair produced is determined by the position of the hair follicle within its group. The age at which a follicle is formed determines its place in the group. At first, during foetal life, there is a succession of four or five generations of central primary hair follicles from the 18th day of gestation. Subsequently, at least two lateral primary follicles will appear on both sides of each central primary follicle. Secondary follicles will establish themselves afterwards within the previously-defined area from the 26th day of gestation. There are two types of secondaries. The epidermal follicles originate from the primitive foetal epidermis as primaries, until the first sign of epidermal keratinization. As the keratinized epidermis is not able around birth to produce more new follicles, the subsequent secondaries are formed by branching from existing epidermal follicles (Hardy and Lyne, 1956). These derived follicles with their original epidermal follicle constitute a compound hair follicle. The hairs of derived follicles pass through the epidermis in the common hair canal developed by the generating epidermal follicle. There are only 4–6 derived follicle buds per follicle group at birth. Thereafter the multiplication of derived follicles is independent of the age but increases in the growing animal to maintain the hair cover up to a weight of about half that of the adult, to reach 25–50 derived follicles per follicle group (Mehner and Koetter, 1965). The follicle group is usually organized in groups of three to five primary hair follicles and a bundle of secondaries (Carter, 1943; Rougeot and Thébault, 1984; Rougeot *et al.*, 1984b). The overhairs are produced by primary hair follicles. The central primaries produce bristles while lateral primaries produce awns. The underhairs are produced by secondary follicles. The secondary derived follicles are more than simple morphological features producing some additional downs in the coat to maintain the hair cover. They play an important role in the efficiency of the underfur cover in growing mammals through infant moults and in adapting the adult coat to seasonal climatic changes through seasonal moulting periods.

7.2.2 Hair structure

The length of the fine undercoat is 20–30 mm whereas the coarse guard hairs are up to 30–40 mm long. The fine downs are square, oval or near-circular in cross-section with a constant diameter along the fibre of 6–20 μm.

The overhairs have a variable diameter from the base to the tip. The change commences at about mid-length and consists of a flattening of the cross-section until it becomes bean-shaped, passing to sausage and dumb-bell. The hair finally ends in a fine point. Diameter is 20–30 μm and 40–60 μm in the basal part, and 60–80 μm and 90–100 μm in the upper part for intermediate and guard hairs, respectively.

The hair shaft is made of three structural keratin-protein components: cuticle, cortex and medulla. The cortex, the main structural component of the hair shaft, consists of a cylindrical array of closely packed cells whose long axes are aligned parallel to the fibre axis. The cortex is covered by a cuticle composed of sheath-like cells which overlay each other in a similar manner to the tiles on a roof. The medulla is unbroken, wide and central, symmetrically-shaped in cross-section. The overall structure of the medulla belongs to the serial type with one or several columns of air-filled cells along the axis of the hair shaft in the downs and overhairs, respectively.

7.2.3 Hair growth and moulting periods

In the normal-coat, the different hair types of rabbits grow for 4–6 weeks, regardless of their nature. Differences in length are the outcome of different growth rates. Each kind of fibre type is replaced periodically. The growth pattern of individual fibres can be divided in three phases: anagen (active growth phase), catagen (follicle regression) and telogen (resting phase) (Chase, 1954). Shedding of the previously-grown fibre tends to occur around the onset of anagen, thus at least one hair is, in general, always present in a follicle. The duration of anagen is constant whereas the duration of telogen can be modified by photoperiodic manipulations, hormones, trauma and plucking. Thus, a hair may be anchored in the skin by its club root for several weeks without growing. Pattern of hair growth and changes in the coat depend on moulting: juvenile moults in the growing animal and seasonal moults in the adult.

In the young rabbit, there are three types of juvenile coats: newborn, infant and sub-adult. The coat of the newborn rabbit stops growing when the animal weighs about 0.4 kg (for an average-size breed). The infant coat is mature at about 8–10 weeks depending on the animal weight, as the number of derived hair follicles in development depends on the surface skin area in the growing animal. The coat of the sub-adult rabbit is mature from about 4.5–5 months (Thébault, 1997).

In the adult, moults are governed by seasonal photoperiodicity and occur in spring and autumn. The spring moults are spectacular, with visible loss of winter coat; but they are slow and irregular and rarely give an entirely stable coat in the summer. The summer coat is short and thin as during spring moult, when the down falls, some secondary derived follicles regress to the hair germ stage or disappear without replacing the hair. By contrast the autumn moult reactivates all the hair follicles in a relatively short time. It gives longer hairs and a thicker coat as all secondary derived follicles produce a down and contribute to the thickness of the winter coat.

7.3 Single Genes Affecting Hair and Coat Structure

In rabbits, several independent and autosomal recessive genes determine characteristics of fibre growth, coat composition and structure. The Angora phenotype is characterized by an abnormally long hair length (Rougeot and Thébault, 1984). The Rex trait suppresses development of long coarse fibres (Letard, 1928; Castle, 1929; Fraser, 1953). The *satin* gene modifies fibre structure (Castle and Law, 1936; Robinson, 1953) and the furless and hairless phenotypes are due to partial or complete suppression of down fibre production, respectively (Castle, 1933; Boucher *et al.*, 1996).

Other autosomal recessive genes ('wuzzy', furnished or waved) affect coat structure but most of them have been observed in association with other recessive genes affecting hair and coat structure (Robinson, 1958).

The Angora and Rex characters are industrially used for fibre and fur production whereas other mutations are mainly used for creating fancy breeds in combination with colour and coat-modifying genes.

7.3.1 Angora

Castle (1903) has firstly demonstrated that the long coat of the Angora, with a length about three times longer than in the normal-coat rabbit, was due to a single recessive gene. Different studies undertaken thereafter demonstrated that this recessive gene acts only on fibre length without any other modifications on coat composition or hair structure such as fibre diameter or coat softness compared to the normal rabbit (Wucherer, 1925; Hardy and Markley, 1944). A woolly coat-type rabbit, having also a long coat but shorter than the Angora was described by Hardy and Markley (1944). However, both rabbit types are owning the same recessive gene, but the Angora possesses coat polygenes increasing hair length and fleece weight (Robinson, 1958). Similarly, the presence of long wool tufts on the tips of the ears, called 'furnishing', was caused by the segregation of modifying polygenes and not due to a dominant gene associated with Angora mutation or correlated to the wool yield (Pickard, 1930).

Extension of the active growth phase of hair (anagen) without changes in growth rate is at the origin of the unusual length of the coat in Angora rabbits (Rougeot and Thébault, 1983, 1984) as in mice (Pennycuick and Raphael, 1984). After rabbits are plucked or defleeced, hair growth stops suddenly at the sixth week in rabbits not carrying the Angora mutation whereas complete hair growth lasts more than thirteen weeks in Angora rabbits. Angora mutation does not prolong the duration of the active phase according to any regular pattern. Some hairs complete their growth cycle at the end of 12 weeks and are shed, whereas others are still growing 5 months later (Rougeot and Thébault, 1984). It seems possible to increase the average duration of the active phase up to a permanent growth pattern with an appropriate selection programme as it happens in the German breed selected for shearing (Rougeot and Thébault, 1984).

Mulsant *et al.* (2004) have reported a tight linkage between the rabbit Angora trait and a polymorphism in exon 3 of fibroblast growth factor 5 (*FGF5*) gene, although no causative mutation has been identified. The *Angora* locus mapped on a 15 cM region of chromosome 15 (OCU15) including the *FGF5* gene as reported by a study that also defined a first-generation genetic map in rabbits using microsatellites (Chantry-Darmon *et al.*, 2006) confirming a previous report (Fox, 1994). It was also confirmed that the *FGF5* gene could be the potential major gene affecting wool yield or that it could be in linkage disequilibrium with the causative mutation(s), supporting the possibility to use markers in this gene for improving wool yield in Angora rabbits (Li *et al.*, 2008). Recently, it was shown that the *FGF5* gene was upregulated in the Angora rabbit (Zhao *et al.*, 2018; Ding *et al.*, 2019).

7.3.2 Rex

The Rex phenotype is characterized by the relative absence of guard hair with a short and soft hair coat which gives a quite different fur from that of the normal-coat rabbit raised for meat production. This coat phenotype was first observed in France in 1919 within a litter of wild gray rabbit (Letard, 1928). Similar phenotypes appeared thereafter in Germany (1926) and France (1927). All these Rex phenotypes showed simple Mendelian recessive inheritance (Castle, 1929, 1933). It was first suggested that these three phenotypes, quite undistinguishable in appearance, were considered genetically different (controlled by three different loci: R^1, R^2 and R^3) as any crosses between these lines produced rabbits with normal coat (Castle and Nachtsheim, 1933). However, as genetic differences between these three phenotypes were never observed thereafter, it was suggested that differences between these Rex phenotypes were mainly due to variations in guard hair length with polygenic additive effects (Lienhart, 1962). It is now accepted that Rex phenotype is due to a single recessive allele (indicated as r^1, following the classical nomenclature of the *Rex* loci) but coat characteristics of the Rex rabbit can be improved through selection programmes to reduce the proportion and dimensions of guard hairs within the Rex coat (Thébault *et al.*, 2000).

In the Rex coat, guard hairs are less numerous and about 10 mm shorter than in normal-coat rabbits. Thus, undercoat and guard hair length are similar, hair is finer and compactness is higher (Lienhart, 1962; Vrillon *et al.*,1988; Diribarne *et al.*, 2011). This abnormally short hair trait with the relative absence of guard hair was thought to originate from a large decrease in fibre growth affecting mainly the outer coat without any modifications of coat composition (Lienhart, 1962) or was due to degenerescence of some primary hair follicles which usually produce guard and intermediate hair (Vrillon *et al.*, 1988). Recent studies, including skin histology methods and fine measurements of fibre length and diameter, state that the lack of guard hair in Rex rabbits was not due to the degenerescence of primary hair follicles. A large part of them remains active and produces hair. But these guard hairs present a modification of their shape and an important decrease of diameter. The hairs produced by the primary follicles seem to have disappeared but their reduction in size and shape, especially for guards, made them look like down (Diribarne *et al.*, 2011). This is the reason why guard and awn hairs have been considered as lacking when fur was only visually assessed. It also explains why Rex fur has characteristics of shortness, fineness, density, smoothness and beautiful velvety texture, and represents a great value for breeders and the industry (Robinson, 1958).

A single polymorphism in the lipase member H (*LIPH*) gene was shown to have a complete linkage disequilibrium and association with the Rex phenotype, confirming the recessive determinism of the trait (Diribarne *et al.*, 2011). This polymorphism was found as a one base pair deletion in exon 9 (c.1362delA) of *LIPH* gene. This deletion modifies transcription and leads to a premature termination of the Rex LIPH protein which affects *LIPH* expression in hair follicle and skin. The higher the relative absence of guard hair, the lower *LIPH* expression is (Diribarne *et al.*, 2012).

7.3.3 Satin

The coat of the Satin rabbit has a characteristic glossy and smooth appearance and feels softer and silkier than the normal rabbit coat without any modification in hair length. This mutation, first observed in the USA in the 1930' within a litter of Havana rabbit, is due to a recessive allele at a single locus (Castle and Law, 1936; Robinson, 1953). Variations in the three structural components of the hair shaft – cuticle, cortex and medulla – give the glossy aspect of the Satin coat. The cortex as well as the medulla and the cuticle layer covering the cortex are thinner in the Satin coat. These characteristics associated to a smoother surface of the cuticle scale increase the visibility of the hair pigment and make in turn the hair colour more brilliant (Boucher, 1993). The cells of the medulla, instead of belonging to the serial type, become more compact and lead to form a spongy-like structure (Castle and Law, 1936) or the absence of medulla (Spendlove and Robinson, 1970). The Satin mutation, as well as the Rex one, in combination with colour type, were often used to create fancy breeds but its commercial interest in the fur industry remains limited.

7.3.4 Waved

The waved (or curly) coat phenotype in the rabbit seems to be due to a recessive allele at a single autosomal locus but is dependent upon the presence of the Rex recessive homozygous genotype. It is only manifested in the fine-coated Rex rabbits (Pickard, 1941). Any attempt to combine the waved trait with a coarse Rex coat or the Angora one failed. This trait is also often called 'astrex' as it can only be manifested in the fine Rex.

7.3.5 'Wuzzy' (hair sticky and matted)

'Wuzzy' rabbits possess a long coat which differs from the Angora. Hairs appear to mat in tufts in a characteristic fashion (Sawin, 1955). A straight hair shaft is observed in the Angora rabbits whereas the tips of hairs have a tendency to mat with each other in a peculiar, seemingly twisted, fashion which becomes progressively more pronounced as hair density increases (Crary and Sawin, 1959). A recessive allele at a single locus was apparently involved but no relation with the Angora has been established (Robinson, 1958).

7.3.6 Naked or furless

The rare occurrences of naked and furless rabbits were first reported in Russia (Kislovsky, 1928) and Great Britain (Castle, 1933), respectively. The furless and naked loci both prevent the normal formation of fine underhairs, leaving primarily the long and coarse guard hairs (Drapeau, 1933). It was then suggested that the naked or furless traits were due to a recessive allele at the same locus (Castle, 1933; Jackson *et al.*, 2006). Other mutations leading to the absence of underhairs were also described: pelt loss (Nachtsheim, 1937, 1958), wirehair (Sawin, 1955) and hairless (Boucher *et al.*, 1996) but all these mutations appeared to be closely similar to the naked or furless trait. The naked rabbit resembles the furless one but with less residual hairs due to a variation in gene expression (Crary and Sawin, 1959). Naked and furless rabbits produce approximately 90% less fur than normal furred rabbits (Rogers *et al.*, 2006). It could be assumed that all the mutations describing the loss of underfur exhibit variations in the absence of downs or any hair over the body as reported by Castle (1933) and Boucher *et al.* (1996), probably due to modifier genes. This mutation was reported to be lethal (Kislovsky, 1928), but contradictory results were described (Castle, 1933;

Jackson *et al.*, 2006) although furless rabbits were found to be difficult to keep alive and have low fertility (Boucher *et al.*, 1996).

7.4 Genetic Improvement of Fibre and Fur Production

Fibre and fur production schemes in rabbits are based on purebred selection schemes with specialized breeds. Genetic improvement of fibre and fur production in rabbits is aimed at:

(i) increasing the production of fibre or fur to give a greater economic return per animal and production unit; and
(ii) improving the quality of the fibre or fur so that it can be processed into superior end products thus attracting a higher unit value.

The development of efficient breeding programmes requires to:

(i) identify the traits of importance;
(ii) know the phenotypic variation levels and genetic parameters among the traits to be improved so that breeding value predictions can be made using genetic model with appropriate adjustments for environmental and non-genetic factors; and
(iii) evaluate selection strategies to achieve the breeding objective.

This part focuses on genetic improvement for the production of fibre and fur in rabbit. Other traits relating to functional and adaptation traits (reproduction, health, growth and maternal traits) have also to be taken into consideration with less or more importance. The genetics of these traits is the subject of other chapters of this book.

7.4.1 Genetic improvement of Angora rabbit for fibre production

Angora wool is almost exclusively used in the carded processing chain, either in drapery or in hosiery. These two uses led to the selection of two different breeds or lines of Angora rabbit which are produced under different management systems.

The French breed, raised mainly in France, intended to produce bristly Angora with a long and hairy coat, is harvested by defleecing with a depilatory treatment. Long guard hairs within well-structured staples are desirable fibres for making luxurious garments and fashion knitwear having a brush appearance.

In the French breed, fibres are collected every 90–100 days by defleecing when the hair follicles reach the resting phase and before hairs start falling. Defleecing does not affect the behaviour of hair follicles and the hairs of Angora rabbits continue to grow in appropriate synchronism, obeying the typical growth rate of each type of hair (Rougeot and Thébault, 1984). At the time of defleecing, all hairs have the same duration of growth, but due to different growth rates, there are significant differences in fibre length between long guard hairs (10–12 cm) and fine downs (6–7 cm). Thus, a well-structured staple is produced. French Angora rabbits are usually defleeced at 8 weeks of age for the first time, then at intervals of 13–14 weeks (Rougeot and Thébault, 1984).

The German breed, mainly raised in China, South America and central Europe, intended to produce a woolly fleece, has a long, soft and fine woolly coat with a minimum of bristles or guard hairs. The German breed is selected for a woolly and finer fibre to make soft and very fine yarns for fashion knitwear and thermal underwear.

In the German breed, the hair is collected by shearing or cutting every 10–13 weeks. As shearing do not appear to encourage hair to grow, an absence of synchronism in hair growth is observed between the different type of fibres; some hairs follow their established growth pattern, whereas others start their growth cycle randomly, so that hairs are cut at different stages of their growth, giving the coat a woolly look. In central Europe and China, Angora rabbits are sheared at 13-week intervals whereas in South America, the first shearing takes place when the rabbits are 9 weeks old, the second at 20 weeks, and subsequent shearing at intervals of 10 weeks (Garcia and Magofke, 1982). The adoption of such a pattern would have the effect of increasing the annual wool output, but with a reduction in the length of the hairs gathered each time.

7.4.1.1 Relevant traits

In both French and German breeds, the main selection criterion is the total fleece weight. Angora rabbits produces each year 1–1.5 kg, even up to 2 kg, of fibre, which is more than a quarter of their body weight.

The fleece of Angora rabbits is usually sorted in five different grades during harvest according to length, type of wool, cleanness and absence of felting. For the French breed, these grades are (Vrillon and Thébault, 1992):

(i) WAJ1 for bristly wool harvested from the back and side of the rabbit, which includes both bristles and down longer than 6 cm and represents about 70% of the total fleece weight;
(ii) WAW1 for woolly wool harvested from the throat and the belly longer than 6 cm;
(iii) WAW2 includes short wool less than 6 cm long harvested mainly from young rabbits at the first and second harvest;
(iv) CFW for clean felted wool; and
(v) DW for dirty wool.

For the German breed, there are no nationwide standard classes. However, Angora wool is graded according to the length (2 or 3 grades from 3 to 6 cm) and the type (woolly or single coat and bristly or double coat) of hair.

By contrast to other fibre-producing animals such as sheep, goats and South-American camelids, fibre diameter and other fleece characteristics (fineness variability, curvature, crimp, staple strength, undesirable fibres) are not considered as traits of importance. Objective measurements based on optical fiber diameter analysis (OFDA) methodology for determining fineness, curvature and proportion of coarse fibre are available (Rafat *et al.*, 2007b) but not widely used. Angora wool is a fine fibre (13–16 μm) like fine cashmere with about 0.5–5% of coarse fibres (diameter >30 μm) without undesirable fibres and a length of 6–10 cm depending upon the interval between harvests.

Other production traits are of less importance in the husbandry of Angora rabbits. Both French and German Angora rabbits are raised to produce fibres with an adult animal, having a long lifetime (4–6 years) depending upon management system and without any meat valorization. Thus, fertility and maternal traits are of less importance as reproduction is only used for animal replacement and selection.

7.4.1.2 Non-genetic factors affecting fibre production in Angora rabbits

Despite the harvest method which affects the type of fleece produced by the German and French breeds, wool production is affected by age and sex of the animals, reproduction and season.

Age or harvest number is important (at least at the first harvest) for all Angora rabbit breeds as the development of secondary follicles producing downs is continuing after birth until animals reach about 50% of their adult weight. Young rabbits produce woolly fleece at the first and second harvest, even in the French breed after defleecing. Total fleece weight is about 30–40 g at the first harvest and then increases rapidly up to the fourth and the fifth harvest to reach a weight five or six times higher than at the first harvest. As live body weight increases rapidly from about 50 g at birth to 1.3 kg, 3 kg and 4 kg at 8, 20 and 30 weeks, respectively, total fleece weight increases with age and live body weight (Thébault *et al.*, 1992).

Fleece weight is greater in females than in males. This sex effect is more important in the French breed where males produce up to 20% less wool than females whereas in the German breed, males produce 0–15% less than females. This sex effect in Angora rabbits is the reverse of that observed in sheep and goats. Due to a lower fibre production in males, females and castrated males in some management systems are mainly used for Angora wool production. Entire males are only kept for reproduction and animal replacement.

Reproductive status of the doe, and especially gestation and lactation, reduce hair production by about one-third (Thébault and de Rochambeau, 1988; Umesh *et al.*, 2004).

Seasonal effects are important. Fibre output is maximal in winter and minimal in summer in both breeds (Magofke *et al.*, 1982a, 1982b; Rougeot and Thébault, 1983; Allain *et al.*, 1999). The seasonal influence decreases with age or harvest number. The relative difference between winter and summer is around 30% for the first and second harvest, around 15% for the third and the fourth harvest and only 9% for subsequent adult harvest (Rougeot *et al.*, 1984b). Fleece parameters are also influenced by the season in which animals are born. The total yield of adult animals born in winter is significantly greater than those born in summer (Allain *et al.*, 1999). However, a greater seasonal influence has been observed in the French breed, when animals are defleeced, than in the German breed, when animals are sheared (Schlolaut, 1980).

7.4.1.3 Genetic parameters

Few studies have reported genetic and phenotypic parameter estimates for wool traits in Angora rabbits. There was a considerable range in published heritability estimates for fleece weight at different harvests to the adult age (Table 7.1).

TOTAL FLEECE WEIGHT The heritability estimated for total fleece weight ranged from 0.08 ± 0.20 to 0.96 ± 0.29 (Table 7.1). The most probable values estimated under an REML animal model ranged from 0.21 to 0.52 with a mean value of 0.35 ± 0.08 which is similar to the weighted mean heritability for clean fleece weight (0.36 ± 0.02)

Table 7.1. Estimates of heritability (± S. E.) with reference to wool traits in French (F) and German (G) Angora rabbit breeds.

Heritability	Breed[1]	Method[2]	Reference
Total fleece weight			
First harvest			
0.20 ± 0.12	G	–	(Caro *et al.*, 1984)
0.40 ± 0.02	F	REML	(Allain *et al.*, 1999)
0.19	G	ROS	(Magofke *et al.*, 1994)
0.44 ± 0.15	G	REML	(Singh and Jilani, 2006)
0.17	G	–	(Risam *et al.*, 2005)
Second harvest			
0.23 ± 0.12	G	–	(Caro *et al.*, 1984)
0.31 ± 0.03	F	REML	(Allain *et al.*, 1999)
0.27	G	–	(Magofke *et al.*, 1994)
0.40 ± 0.20	G	REML	(Singh and Jilani, 2006)
0.25	G	–	(Risam *et al.*, 2005)
Third harvest			
0.09 ± 0.12	G	–	(Caro *et al.*, 1984)
0.32 ± 0.03	F	REML	(Allain *et al.*, 1999)
0.31	G	ROS	(Magofke *et al.*, 1994)
0.49 ± 0.11	G	REML	(Singh and Jilani, 2006)
0.45	G	–	(Risam *et al.*, 2005)
Fourth harvest			
0.08 ± 0.20	G	–	(Caro *et al.*, 1984)
0.30	G	–	(Lin *et al.*, 1994)
0.23	F	REML	(Allain *et al.*, 1996a)
0.21 ± 0.24	G	REML	(Singh and Jilani, 2006)
0.14	G	–	(Risam *et al.*, 2005)
Fifth harvest			
0.26	G	ROS	(Magofke *et al.*, 1994)
0.26 ± 0.13	G	REML	(Singh and Jilani, 2006)
0.11 ± 0.10	G	ROS	(Katoch *et al.*, 1999)
Homogeneity			
0.18	F	REML	(Allain *et al.*, 1996a)
Structure			
0.17	F	REML	(Allain *et al.*, 1996a)
Bristle length			
0.25	F	REML	(Allain *et al.*, 1996a)
Down length			
0.15	F	REML	(Allain *et al.*, 1996a)
Compression			
0.19	F	REML	(Allain *et al.*, 1996a)
Resilience			
0.12	F	REML	(Allain *et al.*, 1996a)

[1] F, French breed; G, German breed.

[2] ROS, regression of offspring on sire; REML, restricted maximum likelihood estimation.

observed in sheep wool breeds (Safari *et al.*, 2005). Despite the large variation observed in the phenotypic expression of Angora wool production according to the age of the animal, heritability estimates of total fleece weight are of the same magnitude in both young and adult animals (Allain *et al.*, 1999).

Maternal genetics effects for total fleece weight are high and significant in both the young and the adult animals but decrease with age and range from 0.44 in the young to 0.10 in the adult (Allain *et al.*, 1999).

FLEECE QUALITY TRAITS Estimations of heritability of different fleece quality traits in French Angora rabbit females range from 0.15 to 0.25 (Allain *et al.*, 1996a) (Table 7.1).

LIVE BODY WEIGHT Direct heritability estimates of live body weight ranging from 0.28 to 0.63 at the fourth harvest and moderate to high maternal genetic estimates from 0.43 at 8 weeks of age to 0.14 in the adult age were reported in the French Angora rabbits (Allain *et al.*, 1999). Higher heritability estimates of live body weight were reported in the German Angora rabbits but these values were probably overestimated due to confusion between direct additive and maternal effects (Caro *et al.*, 1984).

GENETIC CORRELATIONS Genetic correlations of total fleece weight between the different harvests were moderate to high and positive. The first harvest was not highly correlated to the following harvest with estimates ranging from 0.22 to 0.39 (Allain *et al.*, 1996b). Between the consecutive following harvests, genetic correlations were positive and high and ranged from 0.69 to 0.89 indicating that the second harvest is a good predictor of breeding value for total fleece weight (Magofke *et al.*, 1994; Allain *et al.*, 1999).

Genetic correlations between live body weight and total fleece weight are variable according to the age of the animals and the breed type. With the exception of the first harvest, genetic correlations between body weight and fleece weight are low and not significant, indicating that no improvement of fleece weight production can be obtained by selection in the French breed (Allain *et al.*, 1999). On the contrary, high correlations are observed in the German breed (Caro *et al.*, 1984).

Genetic correlations between total fleece weight and the weight of long and bristly wool, as well as between the length of the two kinds of fibres within the staple or at different locations are high (Allain *et al.*, 1996a).

7.4.1.4 Response to selection

Estimated genetic parameters have given us relatively precise predictions for expected responses to selection. In practice, selection for a single trait, i.e. total fleece weight, has been widely used in both the French and the German Angora rabbit breeds. It would be important to check if the prediction of the selection response for total fleece weight would is really effective. Such work has only been reported for the French breed.

DIRECT RESPONSE TO SELECTION FOR TOTAL FLEECE WEIGHT Rafat *et al.* (2007a) explored the genetic variability of wool production and other production traits through an 8-cohort divergent selection experiment for total fleece weight in the French Angora breed. The lines were selected for high and low total fleece weight at each harvest every three months from the third harvest and the selection response measured as deviations between the two lines. A divergence of three genetic standard deviations was observed between the low and high lines after eight years of divergent selection. The pattern of response was consistent with a heritability estimate of 0.38 and selection intensity at a rate of 2.4% per year (Rafat *et al.*, 2007a). Selection for total fleece weight significantly increased the weight of both long bristly and woolly wool grades of the fleece in similar proportion. An increased live body weight was also obtained by selection for total fleece weight as expected in regard to genetic parameters. It was also shown that response to selection at the second harvest was similar to that observed at the following harvest whereas no response to selection was observed on the first harvest (Rafat *et al.*, 2009) in regard to genetic correlations between following harvests from birth. These results confirm that (i) fleece weight at the first harvest is a different trait as hair follicle development is not complete at 8 weeks of age when the first harvest occurs; and (ii) the second harvest is a good predictor of breeding value for total fleece weight (Magofke *et al.*, 1994; Allain *et al.*, 1999).

CORRELATED RESPONSE TO SELECTION FOR TOTAL FLEECE WEIGHT From the same divergent selection experiment, correlated response to selection for total fleece weight on different fleece-quality traits – length, diameter and proportion of the different fibre types, secondary to primary follicle (S/P) ratio, compression and resilience properties of hair – were evaluated. Selection for total fleece weight increased bristle length, S/P ratio and comfort factor (% of fibres having a diameter <30 μm), and decreased bristle diameter, mean fibre diameter, compression and resilience, whereas no correlated responses were observed in down length (Rafat *et al.*, 2008).

No similar selection experiment on the German Angora rabbits has been published. It could be assumed that a similar response to selection would be expected.

CONSEQUENCES FOR BREEDING PROGRAMMES Selection for increasing total fleece weight has positive effects on fleece characteristics such as bristle length, follicle population and fibre diameter (Rafat *et al.*, 2008). The measurement of total fleece weight is simple and easy to do at farm level. Breeding programmes based only on this criterion result in an improvement of both quantitative and qualitative traits of wool production in the Angora rabbits.

7.4.2 Genetic improvement of rabbit for fur production

There are two kind of skins from rabbits used in the production of various fur items and trim: skins from meat rabbits and skins from Rex rabbits.

Skins from normal rabbits specially bred for meat production are retrieved from the slaughterhouse as a by-product of the rabbit meat industry. Such valorization of skins provided pelts of acceptable quality to the fur industry until the 1960s, but due to the development of the meat rabbit industry using specialized breeding lines selected to improve numerical and weight productivity with an earlier slaughtering age, pelt quality fell dramatically (Thomas, 1977, Rochambeau and Vrillon, 1985). Many skins are still collected from European slaughterhouses and exported as raw skins to China. By sorting, a small percentage corresponding to the best quality are dressed and processed to feed the fur industry. Others are shorn to collect hair for textile use. Today this market is not very stable and large fluctuations of the demand are observed. The price of raw skins can vary from 0% to 30% of the total value of the animal at slaughterhouse level.

The Rex rabbits are specially bred for their fur and used for fur production. Their fur is quite distinct from that of regular. The Rex fur has no prominent guard hairs resulting in a silky, dense, soft and plush type of fur (comparable to velvet). This fur is highly desirable and can obtain high market prices. Its popularity in garment production stems from the various colours available throughout the breed's different varieties. The Rex mutation was first observed in France in 1919 and was considered as a new breed in 1925 in France. Then, development of Rex rabbit farming for fur production was popular up to the 1950s and declined until the 1970s with the demand in fur. Thereafter studies for developing a dual-purpose rabbit for both meat and fur production were developed in France (Vrillon and de Rochambeau, 1982) and China. A specific Rex rabbit industry was then created in France (Thébault *et al.*, 2000), China and some other countries aiming to produce a meat rabbit with a valued pelt which could be better valorized than usual by-product pelt from the meat rabbit slaughterhouse industry. A large panel of coloured Rex furs has been obtained and offered to the fur industry.

As fur production is more often associated with meat production from dual-purpose breeds owning the Rex mutation, there are two kinds of important traits: those related to fur production and fur quality and others to meat production including reproduction and maternal abilities, growth and carcass performances and adaptive traits. This part focuses only on traits relating to fur production. Genetics of other production traits are given in other chapters of this book.

7.4.2.1 Relevant traits in Rex rabbits

More traits are relevant for fur production than for fibre production. The main barrier to quality fur production is slaughter age: the pelt must be large enough and the whole coat mature, i.e. a winter coat having a high fur density. Thus, animals are slaughtered at a later age than meat rabbits, from 20 to 35 weeks of age, when pelts are large enough and mature (Rougeot, 1986).

The two other major defects that make normal rabbit fur a down-market product are the unequal growth of the hairs during moulting periods and the long and fragile guard hairs which break off very easily.

The first defect can be ironed out by production techniques (Vrillon *et al.*, 1988; Uzcategui and Johnston, 1990, 1996) that synchronize moulting and by checking the progress of moulting in all parts over the body.

The Rex rabbits are nearly free of the second defect as their coat contains no, or only a few, short guard hairs. Thus, Rex is the main if not the only breed raised and selected for fur production. However, the presence of short guard hairs called bristliness determines fur softness and remains an important quality criterion.

Other fur characteristics such as size, compacity, colour and other skin faults which determine the quality of the fur and its economic value, are taken into account for grading furs within each breed type.

FUR MATURITY A non-whole mature fur cannot be used in the fur industry because non-prime areas show unequal growth of hairs and important hair losses are observed following the tanning process. The crucial time to reach maturity is moulting: juvenile moults for growing rabbits and seasonal moults for adults. There are three types of juvenile coats: that of the newborn rabbit, infant and sub-adult coats. The first two are unusable because they are too small. The sub-adult coat becomes more interesting but the lengthy (4–6 weeks) moult, which is produced, does not start until the rabbit reaches 1.7–1.9 kg. It matures from 5 to 8 months of age (Petersen, 1992). Despite these limits, it is the first coat that could provide a fur of interest (Rougeot, 1986).

The progress of moulting and fur maturity can be assessed by scoring the extension and intensity of blue pigmentation through separating the hair over the skin which indicates hair follicle activity (Allain and Rougeot, 1980). This pigmentation is due to the renewal of melanogenesis in hair follicles during hair growth. At fur maturity, any pigmentation of the skin was not observed over the whole surface of the body (Thébault *et al.*, 2000).

COAT COMPOSITION AND FUR DENSITY The presence of guard hair and composition of the coat are of importance for determining softness and compacity of the fur. Softness is associated with bristliness (presence of coarse hairs) whereas the density of fibres per skin area unit determines compacity. The Rex type is known to shorten and reduce the proportion of coarse fibres. However, residual short guard hairs, a little longer than downs, are remaining and its content determines the softness of the Rex fur. It can be scored by visual assessment (Thébault *et al.*, 2000) or objectively measured using OFDA methods. The former method is cheap and easy to undertake on a large animal population and allows rapid progress (Thébault *et al.*, 2000). The latter one, developed for measuring fineness in wool, can also be used to determine comfort factor (proportion of fibres having a diameter >30 μ) in the fur rabbits. This method is quite expensive and can be used only when visual assessment does not allow any more improvement in coat composition.

Fur compacity which corresponds to the density of fibres or hair follicles per skin area unit can be assessed by handling the coat or can be measured using a compacimeter (Thébault *et al.*, 2000).

OTHER FUR CHARACTERISTICS Other traits such as size, coat depth or fibre length, clarity and skin faults have an impact on fur grading which determines its use by the industry. Size of the fur which is determined by live body growth at slaughter age, and fibre length or coat depth can be easily measured. Variations of fur colour within a colour genotype can be assessed by scoring through different criteria such as extension of the agouti band along the fibre, intensity and homogeneity of the colour. These traits are usually checked at grading to sort furs in different quality batches before marketing and processing, but they could also be taken into account in breeding programmes.

7.4.2.2 Genetic parameters and response to selection for fur production in Rex rabbits

Data about genetic parameters and selection schemes of Rex rabbits for fur production are very sparse. A selection programme based on the quality of fur and mainly the absence of guard hairs was initiated in France during the 1980s from a castor Rex rabbit in order to obtain a dual-purpose animal, a fur only composed of undercoat associated to meat production (Vrillon *et al.*, 1988). Several fur-quality parameters were introduced in the selection programme and particularly fur priming in order to obtain light and soft pelts. Its main objective was to create and develop a dual-purpose Rex rabbit producing both meat and the orylag® fur, a registered trademark fur different from that of the Rex rabbits, with an incomparable softness and lightness exploited under licence by a French breeder co-operative (Thébault *et al.*, 2000).

The different selection criteria taken into account to improve fur quality were: hair length, compactness, bristliness, fur priming and three-colour criteria (intensity, homogeneity and extension of the agouti band along the fibre) determined by scoring. Live body weight at 8 weeks of age was also introduced in the selection scheme. Quite apart from body growth development, it determines the end of hair follicle multiplication in the growing rabbits and hair growth of the second or juvenile coat (Thébault, 1997) and consequently determines moult onset of the sub-adult coat having a commercial value (Vrillon *et al.*, 1988). Selection goals are to decrease bristliness score, to improve 8-week body weight, compactness and fur priming, and to maintain steady hair length. Genetic parameters observed in the Castor orylag® are reported in Table 7.2 (Thébault *et al.*, 2000).

Heritability estimates of bristliness is moderate (0.28) indicating that despite the major effect of the Rex recessive allele on primary hair-follicle activity, there are still other polygenic effects acting on the development of guard hairs in animals owning the Rex genotype and an improvement in direction to suppress guard hairs could be obtained through a quantitative selection programme (Thébault *et al.*, 2000). Heritability estimates for other fur-quality traits are small to moderate and ranged from 0.06 for fur priming to 0.25 for fur-colour intensity. Genetic correlations reveal various antagonisms. Unfavourable correlations between bristliness score and compactness and hair length and between 8-week body weight and compactness were observed. To solve these antagonisms, a multivariate selection procedure including all fur production parameters and live body weight at 8 weeks of age was proposed. This selection obtained a dual-purpose animal producing a carcass which represents about 40% of the total income and a fur of high quality which is valorized in France through a breeder co-operative.

7.5 Concluding Remarks

Fibre and fur production are based on specialized breeds and pure-bred selection schemes. This could result in lower reproductive and productivity performances observed in both Angora and Rex rabbits than in modern meat-rabbit lines.

Hair is mainly collected from an adult Angora rabbit with harvest at about 2–3-month intervals from birth until up to 5–6 years of age. Fibre harvesting is the main competitiveness criterion in Angora rabbit farming whereas reproduction, which is used only for animal replacement and numerical productivity, has a low impact on the selection scheme and fibre production. Angora rabbit is a very efficient animal producing annually 1.2–1.5 kg of

Table 7.2. Heritability and genetic correlation between fur production traits in *Orylag®* Castor (Thébault *et al.*, 2000).

Traits	LBW8	BR	HL	CO	FP	ABE	FCI	FCH
Live body weight at 8 weeks (LBW8)	**0.40**	0.09	0.15	−0.11	0.10	0.20	−0.02	0.32
Bristliness (BR)		**0.29**	−0.38	−0.37	−0.39	−0.21	−0.36	0.33
Hair length (HL)			**0.19**	0.33	−0.16	−0.36	−0.07	0.14
Compactness (CO)				**0.20**	−0.09	−0.21	0.14	−0.09
Fur priming (FP)					**0.09**	−0.02	−0.67	0.69
Agouti band extension (ABE)						**0.16**	0.35	0.09
Fur colour intensity (FCI)							**0.28**	−0.58
Fur colour homogeneity (FCH)								**0.09**

wool that is more than a quarter of its body weight. Selection schemes based only on the measurement of total fleece weight from the second harvest, a moderate to high heritable trait easy to measure at the farm, result in an improvement of both quantity and quality of wool produced by the Angora rabbits.

In fur production, numerical productivity has a larger impact. The income from Rex rabbit farming is balancing between meat and fur production. Animals are slaughtered to produce both meat and furs at 5–8 months when the pelt is large enough and the whole coat mature. If meat can be marketed easily, any fur defect or low-quality grade has a large impact on pelt value. The Rex rabbit, which has a low numerical productivity, cannot compete with meat-production systems based on crossbreeding or multi-purpose line schemes, which avoid adverse effects between reproductive and growth performances. Breeding objectives must lead to produce a high-quality pelt to compensate for the low numerical productivity.

Any deviation from the breeding objectives due to economic reasons could have negative impacts. Any selection to increase numerical productivity must be limited to avoid adverse effects on growth rate of kits from birth. Hair follicle development determines moulting periods and fur density in the growing animals and consequently final fur quality. Any improvement of meat-production income through a higher growth rate or an earlier slaughter age has adverse effects on coat maturity and fur quality. Thus, breeding objectives in fur-production schemes should obtain a dual-purpose Rex line having reasonable numerical productivity, high-quality pelt, taking into account the negative relationship between fur quality criteria, and allowing a high value of the fur, whilst keeping carcass weight within commercial limits.

References

Allain, D. and Rougeot, J. (1980) Induction of autumn moult in mink with melatonin. *Reproduction Nutrition Development* 20, 197–201.

Allain, D., de Rochambeau, H., Thébault, R.G. and Vrillon, J.L. (1996a) Angora rabbit wool production: the inheritance of wool quantity and different fleece characteristics in adult does. *Proceedings of the 6th World Rabbit Congress No. 1.*, pp. 313–319, Toulouse, France.

Allain, D., de Rochambeau, H., Thébault, R.G. and Vrillon, J.L. (1996b) Angora rabbit wool production: heritability and genetic correlations for fleece weight according to age. *Proceedings of the 6th World Rabbit Congress No. 1.*, pp. 321–324, Toulouse, France.

Allain, D., de Rochambeau, H., Thébault, R.G. and Vrillon, J.L. (1999) The inheritance of wool quantity and live body weight in the French Angora rabbit. *Animal Sciences* 68, 441–447.

Boucher, S. (1993) *Les robes du lapin (Oryctolagus cuniculus)*. Etude génétique des patrons 'Blancs aux yeux bleus', 'Agouti gris' et 'Chamois'. Thèse de Doctorat Vétérinaire, Ecole Nationale Vétérinaire, Nantes, France.

Boucher, S., Thébault, R.G., Plassiart, G., Vrillon, J.L. and de Rochambeau, H. (1996) Phenotypical description of hairless rabbits appeared in three different herds. *Proceedings of the 6th World Rabbit Congress No. 1.*, pp. 333–338.

Caro, W., Magofke, J.C., Garcia, F.X., Garcia, G., Carvajal, S. *et al.* (1984) Genetic parameters for fleece production and body weight in Angora rabbits. *Investigactiones del Departamento de Produccion Animal* 3, 210–214.

Carter, H.B. (1943) 1. The development and general histology of the follicle group in the skin of the merino. *Bulletin of the Council for Scientific and Industrial Research (Australia)* 164, 7–21.

Castle, W.E. (1903) The heredity of 'Angora' coat in mammals. *Science* 18, 760–761.

Castle, W.E. (1905) Heredity of coat characters in guinea-pigs and rabbits. *Carnegie Institution of Washington, vol.* 1.

Castle, W.E. (1929) The rex rabbit. *Journal of Heredity* 20, 193–199.

Castle, W.E. (1933) The furless rabbit. *Journal of Heredity* 24, 81–86.

Castle, W.E. and Law, L.W. (1936) Satin, a new hair mutation of the rabbit: with a chromosome map of the rabbit. *Journal of Heredity* 27, 235–240.

Castle, W.E. and Nachtsheim, H. (1933) Linkage interrelations of three genes for Rex (short) coat in the rabbit. *Proceedings of the National Academy of Sciences USA* 19, 1006–1011.

Castle, W.E., Cobbs, S., Mullenix, R.C. and Walter, H.E. (1909) Studies of inheritance in rabbits. *Carnegie Institution of Washington*, vol. 114.

Chantry-Darmon, C., Urien, C., de Rochambeau, H., Allain, D., Pena, B. *et al.* (2006) A first-generation microsatellite-based integrated genetic and cytogenetic map for the European rabbit (*Oryctolagus cuniculus*) and localization of Angora and albino. *Animal Genetics* 37, 335–341.

Chase, H.B. (1954) Growth of hair. *Physiological Review* 34, 113–126.

Crary, D.D. and Sawin, P.B. (1959) Inheritance and hair morphology of the 'Wuzzy' mutation in the rabbit. *Journal of Heredity* 50, 31–34.

Ding, H., Zhao, H., Cheng, G., Yang, Y., Wang, X. *et al.* (2019) Analyses of histological and transcriptome differences in the skin of short-hair and long-hair rabbits. *BMC Genomics* 20, 140.

Diribarne, M., Mata, X., Chantry-Darmon, C., Vaiman, A., Auvinet, G. *et al.* (2011) A deletion in exon 9 of the LIPH gene is responsible for the rex hair coat phenotype in rabbits (*Oryctolagus cuniculus*). *PLOS ONE* 6, e19281.

Diribarne, M., Mata, X., Rivière, J., Bouet, S., Vaiman, A. *et al.* (2012) LIPH expression in skin and hair follicles of normal coat and rex rabbits. *PLOS ONE* 7, e30073.

Drapeau, E.E. (1933) Histological basis of hairlessness in the rabbit. *Journal of Morphology* 54, 365–388.

Fox, R.R. (1994) Taxonomy and genetics. In: Manning, P.J., Ringler, D.H. and Newcomer, C.E. (eds) *The Biology of the Laboratory Rabbits*, 2nd edn. Academic Press, San Diego, California, pp. 1–25.

Fraser, A.S. (1953) A note on the growth of the Rex and Angora coats. *Journal of Genetics* 51, 237–242.

Garcia, F.X. and Magofke, J.C. (1982) Parametros geneticos para producci6n de pelo y peso vivo en conejos Angora. [Genetic parameters for the production of fleeces and body weight in Angora rabbits]. *Avances en Producciôn Animal* 7, 81–90.

Hardy, M.H. and Lyne, A.G. (1956) The prenatal development of wool follicles in Merino sheep. *Australian Journal of Biological Science* 9, 423–441.

Hardy, T.M.P. and Markley, M.H. (1944) A microscopic study of coat variations in white New Zealand and Angora rabbits. *Journal of Heredity* 35, 183–192.

Hurst, C.C. (1905) Experimental studies on heredity in rabbits. *Journal of the Linnean Society of London, Zoology* 29, 283–324.

Jackson, R.A., Rogers, D. and Lukefahr, S.D. (2006) Inheritance of the naked gene and associations with postweaning performance and thermotolerance characters in fryer rabbits from an F2 generation. *World Rabbit Science* 14, 147–155.

Katoch, S., Sambher, V.K, Manuja, N.K., Thakur, Y.P. and Gupta, K. (1999) Studies on genetic and phenotypic parameters for wool production traits in Angora rabbits. *Indian Journal of Animal Research* 33, 126–128.

Kislovsky, D.A. (1928) Naked – a recessive mutation in the rabbit. *Journal of Heredity* 19, 438–439.

Letard, E. (1928) Le lapin Castorrex, L'histoire d'une mutation. *Revue de Médecine Vétérinaire de Toulouse* 80, 136–143.

Li, C.-X., Jiang, M.-S., Chen, S.-Y. and Lai, S.-J. (2008) Correlation analysis between single nucleotide polymorphism of FGF5 gene and wool yield in rabbits. *Yi Chuan [hereditas]* 30, 893–899.

Lienhart, R. (1962) Mise au point de nos connaissances biologiques actuelles sur le lapin Castor Rex. *Bulletin de l'Académie et de la Société Lorraines des Sciences* 2, 84–102.

Lin, Z.H., Shen, Y.Z., Zhang, Z.H., Lin, D.G., Kong, P.L. and Li, G.M. (1994) The estimation of genetic parameters for some quantitative traits in coarse-wool rabbit [in Chinese], *Chinese Journal of Rabbit Farming* 6, 17–19.

Magofke, S.J.C., Garcia, F.X. and Caro, T. (1982a) Factores ambiantales que influyen sobre la producciôn de pelo dei conejo Angora. [Environmental factors influencing Angora rabbit hair production]. *Avances en Production Animal* 7, 113–123.

Magofke, S.J.C., Garcia, F.X. and Caro, T. (1982b) Factores ambiantales que influyen sobre la producciôn de pelo dei conejo Angora. [Environmental factors influencing Angora rabbit hair production]. *Avances en Production Animal* 7, 125–134.

Magofke, S.J.C., Garcia, F.X. and Arinoviche, N.M. (1994) Genetic parameters of Angora rabbits. *Avances en Produccion Animal* 19, 161–171.

Mehner, A. and Koetter, U. (1965) Dichte und Gruppierung des Haarfollikel der Haut des Angorakaninchen. *Zeitschrift fur Tierzüchtung und Züchtungsbiologie [Journal of Animal Breeding and Genetics]* 81, 241–259.

Mulsant, P., de Rochambeau, H. and Thébault, R.G. (2004) A note on linkage between the Angora and fgf5 genes in rabbits. *World Rabbit Science* 12, 1–6.

Nachtsheim, H. (1937) Erbpathologische untersuchungen am kaninchen. Z. induckt. *Abstamm.-u. Vererb. lehre* 73, 463–466.

Nachtsheim, H. (1958) Erbpathologie der Nagetiere. In: Cohrs, P., Jaffe, R. and Meesen, H. (eds) *Pathologie der laboratoriumstiere*. Springer-Verlag, Berlin and New York, pp. 310–452.

Pennycuik, P.R. and Raphael, K.A. (1984) The Angora locus (GO) in the mouse: hair morphology, duration of growth cycle and site of action. *Genetical Research* 44, 283–291.

Petersen, A. (1992) Effect of age on priming and fur quality of the rabbit castor Rex. *Journal of Applied Rabbit Research* 15, 1599–1605.

Pickard, J.N. (1930) Die Vererbung von Haarbuscheln und Wollmenge bei Angora Kaninchen. *Zuchtungskunde* 5, 339–351.

Pickard, J.N. (1941) Waved – a new coat type in rabbits. *Journal of Genetics* 42, 215–222.

Rafat, S.A., Allain, D., Thébault, R.G. and de Rochambeau, H. (2007a) Divergent selection for fleece weight in French Angora rabbits: non-genetic effects, genetic parameters and response to selection. *Livestock Science* 106, 169–175.

Rafat, S.A., de Rochambeau, H., Brims, M., Thébault, R.G., Deretz, S. *et al.* (2007b) Characteristics of Angora rabbit fiber using optical fiber diameter analyzer. *Journal of Animal Science* 85, 3116–3122.

Rafat, S.A., de Rochambeau, H., Thébault, R.G., David, I., Deretz, S. *et al.* (2008) Divergent selection for total fleece weight in Angora rabbits: correlated responses in wool characteristics. *Livestock Science* 113, 8–13.

Rafat, S.A., Thébault, R.G., Bonnet, M., Deretz, S., Pena-Arnaud, B. *et al.* (2009) A note on divergent selection for total fleece weight in adult Angora rabbits: direct response to selection on total fleece weight at first and second harvest. *World Rabbit Science* 17, 39–44.

Risam, K.S., Das, G.K. and Bhasin, V. (2005) Rabbit for meat and wool production in India: a review. *Indian Journal of Animal Science* 75, 365–382.

Robinson, R. (1953) Segregation of the satin gene in the rabbit. *Journal of Heredity* 44, 95–96.

Robinson, R. (1958) Genetic studies of the rabbit. *Bibliographia Genetica* 17, 228–558.

Rochambeau, H. and Vrillon J.L. (1985) Facteurs de variations de la qualité de la fourrure et de la productivité pondérale chez le lapin domestique. *Annales de Zootechmie* 34, 49–76.

Rogers, A.D., Lupton, C.J. and Lukefahr, S.D. (2006) Fiber production and properties in genetically furred and furless rabbits. *Journal of Animal Science* 84, 2566–2574.

Rougeot, J. (1986) Production and marketing of rabbit skins. *World Animal Review* 60, 7–17.

Rougeot, J., and Thébault, R.G. (1983) Variations saisonnières de la composition et de la structure du pelage: exemple de la toison du lapin Angora. *Annals of Zootechnique* 32, 287–314.

Rougeot, J. and Thébault R.G. (1984) *Le lapin Angora sa toison, son élévage*. Les editions du point vétérinaire, Maisons-Alfort, France.

Rougeot, J., Allain, D. and Martinet., L. (1984a) Photoperiodic and hormonal control of seasonal coat changes in mammals with special reference to mink photoperiodic and hormonal control. *Acta Zoologica Fenica* 171, 13–18.

Rougeot, J., Thébault, R.G. and Allain, D. (1984b) The role of the compound hair follicle in seasonal adaptative pelage changes. *Acta Zoologica Fenica* 171, 19–21.

Safari, E., Fogarty, N.M. and Gilmour, A.R. (2005) A review of genetic parameter estimates for wool, growth, meat and reproduction traits in sheep. *Livestock Production Science* 92, 271–289.

Sawin, P.B. (1955) Recent genetics of the domestic rabbit. In: Demerec, M. (ed.) *Advances in Genetics*. Academic Press, pp. 183–226.

Schlolaut, W. (1980) Organisation und Durchführung der Leistungsprüfungen in der Kleintierproduction in der Bundesrepublik Deutschland. [Organisation and procedures of performance testing in small animal breeding in the German Federal Republic]. *Zuchtungskunde* 52, 393–401.

Singh, C.B. and Jilani, M.H. (2006) Genetic studies on the reproductive and productive traits of German Angora rabbits under temperate condition of garhwal himalayas. *Indian Journal of Small Ruminants* 12, 26–32.

Spendlove, W. and Robinson, R. (1970) A linkage test with satin in the rabbit. *Genetica* 41, 635–637.

Thébault, R.G. (1997) Production of rabbit skins and hair for textile. FAO Animal production and Health Series, Rome.

Thébault, R.G. and de Rochambeau, H. (1988) Production data and demographic parameters of a French Angora rabbit strain. *Proceedings of the Congress of the 4th World Rabbit Science Association*, 2, pp. 227–238.

Thébault, R.G., Vrillon, J.L., Allain, D., Farhat, E. and de Rochambeau, H. (1992) Effect of non-genetics factors on quantitative and qualitative features about Angora wool production in French farms. *Proceedings of the 5th World Rabbit Science Congress*. Corvallis, Oregon, pp. 233–241.

Thébault, R.G., Allain, D., de Rochambeau, H. and Vrillon, J.L. (2000) Selection scheme and genetic improvement of Orylag for fur production. *Proceedings of the 7th International Congress in Fur Animal Production*, Kastoria, Greece. *Scientifur* 24, 81–86.

Thomas, C. (1977) Que deviennent les peaux de lapins? *Cuniculture* 14, 86–88.

Umesh, S., Sharma, S.R., Bhatt, R.S., Risam, K.S., Kumar, D. *et al.* (2004) Effect of parity on the production performance of German Angora rabbits. *Indian Journal of Animal Sciences* 74, 550–551.

Uzcategui, M.E. and Johnston, N.P. (1990) Effect of continuous and intermittent photoperiods on the reproductive performance and growth of rabbits. *Journal of Applied Rabbit Research* 13, 215–219.

Uzcategui, M.E. and Johnston, N.P. (1996) Effect of 10-, 12- and 14-hour continuous and intermittent photoperiods on the development of fur prime of Rex rabbits. *Proceedings of the 6th World Rabbit Congress* 1, 381–385.

Vrillon, J.L. and de Rochambeau, H. (1982) Mise au point d'une méthode d'élevage d'un lapin à double fin (viande et fourrure) : résultats d'une première expérimentation. In: *3ème Journées de la recherche cunicole*, Paris

Vrillon, J.L. and Thébault, R.G. (1992) Proposition to classify Angora rabbit fleece. *Journal of Applied Rabbit Science* 15, 1544–1549.

Vrillon, J.L., Thébault, R.G., de Rochambeau, H. and Dardant, P. (1988) Photoperiodism effect on fur maturity and fur quality of rabbits, owning or not Rex gene. In: *4th Congress of the World Rabbit Science Association*, Budapest, pp. 244–252.

Wucherer, E. (1925) Uber den charakter des Angorahaares. *Zeitschrift für Tierzüchtung und Züchtungsbiologie [Journal of Animal Breeding and Genetics]* 4, 119–143.

Zhao, B., Chen, Y., Hao, Y., Yang, N., Wang, M. *et al.* (2018) Transcriptomic analysis reveals differentially expressed genes associated with wool length in rabbit. *Animal Genetics* 49, 428–437.

8 Genetics and Molecular Genetics of Morphological and Physiological Traits and Inherited Disorders in the European Rabbit

Luca Fontanesi
Department of Agricultural and Food Sciences, Division of Animal Sciences, University of Bologna, Italy

8.1 Introduction

Since the rediscovery of Mendel's laws at the beginning of the 20th century, the European rabbit has been used in genetic studies that reflected the multiple purpose of this lagomorph, i.e. as livestock species, fancy animal, animal model, wild animal and pest. The first pioneering genetic works in this species contributed to confirm the validity of Mendelian inheritance in mammals (e.g. Castle, 1905, 1907; Punnett, 1912, 1915). Coat colours and fibre traits were the first targeted phenotypes of these studies (see Chapters 6 and 7) that subsequently were oriented to investigate many other morphological traits and inherited defects. Defects were mainly described in inbred stocks, lines and breeds and were usually related to a high level of inbreeding that increased the frequency of recessive alleles at different loci in a variety of populations (e.g. Chai, 1968, 1969, 1970; Crary and Fox, 1980). Biochemical studies were subsequently introduced to understand basic biological mechanisms of observed variability of blood components, physiological states or enzyme activities determining functional effects at different levels (e.g. Watanabe, 1977).

Mode of inheritance of morphological traits and malformations or other defects ranged from polygenic to monogenic with complete or partial penetrance and different levels of dominance, genetics and environment interactions, complicating the interpretation of the results in several cases. Most of these studies were completed in the pre-DNA era or when genomic analyses could not be applied. They relied only on classical genetic approaches that observed the manifestation of the traits and measured parameters in family-based designs or in approaches that could monitor or infer the inheritance in the investigated rabbit populations and lines. With the advent of molecular genetics and the application of DNA markers in genomic studies, few morphological traits and defects have also been characterized at the molecular level. Results obtained from DNA-based approaches are still limited but they are expected to increase within the next decade by taking advantage of the availability of genomic tools.

This chapter compiles and updates the knowledge on morphological, physiological and biochemical traits under genetic control and on inherited disorders documented in the European rabbit. The chapter is also focused on the recent developments determined by the molecular characterization of these traits and defects.

Contact e-mail address: luca.fontanesi@unibo.it

 The Genetics and Genomics of the Rabbit (ed Luca Fontanesi)
DOI: 10.1079/9781780643342.0008

8.2 Sources of Information

Over the last century, a few authors have reviewed the genetic literature on inherited traits and disorders described in the European rabbit. Castle (1930) wrote the first manual on the genetics of the domestic rabbit that detailed the basic coat colour loci and analysed a few other genetically determined morphological traits, including body size and ear length. Sawin (1955) reviewed the genetic studies in rabbits published at that time and listed loci affecting hair colour and hair morphology, single locus and polygenic variations for morphological and physiological characters, genetic factors affecting the neuromuscular systems and behaviour, serological characteristics and body size, body proportions and growth. An extensive compilation of the genetic works in the rabbit was published by Robinson (1958), which constitutes a milestone on rabbit genetics of the pre-DNA period. His review included genetic knowledge on colour and pelage variation, variability of physiological traits, genetics of gross abnormalities affecting different organs and systems and quantitative genetics of size and growth, ear length, reproduction physiology, skeleton and several other biological aspects. Lindsey and Fox (1974, 1994) listed inherited diseases and variations described in rabbits. Fox (1974, 1975, 1994) reported the latest extensive reviews on rabbit genetics. In the final update (Fox, 1994), he listed a total of 101 recognized mutations (loci). This latest list included loci affecting coat colour and hair (n. 15 loci), morphological traits and inherited defects (n. 44; mainly derived by the works on inbred stocks) and biochemical traits (n. 43). The Online Mendelian Inheritance in Animals (OMIA) database (Nicholas and Hobbs, 2014; https://omia.org/home/, accessed 7 January 2021) includes 97 entries (April 2020) of disorders and other traits for which monogenic or oligogenic inheritance has been claimed or supposed in rabbits.

Together with a comprehensive literature survey, these sources of information have been used to critically compile this chapter and link what was reported by previous reviews, as summarized in Table 8.1.

8.3 Sources of Genetic Variability

Early rabbit breeders developed, by selection, a variety of breeds and lines that can not only be distinguished by the colours of the animals but also by their size, shape and other physiological features. A large number of rabbit breeds are now raised and maintained by fancy breeders who are organized in breeders' associations in many countries (see Chapter 2). Comparative analyses among breeds have also been used to infer genetic predispositions to different defects or to analyse morphological features (e.g. Johnson and Burn, 2019).

Inbred stocks have been developed by research workers to further exploit the usefulness of the rabbit in genetic studies of defects mainly caused by recessive mutations. Investigations of these stocks culminated in the period that ranged from the 1940s to the 1970s when several strains were developed and then maintained at the Jackson Laboratory and in a few other laboratories around the world. Unfortunately, most of these genetic resources could not be fully characterized and were subsequently lost. Others, like the Watanabe heritable hyperlipidemic (WHHL) rabbit and its advanced strains, continued to be used in the subsequent decades in many laboratories and since their development they have been largely analysed as models of relevant human pathophysiological mechanisms (Shiomi, 2020).

8.4 Quantitative Inheritance

Variability of most of the morphological and physiological characteristics in rabbits as well as in all animals are under the control of many genes and the inheritance of these traits can be interpreted using the concepts of quantitative genetics. However, major genes affecting quantitative traits might exist, as demonstrated by QTL and genome-wide association studies carried out in several livestock species, and, more recently, also in rabbits for economically relevant traits (Sternstein *et al.*, 2015; Sosa-Madrid *et al.*, 2020a, 2020b). Other studies applying genomic approaches are expected in the future to better interpret the genetic variability observed for

Table 8.1. List of loci, related defects and morphological traits caused or affected by naturally occurring mutations in the European rabbit with monogenic, oligogenic or polygenic modes of inheritance or with potential major loci affecting the traits (loci affecting coat colour and fibre traits are described in Chapters 6 and 7, respectively).

Loci, defects or traits[1]	OMIA[2]	Genes[3]	Mode of inheritance	Alleles[1]	Descriptions/notes	References
Affecting or involving the eyes:						
Buphthalmia (congenital or infantile glaucoma, hydrophtalmia)[4,5]	000411	–	Autosomal, recessive	Normal (*Bu*) - Buphthalmus or hydropthalmus (*bu*)	Homozygous recessive rabbits developed the defect at 5–6 weeks. The defect has been considered semi-lethal. Several other studies described similar defects probably due to different loci. Buphthalmia and glaucoma have not been clearly distinguished in the rabbit. Rabbits with buphthalmia had also a defect in taurine transport in the ciliary epithelium (Harris *et al.*, 1983).	Vogt (1919); Nachtsheim (1937a); Hanna *et al.* (1962); Fox *et al.* (1969a); Sheppard *et al.* (1971); Holve *et al.* (2011)
Cataract (*Cat* or *Ca*)[4,5]	000168	–	Autosomal, recessive or dominant	First allele series: normal (*Cat-1*)/cataract of lens (*cat-1*), recessive. Second allele series: cataract of lens (*Cat-2*) / normal (*cat-2*), recessive	Two series of juvenile cataract alleles, probably due to different loci have been described. Peng *et al.* (2015) described cataracts in ageing inbred rabbits probably due to other loci.	Nachtsheim (1937b); Nachtsheim and Gürich (1939); Ehling (1957); Nachtsheim (1958); Munger *et al.* (2002)
Coloboma Retinae	–	–	Multiple potential conditions	Some conditions might be due to a single dominant gene or by the segregation of two recessive genes	Several independent cases have been reported. The precise information on the type of inheritance is not known.	Davis and Smith (1930); Grüneberg (1947)
Cyclopia (Otocephaly)[4,5]	000249	–	Autosomal, recessive	Normal (*Cy*) / cyclopian monster (*cy*), recessive	It might be an effect of high inbreeding level.	Menschow (1934); Nachtsheim (1958)

Entropion[5]	000337	–	Unknown	Not described	Conjunctivitis and blepharospasm were observed in a litter of four 2-week-old New Zealand white rabbits. Corneal opacity and neovascularization of the cornea and entropion of the upper eyelids were observed when the rabbits were examined 10 weeks later.	J. G. Fox *et al.* (1979)
Heterochromia iridis	–	–	Unknown	Not described	Reported together with other phenotypic traits or defects.	Punnett (1920); Drews and Picó (1971)
Keratitis associated with French Rex mutation[5]	–	–	Autosomal, recessive	Normal (R_1) / showing keratitis (r_1)	Rabbits homozygous with the pelage mutated allele known as French Rex also have deformed eyelashes. The irritation of the cornea caused by the abnormal hairs has been reported to cause keratitis.	Létard (1929); Nachtsheim (1934b)
Red eye[4,5]	002008	–	Autosomal, recessive	Normal (*Re*) / depigmented red eye (*re*)	The defect was identified in Chinchilla rabbits. Animals had prominent red pupils associated iris depigmentation (from light blue to grey-brown). Affected animals were more susceptible to infections.	Magnussen (1952, 1954)
Affecting the neuromuscular system:						
Acrobat[4,5]	001999	*RORB*	Autosomal, recessive	Normal (*Ak*) / walks on forelegs (*ak* or S^{am})	It is a nervous behaviour affecting the gait. The animals walk normally but the start to proceed only on their forelegs. It is associated with some eye defects. A splice-site mutation in an evolutionary conserved site of *RORB* results in several aberrant transcript isoforms, leading to a drastic reduction of RORB-positive neurons in the spinal cord, as well as defects in differentiation of the DMRT3-expressing population of interneurons.	Létard (1935, 1943); Audigier and Renous (2002); Carneiro *et al.* (2021)

Continued

Table 8.1. Continued.

Loci, defects or traits[1]	OMIA[2]	Genes[3]	Mode of inheritance	Alleles[1]	Descriptions/notes	References
Ataxia[4,5]	000077	–	Autosomal, recessive	Normal (*Ax*) / loss of co-ordination (*ax*), recessive and sub-lethal	This hereditary nervous disorder was described in the strain AX of the Jackson Laboratory. The animals are born apparently normal but develop the defect rapidly (on average at ~70 days of age).	Sawin *et al.* (1942); Tourtellotte *et al.* (1966); Robinson (1970); O'Leary *et al.* (1972, 1974)
Audiogenic seizures (epilepsy)[4,5]	000344 000093	–	Autosomal, recessive	Normal (*Ep*) / epileptic-like seizures (*ep*)	Animals which are subject to epileptic-like seizures occurred occasionally in the Vienna White breed and in the derived Blue-eyed White Beveren strain. The locus might be associated with the Vienna white locus (or could be derived from another allele of this series).	Nachtsheim (1937a, 1939b); Antonitis *et al.* (1954); Sawin (1955); Nellhaus (1958, 1965); Hohenboken and Nellhaus (1970)
Cannabinoid induced behavioural convulsions (Δ^9-tetrahydrocannibinol susceptibility)[4,5]	000267	–	Autosomal, recessive (with full penetrance)	Normal (*Thc*) / seizure causing allele (*thc*)	It was observed in an inbred population of New Zealand White rabbits (Uaz:NZW-thc strain). Seizures were induced by intravenous administration of low doses of Δ^9-tetrahydrocannabinol. Rabbits developed tolerance during long-term administration. It was proposed as model for the study of marijuana psychoactivity.	Martin and Consroe (1976); Fish *et al.* (1981, 1983); Consroe and Fish (1981); Consroe *et al.* (1982); Fish and Consroe (1983)
Hydrocephalus (and cleft palate)[4,5]	000487 000197	–	Apparently autosomal, recessive (incomplete penetrance)	Normal (*Hy*) / high cranial vault, soft and easily depressed, excess of fluid in cranial ventricles (*hy*)	It can occur with several eye defects and is semi-lethal. The defect was first described in an inbred colony of the Henry Phipps Institute. Hydrocephalus and cleft palate seemed determined by the same defective allele. Cases of hydrocephalus were described to occur in inbred rabbit stocks. Environmental factors and incomplete penetrance might affect the occurrence of the defects. It could also occur in several populations.	Da Rosa (1946); Robertson *et al.* (1966); Lukefahr (1982)

Lethal muscle contracture (muscle contracture)[4,5]	000674	–	Autosomal, recessive	Normal (*Mc*) / contracture and atrophy of muscles (*mc*)	Rabbits were stillborn or died soon after birth because they could not ambulate and nurse normally. All legs were held in rigid position. There was atrophy of the musculature. Hind legs were usually less affected. The expression of the anomaly was variable. The affected rabbits had also other malformations (cleft palate and hydrocephaly).	Nachtsheim (1939a); Sawin (1955)
Neuroaxonal dystrophy	000715	–	Autosomal, recessive	Not described	The defect was observed in a C6-deficient rabbit cohort.	Giannini *et al.* (1992)
Paralytic tremor (tremor X linked)[4,5]	000770	*PLP1*	Sex-linked, recessive	Normal (*Pt*) / Parkinson's trembling (*pt*)	The defect was first discovered in Chinchilla rabbits. It seems different from the Tremor defect described by Nachtsheim (1934a). Affected rabbits were recognized during the first week of life by the presence of a coarse tremor which was similar to the involuntary movements seen in Parkinson's disease of humans. The clinical course was variable. The morphological changes included neuroaxonal degeneration. The defect was due to a point mutation in exon 2 of the *PLP1* gene, that results in the substitution of histidine at amino acid 36 by a glutamine, at the end of the first potential transmembrane domain of the protein. This mutation caused the reduction of PLP1 protein and gene expression levels.	Osetowska and Wisniewski (1966); Osetowska (1967); Tosic *et al.* (1993, 1994); Sypecka and Domańska-Janik (2005)
Spastic spinal paralysis (paresis posterior)	000011	–	Autosomal recessive	Normal (*Sp*) / spastic paralysis of the hind legs (*sp*)	The defect was noticed when the animals left the nest.	Nachtsheim (1937b)

Continued

Table 8.1. Continued.

Loci, defects or traits[1]	OMIA[2]	Genes[3]	Mode of inheritance	Alleles[1]	Descriptions/notes	References
Syringomyelia[4,5]	000965	–	Multifactorial, supposed to be autosomal and recessive	Normal (*Sy*) / asymmetrical spastic paralysis (*sy*)	Paralysis of the hind limbs was more frequent than that of the forelegs. The development of the paralysis seemed variable.	Ostertag (1930a, b; 1934); Nachtsheim (1931)
Tremor (shaking palsy)[4,5]	–	–	Autosomal, recessive	Normal (*Tr*) / shaking palsy (*tr*)	It was described in a German breed. The symptoms emerged at 10–14 days of age and were characterized by fine tremors involving the entire body. The tremors became more evident with the passage of time. They were exaggerated in the presence of sudden noises. Flaccid paralysis proceeded with the age from the hind legs to the front legs, followed by complete paralysis and death.	Nachtsheim (1934a, 1958);
Waltzing (circling)[5]	–	–	Unknown	Not described	This disorder was described twice. The first description included just one rabbit in a litter. The second description was in another litter obtained by crossing a New Zealand White buck and a White Beveren doe. Five out of eight obtained rabbits exhibited circling behaviour. Genetic transmission was supposed but could not be tested.	Cole and Steele (1922); Cogan (1943)
Affecting or involving the skin:						
Ehlers-Danlos syndrome (cutaneous asthenia; hyperelastosis cutis)	000327	–	Heterogeneous: dominant or recessive	Not described	Different loci might be involved in determining different types of this syndrome.	Harvey *et al.* (1990); Brown *et al.* (1993); Sinke *et al.* (1997)
Furless[4]	002002	–	Autosomal, recessive	Normal (*F*) / Furless (*f*)	The mutants were described in an inbred rabbit population. This locus is linked to the brachydactyly locus (Castle and Sawin, 1941). It is different from the naked locus.	Castle (1933a); Drapeau (1933)

Furnishing in Angoras (ears)	002001	–	Autosomal, dominant (or partially dominant)	Presence of tufts of wool on the tips of the ears (furnishing) (*Fu* allele) / absence of ear tufts (plain), (*fu* allele)	The trait was observed in Angora rabbits. From breeding records of Angora lines, three groups of rabbits were observed: (i) furnished (putative homozygous *Fu*/*Fu* or heterozygous *Fu*/*fu*); (ii) trace, where only a vestige of ear tuft was present (putative heterozygous *Fu*/*fu*, or *Fu*/*Fu* with modifying factors affecting the expression of the trait); (iii) plain (putative homozygous *fu*/*fu*). High wool yield seems associated with the presence of ear tufts even if conclusive confirmation was not reported.	Pickard (1930); Tegtmeyer (1949); Robinson (1958)
Naked[4]	000700	–	Autosomal, recessive	Normal (*N*) / naked (*n*)	It could also be indicated as Hairless. Similar phenotypes were subsequently described in different rabbit populations (Boucher *et al.*, 1996; Jackson *et al.*, 2006; Rogers *et al.*, 2006).	Kislovsky (1928); David (1932)
Pelt loss[4]	002003	–	Autosomal, recessive	Normal (*Ps-1* and *Ps-2*) / Absence of wool hairs, less pronounced than furless (*ps-1* and *ps-2*)	Two different allele series have been postulated. The allele series was also identified with the symbol *Pl*. Allele *ps-2* might determine absence of underwool hair with a coat thicker than in *ps-1*. It is not completely clear the differences between the two putative loci and the phenotypic difference with the furless locus.	Nachtsheim (1937b, 1954, 1958)
Wire hair[4]	–	–	Autosomal, partially dominant	Absence of wool hair (*Wh*) / Normal (*wh*)	At birth, *Wh*/*Wh* animals had a relatively large proportion of guard to wool hair.	Sawin (1955)

Continued

Table 8.1. Continued.

Loci, defects or traits[1]	OMIA[2]	Genes[3]	Mode of inheritance	Alleles[1]	Descriptions/notes	References
Wuzzy[4]	001078	–	Recessive	Normal (*Wu*) / hair sticky and matted (*wu*)	Gross matted clumping of the hair, ulceration of the eye, and minimal denudation were secondary effects of structural changes in medula, cortex, and particularly of the cuticular scales. Ultimate separation of these scales from the surface of the hair, enhanced by abnormal secretion from the skin, led to entanglement of the hair, irritation of the eye and skin, and secondarily to localized denudation. It was identified in a New Zealand White colony.	Sawin (1955); Crary and Sawin (1959)
Affecting or involving the teeth, the skeleton or related systems:						
Absent incisors (absence of second incisors)[4,5]	–	–	Autosomal, dominant or partially dominant	Absence of secondary incisors (I^2) / presence of secondary incisors (i^2)	Normally, the rabbit has two pairs of incisor teeth in the upper jaw. The second pair of milk teeth erupts normally but either one or both of these may not be succeeded by the permanent teeth. The locus has also been indicated as *Ia*.	Nachtsheim (1936b, 1938, 1958)
Achondroplasia-1[4,5]	001996	–	Autosomal, recessive (lethal)	Normal (*Ac-1*) / Achondroplasia lethal (*ac-1*)	The defect determined a dwarfism characterized by a relative shortening of the extremities with several other secondary defects. Animals that showed these defects were either stillborn or died within a few hours of birth.	Brown and Pearce (1945); Pearce and Brown (1945a, 1945b); Crary and Sawin (1963)
Achondroplasia-2 (Dachs)[4,5]	001997	–	Autosomal, recessive	Normal (*Ac-2*) / Achondroplasia viable (*ac-2*)	Animals were fully viable and appeared almost normal at birth but, when adult, had the characteristic short legs and other anatomical modifications. Nomenclature of the locus changed over publications: the mutated allele was indicated *da* by Crary and Sawin (1952) or *dachs* by Sawin *et al.* (1959a, 1959b).	Crary and Sawin (1952); Sawin (1955); Crary *et al.* (1958); Sawin *et al.* (1959a, 1959b, 1962); Lamb and Sawin (1962, 1963); Crary (1964); Sawin and Trask (1965); Sawin and Hamlet (1970)

Brachydactily (Brachydactylia)[4,5]	000146	–	Autosomal, recessive	Normal (*Br*) / absence of nails, digits, limbs (*br*)	Feet abnormalities can be associated with defects of the ears or other anatomical abnormalities. The defect was also associated with abnormalities of haematopoietic tissue (particularly of foetal liver), including a rarefaction of erythropoietic tissue and a defect of the megakaryocytic series resulting in erythrocyte macrocytosis and thrombopenia. The defect showed variable levels of expression but with almost complete penetrance.	Greene (1935); Greene and Saxton (1939); Inman (1941); Jost *et al.* (1969); Petter *et al.* (1971, 1973, 1977); Bourbon (1976); Ehrensperger *et al.* (1981); Boucher-Ehrensperger and Petter (1984)
Brachygnathia or mandibular prognathism (hypognatia, malocclusion, walrus teeth, buck teeth)[4,5]	000149	–	Autosomal, recessive	Normal (*Bg* or *Mp*) / abnormal development of the upper jaw (*bg* or *mp*)	As a consequence of the retarded development of the upper jaw bones, the incisors of the lower jaw and those of the upper jaw are incorrectly synchronized and are not worn down. Thus, the incisors continue growing and ultimately appear as curved tusks. It was also described with several synonyms.	Nachtsheim (1937b); Robinson (1958); Weisbroth and Ehrman (1967); Fox and Crary (1971c); Huang *et al.* (1981); Korn *et al.* (2016); Harvey *et al.* (2019)
Condrodystrophy[4,5]	000189	–	Autosomal, recessive	Normal (*Cd*) / disproportionate dwarf, lethal (*cd*)	The defect was observed in the III/J strain of rabbits at the Jackson Laboratory. The mutants were viable prenatally but did not survive after birth. The defect is different from dachs and achondroplasia. It is similar to the metatropic dwarf described in humans.	Fox and Crary (1971a, 1975); Webber *et al.* (1981)
Condrodystrophy with femoral luxation	002004	–	Unknown	–	It was described in a sporadic case.	Jepsen and Kinkler (1977)
Congenital luxation (Luxate Femor)[4,5]	–	–	Autosomal, recessive	Normal (*Lu or Lx*) / luxation of the hip (*lu or lx*)	It is not known if it is separated from other similar defects like splay legs.	Da Rosa (1945)

Continued

Table 8.1. Continued.

Loci, defects or traits[1]	OMIA[2]	Genes[3]	Mode of inheritance	Alleles[1]	Descriptions/notes	References
Craniosynostosis	001224	*FGFBP1*, *ITGA3*	Autosomal, dominant with incomplete penetrance and variable expression	Two main loci associated with different factors determining the disease (with alternative alleles)	SNPs achieving a genome-wide significance of $p = 5x10^{-8}$ were identified on chromosome 2 in association with disease occurrence and on chromosomes 14 and 19 in association with disease onset. Genotyping identified a coding variant in fibroblast growth factor binding protein 1 (*FGFBP1*) on chromosome 2 and a non-coding variant upstream of integrin alpha 3 (*ITGA3*) on chromosome 19 that associated with disease occurrence and onset, respectively.	Mooney *et al.* (1994, 1996, 1998a, 1998b, 1998c); Gilbert *et al.* (2018)
Curvature of distal foreleg[5]	001212	–	Autosomal, recessive	Normal (*Fc*) / foreleg curvature (*fc*)	The bowing abnormality occurred only in certain families of purebred Beveren, Belgian, French Silver and Dutch rabbits. The defect was characterized by an inwardly directed curvature of the distal segment of both forelegs. The condition was detected at 2–3 weeks of age, developed rapidly, and reached its final and permanent stage at 2–3 months of age.	Pearce (1960a, 1960b); Pulker *et al.* (2011)
Droopy-ear (abnormal ear carriage)[5]	–		Probably due to a single recessive gene with incomplete penetrance, or to multiple factors	Not described	It was described in inbred stocks at the Jackson Laboratory. The defect was characterized by abnormal ear carriage, i.e., the ears were held downward toward the front and side of the head.	Chai and Clark (1967)

Dwarf[4,5]	000299	*HMGA2*	Autosomal, recessive (or partial recessive)	A few putative series of alleles have been described: Normal (*Dw*) and proportionated dwarf-reduced size (*dw*); Normal (*Nan*) and proportionated dwarf – nanosomia (*nan*); Normal (*Zw*) and proportionated dwarf (*zw*).	It is not completely clear if these series described by different authors correspond to the same locus. Homozygotes (*dw/dw*) are smaller than litter mates, exhibit a swollen head, tiny ears and die within a few days after birth. These animals are called peanuts. Heterozygous (*Dw/dw*) reach about two-thirds the size of the homozygous normal (*Dw/Dw*) rabbits. The *dw* allele is caused by a ~12.1 kb deletion overlapping the promoter region and first three exons of the *HMGA2* gene.	Greene *et al.* (1934); Nachtsheim (1937a); Greene (1940); Castle and Sawin (1941); Latimer and Sawin (1955a, b, c; 1963); Carneiro *et al.* (2017).
Hypoplasia pelvis (spastic paralysis)[4,5]	–	–	Autosomal, recessive	Normal (*Hyp*) / Hypoplasia of ischium (*hyp*)	It was discovered in English Checkered Giant rabbits. The rabbits could not move the hind legs. The lesions involved the pelvis and the femurs.	Nachtsheim (1936a, 1937b, 1940, 1958)
Macrostomus[4,5]	000618	–	Autosomal, recessive (with incomplete penetrance)	Normal (*Mst*) / Defective (*mst*)	Rabbits had deformities of the zygomatic (malar, jugal) bone and related zygomatic processes of adjacent bones. It was proposed as a model for human Treacher Collins syndrome. The defect was described in inbred rabbit strains of the Jackson Laboratory (WH/J and III/DwJ).	Fox and Crary (1979)
Narrow axis[4,5]	001568	–	Autosomal, recessive	Normal (*Nx*) / Narrow axis (*nx*)	The defect causes a marked narrowing of the second cervical vertebra. This condition was first recognizable on X rays at 32–33 days gestation. It was observed in strain X/J rabbits of the Jackson Laboratory.	Crary and Fox (1983)

Continued

Table 8.1. Continued.

Loci, defects or traits[1]	OMIA[2]	Genes[3]	Mode of inheritance	Alleles[1]	Descriptions/notes	References
Osteopetrosis[4,5]	000755	–	Autosomal, recessive	Normal (*Os* or *O*) / abnormal tooth and bone development (*os* or *o*)	The defect is determined by generalized skeletal sclerosis due to reduction in bone resorption and showing abnormal bone and tooth development. Mutants were smaller than normal littermates and died by the fourth or fifth postnatal week. The defect was first described in a Dutch breeding stock and then maintained at the Jackson Laboratory as OS/J strain.	Pearce (1948, 1950); Pearce and Brown (1948); Marks *et al.* (1986); Popoff *et al.* (1989); Lenhard *et al.* (1990); Popoff and Marks (1990, 1991)
Oxycephaly[5]	–	–	Not well characterized (probably recessive)	Not described	It was described as a defect of the growth of the skull bones. It is not known if it was determined by different genetic factors than those affecting other skull disorders (e.g. craniosystosis). Other deviants of the shape of the calvarium were also described in rabbits (trigonocephaly, plagiocephaly and scaphocephaly).	Greene (1932, 1933)
Scoliosis[5]	–	–	Polygenic or oligogenic (two recessive genes)	Not described	Studies in inbred lines confirmed the genetic transmission of the trait even if the results were not conclusive. The character appeared to be sporadic in some cases.	Sawin and Crary (1955, 1962)
Spina bifida[4,5]	000933	–	Autosomal, recessive	Normal (*Sb*) / Spina bifida occulta totalis (*sb*)	Spina bifida has occurred sporadically in the rabbit for many years either as a bifid lumbar spine or as a small meningocele in either cranial or lumbosacral region. The described defect occurred in the AC strain at the Jackson Laboratory.	Crary *et al.* (1966)

Splay legs (hip dysplasia)[5]	000473	–	Complex mode of inheritance	–	It is a defect commonly applied to animals that do not have the ability to adduct one or all legs (usually the rear legs) and have a classical standing position. It is not known if it could be a separate genetic disease from syringomyelia, hypoplasia pelvis, femoral luxation, hip dysplasia and distalis foreleg curvature. It is probably due to imbalanced development of the neural, muscular and skeletal systems of the limb. It is not clear if it could be derived by a recessive gene with reduced expressivity or by more genes and environmental factors.	Innes and O'Steen (1957); Innes (1959); Arendar and Milch (1966); Joosten *et al.* (1981); Pałyga *et al.* (1985)
Supernumerary incisors[4,5]	–		Supposed autosomal, recessive with incomplete penetrance.	Normal (*In* or *Isup*) / presence of extra second incisors (*in* or *isup*)	The defect produces a row of four secondary incisor teeth in the upper jaw.	Nachtsheim (1936b, 1937a, 1958); Robinson (1958)
Syndactyly	002067	–	Unknown	Not described	The defect was also determined by pharmacological treatments (McClain and Langhoff, 1980; Ujhazy *et al.*, 1993).	Gallego and Avedillo (2016)
Vertebral number	–	–	Polygenic	–	Variability in both thoracic and lumbar vertebral number has been reported in several populations.	Sawin (1937, 1945); Sawin and Hull (1946); Greenaway *et al.* (2001); Proks *et al.* (2018)

Continued

Table 8.1. Continued.

Loci, defects or traits[1]	OMIA[2]	Genes[3]	Mode of inheritance	Alleles[1]	Descriptions/notes	References
Affecting or involving other morphological and physiological traits:						
Acromegaly[5]	–	–	Unknown	Not described	It was first described in inbred Dutch rabbits. Acromegalic rabbits had the skin of the neck, the shoulders and the chin that became thickened and reddened at 1–2 weeks of age. The defect of the skin subsequently spread to the entire body. Diseased animals were initially larger than normal animals but then they lost weight and died. Skeletal overgrowth occurred in some animals.	Hu and Greene (1935)
Adrenal hyperplasia[4,5]	000017	*CYP11A1*	Autosomal, recessive (fully penetrant)	Normal (*Ah*) / Hyperplasia and adrenal, lethal (*ah*)	The mutants were maintained at the Jackson Laboratory, strain III VO/*ah*J. Homozygous *ah*/*ah* rabbits were viable prenatally, recognized by gross examination of the adrenal at least as early as 19 days of gestation, but died soon after birth. There was an external feminizing effect of the defective allele (the model was similar to the feminizing congenital lipoid adrenal hyperplasia reported in man). The defect is caused by a gross deletion (not completely characterized yet) of the gene encoding cytochrome P-450 cholesterol side-chain cleavage enzyme, resulting in complete elimination of the adrenal gene expression. The gene was previously named *P450scc* or *P45011A*.	Fox and Crary (1972, 1978); Pang *et al.* (1992); Yang *et al.* (1993); Iwamoto *et al.* (1994)

Aortic arteriosclerosis[5]	–	–	Polygenic	Not described	Aortic arteriosclerosis was independently described by several authors in many different rabbit lines and breeds with differences in the incidence of the disease. Lesions might not be usually comparable to those observed in WHHL rabbits.	Israel (1881); Miles (1907); Levin and Larkin (1910); Kesten (1935); Schenk *et al.* (1966); Gaman *et al.* (1967); Garbarsch *et al.* (1970) Adams *et al.* (1972); Shore and Shore (1976)
Diaphragmatic hernia[4,5]	000459	–	Two autosomal recessive genes, supposed	Normal (*Dh-1, Dh-2*) / Diaphragmatic hernia (*dh-1, dh-2*; both alleles must be present in homozygous condition to cause this lethal defect)	Associated abnormalities included hypoplasia of the ipsilateral lung and an increased incidence of ventricular septal defects. The condition was lethal perinatally due to insufficiency of the respiratory system.	Fox and Crary (1973)
Ear length	–	–	Polygenic	–	It is normally correlated with the size of the animals. Lop rabbits have long ears. Long-eared rabbits have modified skull structure. The trait is associated with increased risk of aural pathologies.	Darwin (1868); Castle and Reed (1936); Dunlop and Hammond (1937); Johnson and Burn (2019), Richardson *et al.* (2019)
Hydrops foetalis[4]	000493	–	Autosomal, dominant	Erythroblastosis fetalis (*Hd*) / Normal (*hd*)	The mutants were identified in a rabbit line (rabbit-hydrops-strain). Immunological aspects were not completely defined.	Nachtsheim (1947); Helmbold (1956)
Hypertension[5]	–	–	Polygenic	Not described	Hypertensive rabbit strains have been developed. Variability in blood pressure was observed among several strains. Heritability of this trait seems moderate.	Alexander *et al.* (1954, 1956); Fox *et al.* (1969b)

Continued

Table 8.1. Continued.

Loci, defects or traits[1]	OMIA[2]	Genes[3]	Mode of inheritance	Alleles[1]	Descriptions/notes	References
Hypogonadia[4,5]	–	–	Autosomal, recessive	Normal (*Hg*) / absence of germ cells (*hg*)	The recessive allele (*hg*) caused sterility in both sexes. Hypogonadics (*hg/hg*) may have either a completely abnormal or a 'mosaic' phenotype. 'Mosaics' may vary from less than 1% to almost 100% abnormal tissue. Hypogonadics had normal body size but infantile external genitalia. 'Mosaics' could mate and produce young.	Sawin and Crary (1962); Fox and Crary (1971b)
Hypospady (split penis)	–	–	Autosomal, recessive	Normal (*Hy*) / defect of the penis (*hy*)	It was as a defect of the penis with variable degree of expression. The extreme cases appeared similar to hermaphrodites.	Robinson (1958)
Left ostium straight[4,5]	000583	–	Autosomal, recessive (with incomplete penetrance)	Normal (*Los*) / left ostium straight (*los*)	The defect had variable expression and resulted in the left oviduct extending lateral to the kidney with the ostium opening upward in otherwise normal animals. The defect was identified in rabbit strains AX/J and AXBU/J of the Jackson Laboratory.	Crary and Fox (1981)
Lymphosarcoma[4,5]	000615	–	Autosomal, recessive	Normal (*Ls*) / susceptibility to lymphosarcoma (*ls*)	Lymphosarcoma occurred often in the WH strain of rabbits. Affected rabbits usually died between the ages of 5 and 13 months. The neoplastic features involved lymphoreticular organs and other organs, especially kidneys. It could be determined by the same locus causing Hemolytic anemia.	Fox *et al.* (1970); Fox, and Meier (1976); Fox *et al.* (1976)

Megacolon (associated with the *English spotting* locus)	000629	*KIT*	Incompletely dominant allele for spotting but recessive for megacolon (*En*) / Self coloured and normal (*en*).	Presence of the megacolon (*En*) / normal (*en*)	Megacolon is probably determined by a regulatory mutation affecting the expression of the *KIT* gene. The *En* allele might have incomplete penetrance for the megacolon defect. The *En/En* rabbits have neuro-interstitial cells of Cajal changes reminiscent of the human non-aganglionic megacolon.	Gerlitz *et al.* (1993); Bödeker *et al.* (1995); Wieberneit and Wegner (1995); Fontanesi *et al.* (2014)
Pelger anomaly[4,5]	000783	–	Autosomal, dominant (or partial dominant)	Chondrodystrophic dwarf, primary effect on nuclei of leukocytes and lethal (*Pg* or *P*) / Normal (*pg* or *p*)	The condition is referred to as either Pelger or Pelger-Huët anomaly. Homozygotes *Pg/Pg* show extreme forms of the defect (nuclear segmentation of the leukocytes completely suppressed and the skeletal system with severe deformities) that results in neonatal death in almost all cases. These animals are called super-Pelgers. Heterozygous *Pg/pg* are indicated as Pelgers. They could not be distinguished from normal except by means of blood tests (nuclei of neutrophils have a reduced number of lobes: one, two or rarely three). Viability of these rabbits is not seriously affected.	Undritz (1939, 1943); Nachtsheim (1950); Holtz *et al.* (1977)
Polycystic kidney disease	000807	–	Unknown	Not described	Clinicopathologic characterization in New Zealand White rabbits revealed similarities to both autosomal-dominant and autosomal-recessive polycystic kidney diseases of humans.	Maurer *et al.* (2004)

Continued

Table 8.1. Continued.

Loci, defects or traits[1]	OMIA[2]	Genes[3]	Mode of inheritance	Alleles[1]	Descriptions/notes	References
Premature senescence (premature aging syndrome)[5]	001213	–	Unknown	Not described	The defect was first reported in a purebred Belgian hare line. The senile appearance was evident in rabbits of one or two years of age which resembled much older rabbits in such respects as an irregular unkempt coat, dry scurfy or scaly skin areas, and an increasingly poor nutritional state. A similar defect was described by Sui *et al.* (2019) in targeted *LMNA*-mutated rabbits.	Pearce and Brown (1960a, 1960b)
Prolapsus vaginae	–	–	Probably due to a recessive allele at one locus with incomplete penetrance or more than one locus	Not described	Rabbits showing the defect were identified in an inbred line of the Jackson Laboratory. The prolapses were primarily caused by an over-expansion of existing blood vessels in the vestibular wall which occurs during a period of increasing sexual activity of the rabbits.	Van Herck *et al.* (1989)
Renal agenesis[4,5]	002007	–	Autosomal, recessive	Normal (*Na*) / absence of one kidney (*na*)	The mutation suppresses the development of either the right or the left kidney at random. The homozygote had reduced general viability whereas the heterozygote appeared normal. The mutated allele was indicated as *ra* by Robinson (1958).	Da Rosa (1943); Robinson (1958)
Renal cysts[4,5]	–	–	Autosomal, recessive (with incomplete penetrance)	Normal (*Rc*) / cortical renal cysts (*rc*)	It was described in the III_{vo} strain of rabbits at the Jackson Laboratory. Cysts were similar to the 'simple cysts' in man. They occurred primarily in adult life and affect the normal structure of the renal cortex, but did not appear to be associated with pathological changes in the normal functioning of the kidney.	Fox *et al.* (1971a)

Retroesophageal right subclavian artery[4,5]	–	–	Autosomal, recessive (with incomplete penetrance)	Normal (*Res*) / retroesophageal right subclavian artery (*res*)	Inheritance of this defect was studied in inbred lines of the Jackson Laboratory. However, its frequency was reported to be around 1% in several other rabbit populations. It is a vascular abnormality classified among the vascular ring malformations.	Sawin and Edmonds (1949); Crary and Fox (1978)
Teat number	–	–	Polygenic	–	High heritability of teat number was estimated in a meat-rabbit population.	Szendrő *et al.* (1991, 1992)
Vena cava morphology[5]	–	–	Polygenic	–	Several patterns of vena cava morphology were identified.	McNutt and Sawin (1943)
Vestigial pulmonary arterial trunk[4,5]	–	–	Two loci, autosomal, recessive with variable expression	Two loci: normal (*Vpt-1, Vpt-2*) / defective expression (*vpt-1, vpt-2*). Both loci should be homozygous for the recessive alleles to obtain the malformation.	The defect was identified in the IIIVO/J rabbits at the Jackson Laboratory as the commonest expression of a malformation involving the heart and great vessels. The animals appeared normal in every other respect. In its extreme forms this mutation was lethal.	Crary and Fox (1975)
Affecting or involving biochemical traits:						
Alkaptonuria	001243	–	Unknown	Not described	It was described in only one doe. The doe died before other information was obtained.	Lewis (1926)
C3 deficiency[4,5]	000155	–	Autosomal, recessive	Normal ($C3^H$) / about 1/10 normal ($C3^L$)	Hereditary hypocomplementemia of the third component of complement (C3) was found in a strain of rabbits in which hereditary C8 alpha-gamma deficiency was also found. The serum C3 concentration, haemolytic C3 activity and total complement haemolytic activity (CH50) of these animals were, respectively, 6–12%, 8–13% and 27–37% of the normal levels. C3 hypocomplemetemic rabbits had reduced survival with sera showing reduced bactericidal activity.	Komatsu *et al.* (1988)

Continued

Table 8.1. Continued.

Loci, defects or traits[1]	OMIA[2]	Genes[3]	Mode of inheritance	Alleles[1]	Descriptions/notes	References
C6 deficiency[4,5]	001624	–	Autosomal, recessive	Normal (*C6*) / Deficient (*C6def*)	C6 deficiency was observed in at least three different rabbit strains. Therefore, they might be caused by different mutations. Delayed xenograft rejection was observed in C6-deficient rabbits. C6-deficient rabbits were more susceptible to ventricular arrhythmias after coronary artery occlusion, independent of reperfusion. C6-deficient rabbits do not have hemolytic, bactericidal and leukocyte chemotactic activity.	Rother and Rother (1961); Biro and Ortega (1966); Rother *et al.* (1966, 1967); Brown *et al.* (1970); Lachmann (1970); Zimmerman *et al.* (1971); Abe *et al.* (1979); O'Regan *et al.* (1979); Goldman *et al.* (1982); Ito *et al.* (1996); Schmiedt *et al.* (1998); Liu *et al.* (2007)
C8 deficiency[4,5]	000156	*C8A*	Autosomal, recessive	Normal ($C8\alpha\text{-}\gamma^{N}$) / Deficient ($C8\alpha\text{-}\gamma^{D}$)	$C8\alpha\text{-}\gamma$-deficiency was characterized by dwarfism (non-pituitary), small litter size, small thymus, a low survival rate, severely reduced serum bactericidal activity and enhanced delayed-type hypersensitivity and normal expression of alpha and gamma genes. Alleles might be determined by abnormal co-translational processing of C8α gene (a mutation of the exon/intron junction of the C8α gene).	Komatsu *et al.* (1985, 1990, 1991); Komatsu (1992)
Carboxylesterase: *Est-1*[4]	–	–	Autosomal, dominant (*Est-1*^{S} is dominant)	*Est-1*^{S} and *Est-1*^{s}.	*Est-1*^{S} gives rise to the three S zones possessing the cocainesterase activity. *Est-1* and *Est-2* are closely linked.	Van Zutphen (1974a, 1974b)

Carboxylesterase: *Est-2* (Atropinesterase deficiency: *As*)[4,5]	002000	–	Autosomal, codominant	*Est-2*^{F}, *Est-2*^{f1}, *Est-2*^{F1}, *Est-2*^{f} [from high atropinesterase activity (*As*) to low atropinesterase activity (*as*)]	The enzyme can also hydrolyze monoacetylmorphine. *Est-2*^{F} gives rise to the three F zones with atropinesterase activity. The presence of the allele can be manifested only with the *Est-1*^{S} allele. *Est-2*^{f} lacks enzyme activity. It is located in the linkage group with the *Extension* locus. Rabbits with low or null atropinesterase activity were used to test different treatments and related response. A correlation of the dietary cholesterol susceptibility with the presence or absence of the esterase zones in the anodal, fast-moving region of the gel, was observed (Van Zutphen and Fox, 1977). Polymorphisms at this locus have been associated with differences in pharmacologic and toxicologic effects. The absence of atropinesterase might not have any harmful effects on the rabbits.	Glick and Glaubach (1941); Sawin and Glick (1943); Margolis and Feigelson (1963, 1964); Stormont and Suzuki (1970); Gasser *et al.* (1973); Ecobichon and Comeau, (1974); Van Zutphen (1974a. 1974b); Van Zutphen, and Den Bieman (1975); Van Zutphen *et al.* (1977, 1981); Forster and Hannafin (1979); Koplovitz, and Stewart (1992, 1994) Stampfli and Quon (1995)
Carboxylesterase: *Est-3*[4]	–	–	Autosomal, two alleles (*Est-1*^{D} s dominant)	*Est-1*^{D} and *Est-1*^{d}	*Est-3*^{D} codes for the D zone. This D esterase reacts with the α-naphthylacetate substrate only in the presence of the F zones.	Van Zutphen (1974a, 1974b)
Diabetes mellitus	–	–	Supposed polygenic	–	It was reported in an inbred New Zealand White colony that was developed from a single affected doe, first observed to be polyuric, polydipsic and hyperglycemic. Diabetes was characterized by fasting hyperglycemia and depressed intravenous glucose stimulated serum insulin levels in some animals and in others it was characterized by glucose disposal with normal or slight elevations in fasting serum glucose levels.	Conaway *et al.* (1980, 1981); Taylor *et al.* (1980); Cannon and Conaway (1981); Boyd *et al.* (1982)

Continued

Table 8.1. Continued.

Loci, defects or traits[1]	OMIA[2]	Genes[3]	Mode of inheritance	Alleles[1]	Descriptions/notes	References
GM2 gangliosidosis	001461	–	Unknown	Not described	The defect was identified in a rabbit with chronic nasal discharge and ataxia and was associated with reduced activity of tissue β-hexosaminidase A in brain and liver tissue.	Rickmeyer *et al.* (2013)
Hemolytic anemia[4,5]	–	–	Autosomal, recessive (lethal)	Normal (*Ha*) / Hemolytic anemia (*ha*)	This condition was rapidly fatal to both sexes with a mean survival time of about 5 months. It could be the same locus of the *Ls* locus conferring susceptibility to lymphosarcoma in strain WH rabbits.	Fox *et al.* (1971b); Fox, and Meier (1976)
Hypercholesterolemia (diet-induced hypercholesterolemia resistance)		–	Polygenic or oligogenic	Not described	Rabbits are more sensitive to a cholesterol diet than other experimental animals. However, there is high variability among rabbit breeds and within breeds on the response to hypercolesterolemic diets. A few authors described the constitution of cholesterol-resistant rabbit line (CRT/mlo) that is resistant to the hypercholesterolemia induced by cholesterol-rich diet. Resistant rabbits showed higher liver expression level of the cholesterol 7α-hydroxylase than normal rabbits.	Pescador (1978); Van Van Zutphen *et al.* (1981); Lindqvist *et al.* (1988); Overturf *et al.* (1989); Loose-Mitchell *et al.* (1991); Poorman *et al.* (1993)
Hypercholesterolemia, Kurosawa-Kusanagi	001567	*LDLR*	Autosomal, recessive	The mutated allele is determined by a 12-base pair deletion in the low-density lipoprotein receptor mRNA	It was described in a rabbit strain (KHC) manifesting inherited and persistent hypercholesterolemia and produced by inbreeding from mutants discovered in 1985. KHC rabbits had abnormally high serum cholesterol, triglyceride, and phospholipid concentrations 8 to 10 times greater than the concentrations in clinically normal Japanese white rabbits. KHC rabbits also had decreased serum high-density lipoprotein cholesterol concentration, about one-third the value in clinically normal rabbits.	Kurosawa *et al.* (1995); Akita *et al.* (2002)

Hyperlipidaemia (Watanabe heritable hyperlipidemia)[4,5]	001160	*LDLR*	Autosomal, recessive	The defect arises from an in-frame deletion of 12 nucleotides that eliminates four amino acids from the cysteine-rich ligand binding domain of the LDL receptor.	The Watanabe heritable hyperlipidemic (WHHL) rabbit, an animal with familial hypercholesterolemia, produces a mutant receptor for plasma low-density lipoprotein (LDL) that is not transported to the cell surface at a normal rate. This model has been extensively studied and a large literature is available, including effects on atherosclerosis and hypercolesterolemia. Only the relevant literature that described this mutation is cited here. Other reviews report additional information (e.g. Shiomi, 2020).	Watanabe (1977, 1980); Watanabe *et al.* (1977); Kita *et al.* (1981); Yamamoto *et al.* (1986)
Lysozyme deficiency	–	–	Autosomal, recessive	Normal (*Ld*) / deficiency of lysozyme (*ld*)	Lysozyme has antibacterial activity. Most of the tissues of lysozyme-deficient rabbits including bone marrow, liver, lung, spleen and bone had levels of lysozyme which were 1% or less of the levels in the corresponding tissues of normal rabbits when measured with the lysoplate method. Levels of lysozyme in the kidney and serum were 6% of controls, but the thymus of the lysozyme-deficient rabbits had normal levels of the enzyme.	Prieur *et al.* (1974); Greenwald *et al.* (1975); Prieur and Camara (1979, 1985); Camara, and Prieur (1980); Camara *et al.* (1990)
N-acetylation rate[4,5]	–	*NAT2*	Autosomal, codominant	Rapid acetylators (*R*) / slow acetylators (*r*)	Slow acetylators have defective drug metabolism. Slow acetylator phenotype in the rabbit results from the deletion of the N-acetyltransferase 2 (*NAT2*) gene. Rabbit liver cytosols of rabbits deleted *NAT2* gene catalyze the N-acetylation of sulfamethazine, benzidine, 4-aminobiphenyl, and the O-acetylation of N-hydroxy-2-amino-1-methyl-6-phenylimidazo[4,5-b]pyridine (N–OH–PhIP) at rates significantly lower than rabbits carrying the *NAT2* gene.	Gordon *et al.* (1973); Weber *et al.* (1976); Glowinski *et al.* (1980); Hein *et al.* (1982); McQueen *et al.* (1982); Blum *et al.* (1989); Sasaki *et al.* (1991); Hein and Doll (2017)

Continued

Table 8.1. Continued.

Loci, defects or traits[1]	OMIA[2]	Genes[3]	Mode of inheritance	Alleles[1]	Descriptions/notes	References
Protease inhibitor (alpha-1-antitrypsin level)[4]	000032	(*LOC1-00328621*)	Autosomal, codominant (sex differences)	High level of alpha-1-antitrypsin (A-I-AT) and trypsin inhibition (*Pi*^{M})/ low level of A-I-AT and trypsin inhibition (*Pi*^{P})	Three electrophoretic phenotypes were distinguished: M, P and MP. M was characterized by a predominant anodal A-1-AT band, and P had a major cathodal component. The MP pattern was observed in heterozygous rabbits. The P pattern was associated with low A-1-AT concentration (about 56% of that in sera with the M phenotype). The levels in MP rabbits were intermediate. No mutations associated with this phenotype have been reported so far. It is not clear if the reduced A-I-AT level could have functional effects.	Kueppers *et al.* (1984)
Red cell esterase: *Es-1*[4,5]	–	–	Autosomal, codominant	*Es-1*^{A}, *Es-1*^{B}, *Es-1*^{C}	It is closely linked to *Es-2*. Rabbits with the *AB* genotype were more resistant to the disease enterities.	Grunder *et al.* (1965, 1968); Bellen *et al.* (1984)
Red cell esterase: *Es-2*[4,5]	–	–	Autosomal, codominant	*Es-2*^{A}, *Es-2*^{B}	It is closely linked to *Es-1*.	Schiff (1970); Schiff and Stormont (1970)
Red cell esterase: *Es-3*[4,5]	–	–	Autosomal, codominant	*Es-3*^{A}, *Es-3*^{B}	It was not assigned to any linkage group.	Schiff (1970); Schiff and Stormont (1970)
Serum paraoxonase activity	–	*PON1*	Autosomal, codominant	Two alleles (PON1A and PON1B) determined by three missense mutations within exon 4 (corresponding to p. P82S, p.K93E and p.S101G)	PON1 was recognized to hydrolyse paraoxon (the active form of the organophosphate insecticide parathion). PON1 is a high-density lipoprotein (HDL)-associated enzyme that hydrolyses aromatic esters, organophosphates and lactones and can protect low-density lipoprotein (LDL) against oxidation. Two PON1 proteins segregated in the investigated rabbit population (PON1A and PON1B) that hydrolysed a variety of substrates at different rates. PON1A was also at least three times more efficient at protecting LDL from oxidation than PON1B.	Zech *et al.* (1999); Watson *et al.* (2001)

Tissue esterase: *Est-4*[4]	–	–	Autosomal, codominant	Absence (*Est-4*^{a}) or presence (*Est-4*^{b}) of two bands of esterase activity	The system was identified after electrophoresis of rabbit tissue homogenates. *Est-4* variants are determined by two bands of esterase activity with intermediate anodal mobility and broad substrate specificity. Expression of *Est-4* was found in liver, small intestine, and spleen but not in kidney, heart and testis.	Van Zutphen *et al.* (1983)
Tissue esterase: *Est-5*[4]	–	–	Autosomal, codominant	*Est-5*^{a} and *Est-5*^{b}	The system was identified after electrophoresis of rabbit tissue homogenates. *Est-5* is coding for cathodally migrating esterases which differ in mobility (Est-5^a and Est-5^b). *Est-5* expression was found only in kidney and testis homogenates. *Est-5* esterases are more active against α-naphthyl acetate than against β-naphthyl acetate and have no activity against α-naphthyl butyrate.	Van Zutphen *et al.* (1983)
Tissue esterase: *Est-6*[4]	–	–	Autosomal, codominant	*Est-6*^{a} and *Est-6*^{b}	*Est-6* is closely linked to the *Est-1,2,4* cluster. Esterase of *Est-6* is found in many organs, particularly in liver and small intestine, but not in erythrocytes and serum. *Est-6* esterase hydrolyses α-naphthyl acetate and butyrate, naphthol AS-D acetate, indoxyl acetate, and butyrate as well as 5-bromoindoxyl acetate, *N*-acetyl-l-alanine-α-naphthyl ester but not 4-methylumbelliferyl acetate and fluorescein diacetate. The enzyme is inhibited by bis-*p*-nitrophenyl phosphate and eserine but not by*p*-chloromercuribenzoate. It is assumed to be homologous with mouse *Es-7*.	Van Zutphen *et al.* (1987)

Continued

Table 8.1. Continued.

Loci, defects or traits[1]	OMIA[2]	Genes[3]	Mode of inheritance	Alleles[1]	Descriptions/notes	References
Von Willebrand disease	001056	–	Autosomal	The presence of a normal allele and a defective allele can be inferred.	The defect was identified in a Flemish Giant Chinchilla colony. Rabbits had moderate to severely reduced levels of factor VIII coagulant activity (FVIII-C). Some have shown prolonged bleeding after venipunctures and gastrointestinal and intramuscular haemorrhages.	Benson and Dodds (1977)
Warfarin resistance	–	–	Unknown	Not described	Some rabbits showed resistance to warfarin and did not become anticoagulated and failed to demonstrate inhibition of peripheral blood monocyte tissue factor expression.	Edwards *et al.* (1986)
Yellow fat[4,5]	001079	*BCO2*	Autosomal, recessive	Normal (*Y*) / Yellow fat (*y*)	The locus was indicated to be linked to the *Albino* and *Brown* loci. The fat of the rabbits is usually white but in some animals it could be yellow. The yellow fat occurs in rabbits that are homozygous *y*/*y* at the *Yellow Fat* locus. These animals accumulate xanthophylls in the fat because they lack the liver enzyme that can metabolize these plant biochemical pigments. The recessive *y* allele is caused by an AAT-deletion in the coding sequence of the *BCO2* gene.	Pease (1928b); Willimott (1928); Castle (1933b); Wilson and Dudley (1946); Jones *et al.* (1965); Strychalski *et al.* (2015, 2016, 2019a, 2019b)

[1] Locus and allele nomenclatures derive from Sawin (1955), Robinson (1958), Lindsey and Fox (1994), Fox (1994), Nicholas and Hobbs (2014) and the cited literature for several traits.
[2] Online Mendelian Inheritance in Animals (OMIA) number. OMIA database was analysed in April 2020.
[3] Gene symbol was reported when gene affecting the trait was identified.
[4] The locus or the trait was also listed by Fox (1994).
[5] The locus or the trait was also listed by Lindsey and Fox (1994).

many traits by the classical genetic works published over the last century.

This paragraph reports information on a few relevant traits for which quantitative inheritance has been shown or inferred.

General body size, body weight, body proportions (or regional body growth), and growth rate are mainly considered under polygenic control and environmental factors. Body size is also used to classify rabbit breeds in three main categories: large breeds, medium-sized breeds and small breeds (see Chapter 2). Examples of extreme breeds in terms of size and weight are the Flemish Giant and the Polish breeds: the adult weight of the first breed is about 3–4 times bigger than that of the second breed. Another category includes the dwarf breeds whose size is largely influenced by the locus (or loci) determining their characteristic dwarfism (considered a monogenic-determined trait; see Table 8.1). The first studies that recognized a polygenic basis of size inheritance in rabbits were published by Castle (1914), MacDowell (1914a, 1914b) and Wright (1918) using data obtained by crossing large and small rabbit strains and recorded in F1, F2 and backcrossed animals. Size inheritance was also approached by Castle (1922) measuring bones (dimension of the skull and length of long leg bones), body weight and ear length. Other early studies were mainly based on weight or growth rate as the prime criterion of size (Punnett and Bailey, 1918; Pease, 1928a; Castle, 1929).

Regional growth was first studied by investigating ear length (Lang, 1910). Among ordinary rabbit breeds, ear length is normally correlated with the size of the animals and this seems to be true also in the purebred lop-eared rabbits (Castle, 1922; Castle and Reed, 1936). However, Lop breeds that have been selected for long and blending ears might have accumulated multiple mutated genes affecting this trait (Castle and Reed, 1936). It seems plausible that, in Lop breeds, one or few major genes might affect ear length and positioning.

Skull shape variability is evident comparing different breeds. In particular, Lop breeds have a typical short skull morphology, resembling brachycephaly, which increases the risk of dental problems (Böhmer and Böhmer, 2017). General skull form was studied by Suchalla (1943) by analyzing F1, F2 and backcrossed rabbits obtained by crossing parental animals of a small breed and of a lop-eared large breed. Heredity of 27 measured skull traits was typically polygenic with complex interactions between the different parts of the skull.

Variation and inheritance of the number of vertebrae in rabbits was investigated in inbred families (Sawin, 1937, 1945; Sawin and Hull, 1946). In these early studies, variability was observed in the number of thoracic vertebrae and ribs (n. 12–13) or considering all presacral vertebrae (n. 26–27), together with variable extent to which the extra element is fully expressed. Family segregation data tended to support a polygenic or an oligogenic mode of inheritance, with the presence of putative major gene(s). However, textbooks reported a certain variability in the number of vertebrae in the general rabbit population and published the following vertebral formula for the European rabbit: cervical (C) 7, thoracic (Th) 12–13, lumbar (L) 6–7, sacral (S) 4 and caudal Cd (15–16) (Capello and Lennox, 2008; Reese and Fehr, 2011; Vella and Donnelly, 2012). More recent studies showed a broader variability of the vertebral number in this lagomorph. Analysing radiographs obtained from a heterogeneous rabbit population, Proks *et al.* (2018) identified seven vertebral formulas with different frequency [C7/Th12/L7/S4 (90.0%), C7/Th12/L6/S4 (3.9%), C7/Th13/L7/S4 (2.9%), C7/Th12/L7/S5 (1.4%), C7/Th12/L8/S4 (1.1%), C7/Th12/L7/S6 (0.35%) and C7/Th11/L7/S4 (0.35%)]. Frequent congenital spinal abnormalities were also reported in rabbit populations (Proks *et al.*, 2018).

Polygenic type of inheritance was also suggested for a few other vertebral-related traits (number and position of ventral spinous processes, number and length of sternebrae and rib ossification; Sawin, 1946; Hull, 1947; Peck and Sawin, 1950). Scoliosis, as investigated in inbred lines, might have been genetically determined based on polygenic or oligogenic actions (Sawin and Crary, 1955). The same could be for several other skeletal deformities (Sawin and Crary, 1962).

The number of teats is another morphological trait that is partially under genetic control. Heritability for this trait was estimated in meat rabbit lines (Szendrő *et al.*, 1991, 1992). Bovo *et al.* (2021) reported the first genome-wide association study for this trait in a meat-rabbit line and identified a few potential major genes affecting the number of teat in this species.

Polygenic variations could explain different vascular-related patterns affecting vena cava (McNutt and Sawin, 1943), the origin of the ilio-lumbar arteries (Sawin and Nace, 1948) and aortic arch patterns (Sawin and Edmonds, 1949), even if the studies that investigated these traits could not be always conclusive and several other observed abnormalities should also be analysed from a genetic point of view (Angell-James, 1974).

Table 8.1 includes some of the mentioned traits for which polygenic inheritance with the potential effects of some major loci/genes could explain the observed genetic variability.

8.5 Single-locus or Oligogenic Determined Traits and Defects

Description of monogenic determined morphological traits in rabbit started after the genetic characterization of loci affecting coat colour and hair structure. Most of the morphological traits that resulted in being determined by single loci emerged in the rabbit populations after inbreeding was used in farming conditions or in laboratory strains. Inbreeding caused the manifestation of many recessive alleles that, in several cases, determined morphological defects. For some traits, the mode of inheritance was not reported in the literature that described the cases and the related phenotypes or this information was not conclusive. Retrospective analysis of the literature suggested that variability for some morphological defects could be better explained supposing the effects of more than one locus. A list of the loci affecting morphological, physiological and biochemical traits, with a short description, the mode of inheritance and the related references is reported in Table 8.1. The list was divided in loci mainly affecting (i) the eyes, (ii) the neuromuscular system, (iii) the skin, (iv) the teeth, the skeleton or related systems, (v) other morphological and physiological traits, and (vi) biochemical traits. Loci described using biochemical variants that were mainly used as genetic markers or that described immunological parameters are discussed in Chapters 4 and 5, respectively. A few loci had pleiotropic effects on several functions, systems and morphological features and are included within the category that mainly defined the described defect.

Among the listed loci and genetically characterized traits, only a few have been investigated at the molecular level and DNA mutations have been reported for 11 of them (Table 8.1). Most of them describe the fine molecular mechanisms that value the rabbit as the animal model for several human diseases.

One of the most important uses of the rabbit as an animal model is for the study of human lipid metabolism and atherosclerosis. Two naturally occurring hypercholesterolemic rabbit models have been frequently used for this purpose (Fan *et al.*, 2015, 2018). One is a cholesterol-fed wild-type rabbit model in which feeding is used to induce severe hypercholesterolemia (Finking and Hanke, 1997) and the other is the Watanabe heritable hyperlipidemic (WHHL) rabbit, which is genetically deficient in low-density lipoprotein (LDL) receptor function (Watanabe, 1977, 1980). Both rabbit models develop aortic and coronary atherosclerosis, but the elevated plasma cholesterol levels are caused by different mechanisms. Wild-type rabbits can be easily induced to develop severe hypercholesterolemia with a cholesterol-rich diet due to the marked increase in hepatically- and intestinally-derived remnant lipoproteins, called β-very low density lipoproteins (VLDL), which are rich in cholesteryl esters. There is quite a large variability on the sensitivity among rabbit breeds and within breeds on fed-induced hypercholesterolemia (e.g. Van Zutphen et al., 1981; Lindqvist et al., 1988). Some rabbits, termed 'low responders', do not exhibit marked hypercholesterolemia (Overturf et al., 1989). Transcriptomic and whole-genome resequencing analyses of hypercholesterolemic fed-induced rabbits have recently been carried out but not on extreme responders (Wang et al., 2016; Loke et al., 2017). The Watanabe heritable hyperlipidemic (WHHL) mutant was the first example of a defective trait described at the DNA level in the rabbit. The defect arises from an in-frame deletion of 12 nucleotides that eliminates four amino acids from the cysteine-rich ligand binding domain of the low-density lipoprotein (LDL) receptor (Yamamoto et al., 1986). This mutation reduces the transportation rate of the plasma LDL to the cell surface and WHHL rabbits are characterized by elevated plasma LDL levels on a standard chow diet, which resembles human familial hypercholesterolemia. Whole-genome resequencing of WHHL rabbits also identified a few other deleterious mutations in genes with potential additional effects on atherosclerosis (*ALDH2*, *ORL1* and *VWF*) that accumulated in the inbred line since the development of this model (Wang *et al.*, 2016).

Other biochemical- and physiological-related defects have been described with the final characterization of the causative mutations. Reduced N-acetylation rate that affects drug metabolism response is caused by the deletion of the N-acetyltransferase 2 (*NAT2*) gene (Blum *et al.*, 1989; Sasaki *et al.*, 1991). Serum paraoxonase activity is affected by three missense mutations in the *PON1* gene (Watson *et al.*, 2001). C8-deficiency, also causing several pleotropic effects at the morphological level, is determined by a mutation affecting *C8A* gene expression (Komatsu *et al.*, 1991; Komatsu, 1992). A three-nucleotide deletion in the *BCO2* gene determines the yellow fat defect caused by increased concentrations of β-carotene and α-tocopherol in the adipose tissue of homozygous mutated rabbits (Strychalski *et al.*, 2015, 2019a).

Mutations identified in a few genes determine morphological defects with several other pleiotropic effects, including metabolic disturbances. Adrenal hyperplasia is determined by a gross deletion (not characterized in detail yet) of the *CYP11A1* gene, resulting in the complete elimination of the adrenal expression of this gene (Yang *et al.*, 1993). A polymorphism in the *KIT* gene is associated with the megacolon and spotted phenotypes (Fontanesi *et al.*, 2014). The dwarfism is determined by the presence of a copy of a mutated *HMGA2* gene lacking the first three exons (Carneiro *et al.*, 2017). A genome-wide association study showed that the occurrence of the craniosynostosis defect is affected by variability in the *FGFBP1* gene and the onset of this disorder is influenced by polymorphisms in the *ITGA3* gene (Gilbert *et al.*, 2018).

Causative mutations were also identified for two neuromuscular defects. A missense mutation in the *PLP1* gene was shown to determine the X linked paralytic tremor disorder (Tosic *et al.*, 1993). The acrobat behavioural disorder affecting the gait of the mutated rabbits is caused by a splice site mutation in the *RORB* gene determining a large reduction of *RORB*-positive neurons (Carneiro *et al.*, 2021). Additional studies have recently reported on the production of several gene-edited rabbits, targeting specific genes to obtain new animal models. These works are not covered in this review.

8.6 Conclusion

This chapter presented a list of inherited morphological, physiological and biochemical traits and parameters and included inherited defects described so far in the European rabbit. Inheritance was demonstrated for a broad range of disorders affecting different systems and physiological mechanisms. The large spectrum of inherited traits and defects demonstrates the usefulness of the rabbit as a valuable animal model for many biological aspects. However, thus far, molecular characterization of most of the described traits has not been reported and, considering the paucity of the available molecular information, the rabbit has a large unexplored potential for further studies in this field. Genomics is expected to facilitate the characterization of most inherited traits and new defects that could appear in the future in many breeding lines and populations with the possibility of developing novel animal models. It would be important to establish close collaborations between breeders, scientists and international networks to collect and maintain new mutants and to provide research infrastructures and expertise for the genomic characterization of the variability present in the rabbit and that could emerge in the future.

References

Abe, T., Koraatsu, M., Oishi, T. and Yamamoto, K.I. (1979) Development and genetic differences of complement activity in rabbits. *Animal Blood Groups and Biochemical Genetics* 10, 19–26.

Adams, W.C., Gaman, E.M. and Feigenbaum, A.S. (1972) Breed differences in the responses of rabbits to atherogenic diets. *Atherosclerosis* 16, 405–411.

Akita, M., Ishii, K., Kuwahara, M. and Tsubone, H. (2002) The daily pattern of cardiovascular parameters in Kurosawa and Kusanagi-Hypercholesterolemic (KHC) rabbits. *Experimental Animals* 51, 353–360.

Alexander, N., Hinshaw, L.B. and Drury, D.R. (1954) Development of a strain of spontaneously hypertensive rabbits. *Proceedings of the Society for Experimental Biology and Medicine* 86, 855–858.

Alexander, N., Hinshaw, L.B. and Drury, D.R. (1956) Further observations on development of a colony of spontaneously hypertensive rabbits. *Proceedings of the Society for Experimental Biology and Medicine* 92, 249–253.

Angell-James, J.E. (1974) Variations in the vasculature of the aortic arch and its major branches in the rabbit. *Acta Anatomica* 87, 283–300.

Antonitis, J.J., Crary, D.D., Sawin, P.B. and Cohen, C. (1954) Sound-induced seizures in rabbits. *Journal of Heredity* 45, 279–284.

Arendar, G.M. and Milch, R.A. (1966) Splay-leg – a recessively inherited form of femoral neck anteversion, femoral shaft torsion and subluxation of the hip in the laboratory lop rabbit: its possible relationship to factors involved in so-called 'Congenital dislocation' of the hip. *Clinical Orthopaedics and Related Research* 44, 221–230.

Audigier, I. and Renous, S. (2002) Les allures du lapin normal peuvent-elles expliquer la marche acrobatique du «lapin sauteur d'Alfort»? *Mammalia* 66, 563–578.

Bellen, H., Van de Weghe, A., Bouquet, Y. and Van Zutphen, L.F.M. (1984) Heterogeneity of Es-1 esterases in the rabbit (*Oryctolagus cuniculus*). *Biochemical Genetics* 22, 853–870.

Benson, R.E. and Dodds, W.J. (1977) Autosomal factor VIII deficiency in rabbits: size variations of rabbit factor VIII. *Thrombosis and Haemostasis* 38, 380.

Biro, C.E. and Ortega, M.L. (1966) Algunas caracteristicas del sexto componente del complemento. *Archivos del Instituto de Cardiología de México* 36, 166–168.

Blum, M., Grant, D.M., Demierre, A. and Meyer, U.A. (1989) N-acetylation pharmacogenetics: a gene deletion causes absence of arylamine N-acetyltransferase in liver of slow acetylator rabbits. *Proceedings of the National Academy of Sciences of the USA* 86, 9554–9557.

Bödeker, D., Türck, O., Lovén, E., Wieberneit, D. and Wegner, W. (1995) Pathophysiological and functional aspects of the megacolon-syndrome of homozygous spotted rabbits. *Journal of Veterinary Medicine Series A* 42, 549–559.

Böhmer, C. and Böhmer, E. (2017) Shape variation in the craniomandibular system and prevalence of dental problems in domestic rabbits: a case study in Evolutionary Veterinary Science. *Veterinary Sciences* 4, 5.

Boucher, S., Thebault, R.G., Plassiart, G., Vrillon, J.L. and de Rochambeau, H. (1996) Phenotypical description of hairless rabbits appeared in three different herds. *Proceedings of the 6th World Rabbit Congress, Toulouse, France* Vol. 1, pp. 333–338.

Boucher-Ehrensperger, M. and Petter, C. (1984) Prevention of thrombocytic defects in the br/br rabbit with folic acid and vitamin B12: analogy with the TAR syndrome in humans. *International Journal for Vitamin and Nutrition Research* 54, 199–203.

Bourbon, J.R. (1976) Dihydrofolate reductase and thymidylate synthetase activities in the liver of rabbit foetuses during the haemapoietic period. *Biochemical Journal* 155, 713–715.

Bovo, S., Schiavo, G., Utzeri, V. J., Ribani, A., Schiavitto, M., Buttazzoni, L., Negrini, R. and Fontanesi, L. (2021) A genome-wide association study for the number of teats in European rabbits (*Oryctolagus cuniculus*) identifies several candidate genes affecting this trait. *Animal Genetics* 52, 237–243.

Boyd, A.E., Bolton, W.E., Conaway, H.H. and Brinkley, B.R. (1982) Identification of microtubules in normal and diabetic rabbit islets grown in monolayer culture. *Metabolism Clinical and Experimental* 31, 154–157.

Brown, D.L., Lachmann, P.J. and Dacie, J.V. (1970) The in vivo behaviour of complement-coated red cells: studies in C6-deficient, C3-depleted and normal rabbits. *Clinical and Experimental Immunology* 7, 401–422.

Brown, P.J., Young, R.D. and Cripps, P.J. (1993) Abnormalities of collagen fibrils in a rabbit with a connective tissue defect similar to Ehlers-Danlos syndrome. *Research in Veterinary Science* 55, 346–350.

Brown, W.H. and Pearce, L. (1945) Hereditary achondroplasia in the rabbit: I. Physical appearance and general features. *Journal of Experimental Medicine* 82, 241–260.

Camara, V.M. and Prieur, D.J. (1980) Lysozyme activity in gastrointestinal and lymphoreticular tissues of rabbits genetically deficient in lysozyme. *Laboratory Investigation: A Journal of Technical Methods and Pathology* 43, 352–357.

Camara, V.M., Harding, J.W. and Prieur, D.J. (1990) Inherited lysozyme deficiency in rabbits: the absence of a primary isozyme of lysozyme as the cause of the condition. *Laboratory Investigation: A Journal of Technical Methods and Pathology* 63, 544–550.

Cannon, D.J. and Conaway, H.H. (1981) Glycosylated haemoglobin levels in a colony of spontaneously diabetic rabbits. *Diabetologia* 20, 242–243.

Capello, V. and Lennox, A.M. (2008) *Clinical Radiology of Exotic Companion Mammals.* Wiley Blackwell, Ames, Iowa.

Carneiro, M., Hu, D., Archer, J., Feng, C., Afonso, S. *et al.* (2017) Dwarfism and altered craniofacial development in rabbits is caused by a 12.1 kb deletion at the HMGA2 locus. *Genetics* 205, 955–965.

Carneiro, M., Vieillard, J., Andrade, P., Boucher, S., Afonso, S. *et al.* (2021) A loss-of-function mutation in RORB disrupts saltatorial locomotion in rabbits. *PLOS Genetics* 17, e1009429.

Castle, W.E. (1905) Heredity of coat characters in guinea pigs and rabbits. *Carnegie Institute of Washington Publication* 23, 1–78.

Castle, W.E. (1907) Colour varieties of the rabbit, and of other rodents; their origin and inheritance. *Science* 26, 287–291.

Castle, W.E. (1914) Nature of size factors as indicated by a study of correlation. *Carnegie Institute of Washington Publication* 196, 51–55.

Castle, W.E. (1922) Genetic studies of rabbits and rats. *Carnegie Institute of Washington Publication* 320, 1–57.

Castle, W.E. (1929) A further study of size inheritance in rabbits, with special reference to the existence of genes for size characters. *Journal of Experimental Zoology* 53, 421–454.

Castle, W.E. (1930) *The Genetics of Domestic Rabbit*. Cambridge Harvard University Press, London.

Castle, W.E. (1933a) The furless rabbit. *Journal of Heredity* 24, 81–86.

Castle, W.E. (1933b) The linkage relations of yellow fat in rabbits. *Proceedings of the National Academy of Sciences of the USA* 19, 947–950.

Castle, W.E. and Reed, S.C. (1936) Studies of inheritance in Lop-eared rabbits. *Genetics* 21, 297–309

Castle, W.E. and Sawin, P.B. (1941) Genetic linkage in the rabbit. *Proceedings of the National Academy of Sciences of the USA* 27, 519–523.

Chai, C.K. (1968) The effect of inbreeding in rabbits: skin transplantation. *Transplantation* 6, 689–693.

Chai, C.K. (1969) Effects of inbreeding in rabbits: inbred lines, discrete characters, breeding performance, and mortality. *Journal of Heredity* 60, 64-70.

Chai, C.K. (1970) Effect of inbreeding in rabbits: skeletal variations and malformations. *Journal of Heredity* 61, 2–8.

Chai, C.K. and Clark, E.M. (1967) Droopy-ear, a genetic character in rabbits. *Journal of Heredity* 58, 149–152.

Cogan, D.G. (1943) The waltzing (circling) phenomenon in rabbits. *Journal of Comparative Psychology* 35, 111–117.

Cole, L.J. and Steele, D.G. (1922) A waltzing rabbit. *Journal of Heredity* 13, 291–294.

Conaway, H.H., Brown, C.J., Sanders, L.L., Cernosek, S.F., Farris, H.E. and Roth, S.I. (1980) Spontaneous diabetes mellitus in the New Zealand white rabbit: history, classification, and genetic analysis. *Journal of Heredity* 71, 179–186.

Conaway, H.H., Faas, F.H., Smith, S.D. and Sanders, L.L. (1981) Spontaneous diabetes mellitus in the New Zealand white rabbit: physiologic characteristics. *Metabolism* 30, 50–56.

Consroe, P. and Fish, B.S. (1981) Rabbit behavioral model of marijuana psychoactivity in humans. *Medical Hypotheses* 7, 1079–1090.

Consroe, P., Martin, A.R. and Fish, B.S. (1982) Use of a potential rabbit model for structure-behavioral activity studies of cannabinoids. *Journal of Medicinal Chemistry* 25, 596–599.

Crary, D.D. (1964) Development of the external ear in the dachs rabbit. *The Anatomical Record* 150, 441–447.

Crary, D.D. and Fox, R.R. (1975) Hereditary vestigial pulmonary arterial trunk and related defects in rabbits. *Journal of Heredity* 66, 50–55.

Crary, D.D. and Fox, R.R. (1978) Retroesophageal right subclavian artery in rabbits. Genetics and pathology. *Journal of Heredity* 69, 19–21.

Crary, D.D. and Fox, R.R. (1980) Frequency of congenital abnormalities and of anatomical variations among JAX rabbits. *Teratology* 21, 113–121.

Crary, D.D. and Fox, R.R. (1981) Left ostium straight: an inherited anomaly of the female genital tract in the rabbit. *Journal of Heredity* 72, 362–363.

Crary, D.D. and Fox, R.R. (1983) Narrow axis: an inherited anomaly of the second cervical vertebra in the rabbit. *Journal of Heredity* 74, 47–50.

Crary, D.D. and Sawin, P.B. (1952) A second recessive achondroplasia in the domestic rabbit. *Journal of Heredity* 43, 255–259.

Crary, D.D. and Sawin, P.B. (1959) Inheritance and hair morphology of the 'wuzzy' mutation in the rabbit. *Journal of Heredity* 50, 31–34.

Crary, D.D. and Sawin, P.B. (1963) Morphogenetic studies of the rabbit. XXXII. Qualitative skeletal variations induced by the AC gene (Achondroplasia). *American Journal of Anatomy* 113, 9–23.

Crary, D.D., Sawin, P.B. and Atkinson, N. (1958) Morphogenetic studies of the rabbit. XXI. The nature of disproportionate dwarfism induced by the Da gene revealed by the early fetal ossification pattern. *American Journal of Anatomy* 103, 69–97.

Crary, D.D., Fox, R.R. and Sawin, P.B. (1966) Spina bifida in the rabbit. *Journal of Heredity* 57, 236–243.

Da Rosa, F.M. (1943) Agenesia de um rim, uma nova mutação no coelho. *Revista de Medicina Veterinaria (Lisboa)* 38, 349–363.

Da Rosa, F.M. (1945) Uma nova mutação, luxação congenita da anca o coelho. *Revista de Medicina Veterinaria (Lisboa)* 40, 1–23.

Da Rosa, F.M. (1946) Hidrocefalia, uma nova mutação no coelho. *Revista de Medicina Veterinaria (Lisboa)* 41, 1–55.

Darwin, C. (1868) *The Variation of Animals and Plants*. Orange Judd & Company, New York.

David, L.T. (1932) External expression and comparative dermal histology of hereditary hairlessness in mammals. *Zeitschrift für Zellforschung und Mikroskopische Anatomie* 14, 616–719.

Davis, F.A. and Smith, H.M. (1930) The role of lens antigens and uveal pigment in the production of hereditary anomalies of the eyes. *Archives of Ophthalmology* 4, 672–690.

Drapeau, E.E. (1933) An anatomical study of the furless condition in rabbits. *Journal of Morphology* 54, 365–388.

Drews, R.C. and Picó, G. (1971) Heterochromia associated with coloboma of the iris. *American Journal of Ophthalmology* 72, 827.

Dunlop, G. and Hammond, J. (1937) The growth and proportions of the rabbit's ear in relation to body weight. *Journal of Genetics* 34, 463–477.

Ecobichon, D.J. and Comeau, A.M. (1974) Genetic polymorphism of plasma carboxylesterases in the rabbit: correlation with pharmacologic and toxicologic effects. *Toxicology and Applied Pharmacology* 27, 28–40.

Edwards, R.L., Schreiber, E. and Brande, W. (1986) The effect of sodium warfarin on rabbit monocyte tissue factor expression. *Thrombosis Research* 42, 125–137.

Ehling, U. (1957) Untersuchungen zur kausalen Genese erblichter Katarakte beim Kaninchen. *Zeitschrift für Menschliche Vererbungs-und-Konstitutionslehre* 34, 77–104.

Ehrensperger, M., Petter, C. and Keuky, L. (1981) Foetal megakaryoblastic abnormality associated with congenital limb amputation in the br/br rabbit. *Development* 62, 109–115.

Fan, J., Kitajima, S., Watanabe, T., Xu, J., Zhang, J. *et al.* (2015) Rabbit models for the study of human atherosclerosis: from pathophysiological mechanisms to translational medicine. *Pharmacology and Therapeutics* 146, 104–119.

Fan, J., Chen, Y., Yan, H., Liu, B., Wang, Y. *et al.* (2018) Genomic and transcriptomic analysis of hypercholesterolemic rabbits: progress and perspectives. *International Journal of Molecular Sciences* 19, E3512.

Finking, G. and Hanke, H. (1997) Nikolaj Nikolajewitsch Anitschkow (1885–1964) established the cholesterol-fed rabbit as a model for atherosclerosis research. *Atherosclerosis* 135, 1–7.

Fish, B.S. and Consroe, P. (1983) The ontogeny of delta-9-tetrahydrocannabinol responsiveness in the rabbit. *Developmental Psychobiology* 16, 147–158.

Fish, B.S., Consroe, P. and Fox, R.R. (1981) Inheritance of Δ^9-tetrahydrocannabinol seizure susceptibility in rabbits. *Journal of Heredity* 72, 215–216.

Fish, B.S., Consroe, P. and Fox, R.R. (1983) Convulsant-anticonvulsant properties of delta-9-tetrahydrocannabinol in rabbits. *Behavior Genetics* 13, 205–211.

Fontanesi, L., Vargiolu, M., Scotti, E., Latorre, R., Faussone Pellegrini, M.S. *et al.* (2014) The *KIT* gene is associated with the *English spotting* coat color locus and congenital megacolon in Checkered Giant rabbits (*Oryctolagus cuniculus*). *PLOS ONE* 9, e93750.

Forster, R.P. and Hannafin, J.A. (1979) Influence of a genetically determined atropinesterase on atropine inhibition of the 'smoke (dive) reflex' in rabbits. *General Pharmacology* 10, 41–46.

Fox, J.G., Shalev, M., Beaucage, C.M. and Smith. M. (1979) Congenital entropion in a litter of rabbits. *Laboratory Animal Science* 29, 509–511.

Fox, R.R. (1974) Taxonomy and genetics. In: Weisbroth, S.H., Flatt, R.E. and Kraus, A.L. (eds) *The Biology of the Laboratory Rabbit*. Academic Press, San Diego, California, pp. 1–22.

Fox, R.R. (1975) The rabbit, *Oryctolagus cuniculus*. In: King, R.C. (ed.) *Handbook of Genetics –Volume 4: Vertebrates of Genetic Interest*. Springer, Boston, Massachusetts, pp. 309–328.

Fox, R.R. (1994) Taxonomy and genetics. In: Manning, P.J., Ringler, D.H. and Newcomer, C.E. (eds) *The Biology of the Laboratory Rabbit*, 2nd edn. Academic Press, San Diego, California, pp. 1–26.
Fox, R.R. and Crary, D.D. (1971a) A new recessive chondrodystrophy in the rabbit. *Teratology* 4, 215–246.
Fox, R.R. and Crary, D.D. (1971b) Hypogonadia in the rabbit: genetic studies and morphology. *Journal of Heredity* 62, 163–170.
Fox, R.R. and Crary, D.D. (1971c) Mandibular prognathism in the rabbit: genetic studies. *Journal of Heredity* 62, 23–27.
Fox, R.R. and Crary, D.D. (1972) A lethal recessive gene for adrenal hyperplasia in the rabbit. *Teratology* 5, 255.
Fox, R.R. and Crary, D.D. (1973) Hereditary diaphragmatic hernia in the rabbit: genetics and pathology. *Journal of Heredity* 64, 333–336.
Fox, R.R. and Crary, D.D. (1975) Hereditary chondrodystrophy in the rabbit: genetics and pathology of a new mutant, a model for metatropic dwarfism. *Journal of Heredity* 66, 271–276.
Fox, R.R. and Crary, D.D. (1978) Genetics and pathology of hereditary adrenal hyperplasia in the rabbit: a model for congenital lipoid adrenal hyperplasia. *Journal of Heredity* 69, 251–254.
Fox, R.R. and Crary, D.D. (1979) Hereditary macrostomus in the rabbit: a model for Treacher Collins syndrome, one form of mandibulofacial dysostosis. *Journal of Heredity* 70, 369–372.
Fox, R.R. and Meier, H. (1976) Identical genetic basis for lymphosarcoma and hemolytic anemia in the rabbit. *Journal of Heredity* 67, 99–102.
Fox, R.R., Crary, D.D., Babino, E.J., Jr. and Sheppard, L.B. (1969a) Buphthalmia in the rabbit: pleiotropic effects of the *bu* gene and a possible explanation of mode of gene action. *Journal of Heredity* 60, 206–212.
Fox, R.R., Schlager, G. and Laird, C.W. (1969b) Blood pressure in thirteen strains of rabbits. *Journal of Heredity* 60, 312–314.
Fox, R.R., Meier, H., Crary, D.D., Myers, D.D., Norberg, R.F. and Laird, C.W. (1970) Lymphosarcoma in the rabbit: genetics and pathology. *Journal of the National Cancer Institute* 45, 719–729.
Fox, R.R., Krinsky, W.L. and Crary, D.D. (1971a) Hereditary cortical renal cysts in the rabbit. *Journal of Heredity* 6, 105–109.
Fox, R.R., Meier, H., Crary, D.D., Norberg, R.E. and Myers, D.D. (1971b) Hemolytic anemia associated with thymoma in the rabbit: genetic studies and pathological findings. *Oncology* 25, 372–382.
Fox, R.R., Norberg, R.F. and Meier, H. (1976) Clinical hematological progression of hereditary lymphosarcoma in rabbits. *Journal of Heredity* 67, 376–380.
Gallego, M. and Avedillo, L. (2016) Case report of bilateral 3-4 metatarsal syndactyly in a pet rabbit. *Case Reports in Veterinary Medicine* 2016, 6957101.
Gaman, E.M., Feigenbaum, A.S. and Schenk, E.A. (1967) Spontaneous aortic lesions in rabbits: Part 3. Incidence and genetic factors. *Journal of Atherosclerosis Research* 7, 131–141.
Garbarsch, C., Matthiessen, M.E., Helin, P. and Lorenzen, I. (1970) Spontaneous aortic arteriosclerosis in rabbits of the Danish country strain. *Atherosclerosis* 12, 291–300.
Gasser, D.L., Silvers, W.K., Reynolds, H.M., Black, G. and Palm, J. (1973) Serum esterase genetics in rats: two new alleles at Es-2, a new esterase regulated by hormonal factors, and linkage of these loci to the Ag-C blood group locus. *Biochemical Genetics* 10, 207–217.
Gerlitz, S., Wessel, G., Wieberneit, D. and Wegner, W. (1993) Zur Problematik der Scheckenzucht bei Kaninchen. 3. Mitteilung: Variabilität des Pigmentierungsgrades, ganglionäre Darmwandversorgung, Beziehung zur Pathogenese–tierzüchterische und tierschützerische Aspekte. *Deutsche tierärztliche Wochenschrift* 100, 237–239.
Giannini, C., Monaco, S., Kirschfink, M., Rother, K.O., Lorbacher de Ruiz, H. *et al.* (1992) Inherited neuroaxonal dystrophy in C6 deficient rabbits. *Journal of Neuropathology and Experimental Neurology* 51, 514–522.
Gilbert, J.R., Losee, J.E., Mooney, M.P., Cray, J.J., Gustafson, J. *et al.* (2018) Genetic associations and phenotypic heterogeneity in the craniosynostotic rabbit. *PLOS ONE* 13, e0204086.
Glick, D. and Glaubach, S. (1941) The occurrence and distribution of atropinesterase, and the specificity of tropinesterases. *Journal of General Physiology* 25, 197–205.
Glowinski, I.B., Weber, W.W., Fysh, J.M., Vaught, J.B. and King, C.M. (1980) Evidence that arylhydroxamic acid N, O-acyltransferase and the genetically polymorphic N-acetyltransferase are properties of the same enzyme in rabbit liver. *Journal of Biological Chemistry* 255, 7883–7890.
Goldman, M.B., Cohen, C., Stronski, K., Bangalore, S. and Goldman, J.N. (1982) Genetic control of C6 polymorphism and C6 deficiency in rabbits. *Journal of Immunology* 128, 43–48.
Gordon, G.R., Shafizadeh, A.G. and Peters, J.H. (1973) Polymorphic acetylation of drugs in rabbits. *Xenobiotica* 3, 133–150.

Greenaway, J.B., Partlow, G.D., Gonsholt, N.L. and Fisher, K.R. (2001) Anatomy of the lumbosacral spinal cord in rabbits. *Journal of the American Animal Hospital Association* 37, 27–34.

Greene, H.S. (1932) Hereditary variations in the skull of the rabbit. *Science* 76, 421–422.

Greene, H.S. (1933) Oxycephaly and allied conditions in man and in the rabbit. *Journal of Experimental Medicine* 57, 967–976.

Greene, H.S. (1935) Hereditary brachydactylia and associated abnormalities in the rabbit. *Science* 405–407.

Greene, H.S. (1940) A dwarf mutation in the rabbit: the constitutional influence on homozygous and heterozygous individuals. *Journal of Experimental Medicine* 71, 839–856.

Greene, H.S. and Saxton, J.A. (1939) Hereditary brachydactylia and allied abnormalities in the rabbit. *Journal of Experimental Medicine* 69, 301–314.

Greene, H.S., Hu, C.K. and Brown, W.H. (1934) A lethal dwarf mutation in the rabbit with stigmata of endocrine abnormality. *Science* 79, 487–488.

Greenwald, R.A., Cantor, J.O., Prieur, D.J. and Young, D.M. (1975) Composition of cartilage from lysozyme-deficient rabbits. *Biochimica et Biophysica Acta* 385, 435–437.

Grunder, A.A., Sartore, G. and Stormont, C. (1965) Genetic variation in red cell esterases of rabbits. *Genetics* 52, 1345–1353.

Grunder, A.A., Rollins, W.C., Stormont, C. and Casady, R.B. (1968) A note on differential mortality rates in young rabbits of esterase phenotypes A, AB, and B. *Animal Science* 10, 221–222.

Grüneberg, H. (1947) *Animal Genetics and Medicine*. Hamish Hamilton, London.

Hanna, B.L., Sawin, P.B. and Sheppard, L.B. (1962) Recessive buphthalmos in the rabbit. *Genetics* 47, 519–529.

Harris, T.M., Nance, C.S., Sheppard, L.B. and Fox, R.R. (1983) Evidence for an hereditary defect in taurine transport in the ciliary epithelium of an inbred strain of rabbits. *Journal of Inherited Metabolic Disease* 6, 163–166.

Harvey, N.D., Oxley, J.A., Miguel-Pacheco, G., Gosling, E.M. and Farnworth, M. (2019) What makes a rabbit cute? Preference for rabbit faces differs according to skull morphology and demographic factors. *Animals* 9, 728.

Harvey, R.G., Brown, P.J., Young, R.D. and Whitbread, T.J. (1990) A connective tissue defect in two rabbits similar to the Ehlers-Danlos syndrome. *The Veterinary Record* 126, 130–132.

Hein, D.W. and Doll, M.A. (2017) Rabbit N-acetyltransferase 2 genotyping method to investigate role of acetylation polymorphism on N-and O-acetylation of aromatic and heterocyclic amine carcinogens. *Archives of Toxicology* 91, 3185–3188.

Hein, D.W., Smolen, T.N., Fox, R.R. and Weber, W.W. (1982) Identification of genetically homozygous rapid and slow acetylators of drugs and environmental carcinogens among established inbred rabbit strains. *Journal of Pharmacology and Experimental Therapeutics* 223, 40–44.

Helmbold, W. (1956) Über Blutkörperchenantigene des Kaninchens in Zusammenhang mit dem Hydropsproblem Die Isolierung von 10 spezifischen Antikorpern asu Isoimmunseren und die Bestimmung des korrespondierenden Antigene. *Blut* 2, 9–31.

Hohenboken, W.D. and Nellhaus, G. (1970) Inheritance of audiogenic seizures in the rabbit. *Journal of Heredity* 61, 107–112.

Holtz, C.R., Pfaff, J. and Gärtner, K. (1977) *Zentralblatt fur Veterinarmedizin. Reihe A. [Studies on the frequency of the Pelger anomaly in the domestic rabbit population]* 24, 409–412.

Holve, D.L., Mundwiler, K.E. and Pritt, S.L. (2011) Incidence of spontaneous ocular lesions in laboratory rabbits. *Comparative Medicine* 61, 436–440.

Hu, C.K. and Greene, H.S.N. (1935) A lethal acromegalic mutation in the rabbit. *Science* 81, 25–26.

Huang, C.M., Mi, M.P. and Vogt, D.W. (1981) Mandibular prognathism in the rabbit: discrimination between single-locus and multifactorial models of inheritance. *Journal of Heredity* 72, 296–298.

Hull, I.B. (1947) Morphogenetic studies in the rabbit. IV. The inheritance of developmental patterns of rib ossification. *Journal of Experimental Zoology* 105, 173–197.

Inman, O.R. (1941) Embryology of hereditary brachydactyly in the rabbit. *The Anatomical Record* 79, 483–505.

Innes, J.R.M. (1959) 'Inherited dysplasia' of the hip joint in dogs and rabbits. *Laboratory Investigation* 8, 1170–1177.

Innes, J.R.M. and O'Steen W.K. (1957) Splayleg in rabbits: an inherited disease analogous to joint dysplasia in children and dogs. *Laboratory Investigation* 6, 171–186.

Israel, O. (1881) Experimentelle Untesuchung über den Zusammenhang zwischen Nierenkrankheiten und Secundären Veränderungen des zirkulations-systems. *Virchows Archiv für pathologische Anatomie und Physiologie und für klinische Medizin* 86, 299–321.

Ito, W., Schäfer, H.J., Bhakdi, S., Klask, R., Hansen, S. *et al.* (1996) Influence of the terminal complement-complex on reperfusion injury, no-reflow and arrhythmias: a comparison between C6-competent and C6-deficient rabbits. *Cardiovascular Research* 32, 294–305.

Iwamoto, K., Yang, X., Rogerson, F.M., Mason, J.I., Artwohl, J. *et al.* (1994) Evidence of a steroidogenic enzyme gene dose effect on adrenal gene expression in hereditary rabbit congenital adrenal hyperplasia. *Pediatric Research* 36, 660–666.

Jackson, R., Rogers, A.D. and Lukefahr, S.D. (2006) Inheritance of the naked gene and associations with postweaning performance and thermotolerance characters in fryer rabbits from an F2 generation. *World Rabbit Science* 14, 147–155.

Jepsen, P.L. and Kinkler, R.J. (1977) Chondrodystrophy in a rabbit. *Veterinary Medicine, Small Animal Clinician* 72, 1374–1375.

Johnson, J.C. and Burn, C.C. (2019) Lop-eared rabbits have more aural and dental problems than erect-eared rabbits: a rescue population study. *Veterinary Record* 85, 758.

Jones, D., Gresham, G.A., Lloyd, H.G. and Howard, A.N. (1965) 'Yellow fat' in the wild rabbit. *Nature* 207, 205–206.

Joosten, H.F., Wirtz, P., Verbeek, H.O. and Hoekstra, A. (1981) Splayleg: a spontaneous limb defect in rabbits: genetics, gross anatomy, and microscopy. *Teratology* 24, 87–104.

Jost, A., Roffi, J. and Courtat, M. (1969) Congenital amputations determined by the BR gene and those induced by adrenalin injection in the rabbit fetus. In: Swinyard, C.A. (ed.) *Limb Development and Deformity: Problems of Evaluation and Rehabilitation*. Thomas, Springfield, pp. 187–199.

Kesten, H.D. (1935) Early incidence of spontaneous medial degeneration ('arteriosclerosis') in the aorta of the rabbit. *Archives of Pathology* 20, 1–8.

Kislovsky, D.A. (1928) Naked – a recessive mutation in the rabbit. *Journal of Heredity* 19, 438–439.

Kita, T., Brown, M.S., Watanabe, Y. and Goldstein, J.L. (1981) Deficiency of low-density lipoprotein receptors in liver and adrenal gland of the WHHL rabbit, an animal model of familial hypercholesterolemia. *Proceedings of the National Academy of Sciences of the USA* 78, 2268–2272.

Komatsu, M. (1992) Molecular biology for genetic deficiencies of complement components in rabbits: C8alpha-gamma deficiency and C3 hypocomplementemia. *Japan Agricultural Research Quarterly* 26, 48–54.

Komatsu, M., Yamamoto, K., Kawashima, T. and Migita, S. (1985) Genetic deficiency of the alpha-gamma-subunit of the eighth complement component in the rabbit. *Journal of Immunology* 134, 2607–2609.

Komatsu, M., Yamamoto, K., Nakano, Y., Nakazawa, M. and Ozawa, A. (1988) Hereditary C3 hypocomplementemia in the rabbit. *Immunology* 64, 363–368.

Komatsu, M., Imaoka, K., Satoh, M. and Mikami, H. (1990) Hereditary C8α-γ deficiency associated with dwarfism in the rabbit. *Journal of Heredity* 81, 413–417.

Komatsu, M., Yamamoto, K.I., Mikami, H. and Sodetz, J.M. (1991) Genetic deficiency of complement component C8 in the rabbit: Evidence of a translational defect in expression of the α-γ subunit. *Biochemical Genetics* 29, 271–274.

Koplovitz, I. and Stewart, J.R. (1992) Efficacy of oxime plus atropine treatment against soman poisoning in the atropinesterase-free rabbit. *Drug and Chemical Toxicology* 15, 117–126.

Koplovitz, I. and Stewart, J.R. (1994) A comparison of the efficacy of HI6 and 2-PAM against soman, tabun, sarin, and VX in the rabbit. *Toxicology Letters* 70, 269–279.

Korn, A.K., Brandt, H.R. and Erhardt, G. (2016) Genetic and environmental factors influencing tooth and jaw malformations in rabbits. *Veterinary Record* 10.1136/vr.103293.

Kueppers, F., Lee, C.C., Fox, R.R. and Mills, J.K. (1984) Genetic heterogeneity of rabbit alpha-1-antitrypsin. *Genetics* 106, 695–703.

Kurosawa, T., Kusanagi, M., Yamasaki, Y., Senga, Y. and Yamamoto, T. (1995) New mutant rabbit strain with hypercholesterolemia and atherosclerotic lesions produced by serial inbreeding. *Laboratory Animal Science* 45, 385–392.

Lachmann, P.J. (1970) C6-deficiency in rabbits. In: Peeters, H. (ed.) *Protides of the Biological Fluids*. Pergamon, Oxford, pp. 301–309.

Lamb, N.P. and Sawin, P.B. (1962) Morphology of the external ear of the dachs (chondrodystrophy) rabbit. *American Zoologist* 2, 421–422.

Lamb, N.P. and Sawin, P.B. (1963) Morphogenetic studies of the rabbit. XXXIII. Cartilages and muscles of the external ear as affected by the dachs gene (Da). *American Journal of Anatomy* 113, 365–388.

Lang, A. (1910) Die Erblichkeitsverhältnisse der Ohrenlänge der Kaninchen nach Castle und das Problem der intermediären Vererbung und Bildung konstanter Bastardrassen. *Zeitschrift für Induktive Abstammungs- und Vererbungslehre* 4, 1–23.

Latimer, H.B. and Sawin, P.B. (1955a) Morphogenetic studies of the rabbit. XII. Organ size in relation to body weights in adults of small sized race X. *The Anatomical Record* 123, 81–102.

Latimer, H.B. and Sawin, P.B. (1955b) Morphogenetic studies of the rabbit. XIII. The influence of the dwarf gene upon organ size and variability in race X. *The Anatomical Record* 123, 447–466.

Latimer, H.B. and Sawin, P.B. (1955c) The weight of the brain, of its parts and the weight and length of the spinal cord in the rabbit (race X). *Journal of Comparative Neurology* 103, 513–539.

Latimer, H.B. and Sawin P.B. (1963) Morphogenetic studies of the rabbit. XXXV. Comparison of the weights and linear measurements in normal and heterozygous dwarf rabbits of race X. *The Anatomical Record* 146, 85–92.

Lenhard, S., Popoff, S.N. and Marks Jr., S.C. (1990) Defective osteoclast differentiation and function in the osteopetrotic (os) rabbit. *American Journal of Anatomy* 188, 438–444.

Létard, E. (1929) La keratite des lapins Castorrex et Rex de couleur. *Revue de Vétérinaire (Toulouse)* 81, 419–425.

Létard, E. (1935) Une mutation nouvelle chez le lapin. *Bulletin de l'Académie Vétérinaire de France* 8, 608–610.

Létard, E. (1943) Troubles de la locomotion et troubles de la vison chez la lapin. *Liaison hereditaire. Bulletin de l'Académie Vétérinaire de France* 16, 184–192.

Levin, I. and Larkin, J.H. (1910) Spontaneous lesions in the rabbit: the early stages of the spontaneous arterial lesions in the rabbit. *Proceedings of the Society for Experimental Biology and Medicine* 7, 109–110.

Lewis, J.H. (1926) Alcaptonuria in a rabbit. *Journal of Biological Chemistry* 70, 659–661.

Lindqvist, P., Olsson, G., Nordborg, C., Bondjers, G., Bräutigam, J. and Östlund-Lindgvist, A.M. (1988) Atherosclerosis in rabbits identified as high and low responders to an atherogenic diet and the effect of treatment with a β1-blocker. *Atherosclerosis* 72, 163–172.

Lindsey, J.R. and Fox, R.R. (1974) Inherited diseases and variations. In: Weisbroth, S.H., Flatt, R.E. and Kraus, A.L. (eds) *The Biology of the Laboratory Rabbit*. Academic Press, San Diego, California, pp. 377–401.

Lindsey, J.R. and Fox, R.R. (1994) Inherited diseases and variations. In: Manning, P.J., Ringler, D.H. and Newcomer, C.E. (eds) *The Biology of the Laboratory Rabbit*, 2nd edn. Academic Press, San Diego, California, pp. 293–319.

Liu, E., Kitajima, S., Wiese, E., Reifenberg, K., Morimoto, M. *et al.* (2007) Re-establishment of complement C6-deficient rabbit colony by cryopreserved sperm transported from abroad. *Experimental Animals* 56, 167–171.

Loke, S.Y., Wong, P.T. and Ong, W.Y. (2017) Global gene expression changes in the prefrontal cortex of rabbits with hypercholesterolemia and/or hypertension. *Neurochemistry International* 102, 33–56.

Loose-Mitchell, D.S., Poorman, J.A., Smith, S.A., Overturf, M.L., Morrisett, J.D. *et al.* (1991) Cholesterol metabolism in hypercholesterolemia-resistant rabbits. *Atherosclerosis* 87, 169–181.

Lukefahr, S. (1982) Inheritance and breeding plans for selective elimination of hydrocephalus in the rabbitry. *Journal of Applied Rabbit Research* 5, 10–12.

MacDowell, E.C. (1914a) Size inheritance in rabbits. *Carnegie Institute of Washington Publication* 196, 7–49.

MacDowell, E.C. (1914b) Multiple factors in Mendelian inheritance. *Journal of Experimental Zoology* 16, 177–194.

Magnussen, K. (1952) Beitrag zur Genetik und Histologie eines isolierten Augen-albinismus beim Kaninchen. *Zeitschrift für Morphologie und Anthropologie* 44, 127–135.

Magnussen, K. (1954) Beitrag zur Genetik eines isolierten Augenalbinismus beim Kaninchen. *II. Zeitschrift für Morphologie und Anthropologie* 46, 24–29.

Margolis, F. and Feigelson, P. (1963) Purification and characterization of a genetically determined rabbit serum esterase. *Journal of Biological Chemistry* 238, 2620–2627.

Margolis, F. and Feigelson, P. (1964) Genetic expression and developmental studies with rabbit serum atropinesterase. *Biochimica et Biophysica Acta (BBA)-General Subjects* 90, 117–125.

Marks Jr., S.C., Seifert, M.F. and Fox, R.R. (1986) The osteopetrotic rabbit: general and skeletal features of a new outbred stock. *Bone* 7, 359–364.

Martin, P. and Consroe, P. (1976) Cannabinoid induced behavioral convulsions in rabbits. *Science* 194, 965–967.

Maurer, K.J., Marini, R.P., Fox, J.G. and Rogers, A.B. (2004) Polycystic kidney syndrome in New Zealand White rabbits resembling human polycystic kidney disease. *Kidney International* 65, 482–489.

McClain, R.M. and Langhoff, L. (1980) Teratogenicity of diphenylhydantoin in the New Zealand white rabbit. *Teratology* 21, 371–379.
McNutt, C.W. and Sawin, P.B. (1943) Hereditary variations in the vena cava inferior of the rabbit. *American Journal of Anatomy* 72, 259–289.
McQueen, C.A., Maslansky, C.J., Glowinski, I.B., Crescenzi, S.B., Weber, W.W. and Williams, G.M. (1982) Relationship between the genetically determined acetylator phenotype and DNA damage induced by hydralazine and 2-aminofluorene in cultured rabbit hepatocytes. *Proceedings of the National Academy of Sciences of the USA* 79, 1269–1272.
Menschow, G.B. (1934) Fattori letali nel coniglio cincilla. *Rivista di Coniglicoltura* 6, 8–9.
Miles, A.B. (1907) Spontaneous arterial degeneration in rabbits. *Journal of American Medical Association* 49, 1173–1176.
Mooney, M.P., Wolfgang, H.L., Siegel, M.I., Lalikos, J.F., Losken, A. *et al.* (1994) Development of a strain of rabbits with congenital simple nonsyndromic coronal suture synostosis. Part I: breeding demographics, inheritance pattern, and craniofacial anomalies. *Cleft Palate-Craniofacial Journal* 31, 1–7.
Mooney, M.P., Aston, C.E., Siegel, M.I., Losken, H.W., Smith, T.D. *et al.* (1996) Craniosynostosis with autosomal dominant transmission in New Zealand white rabbits. *Journal of Craniofacial Genetics and Developmental Biology* 16, 52–63.
Mooney, M.P., Siegel, M.I., Burrows, A.M., Smith, T.D., Losken, H.W. *et al.* (1998a) A rabbit model of human familial, nonsyndromic unicoronal suture synostosis. II. Intracranial contents, intracranial volume, and intracranial pressure. *Child's Nervous System* 14, 247–255.
Mooney, M.P., Siegel, M.I., Burrows, A.M., Smith, T.D., Losken, H.W. *et al.* (1998b) A rabbit model of human familial, nonsyndromic unicoronal suture synostosis. I. Synostotic onset, pathology, and sutural growth patterns. *Child's Nervous System* 14, 236–246.
Mooney, M.P., Burrows, A.M., Wigginton, W., Singhal, V.K., Losken, H.W. *et al.* (1998c) Intracranial volume in craniosynostotic rabbits. *The Journal of Craniofacial Surgery* 9, 234–239.
Munger, R.J., Langevin, N. and Podval, J. (2002) Spontaneous cataracts in laboratory rabbits. *Veterinary Ophthalmology* 5, 177–181.
Nachtsheim, H. (1931) Über eine erbliche Nervenkrankheit (Syringomyelie) beim Kaninchen. *Zeitschrift für Pelztier- und Rauchwarenkunde* 3, 254–259.
Nachtsheim, H. (1934a) Schuttellahmung – Ein Beispiel fur ein einfach mendelndes rezessives Nervenleiden beim Kaninchen. *Erbarzt (Leipzig)* 1, 36–38.
Nachtsheim, H. (1934b) Kurzhaarkaninchen – Drei genotypisch verschiedene Mutanten mit dem gleichen Phanotypus. *Erbarzt (Leipzig)* 1, 97–102.
Nachtsheim, H. (1936a) Die Genetik einiger Erbleiden des Kaninchen, verglichen mit ahnlichen Krankheiten des Menschen. *Deutsche tierärztliche Wochenschrift* 44, 742–746.
Nachtsheim, H. (1936b) Erbliche zahnanomalien bein Kaninchen. *Zuchtungskunde* 11, 273–287.
Nachtsheim, H. (1937a) Erbpathologie des Kaninchens. *Erbarzt (Leipzig)* 4, 25–30, 50–55.
Nachtsheim, H. (1937b) Erbpathologische Untersuchungen am Kaninchen. *Zeitschrift für Induktive Abstammungs- und Vererbungslehre* 73, 463–466.
Nachtsheim, H. (1938) Erbpathologie der Haustiere. *I. Organe des ausseren Keimblattes. Fortschritte der Erbpathologie* 2, 58–104.
Nachtsheim, H. (1939a) Erbleiden des Nervensystems bei Saugetieren. In: Bauer, K.H., Hanhart, E. and Lange, J. (eds) *Handbuch der Erbbiologie des Menschen*. Springer-Verlag, Berlin and New York, pp. 1–58.
Nachtsheim, H. (1939b) Krampfbereitschaft und genotypus. I. Die epilipsie der Weissen Wiener Kaninchen. *Zeitschrift für menschliche Vererbungs- und Konstitutionslehre* 22, 781–810.
Nachtsheim, H. (1940) Erbpathologie des Stützgewebes der Säugetieren. In: Bauer, K.H., Hanhart, E. and Lange, J. (eds) *Handbuch der Erbbiologie des Menschen* Vol. 5. Springer-Verlag, Berlin and New York, pp. 1–55.
Nachtsheim, H. (1947) Eine erbliche fetale Erythroblastose beim Tier und ihre Beziehungen zu den Gruppenfaktoren des Blutes. *Klinische Wochenschrift* 24, 37–38, 590–592.
Nachtsheim, H. (1950) The Pelger-anomaly in man and rabbit; a mendelian character of the nuclei of the leucocytes. *Journal of Heredity* 41, 131–137.
Nachtsheim, H. (1954) Die Mutabilitat menschliche Gene. *Proceedings of the 9th International Congress on Genetics* 1953. pp. 139–154.
Nachtsheim, H. (1958) Erbpathologie der Nagetiere. In: Cohrs, P. Jaffe, R. and Meesen, H. (eds) *Pathologie der Laboratoriumstiere*. Springer-Verlag, Berlin and New York, pp. 310–452.

Nachtsheim, H. and Gürich, H. (1939) Erbleiden des Kaninchenanges. I. Erbliche Nahtbandchentrubung der Linse mit nochfalgendem Kernstar. *Zeitschrift für menschliche Vererbungs- und Konstitutionslehre* 23, 463–483.
Nellhaus, G. (1958) Experimental epilepsy in rabbits: development of a strain susceptible to audiogenic seizures. *American Journal of Physiology* 193, 567–572.
Nellhaus, G. (1965) Experimental epilepsy in rabbits: failure to detect a difference in phenylalanine metabolism in convulsant rabbits. *Proceedings of the Society for Experimental Biology and Medicine* 120, 259–260.
Nicholas, F.W. and Hobbs, M. (2014) Mutation discovery for Mendelian traits in non-laboratory animals: a review of achievements up to 2012. *Animal Genetics* 45, 157–170.
O'Leary, J.L., Fox, R.R., Smith, J.M. and Inukai, J. (1972) Ultrastructural alterations in vestibular and cerebellar nuclei of the ataxic rabbit. *Journal of Comparative Neurology* 145, 307–351.
O'Leary, J.L., Fox, R.R., Smith, J.M. and Inukai, J. (1974) Familial ataxia of the rabbit Sawin-Anders type. Ultrastructural analysis of degeneration of the cochlear nuclei. *Acta Neuropathologica* 30, 11–24.
O'Regan, C.C., Robitaille, P., Pinto-Blonde, M. and Chartrand, C. (1979) Delayed rejection of cardiac xenografts in C6-deficient rabbits. *Immunology* 38, 245–248.
Osetowska, E. (1967) Nouvelle maladie hereditaire du lapin de laboratoire. *Acta Neuropathologica* 8, 331–344.
Osetowska, E. and Wisniewski, H. (1966) Ataxie familiale du lapin, differente de la maladie hereditaire de Sawin-Anders. *Acta Neuropathologica* 6, 243–250.
Ostertag, B. (1930a) Weitere Untersuchungen über vererbbare Syringomyelie des Kaninchens. *Deutsche Zeitschrift für Nervenheilkunde* 116, 147–150.
Ostertag, B. (1930b) Die Syringomyelie als erbbiologisches Problem. *Verhandlungen der Deutschen Gesellschaft für Pathologie* 25, 166–174.
Ostertag, B. (1934) Neuere Ergebnisse bei der vererbbaren Syringomyelie des Kaninchens. *Atti del V Cogresso Mondiale di Pollicoltura* 3, 526–532.
Overturf, M.L., Smith, S.A., Hewett-Emmett, D., Loose-Mitchell, D.S., Soma, M.R. *et al.* (1989) Development and partial metabolic characterization of a dietary cholesterol-resistant colony of rabbits. *Journal of Lipid Research* 30, 263–273.
Pałyga, J., Zelník, J., Luboń, H. and Kołątaj, A. (1985) The nuclear proteins in the liver of rabbits with an inherited motorial failure. *Journal of Animal Breeding and Genetics* 102, 49–55.
Pang, S., Yang, X., Wang, M., Tissot, R., Nino, M. *et al.* (1992) Inherited congenital adrenal hyperplasia in the rabbit: absent cholesterol side-chain cleavage cytochrome P450 gene expression. *Endocrinology* 131, 181–186.
Pearce, L. (1948) Hereditary osteopetrosis of the rabbit. *Federation Proceedings* 7(1 Pt 1), 277.
Pearce, L. (1950) Hereditary osteopetrosis of the rabbit. III. Pathologic observations, skeletal abnormalities. *Journal of Experimental Medicine* 92, 591–600.
Pearce, L. (1960a) Hereditary distal foreleg curvature in the rabbit. I. Manifestations and course of the bowing deformity: genetic studies. *Journal of Experimental Medicine* 111, 801–821.
Pearce, L. (1960b) Hereditary distal foreleg curvature in the rabbit. II. Genetic and pathological aspects. *Journal of Experimental Medicine* 111, 823–829.
Pearce, L. and Brown, W.H. (1945a) Hereditary achondroplasia in the rabbit. II. Pathologic aspects. *Journal of Experimental Medicine* 82, 261–280.
Pearce, L. and Brown, W.H. (1945b) Hereditary achondroplasia in the rabbit. III. Genetic aspects, general considerations. *Journal of Experimental Medicine* 82, 281–295.
Pearce, L. and Brown, W.H. (1948) Hereditary osteopetrosis of the rabbit. I. General features and course of disease, genetic aspects. *Journal of Experimental Medicine* 88, 579–596.
Pearce, L. and Brown, W.H. (1960a) Hereditary premature senescence of the rabbit. I. Chronic form, general features. *Journal of Experimental Medicine* 111, 485–504.
Pearce, L. and Brown, W.H. (1960b) Hereditary premature senescence of the rabbit. II. Acute form, general features. *Journal of Experimental Medicine* 111, 505–516.
Pease, M.S. (1928a) Experiments on the inheritance of weight in rabbits. *Journal of Genetics* 20, 261–309.
Pease, M.S. (1928b) Yellow fat in rabbits, a linked character? *Zeitschrift für induktive Abstammungs- und Vererbungslehre* 2(Suppl.), 1153–1156.
Peck, E.D. and Sawin, P.B. (1950) Morphogenetic studies of the rabbit. VIII. Genetic variations in the sternum as determined by the interaction of general and of regionally specific growth factors. *Journal of Experimental Zoology* 114, 335–357.
Peng, X., Roshwalb, S., Cooper, T.K., Zimmerman, H. and Christensen, N.D. (2015) High incidence of spontaneous cataracts in aging laboratory rabbits of an inbred strain. *Veterinary Ophthalmology* 18, 186–190.

Pescador, R. (1978) Rabbit strain differences in plasma lipoprotein pattern and in responsiveness to hypercholesterolemia. *Life Sciences* 23, 1851–1861.

Petter, C., Bourbon, J., Maltier, J.P. and Jost, A. (1971) Prevention des amputations congenitales hereditaires du lapin par une hyperoxie maternelle. *Comptes Rendus de l'Académie des Sciences* 273, 2639–2642.

Petter, C., Bourbon, J., Maltier, J.P. and Jost, A. (1973) Hematies primordiales et amputations congenitales chez les foetus de lapin porteurs du gene BR. *Comptes Rendus de l'Académie des Sciences* 277, 801–803.

Petter, C., Bourbon, J., Maltier, J.P. and Jost, A. (1977) Simultaneous prevention of blood abnormalities and hereditary congenital amputations in a brachydactylous rabbit stock. *Teratology* 15, 149–157.

Pickard, J.N. (1930) DieVererbung von Haarbuscheln und Wollmenge bei Angora Kaninchen. *Zuchtungskunde* 5, 339–351.

Poorman, J.A., Buck, R.A., Smith, S.A., Overturf, M.L. and Loose-Mitchell, D.S. (1993) Bile acid excretion and cholesterol 7 alpha-hydroxylase expression in hypercholesterolemia-resistant rabbits. *Journal of Lipid Research* 34, 1675–1685.

Popoff, S.N. and Marks Jr., S.C. (1990) Relationship of abnormalities in dental and skeletal development in the osteopetrotic (os) rabbit. *Journal of Oral Pathology and Medicine* 19, 5–12.

Popoff, S.N. and Marks Jr., S.C. (1991) Congenitally osteosclerotic (os/os) rabbits are not cured by bone marrow transplantation from normal littermates. *American Journal of Anatomy* 192, 274–280.

Popoff, S.N., McGuire, J.L., Zerwekh, J.E. and Marks Jr., S.C. (1989) Treatment of congenital osteopetrosis in the rabbit with high-dose 1, 25-dihydroxyvitamin D. *Journal of Bone and Mineral Research* 4, 57–67.

Prieur, D.J. and Camara, V.M. (1979) Inheritance of lysozyme deficiency in rabbits. *Journal of Heredity* 70, 181–184.

Prieur, D.J. and Camara, V.M. (1985) Lysozyme activity in female genital tissues of normal and genetically lysozyme-deficient rabbits. *Experientia* 41, 1603–1604.

Prieur, D.J., Olson, H.M. and Young, D.M. (1974) Lysozyme deficiency – an inherited disorder of rabbits. *American Journal of Pathology* 77, 283–298.

Proks, P., Stehlik, L., Nyvltova, I., Necas, A., Vignoli, M. and Jekl, V. (2018) Vertebral formula and congenital abnormalities of the vertebral column in rabbits. *The Veterinary Journal* 236, 80–88.

Pulker, T., Carstens, A. and Williams, J. (2011) Antebrachial chondrodysplasia in New Zealand white rabbits (*Oryctolagus cuniculus*). *Journal of the South African Veterinary Association* 82, 176–178.

Punnett, R.C. (1912) Inheritance of coat colour in rabbits. *Journal of Genetics* 2, 221–238.

Punnett, R.C. (1915) Further experiments on the inheritance of coat colour in rabbits. *Journal of Genetics* 5, 38–50.

Punnett, R.C. (1920) The genetics of the Dutch rabbit: a criticism. *Journal of Genetics* 9, 303–317.

Punnett, R.C. and Bailey, P.G. (1918) Genetic studies in rabbits. I. On the inheritance of weight. *Journal of Genetics* 8, 1–25.

Reese, S. and Fehr, M. (2011) Small mammals: radioanatomy. In: Tully, T. (ed.) *Diagnostic Imaging of Exotic Pets: Birds, Small Mammals, Reptiles.* Schlüttersche, Hannover, Germany, pp. 158–165.

Richardson, J., Longo, M., Liuti, T. and Eatwell, K. (2019) Computed tomographic grading of middle ear disease in domestic rabbits (*Oryctolagus cuniculi*). *Veterinary Record* 184, 679.

Rickmeyer, T., Schöniger, S., Petermann, A., Harzer, K., Kustermann-Kuhn, B. *et al.* (2013) GM2 gangliosidosis in an adult pet rabbit. *Journal of Comparative Pathology* 148, 243–247.

Robertson, J.M., Samankova, L. and Ingalls, T.H. (1966) Hydrocephalus and cleft palate in an inbred rabbit colony. *Journal of Heredity* 57, 142–148.

Robinson, N. (1970) Enzyme changes in the hereditary ataxic rabbit. *Acta Neuropathologica* 14, 326–337.

Robinson, R. (1958) Genetic studies of the rabbit. *Bibliographia Genetica* 17, 229–558.

Rogers, A.D., Lupton, C.J. and Lukefahr, S.D. (2006) Fiber production and properties in genetically furred and furless rabbits. *Journal of Animal Science* 84, 2566–2574.

Rother, K., Rother, U., Müller-Eberhard, H.J. and Nilsson, U.R. (1966) Deficiency of the sixth component of complement in rabbits with an inherited complement defect. *Journal of Experimental Medicine* 124, 773–785.

Rother, U., Ballantyne, D.L. Jr., Cohen, C. and Rother, K. (1967) Allograft rejection in C'6 defective rabbits. *Journal of Experimental Medicine* 126, 565–579.

Rother, V. and Rother, K. (1961) Uber einen angeborenen Komplementdefekt bei Kaninchen. *Zeitschrift für Immunitätsforschung und Experimentelle Therapie* 121, 224–232.

Sasaki, Y., Ohsako, S. and Deguchi, T. (1991) Molecular and genetic analyses of arylamine N-acetyltransferase polymorphism of rabbit liver. *Journal of Biological Chemistry* 266, 13243–13250.

Sawin, P.B. (1937) Preliminary studies of hereditary variation in the axial skeleton of the rabbit. *The Anatomical Record* 69, 407–428.

Sawin, P.B. (1945) Morphogenetic studies of the rabbit. I. Regional specificity of hereditary factors affecting homoeotic variations in the axial skeleton. *Journal of Experimental Zoology* 100, 301–329.

Sawin, P.B. (1946) Morphogenetic studies of the rabbit. III. Skeletal variations resulting from the interaction of gene determined growth forces. *The Anatomical Record* 96, 183–200.

Sawin, P.B. (1955) Recent genetics of the domestic rabbit. *Advances in Genetics* 7, 183-226.

Sawin, P.B. and Crary, D.D. (1955) Congenital scoliosis in the rabbit. *The Anatomical Record* 121, 2 (abstract).

Sawin, P.B. and Crary, D.D. (1962) Inherited hypogonadia in the rabbit. *The Anatomical Record* 142, 325 (abstract).

Sawin, P.B. and Edmonds, H.W. (1949) Morphogenetic studies of the rabbit. VII. Aortic arch variations in relation to regionally specific growth differences. *The Anatomical Record* 105, 377–397.

Sawin, P.B. and Glick, D. (1943) Atropinesterase, a genetically determined enzyme in the rabbit. *Proceedings of the National Academy of Sciences of the USA* 29, 55–59.

Sawin, P.B. and Hamlet, M. (1970) Morphogenetic studies of the rabbit. XL. Growth gradient interaction and function in morphology. *Journal of Morphology* 130, 397–419.

Sawin, P.B. and Hull, I.B. (1946) Morphogentic studies of the rabbit. II. Evidence of regionally specific hereditary factors influencing the extent of the lumbar region. *Journal of Morphology* 78, 1–26.

Sawin, P.B. and Nace, M.A.G. (1948) Morphogenetic studies of the rabbit. V. Inheritance of an asymmetrical vascular pattern. *Journal of Morphology* 82, 331–354.

Sawin, P.B. and Trask, M. (1965) Morphogenetic studies of the rabbit. XXXV. Pleiotropic effects of the dachs gene and the gradient growth pattern. *Journal of Morphology* 117, 87–113.

Sawin, P.B., Anders, M.V. and Johnson, R.B. (1942) 'Ataxia,' a hereditary nervous disorder of the rabbit. *Proceedings of the National Academy of Sciences of the USA* 28, 123–127.

Sawin, P.B., Crary, D.D. and Webster, J. (1959a) Morphogenetic studies of the rabbit. XXIII. The effects of the dachs gene Da (chondrodystrophy) upon linear and lateral growth of the skeleton as influenced in time. *Genetics* 44, 609–624.

Sawin, P.B., Ranlett, M. and Chary, D.D. (1959b) Morphogenetic studies of the rabbit. XXV. The spheno-occipital synchonodrosis of the dachs (chondrostrophy) rabbit. *American Journal of Anatomy* 105, 257–280.

Sawin, P.B., Ranlett, M. and Crary, D.D. (1962) Morphogenetic studies of the rabbit. XXIX. Accessory ossification centers at the occipitovertebral articulation of the dachs (chondrodystrophy) rabbit. *American Journal of Anatomy* 111, 239–257.

Schenk, E.A., Gaman, E. and Feigenbaum, A.S. (1966) Spontaneous aortic lesions in rabbits. I. Morphologic characteristics. *Circulation Research* 19, 80–88.

Schiff, R. (1970) The biochemical genetics of rabbit erythrocyte esterases histochemical classification. *Journal of Histochemistry & Cytochemistry* 18, 709–721.

Schiff, R. and Stormont, C. (1970) The biochemical genetics of rabbit erythrocyte esterases: two new esterase loci. *Biochemical Genetics* 4, 11–23.

Schmiedt, W., Kinscherf, R., Deigner, H.P., Kamencic, H., Nauen, O. *et al.* (1998) Complement C6 deficiency protects against diet-induced atherosclerosis in rabbits. *Arteriosclerosis, Thrombosis and Vascular Biology* 18, 1790–1795.

Sheppard, L.B., Shanklin, W.M., Harris, T.H. and Fox, R.R. (1971) A histologic study of regenerating epithelium of normal and buphthalmic rabbit cornea. *Ophthalmic Research* 2, 116–125.

Shiomi, M. (2020) The history of the WHHL rabbit, an animal model of familial hypercholesterolemia (I) – contribution to the elucidation of the pathophysiology of human hypercholesterolemia and coronary heart disease. *Journal of Atherosclerosis and Thrombosis* 27, 105–118.

Shore, B. and Shore, V. (1976) Rabbits as a model for the study of hyperlipoproteinemia and atherosclerosis. In: Day, C.E. (ed.) *Atherosclerosis Drug Discovery. Advances in Experimental Medicine and Biology*, vol 67. Springer, Boston, Massachusetts, pp. 123–141.

Sinke, J.D., Van Dijk, J.E. and Willemse, T. (1997) A case of Ehlers-Danlos-like syndrome in a rabbit with a review of the disease in other species. *Veterinary Quarterly* 19, 182–185.

Sosa-Madrid, B.S., Hernández, P., Blasco, A., Haley, C.S., Fontanesi, L. *et al.* (2020a) Genomic regions influencing intramuscular fat in divergently selected rabbit lines. *Animal Genetics* 51, 58–69.

Sosa-Madrid, B.S., Santacreu, M.A., Blasco, A., Fontanesi, L., Pena, R.N. and Ibáñez-Escriche, N. (2020b) A genome-wide association study in divergently selected lines in rabbits reveals novel genomic regions associated with litter size traits. *Journal of Animal Breeding and Genetics* 137, 123–138.

Stampfli, H.F. and Quon, C.Y. (1995) Polymorphic metabolism of flestolol and other ester containing compounds by a carboxylesterase in New Zealand white rabbit blood and cornea. *Research Communications in Molecular Pathology and Pharmacology* 88, 87–97.

Sternstein, I., Reissmann, M., Maj, D., Bieniek, J. and Brockmann, G.A. (2015) A comprehensive linkage map and QTL map for carcass traits in a cross between Giant Grey and New Zealand White rabbits. *BMC Genetics* 16, 16.

Stormont, C. and Suzuki, Y. (1970) Atropinesterase and cocaiesterase of rabbit serum: localization of the enzyme activity in isozymes. *Science* 167, 200–202.

Strychalski, J., Brym, P., Czarnik, U. and Gugołek, A. (2015) A novel AAT-deletion mutation in the coding sequence of the BCO2 gene in yellow-fat rabbits. *Journal of Applied Genetics* 56, 535–537.

Strychalski, J., Gugołek, A., Antoszkiewicz, Z., Kowalska, D. and Konstantynowicz, M. (2016) Biologically active compounds in selected tissues of white-fat and yellow-fat rabbits and their production performance parameters. *Livestock Science* 183, 92–97.

Strychalski, J., Gugołek, A., Brym, P., Antoszkiewicz, Z. and Chwastowska-Siwiecka, I. (2019a) Polymorphism of the BCO2 gene and the content of carotenoids, retinol, and α-tocopherol in the liver and fat of rabbits. *Revista Brasileira de Zootecnia* 48, e20180243.

Strychalski, J., Gugołek, A., Brym, P. and Antoszkiewicz, Z. (2019b) Effect of the β-carotene oxygenase 2 genotype on the content of carotenoids, retinol and α-tocopherol in the liver, fat and milk of rabbit does, reproduction parameters and kitten growth. *Journal of Animal Physiology and Animal Nutrition* 103, 1585–1593.

Suchalla, H. (1943) Variabilität und Erblichkeit von Schädelmerkmalen bei Zwerg-und Riesenrassen; dargestellt an Hermelin-und Widderkaninchen. *Zeitschrift für Morphologie und Anthropologie* (H. 2/3), 274–333.

Sui, T., Liu, D., Liu, T., Deng, J., Chen, M. *et al.* (2019) LMNA-mutated rabbits: a model of premature aging syndrome with muscular dystrophy and dilated cardiomyopathy. *Aging and Disease* 10, 102–115.

Sypocka, J. and Domańska-Janik, K. (2005) Rabbit paralytic tremor phenotype a plp I gene mutation as a model of human Pelizaeus-Merzbacher. *Acta Neurobiologiae Experimentalis* 65, 221–230.

Szendrő, Zs., Mohamed, M.M.A. and Biró-Németh, E. (1991) Teat number of new-born rabbits depending on the teat number of their parents. *Journal of Applied Rabbit Research* 14, 133–135.

Szendrő, Zs., Mohamed, M.M.A., Biró-Németh, E. and Radnai, I. (1992) Heritability of teat number of rabbits. *Journal of Applied Rabbit Research* 15, 174–180

Taylor, R.L., Conaway, H.H. and Roth, S.I. (1980) Depression of leucine and isoproterenol induced insulin secretions in the spontaneously diabetic New Zealand white rabbit. *Endocrine Research Communications* 7, 121–136.

Tegtmeyer, M. (1949) Entwicking, Aufbau, und Stand der Angorakaninchenleistungszucht. *Kühn-Archiv* 62, 341–493.

Tosic, M., Dolivo, M., Amiguet, P., Domanska-Janik, K. and Matthieu, J.M. (1993) Paralytic tremor (pt) rabbit: a sex-linked mutation affecting proteolipid protein-gene expression. *Brain Research* 625, 307–312.

Tosic, M., Dolivo, M., Domańska-Janik, K. and Matthieu, J.M. (1994) Paralytic tremor (pt): a new allele of the proteolipid protein gene in rabbits. *Journal of Neurochemistry* 63, 2210–2216.

Tourtellotte, W.W., Lowry, O.H., Passonneau, J.V., O'Leary, J.L., Harris, A.B. and Rowe, M.J. (1966) Carbohydrate metabolites in rabbit hereditary ataxia. *Archives of Neurology* 15, 283–288.

Ujhazy, E., Balonova, T., Durisova, M., Gajdosík, A., Jansák, J. and Molnárová, A. (1993) Teratogenicity of cyclophosphamide in New Zealand white rabbits. *Neoplasma* 40, 45–49.

Undritz, E. (1939) Das Pelger-Huetsche Blutbild beim Tier und seine Bedeutung fiir die Entwicklungsgeschichte des Blutes. *Schweizerische medizinische Wochenschrift* 69,1177–1186.

Undritz, E. (1943) Das ausschliessliche Vorkommen reifer rundkerniger Leukozyten bei der reingezüchteten Pelger–Huëtschen Anomalie des Kaninchens und die Bedeutung der Pelger–Leukozyten in der vergleichenden Hämatologie. *Folia Haematatologica* 67, 249–291.

Van Herck, H., Hesp, A.P.M., Versluis, A., Zwart, P. and Van Zutphen, L.F.M. (1989) Prolapsus vaginae in the IIIVO/JU rabbit. *Laboratory Animals* 23, 333–336.

Van Zutphen, L.F.M. (1974a) Serum esterase genetics in rabbits. I. Phenotypic variation of the prealbumin esterases and classification of atropinesterase and cocainesterase. *Biochemical Genetics* 12, 309–326.

Van Zutphen, L.F.M. (1974b) Serum esterase genetics in rabbits. II. Genetic analysis of the prealbumin esterase system, including atropinesterase and cocainesterase polymorphism. *Biochemical Genetics* 12, 327–343.

Van Zutphen, L.F.M. and Den Bieman, M.G.C.W. (1975) Serum esterase genetics in rabbits. III. A third allele on the Est-2 locus. *Biochemical Genetics* 13, 19–28.

Van Zutphen, L.F.M. and Fox, R.R. (1977) Strain differences in response to dietary cholesterol by JAX rabbits: correlation with esterase patterns. *Atherosclerosis* 28, 435–446.

Van Zutphen, L.F.M., Den Bieman, M.G.C.W. and Bouw, J. (1977) Serum esterase genetics in rabbits. IV. The prealbumin and β-globulin systems. *Biochemical Genetics* 15, 989–1000.

Van Zutphen, L.F.M., Bieman, M.G.C.W., Hülsmann, W.C. and Fox, R.R. (1981) Genetic and physiological aspects of cholesterol accumulation in hyperresponding and hyporesponding rabbits. *Laboratory Animals* 15, 61–68.

Van Zutphen, L.F.M., Fox, R.R. and Den Bieman, M.G.C.W. (1983) Genetics of two tissue esterase polymorphisms (Est-4 and Est-5) in the rabbit. *Biochemical Genetics* 21, 773–780.

Van Zutphen, L.F.M., Den Bieman, M.G.C.W., Von Deimling, O. and Fox, R.R. (1987) Genetics of a tissue esterase polymorphism (Est-6) in the rabbit (*Oryctolagus cuniculus*). *Biochemical Genetics* 25, 335–344.

Vella, D. and Donnelly, T.M. (2012) Basic anatomy, physiology and husbandry. In: Queensberry, K. and Carpenter, J.W. (eds) *Ferrets, Rabbits and Rodents: Clinical Medicine and Surgery*. Saunders, Philadelphia, pp. 157–173.

Vogt, A. (1919) Vererbter Hydrophthalmus beim Kaninchen. *Klinische Monatsblätter für Augenheilkunde* 63, 233.

Wang, Z., Zhang, J., Li, H., Li, J., Niimi, M. *et al.* (2016) Hyperlipidemia-associated gene variations and expression patterns revealed by whole-genome and transcriptome sequencing of rabbit models. *Scientific Reports* 6, 26942.

Watanabe, Y. (1977) Studies on characteristics of spontaneously hyperlipidemic rabbit and development of the strain with such property. *The Bulletin of the Azabu Veterinary College* 2, 99–124.

Watanabe, Y. (1980) Serial inbreeding of rabbits with hereditary hyperlipidemia (WHHL-rabbit): incidence and development of atherosclerosis and xanthoma. *Atherosclerosis* 36, 261–268.

Watanabe, Y., Ito, T. and Kondo, T. (1977) Breeding of a rabbit strain of hyperlipidemia and characteristics of these strains. *Experimental Animals* 26, 35–42.

Watson, C.E., Draganov, D.I., Billecke, S.S., Bisgaier, C.L. and La Du, B.N. (2001) Rabbits possess a serum paraoxonase polymorphism similar to the human Q192R. *Pharmacogenetics and Genomics* 11, 123–134.

Webber, R.J., Fox, R.R. and Sokoloff, L. (1981) In vitro culture of rabbit growth plate chondrocytes. 2. Chondrodystrophic mutants. *Growth* 45, 269–278.

Weber, W.W., Miceli, J.N., Hearse, D.J. and Drummond, G.S. (1976) N-acetylation of drugs. Pharmacogenetic studies in rabbits selected for their acetylator characteristics. *Drug Metabolism and Disposition* 4, 94–101.

Weisbroth, S.H. and Ehrman, L. (1967) Malocclusion in the rabbit: a model for the study of the development, pathology and inheritance of malocclusion. *I. Preliminary note. Journal of Heredity* 58, 245–246.

Wieberneit, D. and Wegner, W. (1995) Albino rabbits can suffer from megacolon-syndrome when they are homozygous for the 'English Spot' gene (EnEn). *World Rabbit Science* 3, 19–26.

Willimott, S.G. (1928) On the pigment of the fat of certain rabbits. *Biochemical Journal* 22, 1057–1059.

Wilson, W.K. and Dudley, F.J. (1946) Fat colour and fur colour in different varieties of rabbit. *Journal of Genetics* 47, 290–294.

Wright, S. (1918) On the nature of size factors. *Genetics* 3, 367–374.

Yamamoto, T., Bishop, R.W., Brown, M.S., Goldstein, J.L. and Russell, D.W. (1986) Deletion in cysteine-rich region of LDL receptor impedes transport to cell surface in WHHL rabbit. *Science* 232, 1230–1237.

Yang, X., Iwamoto, K., Wang, M., Artwohl, J., Mason, J.I. and Pang, S. (1993) Inherited congenital adrenal hyperplasia in the rabbit is caused by a deletion in the gene encoding cytochrome P450 cholesterol side-chain cleavage enzyme. *Endocrinology* 132, 1977–1982.

Zech, R., Severin, R.M., Chemnitius, J.M. and Nebendahl, K. (1999) Paraoxonase polymorphism in rabbits. *Chemico-Biological Interactions* 119–120, 283–288.

Zimmerman, T.S., Arroyave, C.M. and Müller-Eberhard, H.J. (1971) A blood coagulation abnormality in rabbits deficient in the sixth component of complement (C6) and its correction by purified C6. *Journal of Experimental Medicine* 134, 1591–1600.

9 Genetics of Disease Resistance in the European Rabbit

Herve Garreau[1*], Mélanie Gunia[1], Ana Lemos de Matos[2], Joana Abrantes[3] and Pedro José Esteves[3,4,5]

[1]GenPhySE, INRAE, Université de Toulouse, Castanet-Tolosan, France; [2]Biodesign Center for Immunotherapy, Vaccines, and Virotherapy (B-CIVV), The Biodesign Institute, Arizona State University, Tempe, USA; [3]InBIO-Research Network in Biodiversity and Evolutionary Biology, CIBIO, Universidade do Porto, Vairão, Portugal; [4]Departamento de Biologia, Faculdade de Ciências, Universidade do Porto, Portugal; [5]CITS-Centro de Investigação em Tecnologias de Saúde, CESPU, Gandra, Portugal

9.1 Introduction

Diseases remain major profit-limiting factors in rabbit production. Two main syndromes are generally characterized in rabbits: respiratory syndromes, which prevail in adults, and digestive syndromes, more frequent in growing rabbits (Maertens and Coudert, 2006). Pasteurellosis is the most common bacterial disease affecting the respiratory system. Myxomatosis is a major viral disease for which secondary infections can also cause nasal congestion and the clinical signs of the amyxomatous form are mainly respiratory. Another viral pathology, Rabbit Haemorrhagic Disease (RHD), is characterized by acute pulmonary and hepatic infection. Digestive disorders are responsible for mortality and significant morbidity characterised by reduced growth and poor feed conversion. Digestive disorders can be caused by many infectious agents: parasites, bacteria and viruses. In the case of Epizootic Rabbit Enteropathy (ERE), the aetiological agent is not identified.

The fundamental definition of disease resistance includes freedom from clinical signs of disease after challenge (Raadsma *et al.*, 1997). Disease resistance is used generically to cover resistance to infection, i.e. a host's ability to moderate the pathogen or parasite lifecycle, and also resistance to the disease consequence of infection (Bishop and Morris, 2007). The expression of disease resistance as a binomial trait holds for most diseases in animals where clinical signs are well defined (Raadsma *et al.*, 1997). In the analysis of binomial traits, such as disease resistance, it is also common practice to adopt a threshold model, as described by Dempster and Lerner (1950) and Falconer (1965). The broad concept of disease resistance might also include more specific concepts of tolerance, resilience and resistance that are discussed in other specialized texts (e.g. Bishop and Woolliams, 2014).

To establish a successful commercial breeding programme to improve disease resistance, it is necessary to identify a well-defined and measurable trait that can indicate disease resistance. The indicator trait must have a genetic component of variation and should be genetically correlated with the disease trait of economic importance in

Contact e-mail address: herve.garreau@inrae.fr

DOI: 10.1079/9781780643342.0009

the breeding objective. To be adopted by the breeding companies, this trait should be easy to measure in commercial production systems.

9.2 Pasteurellosis

9.2.1 Disease description

Pasteurellosis is one of the most common bacterial diseases in rabbit commercial breeding farms as well as in laboratory facilities. It is responsible for at least 50% of the major causes of culling of breeding does (Coudert *et al.*, 2006). This disease may take any form in the rabbit: rhinitis, abscesses, pneumonia, mastitis, diarrhoea, metritis, wryneck, septicaemia. In Europe, approaches to disease control have relied on environmental hygiene (Lebas *et al.*, 1997), requiring large capital investment in housing and routine inclusion of antibiotics in rabbit feed (Goñi *et al.*, 2004). Even with these controls, *Pasteurella multocida* infection in meat rabbits still occurs, albeit at low incidence and severity (Rosell, 2003).

9.2.2 Selection for resistance to bacterial infection

Baselga *et al.* (1988) estimated the heritability of lung damage caused by infection to be between 0.12±0.05 and 0.28±0.14. The evaluation of this trait involved post mortem examination of lung tissue, so it could not be a practical option for a breeding programme. Eady *et al.* (2007) studied the genetic component of observable signs of bacterial infection in an experimental population set up to establish a national breeding programme for meat rabbits in Australia (Eady, 2003). Incidence of visual signs of bacterial infection and mortality, from causes related to bacterial infection, were recorded on a weekly basis in growing meat rabbits from 5 to 10 weeks of age. At the beginning of the breeding programme, and at 12-monthly intervals, some dead and sick rabbits were sent to an animal health diagnostic laboratory to confirm the species of bacteria present. Where the signs were abscesses, the predominant bacterium was *Staphylococcus aureus*. Where the signs were nasal discharge (snuffles), the predominant bacterium was *Pasteurella multocida*. Rabbits sent for post mortem examination often showed both signs. Heritability of weekly incidence of disease was the highest at weeks 9 and 10 (0.05±0.02 and 0.06±0.02, respectively, with linear model, and 0.10±0.06 and 0.12±0.05, respectively, with a threshold model). The genetic correlation between Weekly Incidence was high for weeks 6–10 with incidence in week 5 appearing to be under a different genetic control. This is probably not surprising as rabbits at week 5 are freshly weaned and would have some carryover resistance from their dams. Therefore, choosing resistance at week 10 as the selection trait should result in a correlated response for the other weeks, with the exception of week 5. This would provide improvements in the weeks of higher disease incidence. Heritability of mortality (dead and euthanized due to infection) in this study was not significantly different from zero (0.02±0.05). The low genetic variability of mortality could be explained by the low proportion of dead animals and, consequently, by the low variance of mortality. Genetic correlations between disease incidence and mortality were imprecise and not different from zero. However, the results showed a phenotypic relationship between observed disease signs and subsequent death, with the risk of dying being increased between 10- and 50-fold if the rabbit had a positive disease score in the prior week. Selection for reduced disease incidence may not result in a genetic improvement in mortality but still has an economic value in a breeding programme as it appears that mortality will be phenotypically reduced with an improvement in disease resistance. This study demonstrated that observed signs of bacterial infection in rabbits can be used as an indicator trait for resistance to bacterial infection, and the heritability of the trait is high enough to warrant its inclusion in a breeding programme. The disease binary trait was included in the breeding objective of the Australian national breeding programme in 2002.

Previous estimates of heritability for a single measure of bacterial infection, from two commercial populations of meat rabbits in France, suggested useful genetic variation (0.04±0.01 and 0.03±0.01) but heritability estimates were lower (Eady *et al.*, 2004). The lower heritability estimates compared to the previous study may be due to a different recording system. In those commercial populations, the purpose of data collection was not specifically to score for the presence or absence of the disease and the predominant symptom was recorded, rather than all symptoms.

9.2.3 Genetic resistance to experimental *Pasteurella multicoda* infection

A research project dedicated to the genomic evaluation of the genetic resistance of the rabbit to pasteurellosis as a new tool for the control of this disease has been set up by the INRA (Helloin *et al.*, 2009, 2015). The question of the number and diversity of *Pasteurella* strains that can be isolated in breeding units represents an essential preliminary step to evaluate the potential of such a genetic approach. A collection of about 200 strains was established from isolates sampled in rabbit flocks since 2003. These strains were characterized and their genetic diversity evaluated through the development of a Multi-Locus Variable-number-tandem-repeat analysis (MLVA technique), and validated with the already published but less discriminating Multi Locus Sequence Type (MLST) method. Virulence in rabbits of five strains selected as representatives of the main genetic groups of *Pasteurella* was evaluated *in vivo* after intra-nasal or subcutaneous inoculation. The subcutaneous infection route was chosen to compare 20 representative strains from the collection and evidenced their virulence variability, as well as an individual variability in the rabbit's ability to control the infection. Based on these results, a strain of *Pasteurella* was chosen to inoculate a rabbit population representative of the French rabbit lines. In total, 964 inoculated rabbits were phenotyped for disease resistance and genotyped to identify chromosomal segments (QTL) involved in resistance to pasteurellosis. The final objective was to propose genomic-based strategies to select for pasteurellosis resistance. The heritability of pasteurellosis resistance traits was low to moderate. Heritability was 0.13±0.07 for the abscess dissemination score, 0.08±0.05 for bacteria score, 0.29±0.07 for growth at the first week after inoculation, 0.20±0.06 for growth at the second week after inoculation, and 0.16±0.06 for the pasteurellosis resistance score, which combines survival, abscess and bacteria scores. These results show that there is a potential to decrease prevalence of pasteurellosis by selecting resistant rabbits on any of these traits (Shrestha *et al.*, 2018).

9.3 Epizootic Rabbit Enterocolitis and Digestive Disorders

9.3.1 Disease description

Intestinal pathology is one of the most important problems encountered in rabbit breeding and occurs mainly in young rabbits after weaning (Marlier *et al.*, 2003). Digestive disorders are responsible for significant mortality and morbidity (depression of growth and poor feed conversion). Diarrhoea is the main clinical symptom of enteropathy. It is possible to distinguish enteropathies originating from a specific pathogenic agent from those where no clear pathogenic origin is detected. The reduction of the dietary fibre level is a major factor that increases the incidence of non-specific enteropathy in the growing rabbit (Gidenne, 2003). Viruses, bacteria and intestinal parasites can induce enteropathy. Coccidia are the major parasitic agents of digestive affections in the rabbit (Coudert, 1989).

The outbreak of epizootic rabbit enteropathy (ERE) occurred in 1997 (Licois *et al.*, 2000; Jones and Duff, 2001). This new gastrointestinal syndrome induces high mortality in young rabbits after weaning. ERE is generally associated with a specific clinical pattern: weak diarrhoea and pronounced distension of the abdomen. At necropsy, ERE is characterized by the dilatation of all segments of the gastrointestinal tract but there is no digestive inflammation or congestion (Licois *et al.*, 2005). The disease is reproducible and contagious. One or several infectious pathogens could be involved, but those pathogens have not yet been identified (Marlier *et al.*, 2003).

9.3.2 Genetic variability of the resistance to digestive disorders of various origins

Rochambeau *et al.* (2006) analysed the genetic variability for the resistance to three experimental digestive stresses in the growing rabbit: after inoculation of coccidia (trial 'coccidian'), with a fibre-deficient diet (trial 'FD'), and after experimental reproduction of epizootic rabbit enteropathy (trial 'ERE'). Genetic variability was analysed from a sample of 48 bucks, which produced the experimental young rabbits. Two binary indexes were defined to describe the rabbit individual response. The first one ('Alive') dealt with mortality. The second ('Tolerant') dealt with mortality and morbidity. The buck effect was significant for each index in the 'coccidian' and the 'FD' trials. The buck effect was significant for the 'tolerant' index in the 'ERE' trial. Correlations between buck rankings for the two indexes of one trial were often significant. Correlations between buck rankings for indexes of 'coccidian' and 'FD' trials were weakly significant. These results showed that genetic variability for the resistance to ERE is low and independent from the genetic variability for resistance to coccidia or to a fibre-deficient trial.

Genetic parameters of resistance to ERE were estimated in another experiment (Garreau *et al.*, 2006). Two batches of 330 young rabbits were inoculated after weaning using the inoculum INRA TEC3 (Licois *et al.*, 2005). Four indexes were defined – mortality, diarrhoea, abnormal growth, resilience – to assess individual response of the young rabbits from day 5 to day 33. The application of a logistic regression highlighted a significant effect of the sire for the diarrhoea, abnormal growth and resilience indexes but not for mortality index. The genetic parameters of these indexes were estimated using a threshold model and REML methodology. Heritability estimates for mortality, diarrhoea, abnormal growth and resilience indexes were, respectively, 0.05±0.05, 0.21±0.16, 0.38±0.21 and 0.08±0.06.

9.3.3 Selection for the resistance to digestive disorders

To design a meat-rabbit breeding programme aimed to improve resistance to digestive disorders, the first steps are: (i) to identify a measurable and heritable trait related to disease incidence; (ii) to estimate the genetic correlations of this trait with the other traits of economic importance in the breeding objective; and then (iii) to set up an adequate selection strategy. In commercial populations, clinical symptoms of enteropathy (diarrhoea, bloated abdomen) can be routinely observed and recorded without identifying any clear pathogenic origin. Garreau *et al.* (2008) carried out a Bayesian analysis to estimate the genetic parameters of a binary trait based on the observed signs of enteropathy, 63-day body weight, carcass yield and perirenal fat percentage in a commercial paternal line. The incidence of digestive disorder was 8% of the population and 66% of these rabbits died prior to 63 days weighing. Estimated heritability was equal to 0.08±0.02 for the resistance to digestive disorders. The genetic correlations between the disease trait, on the one hand, and the 63 days body weight and the carcass yield on the other, were negative (–0.19±0.10 and –0.34±0.17, respectively) and so favourable. The genetic correlation between the disease score and the perirenal fat percentage was close to zero (–0.07±0.12). Since then the disease binary trait was included in the breeding scheme of two paternal lines (AGP39 and AGP59) of the rabbit breeder Hypharm in order to reduce the enteropathy incidence.

9.3.4 Resistance to an experimental *Escherichia coli* infection

In order to assess selection efficiency against digestive disorders, a one-generation divergent selection experiment was carried out in the lines AGP39 and AGP59 previously mentioned. Animals were inoculated with an enteropathogenic *E. coli 0103* strain (Garreau *et al.*, 2012). Two successive batches of 178 crossbred animals, 89 resistant (R) and 89 sensitive (S), were produced by mating AGP39 resistant bucks with AGP59 resistant does and AGP39 sensitive bucks with AGP59 sensitive does, respectively. Individual Estimated Breeding Values (EBVs) of animals of the R group and of the S group were calculated as the mean of their sire and dam's EBVs. These EBVs were genetic scores ranking animals based on their resistance to digestive disorders. In each batch, 31 animals with the lowest EBVs in the R group and 31 animals with the highest EBVs in the S group were selected to constitute a very resistant group (RR) and a very sensitive group (SS), respectively. Young rabbits received 10^5 bacteria of the strain LY265, *E. Coli* O103:H2:K, rhamnose negative (Licois *et al.*, 1992), by oral administration, at 37 days of age. Thirteen days after inoculation, nine rabbits of each subgroup were sacrificed to measure pH of the caecal digesta and for caecum and caecum appendix weighing. The impact of the *E. coli* challenge was significant for growth and for the caecum relative weight. Although mortality tended to be higher in sensitive animals than in resistant animals (p=0.12) there was no significant difference between R and S groups for cumulative mortality, daily weight gain and caecum measurements. Mortality was significantly lower (p=0.04) in the SS group than in the RR group at day 11, day 12 and day 13 (31.8%, 34.8% and 36.6% vs. 50.0%, 51.0% and 54.7%, respectively). Several reasons can explain the lack of difference between R and S groups. Sires and dams of experimental rabbits were selected for the incidence of non-specific digestive disorders under natural infection. The selected trait was quite different from the traits measured in this study, i.e. a response to an experimental infection with the specific strain LY265, *E. Coli* O103. In other words, selection for resistance to natural and non-specific infection may not be efficient enough to protect animals against an artificial infection or against this specific strain of *E. coli*. The low genetic response to divergent selection due to a low heritability of the trait and to the single generation of selection may also explain the lack of difference in response to inoculation between resistant and sensitive animals. The significant difference in mortality between very resistant and very sensitive animals reflected, consequently, the increased genetic divergence due to the increased selection intensity realized by the selection of extreme animals according to their EBV within each group. These results justify continuing the selection on these bases, which can be regularly tested with the same infectious model, or may be with another one (e.g. ERE, clostridium, coccidia).

9.3.5 Candidate genes studied to identify markers associated with resistance to digestive disorders

Molecular genetics approaches were recently applied to study genetic resistance to digestive disorders (Zhang *et al.*, 2011, 2013, 2014; Chen *et al.*, 2013; Liu *et al.*, 2013; Yang *et al.*, 2013; Fu *et al.*, 2014). The same or similar experimental methodologies were applied by these authors who investigated the variability in several candidate genes involved in host immunity (*MyD88*, *JAK1*, *STAT3*, *Dectin-1*, *TLR4*, *NOD2*, *NLRP3* and *NLRP12*). Single nucleotide polymorphisms (SNPs) were identified by sequencing coding regions or flanking regions of these genes. Some SNPs were genotyped for association analysis based on a case and control design and evaluating allele and genotype frequency differences between the two groups of rabbits (i.e. cases or affected and controls or not affected). Association analysis revealed that these genes played a potential role in the susceptibility to digestive disorders or protection against them. Some SNPs of these genes had protective effects against digestive disorders. Subsequent analyses of mRNA expression of *MyD88*, *Dectin-1*, *NOD2*, *NLRP3* and *NLRP12* genes were performed after experimentally inducing non-specific digestive disorders. The expression of the studied genes varied significantly between healthy individuals and rabbits suffering from the disorders, which confirmed the possible relationships or association between these genes and digestive disorders. However, these studies can be considered preliminary and need to be further validated and confirmed including different rabbit lines/breeds, a large number of animals and different experimental designs. It is also clear that resistance to digestive disorders might have a polygenic nature and at present, no major genes have been identified.

9.4 Multiple Disease Syndromes

Growing rabbits can suffer from digestive and respiratory syndromes. Instead of analysing them separately, another approach consists of considering all disease syndromes together, and considering more general resistance traits such as poor body condition, unspecific infectious syndromes, unspecific infectious mortality (Gunia *et al.*, 2015; Ragab *et al.*, 2015).

9.4.1 Unspecific disease resistance under two feeding systems

Ragab *et al.* (2015) estimated the heritability of disease traits in a Spanish paternal line under a restricted feeding system and under an *ad libitum* feeding system. The traits were unspecific mortality (incidence of 7%), morbidity and mortality from ERE (incidence of 8%), from respiratory syndromes (incidence of 2%), and from poor body condition (incidence of 3%). The observed heritabilities of unspecific mortality, ERE, respiratory syndromes and poor body condition were 0.07 ± 0.02, 0.05 ± 0.02, 0.03 ± 0.01 and 0.06 ± 0.02, respectively. On the underlying scale, heritabilities were 0.27 ± 0.06, 0.17 ± 0.09, 0.27 ± 0.08 and 0.38 ± 0.09, respectively. Similar heritability estimates were found under *ad libitum* feeding. The genetic correlations between the same syndromes under the two feeding systems were significantly different from one, which indicated a genotype by environment interaction. Genetic correlations between disease traits and average daily gain were moderate and negative.

9.4.2 Resistance to infectious syndromes

Gunia *et al.* (2015) estimated the heritability of disease traits in French paternal lines. The studied traits were morbidity or mortality from digestive syndromes (incidence of 7%), from respiratory syndromes (incidence of 4%), from infectious syndromes (a combination of digestive, respiratory and other infectious syndromes; incidence of 12%), and infectious mortality (incidence of 5%). The observed heritabilities of digestive syndromes, respiratory syndromes, infectious syndromes, and infectious mortality were 0.03, 0.04, 0.03 and 0.04, respectively, with standard errors of 0.003. The genetic correlations between digestive and respiratory syndromes were slightly negative (-0.18). The genetic correlations between the composite infectious disease trait and digestive or respiratory syndromes were moderate (0.65 ± 0.04 and 0.61 ± 0.04, respectively). Most of the genetic correlations between disease traits and live weight were not significantly different from zero.

9.4.3 Resistance to non-specific disease in a selection and challenged environment

Gunia *et al.* (2018) performed the same study in a French maternal line in a selection and a challenged environment. In this study, even the slightest clinical signs of disease were recorded. The incidence of infectious syndromes was 26% in the selection environment and 41% in the challenging environment, and the corresponding heritability for the resistance to infectious syndromes were 0.08 ± 0.02 and 0.04 ± 0.01, respectively. The genetic correlation between these two traits was 0.70 ± 0.13, indicating genotype by environment interactions. Most of the genetic correlations between disease and production traits were not significantly different from zero, except between the weaning weight and resistance to infectious syndromes in the selection environment, with a favourable correlation of −0.34 ± 0.12. Therefore, selection on non-specific disease resistance or tolerance using simple observations seems to be feasible. Given these genetic parameters, for the same level of exposure of rabbits to pathogens, the expected response to selection is a reduction of disease incidence of 4–6% per generation.

All these studies showed that traits accounting for multiple disease syndromes are heritable. There is genetic variability underlying a more general resistance to undescribed diseases. The genetic correlation between these general resistance traits and the production traits are neither null nor favourable. Some genotype by environment interaction occurs, depending on the feeding system or the disease incidence, and should be taken into account.

9.5 Myxomatosis

Myxomatosis is a lethal, generalized viral disease of domestic and wild European rabbits (*Oryctolagus cuniculus*) (Fenner and Ratcliffe, 1965; Fenner, 1983; Kerr and Donnelly, 2013) caused by the myxoma virus (MYXV), family *Poxviridae*, subfamily *Chordopoxvirinae*, genus *Leporipoxvirus* (Fenner, 1979). MYXV naturally circulates in two species of cottontail rabbits (*Sylvilagus* sp.), the South American tapeti (*S. brasiliensis*) and the North American brush rabbit (*S. bachmani*) (Moses, 1911; Aragão, 1927, 1943; Marshall and Regnery, 1960; Fenner and Ratcliffe, 1965; Regnery and Miller, 1972). In the two native long-term *Sylvilagus* hosts, MYXV only causes a cutaneous fibroma restricted to the site of inoculation (Fenner and Ratcliffe, 1965). Both *Oryctolagus* and *Sylvilagus* genera are members of the family Leporidae. While *Sylvilagus* species are extensively distributed throughout North and central America, and the northern half of South America (Chapman and Flux, 1990; Matthee *et al.*, 2004), *Oryctolagus*, a monotypic genus, is the most widely dispersed genus beyond its natural range, the Iberian Peninsula, since it has a worldwide distribution mediated by human action (Chapman and Flux, 1990; Ferrand and Branco, 2007). The combination of several molecular markers allowed establishing a robust phylogeny for the Leporidae family, where the divergence time between *Sylvilagus* and *Oryctolagus* genera was estimated around 10 million years ago (Mya) (Matthee *et al.*, 2004; van der Loo *et al.*, 2009). More details on the Lagomorpha order and diffusion of the European rabbit are reported in Chapter 1.

Following deliberate use of MYXV as a biological control agent for wild European rabbits in Australia and several European countries in the 1950s, the virus is now endemic in wild European rabbit populations and can easily be transmitted to highly susceptible farmed, laboratory and pet European rabbits (Fenner and Fantini, 1999; Kerr, 2012; Kerr and Donnelly, 2013). Clinically, myxomatosis can assume different forms, depending on the symptoms exhibited (Kerr and Donnelly, 2013). Acute myxomatosis is the classical mucocutaneous form of myxomatosis, which includes swellings (secondary lesions also denominated myxomas) on the eyelids, nose, ears, anogenital region, and subsequently over the entire body, conjunctival inflammation (blepharoconjuctivitis), and mucopurulent discharge from the nose and eyes (Best *et al.*, 2000). Later in infection, the severity of the symptoms increases with the obstruction of the nostrils and closure of the eyelids by the mucopurulent discharge, and acute respiratory distress (Best *et al.*, 2000). Finally, European rabbits may develop secondary infections, such as severe bacterial infections of the conjunctivae and upper respiratory tract. Death typically occurs 7–15 days after infection, and case fatality rates are nearly 100% (Best *et al.*, 2000). On the other hand, in the peracute form of myxomatosis, sudden deaths may occur with only mild or even no clinical signs of myxomatosis (Silvers *et al.*, 2006; Kerr and Donnelly, 2013). Neurological signs, such as convulsion, muscle twitching, hypersensitivity to stimulation, depression and coma, have been observed in European rabbits infected with the North American MYXV strain MSW (Fenner and Ratcliffe, 1965; Silvers *et al.*, 2006). Death will also occur 7–15 days after virus inoculation. The subacute form of myxomatosis results from infection with attenuated field strains of MYXV. The emergence of these strains is a consequence of the action of natural selection in MYXV and in wild European rabbit populations from Australia and Europe, resulting in less virulent and more efficiently transmitted strains, and in the appearance of resistant European rabbit populations (Fenner and Marshall, 1957; Kerr, 2012). In this disease form, the virus causes typical myxomatosis but with a protracted course, with deaths occurring from 10 to 30-plus days after infection (Kerr, 2012). Recently, when testing 1990s MYXV Australian field isolates in laboratory European rabbits most viruses induced a highly lethal immune collapse syndrome similar to septic shock, a phenotype that was markedly distinct from classical cutaneous myxomatosis. Rabbits became moribund over a few hours and died between days 10–15, with only minor signs of typical myxomatosis: the viral inoculation site was poorly differentiated from the surrounding skin and secondary cutaneous lesions were absent (Kerr *et al.*, 2017). The aforementioned study is another great evidence that the natural arms race between MYXV and Australian rabbits is a still ongoing process, as increase in host resistance has led to the evolution of strains of MYXV that are more immunosuppressive.

The pathogenesis of MYXV has been extensively described for susceptible laboratory European rabbits after infection with highly virulent strains (Hurst, 1937; Fenner *et al.*, 1953; Best *et al.*, 2000; Kerr, 2012; Jeklova *et al.*, 2008) (Fig. 9.1). The virus initially replicates in the skin at the inoculation site, particularly in MHC-II positive denditric-like cells at the epidermal/dermal junction and deeper in the dermis at 24 hours (Best *et al.*, 2000). Also, within 24 hours of infection, the virus can be detected in the lymph node draining the inoculation site. Here, MYXV replicates to high titres within the lymphoid tissue of the paracortex and cortex resulting in the massive loss of lymphocytes in the draining lymph node (Best *et al.*, 2000). From the lymph node, MYXV then spreads, probably in lymphocytes and monocytes, to distal tissues such as lungs, testis, spleen and other lymphoid

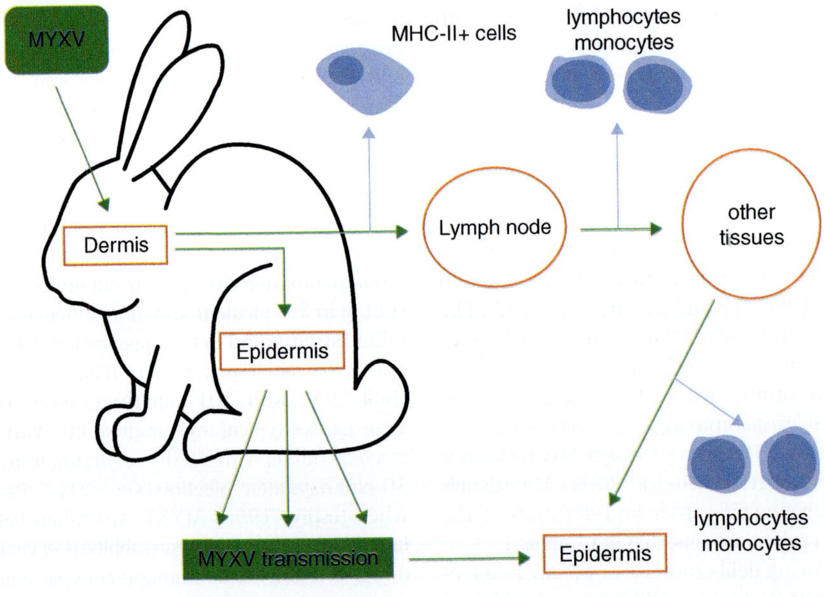

Fig. 9.1. Pathogenesis of myxoma virus (MYXV) in laboratory European rabbits.

tissues, skin and mucocutaneous sites, such as nose, eyelids and anogenital region (Fenner *et al.*, 1953; Best *et al.*, 2000; Kerr, 2012). After several days of infection, the epithelial cells of the epidermis at the inoculation site and in secondary cutaneous lesions, such as eyelids and ears myxomas, become packed with virus and are critical for mosquitoes and fleas transmission (Fenner *et al.*, 1953).

A programme of selection for resistance to myxomatosis in the domestic European rabbit was conducted from 1955 to 1967 in Australia (Sobey, 1969). The aim of this experiment was to estimate the genetic parameters of myxomatosis resistance and build up a line of genetically resistant rabbits for evaluating the effects of new virus strains. Every generation, rabbits were infected with a strain of MYXV. Surviving rabbits or rabbits showing few symptoms of infection were then mated to produce the subsequent generation. Percentage of recovery increased generation after generation (from 50% to 80% for the least virulent strain and from 10% to 20% for the most virulent virus strain).

The successful replication and propagation of MYXV within its natural long-term *Sylvilagus* hosts and in the susceptible European rabbit host requires the evasion and manipulation of the potent and complex host immune defences by a remarkable repertoire of viral proteins (Kerr and McFadden, 2002; Stanford *et al.*, 2007; Spiesschaert *et al.*, 2011; Kerr, 2012). The consequences of this dynamic battle between MYXV and its hosts are expected to be manifested by the presence of selective pressure hallmarks in their genome, with particular incidence in the host immune genes. Once MYXV successfully crossed the species barrier, it had already partially adapted to its natural long-term hosts, the tapeti and the brush rabbit. It is still unclear why MYXV infection has such a strikingly different outcome in the natural hosts and in the susceptible European rabbit; yet it is likely that the severity of the effects of viral proteins depends on the genotype of the host (Zuniga, 2002; Kerr, 2012). For the susceptible *Oryctolagus* host, a level of genetic resistance to MYXV emerged independently in three countries: Australia, UK and France. By conducting a vast study on European rabbit exomes collected before and after the 1950s pandemic for these three geographically distinct populations, a strong pattern of parallel evolution, with selection on standing genetic variation favouring the same alleles was described. Many of these play a regulatory role in the innate immune response, including IFN-a21A, CD200-R, MHC class I, FCRL3, CD96 and MFSD1 (Alves *et al.*, 2019).

The evolutionary and genetic aspects of some other host immune genes likely to play a crucial role in fighting, restricting and inhibiting MYXV infection and replication were studied in different Leporidae genera. The first to be studied was the C-C motif chemokine receptor 5 (*CCR5*) gene in *Oryctolagus*, *Sylvilagus* and *Lepus* genera (Carmo *et al.*, 2006). CCR5 is a seven-transmembrane, specific G protein-coupled receptor that is activated by several low-molecular-weight chemotactic cytokines, named chemokines (Fenner and Ratcliffe, 1965; Zlotnik and Yoshie, 2000). Some of these chemokines are crucial in immune and inflammatory responses when produced by cells during infection or in the presence of a pro-inflammatory stimulus (Fenner and Ratcliffe, 1965; Zlotnik and Yoshie, 2000; Zlotnik and Yoshie, 2012). The study of Leporidae *CCR5* genes revealed the first striking difference between genera and, more importantly, between *Oryctolagus* and *Sylvilagus* (Carmo *et al.*, 2006). Indeed, *Oryctolagus CCR5* experienced a gene conversion event with C-C motif chemokine receptor 2 (*CCR2*) in the second extracellular loop, which was not observed in *Sylvilagus* and *Lepus*, a third Leporidae genus with some evidences of susceptibility to MYXV (Fenner and Ratcliffe, 1965). Indeed, recently, Iberian hares (*Lepus granatensis*) were found dead with lesions consistent with myxomatosis. The sequencing of Myxoma virus detected in Iberian hare tissue samples showed a new MYXV (MYXV Toledo). The genome of this new virus encodes a novel ~2.8 kb recombinant region, which resulted from an insertion of four novel poxviral genes (Agueda-Pinto *et al.*, 2019). The functional significance of this gene conversion event in *Oryctolagus CCR5* remains unknown, but prompted the study of the genetic aspects of four of CCR5 chemokine ligands, CCL3, CCL4, CCL5 and CCL8 (van der Loo *et al.*, 2012; de Matos *et al.*, 2013). The study on Leporidae *CCL8* revealed an *Oryctolagus-Bunolagus* lineage-specific feature, as *CCL8* gene was pseudogenized, while it was intact in the *Lepus-Sylvilagus* lineage (van der Loo *et al.*, 2012, 2016). Selection analyses on CCL3, CCL4 and CCL5 demonstrated that these inflammatory CCR5 ligands are under strong purifying selection. Also, the low number of genetic differences between *Oryctolagus*, *Sylvilagus* and *Lepus* suggested functional binding constraints in CCL3, CCL4 and CCL5, yet it is fundamental to assess whether the species- or genus-specific amino acid substitutions observed in signalling and receptor-binding regions of chemokines influence ligand-receptor binding (de Matos *et al.*, 2013). The sensing of MYXV infection in primary human macrophages by the retinoic acid-inducible gene-I (RIG-I) protein (Wang *et al.*, 2008), a DExD/H box RNA helicase, has driven the search for evidence of positive selection in the three members of the RIG-I-like receptor family in mammals (Lemos de Matos *et al.*, 2013a). As a result, clear evidence of positive selection operating in the host innate sentinel RIG-I protein, melanoma differentiation associated factor 5 (MDA5) protein and laboratory of genetics and physiology 2 (LGP2) protein was found (Lemos de Matos *et al.*, 2013a). When comparing specifically European rabbit and brush rabbit RIG-I proteins, the amino acid differences of positively selected codons between the two species were located in the repressor domain (Lemos de Matos *et al.*, 2013b), the RIG-I region responsible for recognition and binding to RNA from actively replicating viruses in infected cells (Bruns and Horvath, 2012; Schmidt *et al.*, 2012). These insights into the evolution and genetic aspects of some Leporidae host immune genes might help highlight the origins of the species-specific innate responses to MYXV and the different species susceptibility/resistance to myxomatosis.

9.6 Rabbit Haemorrhagic Disease (RHD)

The rabbit haemorrhagic disease (RHD) was first described in 1984 in China (Liu *et al.*, 1984). This disease induces an acute liver necrosis and is responsible for high mortality in wild and domestic rabbits leading to significant ecologic and economic losses (Abrantes *et al.*, 2012). The virus that causes RHD is a Calicivirus named rabbit haemorrhagic disease virus (RHDV). Non-pathogenic and moderately pathogenic strains have also been described (Capucci *et al.*, 1996; Forrester *et al.*, 2007; Bergin *et al.*, 2009; Strive *et al.*, 2009). According to a recently proposed nomenclature, pathogenic strains previously ascribed to genogroups G1-G6 now constitute the *Lagovirus europaeus*/GI.1 genotype, while non-pathogenic strains belong to genotypes GI.3 and GI.4 (Le Pendu *et al.*, 2017).

In 2010, atypical RHD outbreaks were detected in France in rabbits vaccinated against RHDV (Le Gall-Recullé *et al.*, 2011a, 2011b). Molecular characterization of the causative strains showed the emergence of a new genotype, GI.2, that differs by more than 15% from the other genotypes and represents a distinct phylogenetic cluster (Le Pendu *et al.*, 2017). In addition, GI.2 strains have the ability to kill young rabbits (less than 2 months of age). GI.2 strains have been widely dispersed (Rouco *et al.*, 2019) and in some places, such as the Iberian Peninsula, replaced the former circulating GI.1 strains, probably due to a selective advantage (Dalton *et al.*, 2014; Lopes *et al.*, 2014). Emergence of RHDV as a pathogenic virus is controversial, and the hypotheses to explain this emergence include evolution from a non-pathogenic form of the virus or through a species-jump (Esteves *et al.*, 2015). Evolutionary studies showed that the RHDV strains evolved under positive selection and suffered several recombination events (Abrantes *et al.*, 2008; Esteves *et al.*, 2008; Forrester *et al.*, 2008; Alda *et al.*, 2010; Kinnear and Linde, 2010; Lopes *et al.*, 2015, 2017; Hu *et al.*, 2017; Silvério *et al.*, 2018).

Due to the lack of a suitable cell culture system for RHDV, most of the current knowledge on the virus infectious cycle is based on animal experimentation or the employment of indirect systems. These systems demonstrated that the RHDV pathogenic strains have the ability to bind to histo-blood group antigen (HBGAs) present on the surface of host epithelial cells of the respiratory and digestive tracts, most likely co-factors to initiate their infectious cycle (Ruvoen-Clouet *et al.*, 2000; Nyström *et al.*, 2011; Lopes *et al.*, 2018). Expression of HBGAs H/A/B type 2/Le^y is highly variable in rabbits and the RHDV strains have been co-evolving in order to maintain their ability to recognize these host glycans (Nyström *et al.*, 2011; Lopes *et al.*, 2018). Synthesis of these antigens is complex and might require several glycosyltransferase enzymes; for example, synthesis of H type 2 is performed by an α1,2-fucosyltransferase that adds a fucose in α1,2 linkage to a precursor. In rabbits, α1,2-fucosyltransferases are encoded by three genes located in tandem: *FUT1*, *FUT2* and *SEC1*. In mammals, these genes originated from two successive duplications, with *FUT1* being generated first and then *FUT2* and *SEC1* (Apoil *et al.*, 2000). Evolutionary studies further showed that *FUT2* and *SEC1* suffered multiple gene-conversion events, while in the European rabbit, the three genes have suffered gene conversion events (Abrantes *et al.*, 2009). A study on a French rabbit population aiming at finding null alleles at the α1,2-fucosyltransferase genes *FUT2* and *SEC1* that would encode non-functional or weakly functional enzymes constituting (non-secretor) resistant phenotypes, did not detect such alleles (Guillon *et al.*, 2009). However, the authors showed an association of a *SEC1* allele and the increase of survival rate, but this allele was always associated with a functional *FUT2* allele. The authors hypothesized that mutations at regulatory regions rather than at coding regions might be responsible for resistance to RHDV. A follow-up study suggested that in leporids, unlike in most other mammals where it became extinct, *SEC1* evolved a new function and has dominant-negative effect on *FUT1* (Nyström *et al.*, 2015). This negative effect leads to the suppression of the activity of the *FUT1*-encoded α1,2-fucosyltransferases and of the synthesis of RHDV ligands, impairing the virus attachment to the host cells. However, since such glycans are not present in adult rabbit hepatocytes (Ruvoen-Clouet *et al.*, 2000; Nyström *et al.*, 2011), additional hepatic cellular receptor(s) must be involved in virus infection, but remain to be identified.

Although it has been shown that upon infection young rabbit hepatocytes may present damage (Mikami *et al.*, 1999; Prieto *et al.*, 2000; Ferreira *et al.*, 2006b), young and adult rabbits present a marked difference of resistance/susceptibility to GI.1. In young rabbits, enhanced expression of IL-1, IL-6 and IL-8, TNF-α, IFN-α, IFN-γ may have an important role in their natural resistance. Indeed, GI.1 infection in young rabbits leads to the production of these pro-inflammatory cytokines at early times (Marques *et al.*, 2012). An increase in the B and T leukocyte populations is also observed, while in adult rabbits infection induces an early depletion of these populations in the liver and the spleen which critically compromises the immune response (Ferreira *et al.*, 2006a, Marques *et al.*, 2010). It was further demonstrated that by supressing their innate immune response, young rabbits become as fatally susceptible as adult rabbits to GI.1 (Marques *et al.*, 2014). More recently was shown the upregulation of several innate immunity-related genes in young rabbits infected with

GI.1 (e.g. MHC class II, natural cytotoxicity triggering receptor 3, macrophage colony-stimulating factor 1 receptor, CD68) (Neave *et al.*, 2018). Thus, in young rabbits, resistance to RHD caused by GI.1 correlates with a rapid, co-ordinated and efficient innate immune response. As for GI.2, which kills rabbits below 2 months of age that are resistant to GI.1, there is a downregulation of several MHC class I genes, following infection of young rabbits suggesting a suppression of the innate immune response by the virus. However, the mechanisms employed by the virus to surpass such response that could provide additional hints on resistance against GI.2 are currently unknown.

The first step of the immune response is the recognition of foreign antigens. Despite the lack of specificity, innate immunity recognizes a wide variety of pathogens via pattern-recognition receptors (PRRs). PRRs sense generic signals in the pathogens, i.e. the pathogen-associated molecular patterns (PAMPs) and include the toll-like receptors (TLRs) for which little is known in rabbits, in particular on their role in RHDV. In humans, TLR3, 7-9 are expressed intracellularly and can detect viral nucleic acids (Kumar *et al.*, 2009). In rabbits, *TLR7* and *TLR8*, which are involved in the recognition of single-stranded RNA viruses such as RHDV, are absent and pseudogenized, respectively (Astakhova *et al.*, 2009). Since TLR9 senses double-stranded DNA viruses, only TLR3 might be able to detect and signal the presence of viral RNA. A recent study on *TLR3* genetic diversity on rabbits found higher levels of diversity on wild animals (Abrantes *et al.*, 2013), but how this correlates with resistance has yet to be assessed. Other PRRs might be involved in recognition of RHDV.

9.7 Concluding Remarks

Genetic variability on resistance to diseases exists in rabbits. Several studies demonstrated that resistance to specific diseases (such as pasteurellosis, ERE, myxomatosis) as well as unspecific diseases (infectious diseases, respiratory syndromes, digestive syndromes) might exist. Host resistance mechanisms have been more extensively studied in the case of myxomatosis and RHD. Concerning the other diseases, they are largely unknown. More widespread use of genomic tools will provide valuable insights into the genetic mechanisms involved in disease resistance in the European rabbit.

References

Abrantes, J., Esteves, P.J. and van der Loo, W. (2008) Evidence for recombination in the major capsid gene VP60 of the rabbit haemorrhagic disease virus (RHDV). *Archives of Virology* 153, 329–335.

Abrantes, J., Posada, D., Guillon, P., Esteves, P.J. and Le Pendu, J. (2009) Widespread gene conversion of alpha-2-fucosyltransferase genes in mammals. *Journal of Molecular Evolution* 69, 22–31.

Abrantes, J., van der Loo, W., Le Pendu, J. and Esteves, P.J. (2012) Rabbit haemorrhagic disease (RHD) and rabbit haemorrhagic disease virus (RHDV): a review. *Veterinary Research* 43, 12.

Abrantes, J., Areal, H. and Esteves, P.J. (2013) Insights into the European rabbit (*Oryctolagus cuniculus*) innate immune system: genetic diversity of the toll-like receptor 3 (TLR3) in wild populations and domestic breeds. *BMC Genetics* 14, 73.

Águeda-Pinto, A., Lemos de Matos, A., Abrantes, M., Kraberger, S., Risalde, M.A. *et al.* (2019) Genetic characterization of a recombinant myxoma virus leap into the Iberian hare (*Lepus granatensis*). *Viruses* 11, 530.

Alda, F., Gaitero, T., Suárez, M., Merchán, T., Rocha, G. and Doadrio, I. (2010) Evolutionary history and molecular epidemiology of rabbit haemorrhagic disease virus in the Iberian Peninsula and Western Europe. *BMC Evolutionary Biology* 10, 347.

Alves, J.M., Carneiro, M., Cheng, J.Y., Lemos de Matos, A., Rahman, M.M. *et al.* (2019) Parallel adaptation of rabbit populations to myxoma virus. *Science* 363, 1319–1326.

Apoil, P.A., Roubinet, F., Despiau, S., Mollicone, R., Oriol, R and Blancher, A. (2000) Evolution of alpha-2-fucosyltransferase genes in primates: relation between an intronic Alu-Y element and red cell expression of ABH antigens. *Molecular Biology and Evolution* 17, 337–351.

Aragão, H.D.B. (1927) Myxoma dos coelhos. *Memórias do Instituto Oswaldo Cruz* 20, 225–247.

Aragão, H.D.B. (1943) O virus do mixoma no coelho do mato (*Sylvilagus minenses*), sua transmissão pelos *Aedes scapularis* e *aegypti*. *Memórias do Instituto Oswaldo Cruz* 38, 93–99.

Astakhova, N.M., Perelygin, A.A, Zharkikh, A.A., Lear, T.L., Coleman, S.J. *et al.* (2009) Characterization of equine and other vertebrate TLR3, TLR7, and TLR8 genes. *Immunogenetics* 61, 529–539.

Baselga, M., Deltoro, J., Camacho, J. and Blasco, A. (1988). Genetic analysis on lung injury in four strains of meat rabbit. In: *Proceedings of the 4th World Rabbit Congress, Budapest*, vol. 1, pp. 120–127.

Bergin, I.L., Wise, A.G., Bolin, S.R., Mullaney, T.P., Kiupel, M. and Maes, R.K. (2009) Novel calicivirus identified in rabbits, Michigan, USA. *Emerging Infectious Diseases* 15, 1955–1962.

Best, S.M., Collins, S.V. and Kerr, P.J. (2000) Coevolution of host and virus: cellular localization of virus in myxoma virus infection of resistant and susceptible European rabbits. *Virology* 277, 76–91.

Bishop, S.C. and Morris, C.A. (2007) Genetics of disease resistance in sheep and goats. *Small Ruminant Research* 70, 48–59.

Bishop, S.C. and Woolliams, J.A. (2014) Genomics and disease resistance studies in livestock. *Livestock Science* 166, 190–198.

Bruns, A.M. and Horvath, C.M. (2012) Activation of RIG-I-like receptor signal transduction. *Critical Reviews in Biochemistry and Molecular Biology* 47, 194–206.

Capucci, L., Fusi, P., Lavazza, A., Pacciarini, M.L. and Rossi, C. (1996) Detection and preliminary characterization of a new rabbit calicivirus related to rabbit hemorrhagic disease virus but nonpathogenic. *Journal of Virology* 70, 8614–8623.

Carmo, C.R., Esteves, P.J., Ferrand N. and van der Loo W (2006) Genetic variation at chemokine receptor CCR5 in leporids: alteration at the 2nd extracellular domain by gene conversion with CCR2 in *Oryctolagus*, but not in *Sylvilagus* and *Lepus* species. *Immunogenetics* 58, 494–501.

Chapman, J.A. and Flux, J.E.C. (1990) Chapter 1: Introduction and Overview of the Lagomorphs. In: Chapman, A.J. and Flux, J. E. C. (eds) *Rabbits, Hares and Pikas - Status Survey and Conservation Action Plan*. IUCN, Gland, Switzerland, pp. 1–6.

Chen, S.Y., Zhang, W.X., Zhang, G.W., Peng, J., Zhao, X.B. and Lai, S.J. (2013) Case-control study and mRNA expression analysis reveal the MyD88 gene is associated with digestive disorders in rabbit. *Animal Genetics* 44, 703–710.

Coudert, P. (1989) Some peculiarities of rabbit coccidiosis. In: *Proceedings of the 5th International Coccidiosis Conference*. INRA. Tours, France, pp. 481–488.

Coudert, P., Rideau, P., Virag, G. and Cerrone, A. (2006) Pasteurellosis in rabbit. In: Maertens, L. and Coudert, P. (eds) *Recent Advances in Rabbit Sciences*, Ilvo, Melle, Belgium, pp. 147–162.

Dalton, K.P., Nicieza, I., Abrantes, J., Esteves, P.J. and Parra, F. (2014) Spread of new variant RHDV in domestic rabbits on the Iberian Peninsula. *Veterinary Microbiology* 169, 67–73.

de Matos, A.L., Lanning, D.K. and Esteves, P.J. (2013) Genetic characterization of CCL3, CCL4 and CCL5 in leporid genera *Oryctolagus*, *Sylvilagus* and *Lepus*. *International Journal of Immunogenetics* 41, 154–158.

Dempster, E.R. and Lerner, I.M. (1950) Heritability of threshold characters. *Genetics* 2012, 212–236.

Eady, S.J. (2003) *Farmed Rabbits in Australia*. RIRDC Publication No. 02/144. Rural Industries Research and Development Corporation, Canberra.

Eady, S.J., Garreau, H and Hurtaud, J. (2004) Heritability of resistance to bacterial infection in commercial meat rabbit populations. In: *Proceedings of the 8th World Rabbit Congress.* vol. 1, pp. 51–56. Puebla, Mexico.

Eady S.J., Garreau, H. and Gilmour, A.R. (2007) Heritability of resistance to bacterial infection in meat rabbits, *Livestock Science* 112, 90–98.

Esteves, P.J., Abrantes, J., Carneiro, M., Muller A., Thompson, G. and van der Loo, W. (2008) Detection of positive selection in the major capsid protein VP60 of the rabbit haemorrhagic disease virus (RHDV). *Virus Research* 137, 253–256.

Esteves, P.J., Abrantes, J., Bertagnoli, S., Cavadini, P., Gavier-Widén, D. *et al.* (2015) Emergence of pathogenicity in lagoviruses: evolution from pre-existing non-pathogenic strains or through a species jump? *PLOS Pathogens* 11, e1005087.

Falconer, D.S. (1965) The inheritance of liability to certain diseases estimated from the evidence among relatives. *Annals of Human Genetics* 29, 51–76.

Fenner, F. (1979) Portraits of viruses: the poxviruses. *Intervirology* 11, 137–157.

Fenner, F. (1983) Biological control, as exemplified by smallpox eradication and myxomatosis. *Proceedings of the Royal Society B: Biological Sciences* 218, 259–285.

Fenner, F. and Fantini, B. (1999) *Biological Control of Vertebrate Pests: The History of Myxomatosis. An Experiment in Evolution*. CAB International, Wallingford.

Fenner, F. and Marshall, I.D. (1957) A comparison of the virulence for European rabbits (*Oryctolagus cuniculus*) of strains of myxoma virus recovered in the field in Australia, Europe and America. *The Journal of Hygiene (Camb.)* 55, 149–191.

Fenner, F. and Ratcliffe, F. (1965) *Myxomatosis*. Cambridge University Press, Cambridge.

Fenner, F., Marshall, I.D. and Woodroofe, G.M. (1953) Studies in the epidemiology of infectious myxomatosis of rabbits. I. Recovery of Australian wild rabbits (*Oryctolagus cuniculus*) from myxomatosis under field conditions. *The Journal of Hygiene (Camb.)* 51, 225–244.

Ferrand, N. and Branco, M. (2007) The evolutionary history of the European rabbit (*Oryctolagus cuniculus*): major patterns of population differentiation and geographic expansion inferred from protein polymorphism. In: Weiss, S. and Ferrand, N. (eds) *Phylogeography of Southern European Refugia*. Springer, Dordecht, The Netherlands, pp. 207–235.

Ferreira, P.G., Costa-e-Silva, A., Oliveira, M.J., Monteiro, E., Cunha, E.M. and Aguas, A.P. (2006a) Severe leukopenia and liver biochemistry changes in adult rabbits after calicivirus infection. *Research in Veterinary Science* 80, 218–225.

Ferreira, P.G., Costa-e-Silva, A. and Aguas, A.P. (2006b) Liver disease in young rabbits infected by calicivirus through nasal and oral routes. *Research in Veterinary Science* 81, 362–365.

Forrester, N.L., Trout, R.C. and Gould, E.A. (2007) Benign circulation of rabbit haemorrhagic disease virus on Lambay Island, Eire. *Virology* 358, 18–22.

Forrester, N.L., Moss, S.R., Turner, S.L., Schirrmeier, H. and Gould, E.A. (2008) Recombination in rabbit haemorrhagic disease virus: possible impact on evolution and epidemiology. *Virology* 376, 390–396.

Fu, L., Yang, Z.J., Chen, S.Y., Wang, J. and Lai, S.J. (2014) Investigation of JAK1 and STAT3 polymorphisms and their gene-gene interactions in nonspecific digestive disorder of rabbits. *Gene* 543, 8–14.

Garreau, H., Licois, D., Rupp, R. and Rochambeau, H. de (2006) Genetic variability of the resistance to epizootic rabbit enteropathy (ERE): new results. In: *Proceedings of the 8th World Congress for Genetics Applied to Livestock Production*, Belo Horizonte, Brazil, pp. 15–20.

Garreau, H., Eady, S.J., Hurtaud, J. and Legarra, A. (2008) Genetic parameters of production traits and resistance to digestive disorders in a commercial rabbit population. In: *Proceedings of the 9th World Rabbit Congress*, Verona, Italy, pp. 103–107.

Garreau, H., Brard, S., Hurtaud, J., Guitton, E., Cauquil, L. *et al.* (2012) Divergent selection for digestive disorders in two commercial rabbit lines: response of crossbred young rabbits to an experimental inoculation of *Escherichia coli* 0103. In: *Proceedings of the 10th World Rabbit Congress*, Sharm El-Sheikh, Egypt, pp. 217–221.

Gidenne, T. (2003) Fibres in rabbit feeding for digestive troubles prevention: respective role of low-digested and digestible fibre. *Livestock Production Science* 81, 105–117.

Goñi, P., Vergara, Y., Ruiz, J., Albizu, I., Vila, J. and Gomez-Lus, R. (2004) Antibiotic resistance and epidemiological typing of *Staphylococcus aureus* strains from ovine rabbit mastitis. *International Journal of Antimicrobial Agents* 23, 268–272.

Guillon, P., Ruvoen-Clouet, N., Le Moullac-Vaidye, B., Marchandeau, S. and Le Pendu, J. (2009) Association between expression of the H histo-blood group antigen, alpha1,2fucosyltransferases polymorphism of wild rabbits, and sensitivity to rabbit hemorrhagic disease virus. *Glycobiology* 19, 21–28.

Gunia, M., David, I., Hurtaud, J., Maupin, M., Gilbert, H. and Garreau, H. (2015) Resistance to infectious diseases is a heritable trait in rabbits. *Journal of Animal Science* 93, 5631–5638.

Gunia, M., David, I., Hurtaud, J., Maupin, M., Gilbert, H. and Garreau, H. (2018) Genetic parameters for resistance to non-specific diseases and production traits measured in challenging and selection environments; application to a rabbit case. *Frontiers in Genetics* 9, 467.

Helloin, E., Garreau, H., Slugocki, C., Hilgenberg, M., Le Beux, C. *et al.* (2009) Un projet pour améliorer la résistance du lapin à la pasteurellose: travaux en cours et perspectives. In: *Proceedings of the 13èmes Journées de la Recherche Cunicole*, Le Mans, France, pp. 223–226.

Helloin, E., Lantier, I., Slugocki, C., Chambellon, E., Le Roux, H. *et al.* (2015) Vers une amélioration de la résistance du lapin à la pasteurellose. In: *Proceedings of the 16èmes Journées de la Recherche Cunicole*, Le Mans, France, pp. 43–46.

Hu, B., Wang, F., Fan, Z., Song, Y., Abrantes, J. *et al.* (2017) Recombination between G2 and G6 strains of rabbit hemorrhagic disease virus (RHDV) in China. *Archives of Virology* 162, 269–272.

Hurst, E.W. (1937) Myxoma and the Shope fibroma. I. The histology of myxoma. *British Journal of Experimental Pathology* 18, 1–15.

Jeklova, E., Leva, L., Matiasovic, J., Kovarcik, K., Kudlackova, H. *et al.* (2008) Characterisation of immunosuppression in rabbits after infection with myxoma virus. *Veterinary Microbiology* 129, 117–130.

Jones, J.R. and Duff, J.P. (2001) Rabbit epizootic enterocoltis. *Veterinary Record* 149, 532–532.

Kerr, P.J. (2012) Myxomatosis in Australia and Europe: a model for emerging infectious diseases. *Antiviral Research* 93, 387–415.

Kerr, P.J. and Donnelly, T.M. (2013) Viral infections of rabbits. *Veterinary Clinics of North America: Exotic Animal* 16, 437–468.

Kerr, P.J. and McFadden, G. (2002) Immune responses to myxoma virus. *Viral Immunology* 15, 229–246.

Kerr, P.J., Cattadori, I.M., Liu, J., Sim, D.G., Dodds, J.W. *et al.* (2017) Next step in the ongoing arms race between myxoma virus and wild rabbits in Australia is a novel disease phenotype. *Proceedings of the National Academy of Sciences of the USA* 114, 9397–9402.

Kinnear, M. and Linde, C.C. (2010) Capsid gene divergence in rabbit hemorrhagic disease virus. *Journal of General Virology* 91, 174–181.

Kumar, H., Kawai, T. and Akira, S. (2009) Toll-like receptors and innate immunity. *Biochemical and Biophysical Research Communications* 388, 621–625.

Lebas, F., Couldert, P., De Rochambeau, H. and Thebault, R.G. (1997) *The Rabbit - Husbandry, Health and Production*. FAO Animal Production and Health, FAO, Rome.

Le Gall-Reculé, G., Zwingelstein, F., Fages, M.P., Bertagnoli, S., Gelfi, J. *et al.* (2011a) Characterisation of a non-pathogenic and non-protective infectious rabbit lagovirus related to RHDV. *Virology* 410, 395–402.

Le Gall-Reculé, G., Zwingelstein, F., Boucher, S., Le Normand, B., Plassiart, G. *et al.* (2011b) Detection of a new variant of rabbit haemorrhagic disease virus in France. *Veterinary Records* 168, 137–138.

Lemos de Matos, A., McFadden, G. and Esteves, P.J. (2013a) Positive evolutionary selection on the RIG-I-like receptor genes in mammals. *PLOS ONE* 8, e81864.

Lemos de Matos, A., McFadden, G. and Esteves, P.J. (2013b) Evolution of viral sensing RIG-I-like receptor genes in Leporidae genera *Oryctolagus, Sylvilagus*, and *Lepus*. *Immunogenetics* 66, 43–52.

Le Pendu. J., Abrantes, J., Bertagnoli, S., Guitton, J.S., Le Gall-Reculé, G. *et al.* (2017) Proposal for a unified classification system and nomenclature of lagoviruses. *Journal of General Virology* 98, 1658–1666.

Licois, D., Guillot, J.F., Mouline C. and Reynaud, A. (1992) Susceptibility of the rabbit to an enteropathogenic strain of Escherichia coli 0103: effects of animals' age. *Annales de Recherche Vétérinaires* 23, 225–232.

Licois, D., Coudert, P., Ceré, N. and Vautherot, J.F. (2000) Epizootic enterocolitis of the rabbit: review of current research. In: *Proceedings of the 7th World Rabbit Congress*, Valencia, Spain, pp. 187–194.

Licois, D., Wyers, M. and Coudert, P. (2005) Epizootic rabbit enteropathy: experimental transmission and clinical characterization. *Veterinary Research* 36, 601–613.

Liu, S.J., Xue, H.P., Pu, B.Q. and Qian, N.H. (1984) A new viral disease in rabbit. *Animal Husbandry, Veterinary Medicine* 16, 253–255.

Liu, Y.F., Zhang, G.W., Xiao, Z.L., Yang, Y., Deng, X.S. *et al.* (2013) Single nucleotide polymorphisms of NLRP12 gene and association with non-specific digestive disorder in rabbit. *Asian-Australasian Journal of Animal Sciences* 26, 1072–1079.

Lopes, A.M., Correia, J., Abrantes, J., Melo, P., Ramada, M. *et al.* (2014) Is the new variant RHDV replacing genogroup 1 in Portuguese wild rabbit populations? *Viruses* 7, 27–36.

Lopes, A.M., Dalton, K.P., Magalhães, M.J., Parra, F., Esteves, P.J. *et al.* (2015) Full genomic analysis of new variant Rabbit Hemorrhagic Disease Virus (RHDVb) revealed multiple recombination events. *Journal of General Virology* 96, 1309–1319.

Lopes, A.M., Silvério, D., Magalhães, M.J., Areal, H., Alves, P.C. *et al.* (2017) Characterization of old RHDV strains by complete genome sequencing identifies a novel genetic group. *Scientific Reports* 7, 13599.

Lopes, A.M., Breiman, A., Lora, M., Le Moullac-Vaidye, B. and Galanina, O. (2018) Host-specific glycans are correlated with susceptibility to infection by Lagoviruses, but not with their virulence. *Journal of Virology* 92, e01759–17.

Maertens, L. and Coudert, P. (2006) *Recent Advances in Rabbit Sciences*. Plot-it bvba. Merelbeke, Belgium.

Marlier, D., Dewrée, R., Delleur, V., Licois, D., Lassence, C. *et al.* (2003) Description des principales étiologies des maladies digestives chez le lapin européen (*Oryctolagus cuniculus*). *Annales de Médecine Vétérinaire* 147, 385–392.

Marques, R.M., Costa, E.S.A., Aguas, A.P., Teixeira, L. and Ferreira, P.G. (2010) Early acute depletion of lymphocytes in calicivirus-infected adult rabbits. *Veterinary Research Communications* 34, 659–668.

Marques, R.M., Costa, E.S.A., Aguas, A.P., Teixeira, L. and Ferreira, P.G. (2012) Early inflammatory response of young rabbits attending natural resistance to calicivirus (RHDV) infection. *Veterinary Immunology and Immunopathology* 150, 181–188.

Marques, R.M., Teixeira, L., Aguas, A.P., Ribeiro, J.C., Costa-e-Silva, A. and Ferreira, P.G. (2014) Immunosuppression abrogates resistance of young rabbits to Rabbit Haemorrhagic Disease (RHD). *Veterinary Research* 45, 14.

Marshall, I.D. and Regnery, D.C. (1960) Myxomatosis in a California brush rabbit (*Sylyilagus bachmani*). *Nature* 188, 73–74.

Matthee, C.A., van Vuuren, B.J., Bell, D. and Robinson, T.J. (2004) A molecular supermatrix of the rabbits and hares (Leporidae) allows for the identification of five intercontinental exchanges duringthe Miocene. *Systematic Biology* 53, 433–447.

Mikami, D., Park, J.H., Kimura, T., Ochiai, K. and Itakura, C. (1999) Hepatic lesions in young rabbits experimentally infected with rabbit haemorrhagic disease virus. *Research in Veterinary Science* 66, 237–242.

Moses, A. (1911) O virus do mixoma dos coelhos. *Memórias do Instituto Oswaldo Cruz* 3, 46–53.

Neave, M.J., Hall, R.N., Huang, N., McColl, K.A., Kerr, P. *et al.* (2018) Robust innate immunity of young rabbits mediates resistance to rabbit hemorrhagic disease caused by Lagovirus Europaeus GI.1 but not GI.2. *Viruses* 10, E512.

Nyström, K., Le Gall-Reculé, G., Grassi, P., Abrantes, J. and Ruvoën-Clouet, N. (2011) Histo Blood Group Antigens act as attachment factors of Rabbit Hemorrhagic Disease Virus infection in a virus strain-dependent manner. *PLOS Pathogens* 7, e1002188.

Nyström, K., Abrantes, J., Lopes, A.M., Vaidye, B., Marchandeau, S. *et al.* (2015) Neofunctionalization of the Sec1 α 1,2fucosyltransferases in leporids contributes to glycan polymorphism and resistance to Rabbit Hemorrhagic Disease Virus. *PLOS Pathogens* 11, e1004759.

Prieto, J.M., Fernandez, F., Alvarez, V., Espi, A., Garcia Marin, J.F. *et al.* (2000) Immunohistochemical localisation of rabbit haemorrhagic disease virus VP-60 antigen in early infection of young and adult rabbits. *Research in Veterinary Science* 68, 181–187.

Raadsma, H.W., Gray, G.D. and Woolaston, R.R. (1997) Genetics of disease resistance and vaccine response. In: Piper, L. and Ruvinsky, A. (eds) *The Genetics of Sheep.* CAB International, New York, pp. 201–202.

Ragab, M., Ramon, J., Rafel, O., Quintanilla, R., Piles, M. and Sanchez, J.P. (2015) Paramètres génétiques des phénotypes liés aux maladies chez le lapin en engraissement nourri avec deux régimes alimentaires différents. In: *Proceedings of the 16èmes Journées de la Recherche Cunicole*, Le Mans, France, pp. 69–72.

Regnery, D.C. and Miller, J.H. (1972) A myxoma virus epizootic in a brush rabbit population. *Journal of Wildlife Diseases* 8, 327–331.

Rochambeau, H. de, Licois, D., Gidenne, T., Verdelhan, S., Couldert, P. *et al.* (2006) Genetic variability of the resistance for three types of enteropathy in the growing rabbit. *Livestock Science* 101, 110–115.

Rosell, J.M. (2003) Health status of commercial rabbitries in the Iberian Peninsula: a practitioner's study. *World Rabbit Science* 11, 157–169.

Rouco, C., Aguayo-Adan, J.A., Santoro, S., Abrantes, J. and Delibes-Mateos, M. (2019) Worldwide rapid spread of the novel rabbit haemorrhagic disease virus (GI.2/RHDV2/b). *Transboundary and Emerging Diseases* 66, 1762–1764.

Ruvoen-Clouet, N., Ganiere, J.P., Andre-Fontaine, G., Blanchard D. and Le Pendu, J. (2000) Binding of rabbit hemorrhagic disease virus to antigens of the ABH histo-blood group family. *Journal of Virology* 74, 11950–11954.

Schmidt, A., Rothenfusser, S. and Hopfner, K.P. (2012) Sensing of viral nucleic acids by RIG-I: from translocation to translation. *European Journal of Cell Biology* 91, 78–85.

Shrestha, M., Garreau H., Balmisse, E., Bed'hom B., David, I. *et al.* (2018) Estimation of genetic parameters of pasteurellosis resistance in crossbred rabbits. *Proceedings of the World Congress on Genetics Applied to Livestock Production*, Auckland, New Zealand.

Silvério, D., Lopes, A.M., Melo-Ferreira, J., Magalhães, M.J., Monterroso, P. *et al.* (2018) Insights into the evolution of the new variant rabbit haemorrhagic disease virus (GI.2) and the identification of novel recombinant strains. *Transboundary and Emerging Diseases* 65, 983–992.

Silvers, L., Inglis, B., Labudovic, A., Janssens, P.A., van Leeuwen, B.H.and Kerr, P.J. (2006) Virulence and pathogenesis of the MSW and MSD strains of Californian myxoma virus in European rabbits with genetic resistance to myxomatosis compared to rabbits with no genetic resistance. *Virology* 348, 72–83.

Sobey, W.R. (1969) Selection for resistance to myxomatosis in domestic rabbits (*Oryctolagus cuniculus*). *Journal of Hygiene* 67, 743–754.

Spiesschaert, B., McFadden, G., Hermans, K., Nauwynck, H. and Van de Walle, G.R. (2011) The current status and future directions of myxoma virus, a master in immune evasion. *Veterinary Research* 42, 76.

Stanford, M.M., Werden, S.J. and McFadden, G. (2007) Myxoma virus in the European rabbit: interactions between the virus and its susceptible host. *Veterinary Research* 38, 299–318.

Strive, T., Wright, J.D. and Robinson, A.J. (2009) Identification and partial characterisation of a new Lagovirus in Australian wild rabbits. *Virology* 384, 97–105.

van der Loo, W., Abrantes, J. and Esteves, P.J. (2009) Sharing of endogenous lentiviral gene fragments among leporid lineages separated for more than 12 million years (genera *Lepus, Sylvilagus, Bunolagus* and *Oryctolagus*). *Journal of Virology* 83, 2386–2388.

van der Loo, W., Afonso, S., de Matos, A.L., Abrantes, J. and Esteves, P.J. (2012) Pseudogenization of the MCP-2/CCL8 chemokine gene in European rabbit (genus *Oryctolagus*), but not in species of Cottontail rabbit (*Sylvilagus*) and Hare (*Lepus*). *BMC Genetics* 13, 72.

van der Loo, W., Magalhaes, M.J., Lemos de Matos, A., Abrantes, J., Yamada, F. and Esteves, P.J. (2016) Adaptive gene loss? Tracing back the pseudogenization of the rabbit CCL8 chemokine. *Journal of Molecular Evolution* 83, 12–25.

Wang, F., Gao, X., Barrett, J.W., Shao, Q., Bartee, E. *et al.* (2008) RIG-I mediates the co-induction of tumor necrosis factor and type I interferon elicited by myxoma virus in primary human macrophages. *PLOS Pathogens* 4, e1000099.

Yang, Y., Zhang, G.W., Chen, S.Y., Peng, J. and Lai, S.J. (2013) Polymorphism of NLRP3 gene and association with susceptibility to digestive disorders in rabbit. *Asian-Australasian Journal of Animal Science* 26, 455–462.

Zhang, G.W., Wang, H.Z., Chen, S.Y., Li, Z.C., Zhang, W.X. and Lai, S.J. (2011) A reduced incidence of digestive disorders in rabbits is associated with allelic diversity at the TLR4 locus. *Veterinary Immunology and Immunopathology* 144, 482–486.

Zhang, G.W., Zhang, W.X., Chen, S.Y., Yoshimura, Y., Isobe, N. and Lai, S.J. (2014) Dectin-1 gene polymorphism is associated with susceptibility to nonspecific digestive disorders and cytokine expression in rabbits. *Journal of Animal Science* 91, 4051–4059.

Zhang, W.X., Zhang, G.W., Peng, J., Zhang, J.L., Yang, Y. and Lai. S.J. (2013) A synonymous mutation in NOD2 gene was significantly associated with non-specific digestive disorder in rabbit. *Gene* 516, 193–197.

Zlotnik, A. and Yoshie, O. (2000) Chemokines: a new classification system and their role in immunity. *Immunity* 12, 121–127.

Zlotnik, A. and Yoshie, O. (2012) The chemokine superfamily revisited. *Immunity* 36, 705–716.

Zuniga, M.C. (2002) A pox on thee! Manipulation of the host immune system by myxoma virus and implications for viral-host co-adaptation. *Virus Research* 88, 17–33.

10 Genetics and Genomics of Growth, Carcass and Meat Production Traits in Rabbits

Agustin Blasco[1]*, István Nagy[2] and Pilar Hernández[1]
[1]*Institute for Animal Science and Technology, Universitat Politècnica de València, Spain;* [2]*Department of Animal Genetics and Biotechnology, Kaposvár University, Hungary*

10.1 Introduction

The most common breeding scheme in rabbit industrial production consists of two lines selected for litter size to produce a crossbred dam, and a terminal sire selected for growth rate. Selection for growth rate has consequences on carcass and meat quality. Although meat quality in rabbit is not paid nowadays, it is convenient to examine possible changes produced by selection. In this chapter we review the genetic variability between breeds and within breeds related to growth, carcass and meat quality, as well as on the selection experiments performed. This chapter updates several topics on this subject previously reviewed (Blasco *et al.*, 2018), adding new information about genomics and metagenomics experiments on these traits.

10.2 Economic Importance of Growth and Carcass Traits in Genetic Programmes

Growth has a decisive importance in rabbit meat production. Table 10.1 shows the costs of a typical industrial rabbitry that can be managed by one person. Feeding growing rabbits represents almost 30% of the total costs of each rabbit sold when rabbits are slaughtered at a young age (9 weeks of age in the Spanish case) and more than 30% for rabbits slaughtered at later ages. This implies that food conversion rate and correlated traits (residual feed intake, growth rate and slaughter weight) have a decisive importance in the profits. Economic weights and profit functions of rabbit production have been assessed by Armero and Blasco (1992), Prayaga and Eady (2000) and Cartuche *et al.* (2014).

The most important traits were feed conversion rate and litter size. The first represents the main part of the variable costs of the product sold (the rabbit to the slaughterhouse) whereas litter size is important in reducing the fixed costs, sharing them among more rabbits if litter size increases (Cartuche *et al.*, 2014). Other considered traits had lower relevance. Costs derived from traits affecting the reproductive stock, for example longevity or traits related to artificial insemination (AI), are shared by all rabbits produced by a doe or a sire, therefore the cost by rabbit sold becomes very small. Other traits can be important but are near their optimum, as survivals. As the only relevant return is the one coming from the liveweight of rabbits sold, feed efficiency and litter size become the keys of the profit in rabbit

Contact e-mail address: ablasco@dca.upv.es

DOI: 10.1079/9781780643342.0010

Table 10.1. Costs of an industrial rabbitry with 750 reproductive does. Weaning at 35 days and slaughter at 63 days (2.2 kg of live weight). Management in batches with artificial insemination. Costs are based on Cartuche *et al.* (2014).

	€ / kg live weight	€ / doe year	% total
Feeding does	0.28	32.7	15.9
Feeding rabbits for slaughter	0.53	60.5	29.4
Artificial insemination	0.08	8.69	4.2
Replacement reproductive stock	0.10	11.8	5.7
Health	0.12	14.3	6.9
Labour	0.32	37.3	18.1
Amortization	0.18	20.2	9.8
Others	0.19	20.7	10.1

meat production. Feed conversion rate measure needs special facilities and higher labour costs; thus, it is intended in selection programmes to measure correlated traits like growth rate with the objective of improving feed conversion rate. Sometimes feed intake or residual feed intake have been proposed to substitute feed conversion rate, but they have the same problem of expensive recording. Research is now being carried out to design electronic individual feed recording, as in pigs (Sánchez *et al.*, 2021). An alternative using food consumption of the cage with individual growth rate has been proposed by Piles *et al.* (2019) based on a model of Su *et al.* (2018) for pigs; this simplifies food recording and reduces its cost, although prediction of breeding values for feed efficiency have a lower accuracy than individual feed conversion.

In rabbits, selection programmes are based on three-way crosses. A crossbred commercial doe is produced crossing two lines selected for litter size, and one terminal sire comes from a line selected for average daily gain (ADG) (Baselga, 2004). This scheme is similar to that commonly used in pigs. Selection for lean content is applied in pigs, and often conformation and some meat-quality characteristics (e.g., Pale Soft Exudative meat: PSE) play important roles in this species but they are not relevant in rabbit. Rabbit has a very lean carcass when it is sold (Dalle Zotte, 2002; Hernández and Gondret, 2006), thus meat content is not hitherto an important trait. Conformation is not important in rabbit carcasses, and rabbit meat does not have problems like PSE meat, thus these traits are not included in commercial genetic programmes. Feed consumption is often measured in pigs using electronic tags, but they are still not available in rabbits, although this system is being developed. Moreover, labour costs and amortization costs are higher in rabbits (Table 10.1) and measuring feed conversion rate or feed consumption increases labour costs, which is not feasible in commercial rabbit programmes. Feed conversion rate is improved through selection for growth rate, but genetic correlations are not high and success in this field has been limited, as discussed later. An undesirable consequence of selection for growth rate is augmenting adult liveweight (Blasco *et al.*, 2003), but in commercial programmes this has limited effect, since artificial insemination is widely implemented and dams are not selected for increasing their growth.

Selection can also target dressing-out percentage or percentage of muscle in the carcass, but this cannot be performed in the candidate, thus sibs or progeny should be recorded; however, both ways are costly and less efficient than direct selection on the candidate. Selection schemes based on progeny testing of terminal sires for dressing-out percentage were carried out by Varewyck *et al.* (1986) and Szendrő *et al.* (1988), but the programmes were abandoned due to their high cost and larger generation interval. Nevertheless, application of non-invasive techniques such as computer tomography (CT) makes selection for slaughter traits using live animals possible, as shown by Szendrő *et al.* (2010) and Nagy *et al.* (2013), although this technique is also limited by its cost.

10.3 Genetics of Growth Traits

10.3.1 Between-breed genetic variability

Adult size is very variable between breeds in rabbits, from dwarf (around 1 kg adult weight) to giant (around 7 kg adult weight) lines. Medium-size breeds are used for reproduction in commercial production because of their high prolificacy, and large breeds are preferred for terminal sires because of their high growth rate. Using medium-size breeds for dams, not selected for growth, facilitates management, also lowering maintenance cost, and using large-size breeds selected for growth rate for terminal sires leads to commercial rabbits having a high growth rate.

Differences between lines or breeds due to genetic effects should be found by comparing animals at the same stage of maturity. Some lines grow quicker than others, thus they arrive to a commercial slaughter weight being younger. Therefore, meat quality of different breeds is different at the same commercial weight for two reasons: (i) the age; and (ii) the genetic background. Both effects are confused, thus they should be compared at the same stage of maturity; i.e. at the same proportion of adult weight, if the interest is knowing genetic differences between lines or breeds (Taylor, 1985). When breeds under different treatments, or other groups of animals, are compared at the same stage of maturity, many of the differences previously found disappear or are substantially reduced. When adult weight is not available, Pascual *et al.* (2015) showed in commercial rabbit lines that comparisons at the same age can be good approximations, but even at the same age, the stage of maturity of very different lines can also be different (Ouhayoun and Rouvier, 1973).

Bolet *et al.* (2000) and Ouhayoun and Poujardieu (1978) have published comparisons among breeds of several adult sizes. Large differences in average daily gain (ADG) between breeds were found as expected (more than 15 g/d between 4 and 11 weeks of age). It seems to be a negative favourable between-breeds relationship between growth rate and feed conversion ratio (FCR) (Ouhayoun and Poujardieu, 1978). FCR between 4 and 11 weeks varied from 3.61 in Flemish Giant to 4.52 in Small Russian. McCarthy (1980) has explained this as a consequence of better thermoregulation of heavy breeds; per kg of body weight (BW), maintenance energy is lower in giant lines since maintenance energy is proportional to metabolic weight ($BW^{0.75}$ for adult BW). Therefore, giant lines have more energy available for growing.

10.3.2 Genetic parameters of growth traits

Several estimates of heritability have been reported in the literature for weight at a fixed age, normally at slaughter time (SW), which is different in several countries (from 9 weeks in Spain to 13 weeks in the north of Italy), but also for weaning weight (WW), ADG and daily gain (SW-WW). Heritability of SW ranges from 0.12 to 0.67 with a mean of estimates of 0.27 as deduced from 17 publications summarized by Hernández and Gondret (2006). However, caution should be taken when examining average estimates of many studies. First, their quality is not the same, some of them having large standard errors or being biased because of the method of estimation or because the models were incomplete. Second, there are differences among farms in environmental variability, affecting not only the estimates but the actual heritabilities. Third, negative estimates are forced to be positive, leading to biased averages of estimates. Fourth, non-additive variability can be contained in some of the relationships used in the estimation (for example, maternal effects, dominance, epistasis, etc.). In general, estimates tend to be over-optimistic, therefore it is advisable to consult the estimates derived from selection experiments, particularly the ones having additional evidence from control populations or divergent selected populations, which do not depend on estimates derived from relationships.

Another problem related to heritability estimates comes from the type of feeding regime, restricted or *ad libitum*. If the commercial farms are under restricted feeding, Piles *et al.* (2017) have shown that selection for daily gain under *ad libitum* feeding can be inefficient due to the behaviour of rabbits competing for feed. Moreover, Piles and Sánchez (2019) found only a moderate correlation (0.48) between growth rate under restricted or *ad libitum* feeding. Restricted feeding is less efficient for growth than *ad libitum*, but it is common in France and Italy in order to avoid digestive disorders. This practice is not common in other countries like Spain, in which rabbits are slaughtered earlier.

Measuring feed intake is expensive, but if genetic correlation between feed efficiency and growth rate is high, selecting for growth rate can be used for indirectly improving efficiency. However, FCR seems to be less related to growth rate in rabbits than in other species. There are three estimates of the genetic correlation between growth rate and FCR in the literature; −0.49 and −0.47 for two different lines (Piles *et al.*, 2004) and −0.38 (Drouilhet *et al.*, 2013), all of them with large confidence intervals. As the heritabilities of ADG and FCR are similar (from 0.22 to 0.31) (Piles *et al.*, 2004; Drouilhet *et al.*, 2013), for a true genetic correlation around −0.4 or −0.5, selection for growth rate will be inefficient for improving FCR.

Other measures of feed efficiency have been proposed by Koch *et al.* (1963): actual feed intake adjusted for growth and requirements for maintenance (called nowadays residual feed intake, RFI), or actual growth adjusted for feed consumption and maintenance requirements (residual growth rate, RGR). The aim of RFI is to select animals arriving to the same weight with lower feed consumption, and the aim of RGR is to obtain animals growing quicker for the same amount of feed consumed. RFI is often estimated as the residual of feed intake (FI) adjusted for ADG and maintenance requirements (in rabbits, maintenance requirements have been estimated as the average live weight (LW) between 30 and 63 days to the power 0.75). RFI has a low heritability (0.10–0.16) (Drouilhet *et al.*, 2013). A higher value (0.45) was found by Larzul and Rochambeau (2005) in a one-generation divergent selection experiment in which BW, ADG and RFI were unusually high. For this reason, their results should be considered with caution. Genetic correlation between FCR and RFI is very high (0.96) (Drouilhet *et al.*, 2013). As the heritability of FCR is much higher, the advantage of using RFI instead of FCR in selection is unclear in rabbits. Using Bayesian techniques (Blasco, 2017), Piles *et al.* (2007) estimated the heritabilities of the partial regression coefficients used to define RFI. In this study, the heritabilities of these coefficients are similar to the ones of ADG. Piles and Sánchez (2019) have estimated genetic parameters of RFI using group (cage) food intake instead of individual food intake, using the phenotypic parameters in the regression coefficient, as usual, or using the genetic parameters, as recommended by Kennedy *et al.* (1993) to ensure genetic independence between growth rate and feed intake in order to reduce food consumption at the same growth rate. Heritability of RFI using genetic parameters was lower (0.07) than the usual RFI (0.22), and the genetic correlation of the latter with growth rate was positive (0.58). Another problem of RFI calculated with genetic parameters is that it depends on the accuracy of these parameters, which is generally low due to the difficulty of having a high number of data for feed intake. Piles and Sánchez (2019) also estimated residual growth rate after correcting individual growth rate by cage food intake and metabolic weight. This is a more attractive measure of feed efficiency because RFI obtains feed efficiency by reducing feed intake for the same weight and RGR by increasing growth for the same amount of feed. As farmers do not control feed intake but observe growth rate, RGR seems more attractive for commercial terminal sire lines. Again, they found lower heritabilities when genetic parameters were used for estimating the residuals.

10.4 Genetics of Carcass Traits

10.4.1 Between-breed genetic variability

Comparisons of different breeds and crosses show different tendencies when estimated at the same BW (Pla *et al.*, 1996, 1998; Gómez *et al.*, 1998; Hernández *et al.*, 2006) or at the same age (Rouvier, 1970; Lukefahr *et al.*, 1982; Brun and Ouhayoun, 1989, 1994; Ozimba and Lukefahr, 1991; Metzger *et al.*, 2006a, 2006b; Szendrő *et al.* 2009; Ouyed *et al.*, 2011).

Breeds with lower adult BW are in a higher maturity stage at slaughter (Pla *et al.*, 1996, 1998; Gómez *et al.*, 1998; Hernández *et al.*, 2006). They had better dressing-out percentage, higher hind part and lower fore part and greater fat depots.

There are few studies that estimated heterosis from different crosses (Brun and Ouhayoun, 1989, 1994; Ouyed *et al.*, 2011). Carcass yield and carcass fatness showed favourable results in some cases, but individual or maternal heterosis had no effect on carcass composition.

10.4.2 Genetic parameters of carcass traits

Few studies estimated carcass traits genetic parameters in rabbits because of the large samples needed for obtaining an appropriate precision.

Heritability estimates of retail cuts are generally moderate, as common litter effects (Ferraz *et al.*, 1991,1992; Al-Saef, *et al.*, 2008). Carcass ratios have higher heritabilities, varying from moderate to high. Perirenal fat percentage has shown the highest heritability; however, muscle percentage, a trait much more important for consumers, has only a moderate heritability. Thigh muscle volume measured *in vivo* by computer tomography (CT) showed low heritabilities (Gyovai *et al.*, 2008, 2012; Nagy *et al.*, 2010, 2013), but the average surface of the CT estimation of *Longissimus thoracic and lumborum* (LTL) muscle had a substantially higher heritability. Heritability of *Longissimus lumborum* muscle surface between the second and third lumbar vertebrae, estimated by using ultrasound, was moderate (Lenoir and Morien, 2015, 2016). Maternal effects were generally low for carcass components (Krogmeier *et al.*, 1994) and also for muscle traits measured *in vivo* (Gyovai *et al.*, 2008, 2012; Nagy *et al.* 2010, 2013).

Dressing-out percentage has a moderate heritability, although the results of the studies that estimated it cannot always be compared because slaughter age was not the same (96 vs. 63 days, Krogmeier *et al.* 1994; Larzul *et al.*, 2005; Garreau *et al.*, 2008), dissection was not made according to World Rabbit Science Association recommendations (Blasco *et al.*, 1993; Blasco and Ouhayoun, 1996), carcasses had no head (Ferraz *et al.*, 1992; Lukefahr *et al.*, 1996), or distal parts of the legs were not removed (Rouvier, 1970). Moreover, hot carcass weight was used sometimes instead of cold carcass weight for dressing-out percentage (Ferraz *et al.*, 1992; Krogmeier *et al.*, 1994; Al-Saef *et al.*, 2008). Colour values had heritabilities ranging from 0.11 (b*, yellowness) to 0.36 (a*, redness), according to the only experiment reporting estimations for these parameters (Martínez-Álvaro *et al.*, 2016a).

Due to its cost, carcass studies have a limited sample size; therefore, estimates of genetic correlations, which need large samples for being precise enough, should be taken with caution, since they often have large standard errors. Krogmeier *et al.* (1994) and Garreau *et al.* (2008) did not find a relevant genetic correlation between perirenal fat percentage and dressing-out percentage. Nagy *et al.* (2006) observed a negative genetic correlation between perirenal fat weight and muscle LTL volume based on CT scans. These authors also found a moderately high genetic correlation between dress-out percentage and the average cross-sectional area of the muscle LTL, which was corroborated using ultrasounds by Lenoir and Morien (2015). No genetic correlation was found between dressing-out percentage and thigh muscle volume, but thigh muscle volume had a relatively high genetic correlation with hind part percentage (Nagy *et al.*, 2010), showing that, using CT, selection for thigh muscle volume might improve hind part percentage.

10.5 Genetics of Rabbit Meat Quality

Meat quality is controlled by metabolic pathways affecting many traits. No single genes have been found for rabbit meat quality. Thus, it seems that meat quality in rabbits depends on many genes with small effects. Traits measured for assessing meat quality are often difficult and expensive to record, since they are commonly measured after slaughter. This is one of the reasons for not including meat quality in selection programmes. Nevertheless, it is convenient to examine the consequences of selection for growth rate on meat quality.

10.5.1 Genetic variability between rabbit lines

Several meat-quality traits have shown differences between lines or crosses. However, in meat-quality studies, sample sizes are small, because traits are difficult and expensive to record, which often leads to relevant differences that could be non-significant. Moreover, differences in the size of the rabbit line can explain some of the differences found in meat quality. There are substantial differences between lines or crosses in meat pH (around one standard deviation of the trait, see review by Hulot and Ouhayoun, 1999). In general, no association has been found between breed size and meat pH. Within lines, moreover, Blasco and Piles (1990) did not find any correlation between carcass weight and meat pH.

Differences in meat colour values (L*a*b*) have been found in lines of different sizes by Bernardini Battaglini *et al.* (1995), Hernández *et al.*

(2006) and Dalle Zotte *et al.* (2015), but they did not show any clear patterns. Comparing lines of different size for water-holding capacity (WHC), no differences were found by Bernardini Battaglini *et al.* (1995) and Ariño *et al.* (2006). Differences between synthetic giant and medium lines in meat-texture traits were reported by Ariño *et al.* (2006) and Lukefahr *et al.* (1982). Heavy lines were more tender in both cases, and in the study of Ariño *et al.* (2006), the heavier line had lower collagen content and higher proteolytic activity. A sensory analysis performed by Ariño *et al.* (2007) found differences in tenderness in the same direction, but they did not find differences in odour or flavour traits.

Ramírez *et al.* (2005) and Hernández *et al.* (2008) compared lipid content and fatty acid composition of rabbit hind leg meat and perirenal fat using the same lines as Ariño *et al.* (2006, 2007). They found higher polyunsaturated fatty acids (PUFA), lower saturated fatty acids (SFA) and no differences in monounsaturated fatty acids (MUFA) percentages in one of the lines selected for litter size; the differences between lines were high, more than one standard deviation. Other authors compared a local breed and a commercial breed and found no differences in SFA, MUFA and PUFA, but their experiment had large standard errors, thus they could not conclude that there were no differences between both breeds (Gasperlin *et al.*, 2006). Hernández *et al.* (2008) also found differences in lipolytic activities between lines selected for litter size and a line selected for growth rate, but no line effect was found for free fatty acids (FA) after refrigerated storage. No differences between lines were found either for oxidative parameters or for the enzyme activity of the muscle energy metabolism, such as aldolase and isocitrate dehydrogenase (ICDH) (Hernández *et al.* 2006).

10.5.2 Genetic parameters of meat-quality traits in rabbits

Reliable estimates of heritability have been published only for carcass and meat colour, pH, intramuscular fat (IMF), meat FA profile and instrumental texture. Larzul *et al.* (2005) gave a heritability estimate of 0.16 for ultimate pH (pH_u) not different from zero (s.e. was 0.09), Martinez-Alvaro *et al.* (2016a) reported a similar result, 0.08 with [0.01, 0.20] confidence interval at 95% probability, but Larzul and Rochambeau (2005) gave a much higher heritability, 0.50 (s.e. 0.16), suggesting that pH_u might be selected with success. Carcass colour (L*a*b*) does not show a clear pattern. Larzul and Rochambeau (2005) and Larzul *et al.* (2005) found heritabilities near zero, but Martinez-Alvaro *et al.* (2016a) found heritabilities between 0.14 and 0.25 for these traits. The latter result is consistent with the correlated response to selection found in an experiment by Hernández *et al.* (2004) described in section 10.6. Intramuscular fat (IMF) showed a high heritability, 0.54, with a probability of 95% of being higher than 0.40 (Martinez-Alvaro *et al.*, 2016a). This result was corroborated, as described later, by the results of a selection experiment. Genetic correlations between carcass fat depots and IMF were positive but low (0.3); they were of the same order as the correlation between IMF and reference carcass weight, meat/bone ratio and pH_u (Martinez-Alvaro *et al.*, 2016a). This means that an increment in IMF by selection will not necessarily deteriorate carcass quality by increasing its fat depots. Martinez-Alvaro *et al.* (2016a) also investigated the genetics of fat composition and found low heritability (0.09) for SFA and high heritabilities for PUFA (0.45), MUFA (0.61), and also for the PUFA:SFA ratio (0.42). Strong genetic correlations were found between IMF and meat FA percentages; negative genetic correlation for PUFA and PUFA:SFA ratio, and positive for MUFA. IMF and SFA had a positive genetic correlation, but with a wide confidence interval. Individual FA percentages of C14:0, C18:0, C16:1, C18:1n-9, C18:2n-6 and C20:4n-6 also showed high heritabilities (Martinez-Alvaro *et al.*, 2018c). Strong negative correlations were found between IMF and C18:0, C18:2n-6, and C20:4n-6, and high positive genetic correlations between IMF and C14:0, C16:1, C18:1n-9. Correlated responses to selection corroborated all these estimates (Martinez-Alvaro *et al.*, 2018c). Larzul *et al.*, (2005) provided a high heritability for instrumental texture, 0.57 (s.e. 0.02, which for a meat-quality trait is surprisingly low). No genetic correlations were found between growth rate and Warner-Bratzler Shear Force (WBSF) or pH_u, which agrees with the null correlated response obtained for pH in selection experiments for growth rate.

10.6 Selection Experiments

10.6.1 Selection for growth

Responses to selection for growth rate with a control population have been reported by Rochambeau *et al.* (1989), Lukefahr *et al.* (1996), Piles and Blasco (2003) and Larzul *et al.* (2005). Responses to selection using divergent selection lines have been reported by Moura *et al.* (1997) and Larzul *et al.* (2005). In France, feed restriction is commonly used in commercial farms for preventing digestive disorders; this has motivated a selection experiment for ADG under feed restriction by Drouilhet *et al.* (2013, 2016). Other studies have less clear interpretation (Ferraz *et al.*, 1992; Rochambeau *et al.*, 1994) or an arguable methodology (Niedzwiadek *et al.*, 1992).

Selection was successful in all these experiments. Rochambeau *et al.* (1989) selected for growth rate between 30 and 77 days of age in eight generations of selection and obtained a response of 0.83 g/d per generation (2% of the mean per generation). Lesser progress was obtained by Piles and Blasco (2003) in seven generations of selection, 0.56 g/d per generation (1.2 % of the mean per generation). In both experiments an increment in slaughter weight (SW) was observed, but no correlated responses were found in weaning weight (WW). Lukefahr *et al.* (1996) selected by live weight at 70 days and obtained a correlated response for ADG (from 28 to 70 days of age) of 2.7 g/d (1.4% of the mean per generation) in five generations of selection. Larzul *et al.* (2005) selected for live weight (LW) at 63 days in an experiment of divergent selection that also had a control population. After five generations they found a difference between lines for ADG between 28 and 63 days of 12 g/d, being the control population intermediate between both lines, which means 6 g/d per line in five generations (1.2% of the mean per generation). Moura *et al.* (1997) reported a difference between the high and low lines of their divergent selection experiment of 8.4 g/d from 56–60 days to 84–88 days of age (2.25% of the mean per generation under the supposition of symmetrical response). However, it seems that greater success was obtained in decreasing than in increasing daily gain. Selecting ADG under restricted feeding, Garreau *et al.* (2015a, 2016) found, after nine generations of selection, a response of 1.9 g/day (0.5% of the mean per generation), lower than the response found without restriction. This is expected because the full potential for growth is not expressed under restriction. In their experiment, it was found a difference of 11.2 g/d between *ad libitum* and restricted feeding.

Actual responses were lower than the expected ones based on heritability estimates of LW or ADG, but as rabbit generation interval can be very small (six months), the responses per year would be between 2% and 4% of the mean, which are reasonably good results when comparing with other species (Smith, 1984).

10.6.2 Selection for feed efficiency

The most common measure of feed efficiency is feed conversion rate (FCR), that is the ratio between feed intake (FI) and body weight gain (BWG) in a fixed range of days. Residual feed intake (RFI) is another popular measure of feed efficiency. Both parameters have advantages and disadvantages. FCR can be improved either by augmenting LW for a fixed amount of FI or by reducing FI at a fixed weight. Selection for FCR has a higher correlated response on the trait with a higher variability and tends to reduce consumption without changing BWG. This has been observed both in growing pigs (Webb, 1989) and in sows (Prunier *et al.*, 2010). In sows it may create some problems if increasing litter size leads to higher nutrition demands. Another issue is that the FCR is a ratio and has the so-called 'spurious correlations' (Pearson, 1897) with its components, it is not normally distributed, and it has some inconvenient properties for predicting responses to selection (Gunsett, 1984). On the other hand, RFI has also received criticisms. Kennedy *et al.* (1993) stressed that genetic parameters should be used when defining RFI instead of the phenotypic ones, in order to ensure independence between feed intake and ADG, but this implies having good estimates of genetic parameters for feed consumption and ADG, and large sample sizes. Besides, RFI seems to have in rabbits a genetic correlation with FCR close to one, but it also shows a lower heritability. Therefore, its use in selecting for RFI would be less successful for improving feed efficiency than directly measuring FCR. A further criticism is that metabolic weight is usually estimated as BW to the power of 0.75 when defining RFI, which is correct only for adult BW and not for BW of growing animals (Brody, 1945; Taylor, 2009). Selection for feed consumption and daily gain in an index is the most efficient way for improving benefits, since traits are weighed to obtain the maximum benefit, but accurate genetic parameters and estimates of economic weights are needed. If these parameters are not available, selection for FCR or RFI may produce better results.

Experiments in rabbits have been performed selecting for FCR (Moura *et al.*, 1997), RFI (Larzul and Rochambeau, 2005; Drouilhet *et al.*, 2013, 2016) and ADG under restricted feeding (Drouilhet *et al.*, 2013, 2016). Moura *et al.* (1997) reported inconsistent results in a divergent selection experiment, with the low line having a higher FCR than the high line at the end of the experiment. However, these authors found a symmetric progress for FCR of 0.6% per generation in each direction in the period ranging from 56–60 days to 84–88 days of age, using mixed-model breeding predictions. The experiment carried out by Larzul and Rochambeau (2005) on divergent selection for RFI was too short to drive conclusions. Drouilhet *et al.* (2013, 2016) carried out a longer experiment on RFI between 30 and 65 days of age, showing similar results when using mixed-model techniques or when comparing with a control population. This study found a response of –39 g of RFI per generation (0.9% of the mean per generation) and a correlated response of –0.20 in FCR (0.8% of the mean per generation), after nine generations of selection (Garreau *et al.*, 2015a, 2016). Selection acted reducing feed consumption, since no correlated response was found for growth rate.

10.6.3 Consequences of selection for growth rate or feed efficiency

10.6.3.1 *Changes in adult weight*

Selection for growth rate has a correlated response increasing adult weight in all species (Taylor, 1980, 1985). This has been shown in rabbits in an experiment of selection for growth rate by Blasco *et al.* (2003); adult weight increased by 1% per generation, a similar rate as the direct response for growth rate. In consequence, lines selected for growth rate would become giant lines, that are more expensive to maintain and manage. Nevertheless, this should not be a serious problem in modern industrial rabbit production, as very few terminal sires are needed due to the wide use of artificial insemination.

10.6.3.2 *Changes in feed conversion rate*

Selection on ADG aims to improve feed conversion rate (FCR) in rabbits. The only direct evidence of this effect comes from Moura *et al.* (1997), who showed lower values for FCR in the line selected for increased ADG, and a response of 3.5% of the mean per generation in each direction of their divergent selection experiment. However, no relevant changes in ADG were found in the line selected for improved FCR, thus the experiment did not show a clear evidence. Larzul *et al.* (2005) did not find, after five generations of selection, changes in FCR when comparing their line selected for ADG with the control line.

Under restricted feeding, Garreau *et al.* (2015b) found a correlated response in FCR of –0.19 (0.8% of the mean per generation), similar to the response obtained for FCR in their line selected for RFI (Garreau *et al.*, 2015a). As both lines had almost the same response to selection for FCR, under restricted feeding, selection for growth rate seems easier to implement. In the long term, this will produce heavier animals (Garreau *et al.*, 2015a, Drouilhet *et al.*, 2016) and rabbits should be slaughtered earlier for the same commercial weight.

10.6.3.3 *Changes in carcass quality*

Selection for growth rate produces younger rabbits that arrive earlier at the slaughter commercial weight. This leads to poorer carcass yield, but also to slightly higher muscle/bone ratio and to retail cuts with slightly different proportion between them. It can also be expected to have lower fat content, since fat deposition increases at later ages; however, selecting for growth rate increases feed consumption, and it is well known that augmenting food intake leads to increased fat deposition at all ages (Whittemore, 1987). Pascual and Pla (2007) and Gondret *et al.*

(2005) have studied the effect of growth rate selection at fixed BW. Higher percentage of dissectible fat of the carcass and lower muscle/bone ratio in the hind leg were found in the high line in both studies. Gondret *et al.* (2005) also found poorer carcass yield in their high line, as expected. As rabbits were slaughtered at fixed body weight, they were at different maturity stage, therefore the effect of maturity and the actual effect of selection are confounded.

Body weight is highly correlated with retail cuts and tissue composition; thus, when comparisons are made at the same maturity stage, most differences found at the same weight disappear or are highly reduced. The effect of growth rate selection on the relative growth of retail cuts and carcass tissues has been studied by Pascual *et al.* (2008) who compared the line selected for growth rate by Piles and Blasco (2003) and a control population. This study correlated response in allometric coefficients of retail cuts and also examined the correlated response in muscle/bone ratio of the hind leg. No effect of selection was evidenced in any of the allometries or in the muscle/bone ratio studied. The correlated response to selection for growth rate on carcass composition has been examined at the same age by Lukefahr *et al.* (1996), Hernández *et al.* (2004) and Larzul *et al.* (2005), and under restricted feeding conditions by Lukefahr *et al.* (1996), Hernández *et al.* (2004) and Garreau *et al.* (2015b) did not find any correlated response in carcass yield, although Larzul *et al.* (2005) found a small favourable difference between the line selected to increase ADG and the control line. Piles *et al.* (2000) and Hernández *et al.* (2004) found less fat in the line selected for increased ADG than in the control line, which disagrees with the results of Larzul *et al.* (2005), who found more dissectible fat in their high line than in their low line. Lukefahr *et al.* (1996) found some advantage for their selected line in muscle/bone ratio of the loin cut. Gondret *et al.* (2005) found that their low line had a better muscle/bone ratio than the control line, although they did not find differences between the high line and the control line. These results disagree with Hernández *et al.* (2004) who found in the line selected for ADG a higher muscle/bone ratio. No differences between selected and control lines in retail cuts and other parts of the carcass (liver, heart, head, lungs and kidneys) were found. Under restricted feeding, Drouilhet *et al.* (2013, 2016) reported a relevant correlated response in perirenal fat (19% reduction in nine generations, 2% of the mean per generation), although scapular fat did not change. This study did not find correlated responses in retail cuts, muscle/bone ratio and carcass yield. Higher correlated response in perirenal fat in their line selected for RFI (33% in nine generations, 3.6% of the mean per generation) and a correlated response in scapular fat (2% of the mean per generation) were observed. Hind leg proportion and muscle/bone ratio showed favourable correlated responses in the same line.

10.6.3.4 Changes in meat quality

The effect of selection for growth rate in meat quality has been investigated by Hernández *et al.* (2004) (study 1) using a control population, and by Larzul *et al.* (2005) (study 2) using divergent selection and also a control population. Study 1 found a correlated response in muscle LTL meat colour and also in carcass colour; selection for growth rate led to higher L*, and lower a* and b* when comparing selected and control lines, whereas in study 2 no correlated responses were found. No correlated responses on pH_u were found, neither in LTL in study 1 (Ramírez *et al.*, 2004) nor in LTL and in m. *Semitendinosus* in study 2 (Larzul *et al.*, 2005). Negative correlated response in WHC was found in study 1, also observed in previous analyses with the same lines for cooked meat (Piles *et al.*, 2000). Study 1 also showed some correlated response of hind leg fat content and fatty acids composition. The strongest effect was in the PUFA/SFA ratio, which in 14 generations of selection changed from 1.06 to 0.95 (Ramírez *et al.*, 2005). Both studies found a correlated response in meat toughness, but no clear pattern appeared for texture parameters. Study 1 had a trained panel test who did not find any correlated response in fibrousness, tenderness or juiciness. Moreover, no differences in muscle fibre analyses, related to tenderness, were found in study 2. Negative correlated response was found in study 1 for some sensory traits, decreasing aniseed odour and flavour and increasing liver flavour. Proteolitic and lypolitic enzyme activities related to meat quality have also been analysed in study 1, but no correlated response to selection for growth rate was found (Gil *et al.*,

2006). Under restricted feeding, some correlated responses in pH_u and L* were obtained, but no correlated response of WHC in cooked meat was found (Molette *et al.*, 2016). In their line selected for RFI, this study found minor unfavourable correlated responses in the latter traits.

As a general conclusion, it seems that there are no clear correlated responses to selection for growth rate in meat-quality traits. The experiments showed some correlated responses, not in the same direction, which can be due to sampling error or genetic drift.

10.6.4 Selection for increasing muscle volume

Computer tomography (CT) scanning of live rabbits has been used for two divergent selection experiments. In the first experiment, with two generations for the low line and three generations for the high lines, males were selected for increasing or decreasing the average surface of the m. *Longissimus lumborum* (between the 2nd and 3rd and 4th and 5th lumbar vertebra) (Szendrő *et al.*, 1996). In the second experiment, with three generations of selection, rabbits (males and females) were selected for increasing thigh muscle volume (Szendrő *et al.*, 2012). Both experiments were successful. In the first experiment, a difference between the high and low lines of 1.3 cm^2 in the surface of the m. *Longissimus lumborum* was observed, and a correlated response for dressing-out percentage of 2% was also found. The intermediate and hind parts of the carcass had a difference of 22 g and 14 g, respectively, whereas the gastrointestinal tract showed a difference of 23 g between high and low lines. The second experiment reported similar results (Szendrő *et al.*, 2012), the difference between the high and low lines for the thigh muscle volumes was 25 cm^3; moreover, correlated responses for FI and FCR were found, having lower FI in the high line (128 g/d in the high line vs. 138 g/d in the low line) and also better FCR in the high line (2.81% vs. 3.01). Correlated response in perirenal fat (2.40% vs.1.90%) and scapular fat (1.07% vs. 0.49%), were also found, as well as differences in retail cuts proportions: fore part of the reference carcass was higher in the low line (30.1% vs. 29.4%), but the high line had higher hind part (36.3% vs. 38.2%) and higher meat of both hind legs (26.9% vs. 28.7%). Therefore, CT-aided selection can improve carcass traits and muscle volume. Matics *et al.* (2014) have summarized the main results of the CT-aided selection breeding programme.

10.6.5 Selection for intramuscular fat content

Intramuscular fat (IMF) content is an important trait affecting meat quality. By increasing IMF, tenderness, juiciness and flavour are improved. The nutritional value of meat is also affected by IMF and its composition. Zomeño *et al.* (2013a, 2013b) and Martinez-Alvaro *et al.* (2016a, 2016b) performed and analysed a divergent selection experiment on IMF of muscle *Longissimus thoracic and lumborum* (LTL) in rabbits. Direct response to selection was 1.09 g/100 g after seven generations of selection (5% of the mean per generation), having symmetrical genetic trends for both lines (Martinez-Alvaro *et al.*, 2016a). Selection also increased IMF in other muscles: *Semimembranosus proprius*, *Supraspinatus* and *Biceps femoris* (Martínez-Álvaro *et al.*, 2018a). pH_u did not show any correlated response in any muscle. No correlated responses were observed for colour of the meat and of the carcass (Martinez-Alvaro *et al.*, 2016a). A correlated response in lipogenic activities was observed in muscles LTL and *Semimembranosus proprius*, liver and perirenal fat, being greater in the high line (Martínez-Álvaro *et al.*, 2017, 2018b). Perirenal fat content was also higher in the high line. Regarding fatty acid composition, the low line had lower MUFA and higher PUFA, whereas no correlated response was found in SFA (Martinez-Alvaro *et al.*, 2016a). This led to unfavourable ratio PUFA/SFA in the high line, but favourable MUFA/SFA ratio in this line. A similar pattern was found for single FA of the MUFA and PUFA groups, with the exception of C18:3n-3, which was greater in the high line (Martínez-Álvaro *et al.* 2017). The other muscles showed the same pattern (Martínez-Álvaro *et al.*, 2018c). From a nutritional point of view, selection for IMF can deteriorate carcass and meat quality due to the worsening in PUFA/SFA ratio and the augmentation of dissectible fat. However, rabbit carcasses are very lean, having a low percentage of dissectible fat (2.5% at 9 weeks and 3.5% at 13 weeks; Hernández *et al.*, 2004). The same happens with IMF (about 1%) (Zomeño *et al.*, 2013a), human health is not going to be compromised by selection for IMF when consuming rabbit meat. Finally, the low line had a 9.9% greater WBSF toughness, whereas other instrumental texture and sensory attributes, and cooking loss were similar in both lines. Selection for IMF did not affect sensory attributes (Martinez-Alvaro *et al.*, 2016b).

10.7 Genomics in Rabbit Growth, Carcass and Meat Quality

Studies using DNA markers in candidate genes or involving a high-density single nucleotide polymorphism (SNP) genotyping platform with 200,000 SNPs covering the whole genome have been applied to identify genes or chromosome regions associated with growth performances, meat quality and carcass traits in rabbits.

The candidate gene approach consists of proposing a gene that may be related to the targeted productive trait because of its biological role in the metabolic or physiological routes related with the trait of interest. Having markers within the gene or closely linked to it, it is possible to examine whether rabbits having the favourable version of a gene allele perform better than other rabbits carrying alternative alleles. Several candidate genes have been proposed for body weight at different ages (Fontanesi *et al.*, 2011, 2012a, 2012b, 2013, 2016; Zhang *et al.*, 2012, 2013; Peng *et al.*, 2013) and for carcass weight (Zhang *et al.*, 2012; Sternstein *et al.*, 2015). Some candidate genes have also been proposed for meat-quality parameters, including pH (Zhang *et al.*, 2012), intramuscular fat (Zhang *et al.*, 2013) and several other traits (Wang *et al.*, 2016, 2017; Migdal *et al.*, 2018). Being useful as exploratory works, the candidate gene approach has some limitations. It is sometimes difficult to identify the candidate genes to be investigated whose variability could explain a relevant proportion of the genetic variability for that specific trait. Analysed markers in the candidate gene could not be in complete linkage disequilibrium with a potential causative mutation. This means that the same markers in other populations, or even in the same population after some generations, can give different results, and the associations may be lost. Moreover, the hypothesis tested in these works use P-values that can easily lead to false positive results.

10.7.1 Genome-wide association studies for growth, carcass and meat quality in rabbits

A genome-wide association study (GWAS) for growth traits has been recently performed by Sánchez *et al.* (2019) who identified some chromosome regions putatively associated with growth rate and residual feed intake and suggested some candidate genes, one of them (*FTO*) previously associated with growth in rabbits (Zhang *et al.*, 2013). The first GWAS in rabbit for meat-quality traits identified some genome regions affecting intramuscular fat (Sosa-Madrid *et al.*, 2020a). A positional candidate gene suggested by this study explained part of the IMF genomic variance (7.34%). Sosa-Madrid *et al.* (2020b) have also found some signatures of selection in genomic regions that include genes involved in energy, fatty acid and lipid metabolic processes in the divergent selection experiment for intramuscular fat described by Martínez-Álvaro *et al.* (2016a). Laghouaouta *et al.* (2020) performed a GWAS on fatty acid composition of the same animals of Sosa-Madrid *et al.* (2020b) and identified a region associated with C16:0 and SFA, explaining, respectively, 11.2% and 11.3% of the genomic variance and another region explaining up to 8% of the genomic variance of MUFA/SFA. These regions harbour several positional and functional candidate genes involved in lipid metabolism. The findings of Sosa-Madrid *et al.* (2020a, 2020b) and Laghouaouta *et al.* (2020) suggest that there is a large component of polygenic effect determining IMF and its composition, as the variance explained by the genomic analysis was low. These studies indicated that most complex traits are determined by many genes of small effects and some associations might be affected by the experimental design and by the number of genotyped and phenotyped animals (Visscher *et al.*; 2012; Blasco and Toro, 2014).

10.7.2 Rabbit metagenomics for growth, carcass and meat-quality traits

Animals live with other genomes, in addition to their own genome, which have not been given attention until recently: the genomes of gastrointestinal microbes. Their importance to regulate traits related to health and well-being has already been establish in humans and animal models. The gastrointestinal microbes and their genes can also play some roles in the regulation of productive traits in livestock such as feed conversion rate, growth and carcass- and meat-quality traits. Microbiota is analysed by classifying microbes into groups (species, gender, order, family, class or phyla). Studies in rabbits have been focused on the characterization of the microbial communities of the caecum, with a primary focus on digestive health (Combes *et al.*, 2017), on the effects of feed restriction (Velasco-Galilea *et al.*, 2018) and on the effects on selection for feed efficiency (Drouilhet *et al.*, 2016). The metagenome is more expensive and difficult to study, because it involves sequencing all the genes of the bacteria present in the samples. The advantage of the metagenome is that it shows the genes that are present in a sample, which can belong to several species of bacteria, leading to a better knowledge of the metabolic pathways involved in the investigated traits. At present, few studies on the relationships between the microbiota or the metagenome and meat quality in any livestock species have been carried out. In rabbits, Zubiri-Gaitán *et al.* (2019), Blasco *et al.* (2019) and Martínez-Alvaro *et al.* (2019) have shown that intramuscular fat selection modifies the microbial genome for metabolic energy pathways in the rabbit intestine. In the divergent selection for intramuscular fat quoted before (Martínez Álvaro *et al.*, 2016a), the genes involved in specific carbohydrate metabolism showed a greater relative abundance in the line selected for higher intramuscular fat content compared to the line selected for lower intramuscular fat. Other bacterial genes, such as those involved in the biosynthesis of lipopolysaccharides, were more abundant in the selected line for lower intramuscular fat content. This study confirms that the microbiota is also determined by interaction with the genes of the individual in which the microbiota is housed.

10.7.3 Genomic selection for meat-quality traits

The low cost of SNPs markers and their availability in platforms with several tens of thousands of SNPs for livestock species, has promoted the use of these markers for better estimating the genetic value of the candidate animals to selection. In dairy cattle, in which selection is based on progeny test, this has permitted to evaluate bulls before having daughters, which has dramatically reduced the generation interval, leading to a spectacular increase of the genetic progress per year. In species like pigs or poultry, in which genomics does not reduce the generation interval, genomic selection has also been useful, allowing improvements of genetic responses around 20–30% of the total economic value (Blasco and Toro, 2014; Blasco and Pena, 2018).

In rabbits, genomic selection may be particularly useful for litter size, a trait that is not expressed in the female when being selected, so that future dams are selected using only the information of their relatives. Meat-quality traits have been proposed for genomic selection, because breeding values for meat quality can be estimated in live animals using their genomic data. However, there are some limitations of genomic selection that prevent its implementation in rabbits. First, a large 'reference population' (several thousand animals) is needed for the construction of the prediction equations for breeding values. These equations need to be reconstructed every two or three generations because recombination breaks the association between SNPs and causal genes (Ibáñez and Blasco, 2011).

Another major problem is the cost of genotyping that is very relevant in the case of rabbits because of the small individual value of the animals. Low-density chips with a few hundred SNPs, inferring the missing SNPs from high-density chips, will be much cheaper and will help to solve the cost problem. This technique, called 'imputation', relies on a low recombination rate in a single generation, and has produced efficient results (Huang *et al.*, 2012; Cleveland and Hickey, 2013). As before, imputation should be estimated again every three or four generations, due to errors of imputation produced by recombination. Nowadays genomic selection with imputation is being examined in rabbits (Mancin *et al.*, 2019) and it may be implemented in the future if costs of genotyping decrease.

References

Al-Saef, A.M., Khalil, M.H., Al-Dobaib, S.N., Al-Homidan, A.H., García, M.L. and Baselga, M. (2008) Comparing Saudi synthetic lines of rabbits with the founder breeds for carcass, lean composition and meat quality traits. *Livestock Research for Rural Development* 20, 1–12.

Ariño, B., Hernández, P. and Blasco, A. (2006) Comparison of texture and biochemical characteristics of three rabbit lines selected for litter size or growth rate. *Meat Science* 73, 687–692.

Arino, B., Hernández, P., Pla, M. and Blasco, A. (2007) Comparison between rabbit lines for sensory meat quality. *Meat Science* 75, 494–498.

Armero, Q. and Blasco, A. (1992) Economic weights for rabbit selection indices. *Journal of Applied Rabbit Research* 15, 637–642.

Baselga, M. (2004) Genetic improvement of meat rabbits. Programmes and diffusion. *8th World Rabbit Congress*. 7–10 September 2004, Puebla, Mexico, pp. 1–13.

Bernardini Battaglini, M., Castellini, C. and Lattaioli, P. (1995) Effect of sire strain, feeding, age and sex on rabbit carcass. *World Rabbit Science* 3, 9–14.

Blasco, A. (2017) *Bayesian Data Analysis for Animal Scientists*. Springer, New York.

Blasco, A. and Ouhayoun, J. (1996) Harmonization of criteria and terminology in rabbit meat research. Revised proposal. *World Rabbit Science* 4, 93–99.

Blasco, A. and Pena, R. (2018) Current status of genomic maps, genomic selection/GBV in livestock. In: Niemann, H. and Wrenzycki C. (eds) *Animal Biotechnology 2.* Springer. New York, pp. 61–80.

Blasco, A. and Piles M. (1990) Muscular pH of the rabbit. *Annales de Zootechnie* 39, 133–136.

Blasco, A. and Toro, M.A. (2014) A short critical history of the application of genomics to animal breeding. *Livestock Science* 166, 4–9.

Blasco, A., Ouhayoun, J. and Masoero, G. (1993) Harmonization of criteria and terminology in rabbit meat research. *World Rabbit Science* 1, 3–10.

Blasco, A., Piles, M. and Varona, L. (2003) A Bayesian analysis of the effect of selection for growth rate on growth curves in rabbits. *Genetics Selection Evolution* 35, 21–42.

Blasco, A., Nagy, I. and Hernández, P. (2018) Genetics of growth, carcass and meat quality in rabbits. *Meat Science* 145, 178–185.

Blasco, A., Martinez-Alvaro, M., Zubiri-Gaitan, A. and Hernandez, P. (2019) Réponse métagénomique à la sélection de la teneur en lipides intramusculaire chez le lapin. *18èmes Journées de la Recherche Cunicole*. 27–28 May 2019, Nantes, France, pp. 65–68.

Bolet, G., Brun, J.M., Monnerot, M., Abeni, F., Arnal, C. *et al.* (2000) Evaluation and conservation of European rabbit (*Oryctolagus cuniculus*). Genetic resources. First results and Inferences. 7th *World Rabbit Congress.* 9–12 July, Valencia, Spain, Vol. A, pp. 281–316.

Brody, S. (1945) *Bioenergetics and Growth*. Reinhold Publishing, New York.

Brun, J.M. and Ouhayoun, J. (1989) Growth performances and carcass traits in the three strains of rabbits and their two-way crosses. *Annales de Zootechnie* 38, 171–179.

Brun, J.M. and Ouhayoun, J. (1994) Qualités bouchères de laperaux issus d'un croisement diallèle de 3 souches: interaction du type génétique et de la taille de portée d'origine. *Annales de Zootechnie* 43, 173–183.

Cartuche, L., Pascual, M., Gómez, E.A. and Blasco, A. (2014) Economic weights in rabbit meat production. *World Rabbit Science* 23, 165–177.

Cleveland, M.A. and Hickey, J.M. (2013) Practical implementation of cost-effective genomic selection in commercial pig breeding using imputation. *Journal of Animal Science* 91, 3583–3592.

Combes, S., Massip, K., Martin, O., Furbeyre, H., Cauquil, L. *et al.* (2017) Impact of feed restriction and housing hygiene conditions on specific and inflammatory immune response, the cecal bacterial community and the survival of young rabbits. *Animal* 11, 854–863.

Dalle Zotte, A. (2002) Perception of rabbit meat quality and major factors influencing the rabbit: carcass and meat quality. *Livestock Production Science* 75, 11–32.

Dalle Zotte, A., Szendrő, K, Gerencsér, Zs., Szendrő, Zs., Cullere, M. *et al.* (2015) Effect of genotype, housing system and hay supplementation on carcass traits and meat quality of growing rabbits. *Meat Science* 110, 126–134.

Drouilhet, L., Gilbert, H., Balmisse, E., Ruesche, J., Tircazes, A. *et al.* (2013) Genetic parameters for two selection criteria for feed efficiency in rabbits. *Journal of Animal Science* 91, 3121–3128.

Drouilhet, L., Achard, C.S., Zemb, O., Molette, C., Gidenne, T. *et al.* (2016) Direct and correlated responses to selection in two lines of rabbits selected for feed efficiency under ad libitum and restricted feeding. I. Production traits and gut microbiota characteristics. *Journal of Animal Science* 94, 38–48.

Ferraz, J.B.S., Johnson, R.K. and Eler, J.P. (1991) Genetic parameters for growth and carcass traits of rabbits. *Journal of Applied Rabbit Research* 14, 187–192.

Ferraz, J.B.S., Johnson, R.K. and Van Vleck, L.D. (1992) Estimation of genetic trends and genetic parameters for reproductive and growth traits of rabbits raised in subtropics with animal models. *Journal of Applied Rabbit Research* 15, 131–142.

Fontanesi, L., Scotti, E., Frabetti, A., Fornasini, D., Piccone, A. and Russo, V. (2011) Identification of polymorphisms in the rabbit (*Oryctolagus cuniculus*) myostatin (*MSTN*) gene and association analysis with finishing weight in a commercial rabbit population. *Animal Genetics* 42, 339.

Fontanesi, L., Dall'Olio, S., Spaccapaniccia, E., Scotti, E., Fornasini, D. *et al.* (2012a) A single nucleotide polymorphism in the rabbit growth hormone (GH1) gene is associated with market weight in a commercial rabbit population. *Livestock Science* 147, 84–88.

Fontanesi, L., Mazzoni, G., Bovo, S., Frabetti, A., Fornasini, D. *et al.* (2012b) Association between a polymorphism in the IGF2 gene and finishing weight in a commercial rabbit population. *Animal Genetics* 43, 651–652.

Fontanesi, L., Scotti, E., Cisarova, K., Di Battista, P., Dall'Olio, S. *et al.* (2013) A missense mutation in the rabbit melanocortin 4 receptor (MC4R) gene is associated with finishing weight in a meat rabbit line. *Animal Biotechnology* 24, 268–277.

Fontanesi, L., Sparacino, G., Utzeri, V.J., Scotti, E., Fornasini, D. *et al.* (2016) Identification of polymorphisms in the rabbit growth hormone receptor (*GHR*) gene and association with finishing weight in a commercial meat rabbit line. *Animal Biotechnology* 27, 77–83.

Garreau, H., Eady, S.J., Hurtaud, J. and Legarra, A. (2008) Genetic parameters of production traits and resistance to digestive disorders in a commercial rabbit population. *9th World Rabbit Congress*, Verona, Italy, pp. 103–108.

Garreau, H., Gilbert, H., Molette, C., Larzul, C., Balmisse, E. *et al.* (2015a) Réponses à la sélection pour deux critères d'efficacité alimentaire chez le lapin. 1. Croissance, ingéré et efficacité alimentaire. *16èmes Journées de la Recherche Cunicole*, Le Mans, France, pp. 161–164.

Garreau, H., Molette, C., Gilbert, H., Larzul, C. and Balmisse, E. (2015b) Réponses à la sélection pour deux critères d'efficacité alimentaire chez le lapin. 2. Caractères de carcasse et de qualité de viande. *16èmes Journées de la Recherche Cunicole*, Le Mans, France, pp. 169–172.

Garreau, H., Gilbert, H, Molette, C., Larzul, C., Balmisse, E. *et al.* (2016) Direct and correlated responses to selection in two lines of rabbits selected for feed efficiency under ad libitum and restricted feeding. *11th World Rabbit Congress.* Qingdao, China, pp. 43–46.

Gasperlin, L., Polak, T., Rajar, A., Skvarea, M. and Lender, B. (2006) Effect of genotype, age, at slaughter and sex on chemical composition and sensory profile of rabbit meat. *World Rabbit Science* 14, 157–166.

Gil, M., Ramírez, J., Pla, M., Ariño, B., Hernández, P. *et al.* (2006) Effect of selection for growth rate on the ageing of myofibrils, the meat texture properties and the muscle proteolytic potential of longissimus from two groups of rabbits. *Meat Science* 72, 121–129.

Gómez, E.A., Baselga, M., Rafel, O. and Ramon, J. (1998) Comparison of carcass characteristics in five strains of meat rabbit selected on different traits. *Livestock Production Science* 55, 53–64.

Gondret, F., Larzul, C., Combes, S. and de Rochambeau, H. (2005) Carcass composition, bone mechanical properties, and meat quality traits in relation to growth rate in rabbits. *Journal of Animal Science* 83, 1526–1535.

Gunsett, F.C. (1984) Linear index selection to improve traits defined as ratios. *Journal of Animal Science* 59, 1185–1193.

Gyovai, P., Nagy, I., Gerencsér, Zs., Metzger, Sz., Radnai, I. and Szendrő, Zs. (2008) Genetic parameters and trends of the thigh muscle volume in Pannon White rabbits. *9th World Rabbit Congress.* Verona, Italy, pp. 115–119.

Gyovai, P., Nagy, I., Gerencsér, Zs., Matics, Zs., Radnai, I. *et al.* (2012) Genetic parameters for litter weight, average daily gain and thigh muscle volume measured by in vivo Computer Tomography technique in Pannon White rabbits. *Livestock Science* 144, 119–123.

Hernández, P. and Gondret, F. (2006) Rabbit meat quality. In: Maertens, L. and Coudert, P. (eds) *Recent Advances in Rabbit Sciences*. ILVO, Melle, Belgium, pp. 269–290.

Hernández, P., Aliaga, S., Pla, M. and Blasco, A. (2004) The effect of selection for growth rate and slaughter age on carcass composition and meat quality traits in rabbits. *Journal of Animal Science* 82, 3138–3143.

Hernández, P., Ariño, B., Grimal, A. and Blasco, A. (2006) Comparison of carcass and meat characteristics of three rabbit lines selected for litter size or growth rate. *Meat Science* 73, 645–650.

Hernández, P., Cesari, V. and Blasco, A. (2008) Effect of genetic rabbit lines on lipid content, lipolytic activities and fatty acid composition of hind leg meat and perirenal fat. *Meat Science,* 78, 485–491.

Huang, Y., Hickey, J.M., Cleveland, M.A. and Maltecca, C. (2012) Assessment of alternative genotyping strategies to maximize imputation accuracy at minimal cost. *Genetics Selection Evolution* 44, 25.

Hulot, F. and Ouhayoun, J. (1999) Muscular pH and related traits in rabbits: a review. *World Rabbit Science* 7, 15–36.

Ibañez, N. and Blasco, A. (2011) Modifying growth curve parameters by multitrait genomic selection. *Journal of Animal Science* 89, 661–668.

Kennedy, B.W., Van der Werf, J.H.J. and Meuwissen, T.H.E. (1993) Genetic and statistical properties of residual feed intake. *Journal of Animal Science* 71, 3239–3250.

Koch, R.M., Swiger, L.A., Chambers, D. and Gregory, K.E. (1963) Efficiency of feed use in beef cattle. *Journal of Animal Science* 22, 486–494.

Krogmeier, D., Dzapo, V. and Mao, I.L. (1994) Additive and maternal effects of postweaning growth and carcass traits in rabbits. *Journal of Animal Breeding and Genetics,* 111, 289–297.

Laghouaouta, H., Sosa-Madrid, B.S, Zubiri-Gaitán, A., Hernández, P. and Blasco, A. (2020) Novel genomic regions associated with intramuscular fatty acid composition in rabbits. *Animal* 10 (11) 2090.

Larzul, C. and de Rochambeau, H. (2005) Selection for residual feed consumption in the rabbit. *Livestock Production Science,* 95, 67–72.

Larzul, C., Gondret, F., Combes, S. and de Rochambeau, H. (2005) Divergent selection on 63-day body weight in the rabbit: response on growth, carcass and muscle traits. *Genetics Selection Evolution* 37, 105–122.

Lenoir, G. and Morien, F. (2015) Intérét de l'évaluation par échographie des caracteristiques de carcasse pour la selection du lapin de chair. *16èmes Journées de la Recherche Cunicole*, Le Mans, France, pp. 177–180.

Lenoir G. and Morien F. (2016) Estimation of genetic parameters for carcass traits evaluated by in vivo real-time ultrasonography in meat rabbit breeding. In: Lin, Y., Li, F. and Gidenne, T. (eds) *11th World Rabbit Congress,* Qingdao, China, pp. 71–75.

Lukefahr, S.D., Hohenboken, W.D., Cheeke, P.R., Patton, N.M. and Kennick, W.H. (1982) Carcass and meat characteristics of Flemish Giant and New Zealand White purebred and Terminal-Cross rabbits. *Journal of Animal Science* 54, 1169–1174.

Lukefahr, S.D., Odi, H.B. and Atakora, J.K.A. (1996) Mass selection for 70-day body weight in rabbits. *Journal of Animal Science* 74, 1481–1489.

Mancin, E., Sosa-Madrid, B.S., Blasco, A., Mantovani, R. and Ibáñez-Escriche, N. (2019) Efficiency of imputation for genomic selection in rabbits. *23rd National Congress of the Animal Science and Production Association*, 11–14 June, Sorrento, Italy.

Martínez-Álvaro, M., Hernández, P. and Blasco, A. (2016a) Divergent selection on intramuscular fat in rabbits: responses to selection and genetic parameters. *Journal of Animal Science* 94, 4993–5003.

Martínez-Álvaro, M., Penalba, V., Hernández, P. and Blasco, A. (2016b) Effect of divergent selection for intramuscular fat on sensory traits and instrumental texture in rabbit meat. *Journal of Animal Science* 94, 5137–5143.

Martínez-Álvaro, M., Blasco, A. and Hernández, P. (2017) Muscle lipid metabolism in two rabbit lines divergently selected for intramuscular fat. *Journal of Animal Science* 95, 2576–2584.

Martínez-Alvaro, M., Hernández, P., Agha, S. and Blasco, A. (2018a) Correlated responses to selection for intramuscular fat in several muscles in rabbits. *Meat Science* 139, 187–191.

Martínez-Álvaro, M., Paucar, Y., Satué, K., Blasco, A. and Hernández, P. (2018b) Liver metabolism traits in two rabbit lines divergently selected for intramuscular fat. *Animal* 12, 1217–1223.

Martínez-Álvaro, M., Blasco, A. and Hernández, P. (2018c) Effect of selection for intramuscular fat on the fatty acid composition of rabbit meat. *Animal* 12, 2002–2008.

Martínez-Álvaro, M., Zubiri-Gaitán, A., Blasco, A., Hernández, P. (2019) Intramuscular fat selection in rabbits modifies microbial genes for energy metabolic routes. *65th International Congress of Meat Science and Technology,* 4–9 August, Berlin.

Matics, Zs., Nagy, I., Gerencsér, Zs., Radnai, I., Gyovai, P. *et al.* (2014) Pannon breeding program at Kaposvár University. *World Rabbit Science* 22, 287–300.

McCarthy, J.C. (1980) Morphological and physiological effects of selection for growth rate in mice. In: Robertson, A. (ed.) *Selection Experiments in Laboratory and Domestic Animals*. Commonwealth Agricultural Bureau, Slough, UK, pp. 100–109.

Metzger, Sz., Odermatt, M., Szendrő, Zs., Mohaupt, M., Romvári, R. *et al.* (2006a) A study of the carcass traits of different rabbit genotypes. *World Rabbit Science* 14, 107–114.

Metzger, Sz., Odermatt, M., Szendrő, Zs., Mohaupt, M., Romvári, R. *et al.* (2006b) Comparison of carcass traits and meat quality of Hyplus hybrid, purebred Pannon White rabbits and their crossbreds. *Archiv Tierzucht* 49, 389–399.

Migdał, K., Pałka, S., Migdał, W., Otwinowska-Mindur, A., Kmiecik, M. *et al.* (2018) Single nucleotide polymorphisms within rabbits (*Oryctolagus cuniculus*) fatty acids binding protein 4 (FABP4) are associated with meat quality traits. *Livestock Science* 210, 21–24.

Molette, C., Gilbert, H., Larzul, C., Balmisse, E., Ruesche, J. *et al.* (2016) Direct and correlated responses to selection in two lines of rabbits selected for feed efficiency under ad libitum and restricted feeding. II. Carcass and meat quality. *Journal of Animal Science* 94, 49–57.

Moura, A., Kaps, M., Vogt, D.W. and Lamberson, W.R. (1997) Two-way selection for daily gain and feed conversion in a composite rabbit population. *Journal of Animal Science* 75, 2344–2349.

Nagy, I., Ibañez, N., Mekkawy, W., Metzger, Sz., Horn, P. and Szendrő, Zs. (2006) Genetic parameters of growth and in vivo computerized tomography-based carcass traits in Pannon White rabbits. *Livestock Science* 104, 46–52.

Nagy, I., Gyovai, P., Radnai, I., Matics, Zs., Gerencsér, Zs. *et al.* (2010) Genetic parameters of growth in vivo CT based and slaughter traits in Pannon white rabbits. *9th World Congress on Genetics Applied to Livestock Production*, Leipzig, Germany, 341.

Nagy, I., Gyovai, P., Radnai, I., Nagyné Kiszlinger, H., Farkas, J. and Szendrő, Zs. (2013) Genetic parameters, genetic trends and inbreeding depression of growth and carcass traits in Pannon terminal line rabbits. *Archiv Tierzucht* 56, 191–199.

Niedzwiadek, S., Fijal, J. and Bielanski, P. (1992) Progress in area of reproduction in a specialized line of rabbits. *5th World Rabbit Congress*, Corvallis, Oregon, Vol. A, pp. 314–321.

Ouhayoun, J. and Poujardieu, B. (1978) Comparative study of rabbit crossbreeding, between-breed and within-breed relationships between traits of the terminal products. *2èmes Journées de la recherche cunicole en France*. Toulouse, France, Comm. 25.

Ouhayoun, J. and Rouvier, R. (1973) Composition corporelle e degré de maturité en poids de lapereaux de plusieurs génotypes. *1ères Journées de recherches avicoles et cunicoles*. Paris, pp. 85–88.

Ouyed, A., Rivest, J. and Brun, J.M. (2011) Heterosis, direct and maternal additive effects on rabbit growth and carcass traits from a Canadian experiment. *World Rabbit Science* 19, 31–41.

Ozimba, C.E. and Lukefahr, S.D. (1991) Evaluation of purebred and crossbred rabbits for carcass merit. *Journal of Animal Science* 69, 2371–2378.

Pascual, M. and Pla, M. (2007) Changes in carcass composition and meat quality when selecting rabbits for growth rate. *Meat Science* 77, 474–481.

Pascual, M., Pla, M. and Blasco, A. (2008) Effect of selection for growth rate on relative growth in rabbits. *Journal of Animal Science* 86, 3409–3417.

Pascual, M., Calle, E.W. and Blasco, A. (2015) Comparison of degrees of maturity of rabbit lines selected for different traits. *World Rabbit Science* 23, 155–161.

Pearson, K. (1897) Mathematical contributions to the theory of evolution. On a form of spurious correlation which may arise when indices are used in the measurement of organs. *Proceedings of the Royal Society of London* 60, 489–498.

Peng, J., Zhang, G.W., Zhang, W.X., Liu,Y.F., Yang, Y. and Lai, S.J. (2013) Rapid genotyping of MSTN gene polymorphism using high-resolution melting for association study in rabbits. *Asian-Australian Journal Animal Science* 26, 30–35.

Piles, M. and Blasco, A. (2003) Response to selection for growth rate in rabbits estimated by using a control cryopreserved population. *World Rabbit Science* 11, 53–62.

Piles, M. and Sánchez J.P. (2019) Use of group records of feed intake to select for feed efficiency in rabbit. *Journal of Animal Breeding and Genetics* 136, 474–483.

Piles, M., Blasco, A. and Pla, M. (2000) The effect of selection for growth rate on carcass composition and meat characteristics of rabbits. *Meat Science* 54, 347–355.

Piles, M., Gomez, E., Rafel, O., Ramon, J. and Blasco, A. (2004) Elliptical selection experiment for the estimation of genetic parameters of the growth rate and feed conversion ratio in rabbits. *Journal of Animal Science* 82, 654–660.

Piles, M., Garcia-Tomas, M., Rafel, O., Ramon, J., Ibanez-Escriche, N. and Varona, L. (2007) Individual efficiency for the use of feed resources in rabbits. *Journal of Animal Science* 85, 2846–2853.

Piles, M., Davi, I., Ramon, J., Canario, L., Rafel, O. *et al.* (2017) Interaction of direct and social genetic effects with feeding regime in growing rabbits. *Genetics Selection Evolution* 49, 58.

Pla, M., Hernández, P. and Blasco, A. (1996) Carcass composition and meat characteristics of two rabbit breeds of different degrees of maturity. *Meat Science* 44, 85–92.

Pla, M., Guerrero, L., Guardia, D., Olivier, M.A. and Blasco, A. (1998) Carcass characteristics and meat quality of rabbit lines selected for different objectives. I. Between lines comparison. *Livestock Production Science* 54, 115–123.

Prayaga, K.C. and Eady, S. (2000) Rabbit farming for meat production in Australia: preliminary estimates of economic values for production traits. *Asian-Australian Journal of Animal Sciences* 13, 357–359.

Prunier, A., Heinonen, M. and Quesnel, H. (2010) High physiological demands in intensively raised pigs: impact on health and welfare. *Animal* 4, 886–898.

Ramírez, J.A., Oliver, M.A., Pla, M., Guerrero, L., Ariño, B. *et al.* (2004) Effect of selection for growth rate on biochemical, quality and texture characteristics of meat from rabbits. *Meat Science* 67, 617–624.

Ramírez, J.A., Díaz, I., Pla, M., Gil, M., Blasco, A. and Oliver, M.A. (2005) Fatty acid composition of leg meat and perirenal fat of two groups of rabbits selected by growth rate. *Food Chemistry* 90, 251–256.

Rochambeau, H., de la Fuente, L.F., Rouvier, R. and Ouhayoun, J. (1989) Sélection sur la vitesse de croissance post-sevrage chez le lapin. *Genetics Selection Evolution* 21, 527–546.

Rochambeau, H., Retailleau, B., Poivey, J.P. and Allain, D. (1994) Selection pour le poids a 70 jours chez le lapin. *6èmes Journées de la Recherche Cunicole en France*, La Rochelle, France, Vol. 1, pp. 235–240.

Rouvier, R. (1970) Variabilité génétique du rendement a l'abattage et de la composition anatomique de lapins de trois races. *Annales de Génétique et de Sélection Animale* 2, 325–346.

Sánchez, J.P., Legarra, A., Velasco-Galilea, M., Piles, M., Rafel, O. *et al.* (2019) Genomic regions associated with individual growth and cage feed efficiency in rabbits under two feeding regimes. *37th International Society for Animal Genetics Conference,* 7–12 July, Lleida, Spain.

Sánchez, J.P., Perucho, O., Pascual, M., Rafel, O. and Piles, M. (2021) Electronic feeder to record individual feed intake on rabbits raised in collective cages. *12th World Rabbit Congress*, 1–3 July 2020, Nantes, France.

Smith, C. (1984) Rates of genetic change in farm animals. *Journal of Agricultural Research and Development* 1, 79–85.

Sosa-Madrid, B.S., Hernández, P., Blasco, A., Haley, C.S., Fontanesi, L. *et al.* (2020a). Genomic regions influencing intramuscular fat in divergently selected rabbit lines. *Animal Genetics* 51, 58–69.

Sosa-Madrid, B.S., Varona, L., Blasco, A., Casto, C. and Ibañez-Escriche, N. (2020b) The effect of divergent selection for intramuscular fat in the domestic rabbit genome. *Animal* 14, 2225–2235.

Sternstein, I., Reissmann, M., Maj, D., Bieniek, J. and Brockmann, G.A. (2015) A comprehensive linkage map and QTL map for carcass traits in a cross between Giant Grey and New Zealand White rabbits. *BMC Genetics* 16, 16.

Su, G., Madsen, P., Nielsen, B., Ostersen, T., Shirali, M. *et al.* (2018) Estimation of variance components and prediction of breeding values based on group records from varying group sizes. *Genetics Selection Evolution*, 50, 42.

Szendrő, Zs., Ballay, A., Ráczkevy, S. and Bíró, E. (1988) Progeny test onstation in Hungary. In: Holdas, S. (ed.) *Proceedings of the 4th WorldRabbit Congress*, Budapest, pp. 289–293.

Szendrő, Zs., Romvári, R., Horn, P., Radnai, I., Bíró-Németh, E. and Milisits, G. (1996) Two-way selection for carcass traits by computerised tomography. *6th World Rabbit Congress*, Toulouse, France, Vol. 2, pp. 371–375.

Szendrő, Zs., Matics, Zs., Gerencsér, Zs., Radnai, I., Lengyel, M. *et al.* (2009) Effect of adult weight and CT based selection on carcass traits of growing rabbits. *Italian Journal of Animal Science* 8, 240–242.

Szendrő, Zs., Matics, Zs., Gerencsér, Zs., Nagy, I., Lengyel, M. *et al.* (2010) Effect of dam and sire genotypes on productive and carcass traits of rabbits. *Journal of Animal Science* 88, 533–543.

Szendrő, Zs., Metzger, Sz., Nagy, I., Szabó, A., Petrási, Zs. *et al.* (2012) Effect of divergent selection for the Computer Tomography measured thigh muscle volume on productive and carcass traits of growing rabbits. *Livestock Science* 149, 167–172.

Taylor, St C. (1980) Genetic size-scaling rules in animal growth. *Animal Production* 30, 161.

Taylor, St C. (1985) Use of genetic size-scaling in evaluation of animal growth. *Journal of Animal Science* 61, Suppl. 2, 118–143.

Varewyck, H., Bouquet, Y. and Van Zeveren, A. (1986) A progeny test for carcass quality in meat rabbits. *Archiv für Geflügelkunde* 50, 26–31.

Velasco-Galilea, M., Piles, M., Viñas, M., Rafel, O. and Sánchez, J.P. (2018) Rabbit microbiota changes throughout the intestinal tract. *Frontiers in Microbiology* 9, 2144.

Visscher, P.M., Brown, M.A., McCarthy, M.I. and Yang, J. (2012) Five years of GWAS discovery. *American Journal of Human Genetics* 90, 7–24.

Wang, J., Elzo, M.A., Yan, L., Chen, S., Jia, X. *et al.* (2016) A single nucleotide polymorphism in CAST gene is associated with meat quality traits in rabbits. *Animal Science Papers and Reports* 34, 269–278.

Wang, J., Shi, Y., Elzo, M.A., Su, Y., Jia, X. *et al.* (2017) Myopalladin gene polymorphism is associated with rabbit meat quality traits. *Italian Journal of Animal Science* 16, 400–404.

Webb, A.J. (1989) Genetics of food intake in the pig. In: Forbes, J.M., Varley, M.A. and Lawrence, T.L.J. (eds) *The Voluntary Food Intake of Pigs*. British Society of Animal Production, Midlothian, UK, pp. 41–50.

Whittemore, C.T. (1987) *Elements of Pig Science*. Longman, Harlow, UK.

Zhang, G.W., Gao, L., Chen, S.Y., Zhao, X.B., Tian, Y.F. *et al.* (2013) Single nucleotide polymorphisms in the FTO gene and their association with growth and meat quality traits in rabbits. *Gene* 527, 553–557.

Zhang, W.X., Zhang, G.W., Peng, J. and Lai, S.J. (2012) The polymorphism of GHR gene associated with the growth and carcass traits in three rabbit breeds. *10th World Rabbit Congress*, Sharm El-Sheikh, Egypt, pp. 75–78.

Zomeño, C., Hernández, P. and Blasco, A. (2013a) Divergent selection for intramuscular fat content in rabbits. I. Direct response to selection. *Journal of Animal Science* 91, 4526–4531.

Zomeño, C., Hernández, P. and Blasco, A. (2013b) Divergent selection for intramuscular fat content in rabbits. II. Correlated responses in meat characteristics. *Journal of Animal Science* 91, 4532–4539.

Zubiri-Gaitán, A., Martínez-Álvaro, M., Blasco, A. and Hernández, P. (2019) Effect of selection for intramuscular fat content in the cecum microbiome content in rabbits. *37th International Society for Animal Genetics Conference*, 7–12 July, Lleida, Spain.

11 Biology of Reproduction and Reproduction Technologies in the Rabbit

Cristiano Boiti[1*], Cesare Castellini[2], Pascale Chavatte-Palmer[3], Peter Chrenek[4,5], Bernd Fischer[6] and Adriana Kolesárová[7]

[1]Department of Veterinary Medicine, University of Perugia, Italy; [2]Department of Agricultural, Food and Environmental Sciences, University of Perugia, Italy; [3]INRAE UMR 1198 Biologie du Développement et Reproduction, Jouy-en-Josas, France; [4]Department of Biochemistry and Biotechnology, Faculty of Biotechnology and Food Sciences, Slovak University of Agriculture in, Nitra, Slovak Republic; [5]NPPC—Research Institute for Animal Production Nitra, Lužianky, Slovak Republic; [6]Department of Anatomy and Cell Biology, Martin Luther University, Faculty of Medicine, Halle (Saale), Germany; [7]Department of Animal Physiology, Faculty of Biotechnology and Food Sciences, Slovak University of Agriculture in Nitra, Slovak Republic

11.1 Introduction

In rabbits, as well as in any other species, progress in the biology of reproduction is fostered by our understanding of the underlying factors that control the entire sequence of events from gametogenesis to ovulation, from fertilization to embryo development, gestation and parturition. Unquestionably, a better knowledge of those mechanisms that drive the emergence of high-quality oocytes and sperms after meiosis as well as embryos following fertilization is also a key requirement for the development and/or refinement of novel reproduction technologies that are better tailored to rabbit reproduction.

Due to the short gestation length and the large number of offspring, the rabbit has high reproductive potential. The profitability of commercial rabbit farms has increased in recent years due primarily to improvements in genetic selection and the management of reproduction (Dal Bosco *et al.*, 2011). As for other animal species, further development of reproductive techniques could lead to relevant changes in management, selection and preservation of rabbit populations.

In addition, due to their optimal size and physiologic similarities to humans, rabbits have proved to be an important translational model for studying cardiovascular diseases, the toxicology of pregnancy, and the developmental origins of health and diseases (Tyl, 2010; Duranthon *et al.*, 2012; Fischer *et al.*, 2012).

In this chapter, we have updated the recent advancements acquired in the field of biology and technology of reproduction that may have significant repercussion in the breeding of rabbits and in biomedical research.

11.2 Gonadogenesis

In both rabbit sexes, the first germ cells are detected from the ninth day *post coitum*. They migrate into the genital crests until day 16 and

Contact e-mail address: boiti.cristiano@gmail.com

 The Genetics and Genomics of the Rabbit (ed Luca Fontanesi)
DOI: 10.1079/9781780643342.0011

thereafter undergo a series of mitosis until final differentiation at meiosis. Histological features and gene expression during gonadal development in the rabbit have been thoroughly described (Daniel-Carlier *et al.*, 2013).

In females, contrary to humans and most mammals, meiosis takes place entirely after birth, starting from birth to 15 days of age and subsequently followed by the onset of folliculogenesis (Hutt *et al.*, 2006). Puberty takes place around 3–4 months of age.

In males, primitive seminiferous tubules are formed and contain Sertoli and germline cells before birth (Jost *et al.*, 1985). Testes and the accessory glands grow rapidly after five weeks of age. Meiotic activity appears around 50 days and becomes abundant 15 days later. The first spermatozoa are present in the ejaculate at about 110 days. Puberty is achieved at 5–6 months of age.

11.3 Reproduction in the Male

The testicles in mature bucks are oval-shaped and, during mating, appear prominent. The scrotum hangs from the body surface, so that the testicles are located at a temperature of about 35°C, necessary for optimal maturation of spermatozoa. At birth, the testicles are located in the abdominal cavity and, at about 2 months of age, they descend into the scrotal sac. At rest, the penis is directed obliquely backwards and is encased in a skin-fold penis; it points forward when erect.

During maturation, sperms are protected from anti-sperm antibodies and toxic substances by the blood–testis barrier. A series of divisions leading to spermatocytes II, spermatids and sperm undergo in the seminiferous tubules spermatogonial stem cells. The sperm cells are then released into the lumen of the seminiferous tubules and move into the efferent channels. It takes 48–55 days to obtain an epididymal sperm from a spermatogonium. The transit from initial to the final part of the epididymis lasts 8–10 days.

11.3.1 Sperm cell

The sperm is a highly differentiated cell with a head, containing DNA, and a tail that allows the movement. During the early stages of spermatogenesis, granules secreted by the Golgi apparatus form the acrosome, which occupies about two thirds of the head and contains glycoproteins and lytic enzymes involved in sperm, attach to the oocyte and penetration of zona pellucida (ZP). The nuclear chromatin progressively condenses for stabilizing and protecting the DNA. The histone proteins are replaced by transition proteins and later by arginine-rich proteins. Associated with these changes there is a reduction in the nuclear volume and a block in gene transcription. The tail in its initial part contains the mitochondrial sleeve with key enzymes ensuring the energy metabolism. Sperm have lost almost all the common cellular functions, lack the reparative processes and are very sensitive to local environmental conditions (seminal plasma, vaginal and uterine fluids, conservation medium).

11.3.2 Sperm membrane

Sperms have a high amount of long-chain polyunsaturated fatty acid (PUFA) of n-3 and n-6 series (Apel-Paz *et al.*, 2003) which are responsible for the membrane fluidity and competence (capacitation, acrosome reaction and oocyte membrane fusion). Metabolic and functional characteristics of the sperm are influenced by dietary PUFA, which improves the kinetic parameters of the cell. On the contrary, high level of dietary cholesterol alters the metabolism of Sertoli cells (Yamamoto *et al.*, 1999) and the normal spermatogenesis.

11.3.3 Sperm capacitation and acrosome reaction

Capacitation and acrosome reaction (AR) are closely linked. Capacitation implies the acquisition by the sperm of the full fertilizing ability due to the contact with the uterine and oviductal fluids. Capacitation depends on the removal of carbohydrate groups that enhance the removal of cholesterol from the membrane. The capacitated sperm also exhibits hyperactivation consisting in the beating of the tail that facilitates the crossing of gel areas and the adhesion to the ZP.

The initiation of capacitation and AR also depends on free radicals. In fact, they activate a molecular cascade resulting in an increase of intracellular Ca^{2+} and an efflux of cholesterol from the membrane that increases membrane fluidity (Darin-Bennett and White, 1977). The main molecular changes involved in the activation of capacitation, besides the loss of cholesterol, are ionic flows due to an altered membrane potential and an increase in tyrosine phosphorylation of proteins. The changes of the post-translational proteins are essential for sperm since sperm are not able to synthesize any protein.

Only capacitated sperm undergo the AR, recognize the ZP and fuse with it. The AR involves the acrosome lysis and the release of proteolytic enzymes. The interaction between sperm and oocytes is mediated by molecules whose receptors are expressed on the plasma membrane of the sperm while the glycans (complementary) are located on the ZP.

11.3.4 Seminal granules (SGs), ejaculate and daily sperm production

Estrous plasma contains many round-shaped granules of different sizes (0.5–6 μm diameter) found also in several other mammalian species (El-Hajj *et al.*, 2004). These granules, which are secreted by the prostate, are surrounded by an electron dense membrane (Castellini *et al.*, 2012) and contain a large amount of cholesterol and sphingomyelin (Castellini *et al.*, 2006).

These granules, together with seminal plasma (Castellini *et al.*, 2013), affect sperm capacitation and AR, their kinetics, immune-response, and transit in the female tract (Johansson *et al.*, 2004). Granules and seminal plasma probably act as donors of sterols that protect spermatozoa against environmental shock and premature acrosome reaction. In addition, seminal granules are very rich in tocopherol (Mourvaki *et al.*, 2008), which contributes to reducing the responsiveness of sperm to exogenous stimuli by lowering the free radicals in the semen.

The role of SGs as modulators of capacitation and AR along the female tract is particularly sound in rabbit because the ovulation in rabbit female is induced by coitus. Indeed, ovulation occurs about 10–16 hours from mating and, during this lag-phase, SGs contribute to avoid premature capacitation and AR of rabbit sperm.

The standard semen volume is about 0.3–0.8 mL, whereas sperm concentration ranges from 150 to 500×10⁶ sperm/mL. During the epididymal transit about 30% of sperm (mainly defective) is reabsorbed while 20% is excreted in the urine. The amount of sperm produced by bucks depends, however, on several factors (individual, age, photoperiod, temperature, collection rhythm, stress, drugs).

11.3.5 Buck and semen

Several factors affect seminal traits (Boiti *et al.*, 2005) and thus it is crucial to define suitable protocols to improve sperm characteristics and the expected fertility. Semen evaluation provides information on the fertilizing ability of spermatozoa: the most relevant parameters correlated with the fertility rate are the number of spermatozoa/AI and their motility. However, the use of a single trait is not able to predict the fertilizing ability of the semen (Lavara *et al.*, 2005).

11.3.5.1 Factors influencing semen production

Individual and genetic variability of semen characteristics in male rabbits is generally high (Mocé *et al.*, 2005). Buck effects were also reported on the fertility and prolificacy after AI with frozen-thawed semen from different selected rabbit lines (Mocé *et al.*, 2003). Lavara *et al.* (2013, 2017) showed that sperm freezability in rabbit is heritable and heritability for some rabbit frozen-thawed semen traits ranged from 0.08 to 0.15.

Frequency of collection has an effect on semen characteristics (Bencheikh, 1995) both in terms of quality and quantity. A too-intensive collection rhythm or very few collections exert a depressive effect on sperm output. Light affects the hypothalamus-pituitary axis, hormonal release and spermatozoa production. A daily, constant 16L:8D light programme increases sperm production. By contrast, light intensity does not affect semen characteristics (Besenfelder *et al.*, 2004).

Sexual maturity occurs approximately at 5 months (depending on the strain) and semen quality generally decreases in old rabbits (>30 months). Some authors showed that the chromatin structure of the rabbit sperm is less stable in young (<5 months) and old bucks (Gogol *et al.*, 2002). Bucks' health has to be regularly controlled because inflammation of male reproductive apparatus impairs testicular function and seminal characteristics.

Specific dietary recommendations for rabbit bucks are not available, and only some requirements have been established. However, a balanced fatty acid composition seems one of the most important aspects of male nutrition. Dietary addition of PUFA modified several traits of rabbit spermatozoa (Castellini *et al.*, 2019). At the same time, the high unsaturation level of spermatozoa membrane renders these cells very susceptible to peroxidation and, therefore, diets should be fortified with antioxidants. High dietary level of alpha-tocopherol is effective in limiting the oxidative damage of semen (Castellini *et al.*, 2003).

11.4 Reproduction in the Female

11.4.1 Estrous synchronization and ovulation

Estrous synchronization increases fertility and prolificacy, while decreasing the number of inseminations required for gestation (Rebollar *et al.*, 2006). High-yield lactating does need effective estrous synchronization methods to improve their reproductive outcome by enhancing ovarian function (Arias-Álvarez *et al.*, 2010). The most known regulators of ovarian functions are hormones, growth factors and mediators of their action, e.g. cyclic nucleotides, protein kinases (Chrenek *et al.*, 2010; Sirotkin *et al.*, 2010; Sirotkin, 2011), and plant molecules (resveratrol, rapamycin), which can have either regulatory or protective properties (Kadasi *et al.*, 2012; Kolesarova *et al.*, 2012). On the other hand, environmental factors such as heavy metals (Kolesarova *et al.*, 2010), mycotoxins (Kolesarova *et al.*, 2012) or high fat–high cholesterol diet (Cordier *et al.*, 2013) can affect ovarian functions.

Ovarian gonadotropin stimulation is routinely used in animal reproductive technologies for estrus synchronization, multiple ovulation, and/or embryo transfer protocols to favour follicular growth and generate more developmentally competent oocytes. Gonadotropin releasing hormone (GnRH) plays a central role in the regulation of reproductive functions as it regulates the release of both LH and FSH gonadotropins (Teplán, 1989). Ovulation in mammals involves pulsatile release of GnRH from the hypothalamus into the hypophyseal portal system with subsequent release of LH from the anterior pituitary into systemic circulation. Elevated circulating concentrations of LH induce a cascade of events within the mature follicle, culminating in follicle rupture and evacuation. The broad classification of species as either spontaneous or induced ovulators is based on the type of stimulus responsible for eliciting GnRH release from the hypothalamus. In spontaneously ovulating species, the release of GnRH from the hypothalamus is triggered when, in the absence of progesterone, systemic estradiol concentrations exceed a threshold. In induced ovulators such as rabbits, the release of GnRH is contingent upon copulatory stimuli; hence, ovulation is not a regular cyclic event (Adams and Ratto, 2013).

Gonadotropins (FSH, LH) and their analogues – equine chorionic gonadotropin (eCG) and human chorionic gonadotropin (hCG) – and GnRH and its analogues are the most known and widely used promoters of rabbit ovarian follicular growth and ovulation (Dal Bosco *et al.*, 2011). The mechanisms of action of these hormones are not fully elucidated. Nevertheless, it was shown that FSH, LH, and GnRH can exert their effects through changes of ovarian cell proliferation, apoptosis, transcription factor CREB-1, release of steroid hormones, oxytocin, vasopressin and IGF-I (Nitray *et al.*, 1992; Laukova and Sirotkin, 2007; Zerani *et al.*, 2010).

Several methods can be used to increase female receptivity, the most common being eCG treatment, although the use of hormones in meat-producing animals should be avoided (Castellini, 1996). In rabbit breeding, eCG is usually applied for improving follicular development and as a method for estrus synchronization in high-yield lactating animals because lactation exerts an inhibitory effect on ovarian function, especially in primiparous does (Rebollar *et al.*, 2008; Theau-Clément *et al.*, 2008). Moreover, gonadotropin stimulation protocols can modify peripheral 17β-estradiol and progesterone concentrations, as well as their ratio, and therefore the environment in which oocytes and embryos are developed (Ertzeid and Storeng, 1992). It is generally accepted that progesterone is a key regulator of reproductive function in mammals (Fair and Lonergan, 2012). It is essential for normal ovarian cycles and contributes in regulating ovarian follicular development. Progesterone also exerts an inhibitory action on follicular development and steroidogenic function in the rabbit ovary (Setty and Mills, 1987; Parillo *et al.*, 2013).

Follicular development and function are regulated by several hormones such as gonadotropins, leptin and insulin-like growth factor I (IGF-I), which either alone or in combination, affect hormonal secretion, proliferation and apoptosis through intracellular regulators (Sirotkin, 2011). Indeed, IGF-I stimulated progesterone and prostaglandin (PGF) but suppressed testosterone and estradiol secretion by cultured rabbit ovarian granulosa cells (Sirotkin *et al.*, 2009). The presence of IGF-I in cultures was associated with a significant increase of 17β-estradiol and androstenedione production by cumulus–oocyte complexes. Moreover, the addition of IGF-I significantly enhanced nuclear maturation in rabbit oocytes. It must be underlined, however, that all these effects were found in oocytes surrounded by cumulus cells (Lorenzo *et al.*, 1997).

11.4.2 Luteal function and pseudopregnancy

The corpus luteum (CL) is a transient endocrine gland that secretes progesterone, a hormone necessary for the development and maintenance of pregnancy. The CL develops rapidly from an ovulated follicle reaching its maximal size and functional capacity within 9–10 days. The growing CL is characterized by active angiogenesis and intense tissue remodelling with luteinization of the granulosa cells caused by LH (Webb *et al.*, 2002). Several factors, including lymphokines, growth factors and prostaglandin (PGE2 and PGF2α) released by different luteal cell types such as endothelial cells, fibroblasts and resident immune cells, as well as progesterone and 17β-estradiol released by luteal and follicular cells (Goodman *et al.*, 1998) act to programme the lifespan of CL. However, the overall balance between luteotrophic and luteolytic actions may change with the age of the CL and/or in the presence of an embryo (Stocco *et al.*, 2007).

In rabbits, the CL is maintained throughout gestation, a characteristic that differentiates this species from many other mammals. If pregnancy fails to occur, the CL and uterus develop as in a normal pregnancy, but for a shorter period, lasting 16–18 days. Thus, pseudopregnancy is the consequence of an ovulation following unsuccessful mating or mating by a sterile buck. It can also be induced by vaginal stimulation, mounting by another doe, proximity to a male rabbit, or treatment with GnRH or hCG, but often no specific cause can be associated with pseudopregnancy. The pseudopregnant doe is unable to conceive up to day 15 (Fischer *et al.*, 1986). At day 16 of pseudopregnancy, a normal pseudo-/pregnant endometrial transformation can be induced again and fertilization and implantation can be achieved at day 15, although full restoration of the epithelium is not completed before day 24 (Busch *et al.*, 1986).

Luteolysis is a dynamic regression process (McCracken *et al.*, 1999) that comes into play at around days 12–14 of pseudopregnancy under the action of PGF2α secreted by the endometrium and ends with the complete functional and structural demise of the CL, a few days later. The presence of the uterus is necessary for a timely luteolysis, although there is no evidence for a counter-current transfer in the utero-ovarian pedicle allowing a local control between the uterine horn and the adjacent ovary. In addition, several distinct intraluteal pathways have emerged as potential candidate mediators of the initial PGF2α triggering luteolytic action, including nitric oxide (NO), endothelial derived factors (Boiti *et al.*, 2007), immune and apoptosis related factors (Maranesi *et al.*, 2010) as well as locally synthesized prostaglandins (Zerani *et al.*, 2007). In addition, receptors for estrogen, progesterone, PGF2α, GnRH, ACTH, leptin, dopamine (Guelfi *et al.*, 2011; Parillo *et al.*, 2013) are localized in the rabbit CL and may mediate the expression of pro- and/or anti-apoptotic genes that control the CL lifespan.

In rabbits, complete functional luteolysis can be induced by exogenous PGF2α administration only when the CL have acquired a full luteolytic capacity, from day 9 after ovulation (Boiti *et al.* 1998). The PGF2α has no effect on CL during the early-luteal phase up to day 6 of pseudopregnancy, a borderline day between early and mid-luteal stages when CL shift from refractoriness to partial responsiveness.

11.4.3 Breeding strategies

Rabbit does are considered fertile at any stage of the reproductive cycle. However, the success of artificial insemination (AI) largely depends on the sexual receptivity of the female which, in turn, interacts with genetic factors, physiological and metabolic state (lactating or not, stage of lactation), and parity (Parigi-Bini and Xiccato, 1993; Fortun-Lamothe *et al.*, 1999). Accordingly, the attainment of high sexual receptivity at AI is a main goal. Such increase could be reached in different ways: use of synchronizing hormones or biostimulations induced by change of cage, does gathering, dam-litter separation, feeding programme, and light stimulations.

11.4.3.1 Induction of ovulation

The rabbit is a reflex-ovulating species in which sensory and neuroendocrine stimuli act together to determine the ovulatory response. In the standard AI technique, a GnRH analogue is intramuscular injected to induce pituitary LH release. However, ovulation can also be induced by the vaginal absorption of GnRH analogues added to the seminal dose (Rebollar *et al.*, 2012).

11.4.3.2 Sexual receptivity

AI also allows the induction of pregnancy in females which refuse mating, but their fertility is significantly lower than in receptive does. The low fertility and prolificacy of non-receptive does are mainly due to ovulation and pregnancy failures (Theau-Clément, 2007). Moreover, non-receptive does have a lower number of pre-ovulatory follicles (Kermabon *et al.*, 1994) and estrogen concentrations (Rebollar *et al.*, 1992). Consequently, the productivity of non-receptive females is 3 to 4 times lower than that of receptive females. Although rabbit does do not have an estrous cycle, immediately after kindling and after weaning, almost all does are receptive. Therefore, the reproductive rhythm should be co-ordinated with this 'natural' trend of receptivity. Unfortunately, *post partum* AI is too stressful for the does, whereas post-weaning AI is considered too extensive for commercial breeding (Cardinali *et al.*, 2008). The main methods for improving sexual receptivity are hormonal treatments or alternative methods called 'biostimulations'.

11.5 Modern Reproductive Technologies

11.5.1 Cryopreservation of embryos

Cryoconservation methods serve for a long-term preservation of the biological material and are important for the gene bank formation, which guarantees genetic material resources. Cryoconservation methods are slow freezing and fast freezing or vitrification.

Slow freezing is based on gradual dehydration of embryos with temperature decreasing gradually, approximately 1°C per minute, but is time-consuming and requires controlled and expensive freezing units.

Vitrification is the solidification of a solution at low temperature through extreme elevation in viscosity during cooling (water assumes a glassy, vitrified state from the liquid phase). This method combines the use of concentrated solutions of cryoprotectants with rapid cooling to avoid ice formation. There are permeating (glycerol, ethylene glycol, dimetyl sulfoxide) and not permeating cryoprotectants (sacharides, proteins, polymers) depending on their ability to enter the cell membranes. Permeating substances have a strong dehydrating effect, are hydrophilic and decrease the solution freezing point delaying ice crystal formation that usually occurs at −10°C and is reached at −35°C until −45°C. Non-permeating substances significantly decrease the concentration of the intracellular cryoprotectants necessary for the successful cryoconservation and so also their toxic effect (Chrenek *et al.*, 2011).

Cryoconservation of embryos includes six steps: (i) primary exposure to cryoprotectant; (ii) cooling (fast or slow) below freezing point; (iii) storage in liquid nitrogen at −196°C; (iv) thawing; (v) dilution and flushing out the cryoprotectant; and (vi) transfer of embryos into the uterus. During freezing, embryos can be damaged by intracellular formation of ice crystals, extracellular crystallization, chemical toxicity of the cryoprotectant, osmotic swelling, osmotic shrivelling, and by rupturing.

11.5.2 Cryopreservation of rabbit spermatozoa

In several livestock species, including rabbit, artificial insemination (AI) is used to optimize reproduction and breeding management and improve selection. In the rabbit-meat industry, AI is performed with fresh or cooled semen. Frozen rabbit semen is not used because of the poor fertility resulting after thawing (Mocé and Vicente, 2009). However, frozen thawed rabbit semen is used for other purposes: *ex situ* conservation of rabbit genetic resources (including rare or endangered breeds or high-value males); international export (semen from selected lines); and for research (Kulíková *et al.*, 2017; Lavara *et al.*, 2017).

Cryopreservation of rabbit semen is applied in similar ways as carried out in other species with some adaptations. After semen collection, the spermatozoa are diluted to get larger semen volume and greater number of insemination doses. The main function of the diluent is to conserve semen and to prolong the viability and fertilizing ability of the gametes.

The preservation length of semen biological features depends on the diluent used. The interval between semen collection and dilution should not be longer than 15 minutes to prevent development of intoxication processes. The diluent must have good buffer ability, small electrolytes content, same osmotic pressure of spermatozoa, suitable pH, and temporary energetic source for spermatozoa. The diluent should not affect spermatozoa, their metabolism and agglutination, as well as fertility, cause inflammatory processes in genitals of the females, contain germs that might induce infections and diseases in human and animals, and contain more than 5000 non-pathogenic germs in the insemination dose.

Semen dilution has several principles that should be respected. First of all, the semen must be assessed macroscopically and microscopically. The diluent is always added to the semen, not contrariwise, and added slowly along with permanent mixing. The diluent temperature has to be the same as that of the ejaculate (± 1°C). The pH of the diluent solution should vary from 6.2 to 7.5, and all tools used for dilution must be sterile and at the right temperature. The most used diluents are yolk added with either citrate or milk, or plant juices and honey as well as mineral waters.

Semen, which is conserved using sodium citrate (pH 6.15) or acids (carbonic, citric) has short fertilizing ability. Therefore, this approach is substituted by deep freezing. Before deep freezing, the diluted ejaculate must be cooled. It is chilled at 2–4°C, so the fertilizing ability is prolonged to 72 hours, but the optimal use is within 24–48 hours. In the fast-freezing method, the whole freezing process lasts 7–12 minutes. The semen samples are conserved in straws (plastic cavities) or in pellets.

11.5.2.1 Semen freezing in the straws (French method)

The process starts with the dilution of the ejaculate in a yolk-citrate diluent for half of the final dilution. In the second step, it is stored for 50 minutes in a refrigerator at 1–3°C. When cooled, the semen is diluted in the same diluent volume enriched with 10–20% glycerol. Glycerol diluents should be added to the semen gradually, for 15 minutes. After final dilution, the ejaculate must show at least 70% of spermatozoa activity before freezing. It is diluted in such a ratio that the insemination dose will contain at least $10–12x10^6$ motile spermatozoa.

Diluted semen is divided into a volume of 0.25 mL using an automatic device and then hermetically sealed into ministraws. One end of the ministraw is closed with polyvinyl alcohol. All these manipulations are done at a temperature of 5°C. In another step, the ministraws are put in a horizontal position into the water bath of the same temperature as the ejaculate. They are transferred to the refrigerator where they equilibrate at 3°C for 2–6 hours. Before deep freezing, diluted semen must be kept at a temperature above 0°C, because during this period the sperm membranes become more resistant to freezing, against the critical temperature from −15°C to −25°C. After equilibration, the semen is frozen in the vapours of liquid nitrogen for 7 minutes and then it is directly immersed into liquid nitrogen.

15.5.2.2 Semen freezing in the PELLETS (Japanese method)

Fresh semen is diluted at a temperature of 35°C in the diluent that includes 75.3% of 11% lactose solution, 20% egg yolk, and 4.7% glycerol. After dilution, the semen is put into the refrigerator at a temperature of 1–3°C for equilibration that lasts about 4–7 hours. In another step, the semen is frozen by the fast method at the rate of 20°C/min on the prism of dry ice with thickness of at least 3 cm. Hollows are 6 mm in diameter and about 4–8 mm depth. The semen freezing takes 7 minutes. Afterwards, pellets are stored in the container with liquid nitrogen. Frozen spermatozoa cannot be thawed at the critical point of −60°C. Diluted semen is frozen into 0.1 mL pellets directly on the dry ice and stored in a container with liquid nitrogen.

There are three main mechanisms during the semen freezing (and consequent thawing) that may damage the gametes: extracellular and intracellular crystallization, formation of concentrated solutions of electrolytes, and rapid decrease in temperature.

During the cell suspension thawing, the ice crystals are firstly formed in the extracellular fluid, and then inside the cells. A prerequisite for the successful storage of the spermatozoa is to prevent, from chemical processes, the change of the free extracellular and intracellular water into solid state. At selected temperature, the re-crystallization processes should be eliminated or significantly restricted; these processes are stopped at the temperature of −160°C and at the temperature of −196°C the storage should be time unlimited.

11.5.3 Embryo development and losses

Most of the data go back to studies performed in the 1950s to 1970s. The extensive knowledge on pre-, peri- and post-implantation embryo development and embryo–maternal interactions together with the rare biological feature of the uterus duplex (two cervices and uteri) make the rabbit an ideal experimental model for many aspects of biomedical research.

Approximately 10–12 hours after mating (*post coitum*, p.c.) or hCG/GnRH injection, ovulation takes place. About 4 hours later the second polar body is visible as proof of fertilization. Twenty-four hours p.c., the first cleavage division to a two-cell embryo occurs. The next cleavages and time intervals are as follows: 4-cell stage at 28 hours, 8-cell stage at 36 hours, 16-cell stage at 44 hours, morula at 60 hours, and blastocyst formation at 72 hours p.c. Cleavage stages up to the morula are usually recovered from the oviduct; later stages are obtained from uterine flushing. Cell numbers in morulae and non-expanded blastocysts are reported to be approximately 140 cells in day 3 morulae and 800–1000 in day 4 blastocysts.

Up to the early blastocyst stage, the size of the cleavage-stage rabbit embryo does not increase within the zona pellucida (ZP); the vitelline (or intrazonal) diameter ranges between 125 μm and 160 μm. The outer size of the embryo, however, does increase due to the overlay of the extracellular covering mucoprotein layer and gloiolemma (Fischer *et al.*, 1991). The total diameter increases from about 200 μm in early cleavage stages to 350 μm for morulae and 500 μm for non-expanded blastocysts. The mucoprotein layer is deposited on the ZP during oviductal passage by tubal epithelial cells. The gloiolemma originates from secretions of endometrial epithelial cells (Denker and Gerdes, 1979; Fischer *et al.*, 1991).

The extracellular embryonic coverings are somewhat specific but not unique to the rabbit (Leiser and Denker, 1988). Notably, the ZP gets dissolved during the third and fourth day of development and is replaced by the neozona from day 5 p.c. onwards (Denker and Gerdes, 1979).

Implantation in the rabbit has to be divided in two distinct processes. The expanded blastocyst attaches at 6 days and 16 hours p.c. to the uterine epithelium at the antimesometrial side with its abembryonic pole, leading to the so-called yolk sac placenta. Final implantation of the blastocyst starts at day 8 p.c. with the embryonic pole at the mesometrial side, resulting in the definite chorioallantoid placenta. A haemochorial contact is established in both cases. The yolk sac placenta is only transiently built and soon replaced by the chorioallantoid mesometrial placenta (Denker 1977). The extracellular coverings are intact until 16 hours at day 6 p.c. indicating that attachment and invasion of the uterine epithelium have not yet started. These processes, however, begin directly afterwards with trophoblast cells invading the endometrium and subsequent fusion of trophoblast and uterine epithelial cells. Implantation sites are restricted to the so-called placental folds. Trophoblast invasion goes down to endometrial blood vessels, forming the haemochorial placenta by erosion of maternal blood vessels.

After blastocyst formation, blastocyst size increases dramatically during the next days, to approximately 1 mm at day 5, 2.5 mm at day 6 and 7 mm at day 7 p.c. Sizes of individual blastocysts vary considerably within litters. The blastocyst cavity is surrounded and protected against the uterine milieu by a trophoblast layer, which is tight and highly active in cellular transport. The remarkably large sizes – the biggest blastocysts among mammalian pre-implantation embryos – make the rabbit blastocyst a unique experimental model with many advantages (Fischer *et al.*, 2012). It allows detailed genetic, molecular and cellular analyses, analysis of the blastocyst cavity fluid as well as observation and counting of the blastocyst numbers after uterine flushing or number of implantation sites per uterus by naked eye and/or light microscopy.

As said, the rabbit model allows molecular analyses in individual blastocysts, an advantage not offered by most other mammalian experimental animals. The cell numbers in the embryonic disc are as high as 2000 and 7000 embryo blast cells in stage 1 and 2, respectively (Ramin *et al.*, 2010). Total cell numbers in 6½ day-old blastocysts are in the range of 100,000 (Fujimoto *et al.*, 1975), indicating that trophoblast cell numbers in these ontogenetic stages are still clearly higher. Furthermore, gastrulation starts prior to implantation (Fischer *et al.*, 2012) and can be reliably staged by fairly simple morphological criteria in rabbit blastocysts. The rabbit is one of the few mammalian species that allow analysis of early gastrulation in intact blastocysts as the free-floating blastocysts can be flushed out of the uterine lumen until day 6.6 p.c.

What are the critical phases of early embryo development and for embryo mortality in the rabbit? C.E. Adams, from the ARC Institute of Animal Physiology in Cambridge (UK), has pioneered these studies and published a broad body of information in the late 1950s to 1970s (Adams 1960, 1961, 1970). A short summary of the main findings indicates that about 5% of female rabbits in which fertilization did occur suffer from a total embryonic loss, about 50% lose some embryos, and the others none. Adams has calculated that the pre-implantation embryonic loss in the rabbit is in the range of 10%. After implantation, about 3% of the does lose all implants and 70% lose some. The percentage of post-implantation losses sums up to about 20%. We also know from the studies of Adams when these losses occur: about 7% peri-implantational, 66% between day 8 and 17, and 27% between day 17 and 23 p.c. In total, prenatal embryo mortality in the rabbit sums up to approximately 30% of the ovulated fertilized oocytes. Due to resorption, embryo loss before day 17 p.c. cannot be observed by the rabbit breeders.

Embryo mortality is closely related to embryo numbers. Only 20% of females bearing one foetus complete pregnancy. However, if the other uterus accommodates embryos, then also the single embryo will survive. In the opposite situation, in the case of superfetation with too many embryos and/or implantations, in general all litters are lost rather than only a few embryos.

11.5.4 Superovulation and collection of rabbit embryos and embryo transfer

Superovulation treatment is used to produce a larger number of embryos and is usually initiated 68–72 hours before AI or mating using repeated injections of FSH or eCG (Besenfelder *et al.*, 2000). To induce ovulation at a known time, hCG or a GnRH agonist can be injected just after AI or mating.

Embryos can be recovered from the uterine horns 3.5 days after ovulation. Embryos are collected *post mortem* by flushing the oviducts and uterine horns of the doe or by laparoscopy (Mehaisen *et al.*, 2004). Laparoscopy can be performed repeatedly on the same does, but to avoid abdominal adhesions it is recommended to wash the reproductive tract with 0.1% EDTA solution in PBS (Mehaisen *et al.*, 2004). Since all embryos may not be recovered and the doe may still be pregnant after flushing, treatment with PGF2α is needed if pregnancy has to be avoided.

In vivo-produced embryos can be transferred by laparotomy or laparoscopy. The embryos are mounted in a small catheter connected to a 1 mL syringe. In the case of transfer of early embryos (less than 3.5 days), the embryos are placed directly in the oviduct through the ampulla. For older embryos, a small incision is made with a needle into the tip of the horn to introduce the catheter and gently push the embryos into the lumen. *In vitro*-produced embryos should preferably be transferred in the oviduct where they will acquire the mucin coat.

11.6 Feto-placental Development, Growth and Parturition

Implantation takes place at day 6.6 p.c. Each foetus forms a placenta with a foetal part (placenta) and a maternal part (decidua). Exponential growth of the foetus starts around 12 days of pregnancy (Bal'magiya and Surovtseva, 1974). The rabbit gestation lasts 31–32 days.

After implantation, the rabbit foetus is mostly nourished through the placenta, although the ingestion of amniotic liquid also contributes to foetal growth in the last third of gestation. The placenta is discoid in shape, made of two closely apposed discs (cotyledons). The rabbit placentation is hemochorial and two cell layers between maternal and foetal blood as in humans. The cellular organization within the placenta is close to that of the human placenta in the exchange zone, which makes it an attractive model for placental studies (McArdle et al., 2009). Maternal and foetal blood circulate counter-current and the feto-placenta weight ratio ranges between 1/5 and 1/10, like in humans.

Foetal growth is dependent on oxygen and nutriment transfers from maternal circulation, which depend mainly on blood flow and presence of transporters. The blood flow to each placenta is about 3–4 mL/min at term (Bruce and Abdul-Karim 1973). Placental vascularization can be evaluated *in vivo* by Power Doppler (Lecarpentier *et al.* 2012). For oxygen exchange, in contrast to humans, there is no foetal haemoglobin in the rabbit. The major placental glucose transporter present in the human placenta, SLC2A1 (previously GLUT1), is expressed in the rabbit placenta, as well as SLC2A3, SLC2A4 and SLC5A1 (Kevorkova *et al.*, 2007). Transporters involved in neutral amino acid transport have been reported in the rabbit placenta. In rabbits, as in humans, the placenta is very permeable to free fatty acids and changes in maternal dietary lipid intake can alter their placental transfers to the foetus (Stephenson *et al.*, 1990; Montoudis *et al.*, 2004), in a sex-dependent manner. Immunoglobulins do not cross the placenta in the rabbit and their transfer is mediated partly through the absorption of amniotic liquid before birth and partly through colostrum after birth (Brambell, 1969).

The concurrent effects of pregnancy and lactation reduce litter size and foetal growth, possibly due to the combined effects of reduced energy balance and hyperprolactinemia (Fortun-Lamothe *et al.*, 1999). Due to a better vascularization of the uterus close to the ovary, the largest foetuses are usually close to the tip of the uterine horn. Foetal and placental growths can be monitored by ultrasound imaging (Chavatte-Palmer *et al.*, 2008).

One to two days before delivery, does prepare a nest for the kits using hair that they pull from their own fur. Parturition lasts 10–20 minutes. Litters size is very variable, with a mean of 8–10 kits. Birth weight ranges from 30–60 g in the New-Zealand breed, with 15–20% variability. The doe rapidly cleans the offspring after birth and eats the placentas within one hour. After parturition, the uterus involutes rapidly and loses about half of its weight within 48 hours.

11.7 The Rabbit Model in the Toxicology of Pregnancy and in the Developmental Origins of Health and Diseases

Before drugs are used in humans, tests are performed in animals to evaluate their potential deleterious effects, in particular during pregnancy. Thalidomide, a teratogenous agent in humans, initially was tested only in rodents where it was not harmful; this led to its use in women, causing very serious foetal malformations. Had it been tested in rabbits, however, its teratogenic effects would have been observed. Since then, the rabbit has become a reference species for testing drug teratogenicity (Tyl, 2010).

The rabbit embryo is a powerful tool to study gastrulation and early effects of maternal environment on development (Schindler *et al.*, 2020; Pendzialek *et al.*, 2017). Maternal diabetes was shown to impair gastrulation (Ramin *et al.*, 2010) and affect mother–embryo communication involving the IGF1 axis (Thieme *et al.*, 2012). The rabbit is also large enough to perform fine surgery. Rabbit models of intra-uterine growth retardation (IUGR) have been developed using placental arterial ligatures (Eixarch *et al.*, 2011) or pharmaceutical approaches (Lecarpentier *et al.*, 2012).

Finally, the rabbit lipid metabolism is close to that of humans, with high LDL concentrations in response to high-fat diets and rapid response to high-cholesterol diets through the development of atherosclerosis. The rabbit has become a species of choice for the study of long-term effects of maternal high-fat diets on offspring development and health. A diet supplemented with cholesterol (+0.1%) and lipids (8%) administered to does from 10 weeks of age onwards advances puberty, reduces the number of antral follicles, and perturbs pituitary response to GnRH (Cordier *et al.*, 2013). Similar cholesterol intake during pregnancy affects placental transfer of fatty acids and glucose (Montoudis *et al.*, 2004; Marseille-Tremblay *et al.*, 2007), inducing IUGR with a sexual dimorphism. Fatty streaks (precursors of atheroscleroslerosis) are also induced in the foetal aorta (Napoli *et al.*, 2000). Adult offspring from does fed a diet supplemented with cholesterol and lipids develop components of the metabolic syndrome and reproductive disorders (Picone *et al.*, 2011). These data provide important insight into mechanisms that may play a role in the current epidemy of obesity, metabolic diseases and reduced fertility observed in the human population.

References

Adams, C.E. (1960) Prenatal mortality in the rabbit *Oryctolagus cuniculus*. *Journal of Reproduction and Fertility* 1, 36–44.

Adams, C.E. (1961) Early embryonic mortality in the rabbit. *Journal of Reproduction and Fertility* 1, 315–316.

Adams, C.E. (1970) Maintenance of pregnancy relative to the presence of few embryos in the rabbit. *Journal of Endocrinology* 48, 243–249.

Adams, G.P. and Ratto, M.H. (2013) Ovulation-inducing factor in seminal plasma: a review. *Animal Reproduction Science* 136, 148–156.

Apel-Paz, M. Vanderlick, T.K. Chandra, N. and Doncel, F.G. (2003) A hierarchy of lipid constructs for the sperm plasma membrane. *Biochemical Biophysical Research Communications* 309, 724–732.

Arias-Álvarez, M., García García, R.M., Torres-Rovira, L., González-Bulnes, A., Rebollar, P.G. and Lorenzo, P.L. (2010) Influence of hormonal and nonhormonal estrus synchronization methods on follicular and oocyte quality in primiparous lactating does at early postpartum period. *Theriogenology* 73, 26–35.

Bal'magiya, T.A. and Surovtseva, Z.F. (1974) Growth pattern of rabbit fetuses during normal pregnancy and with inhibition of the 'gestation dominant'. *Byulleten' Eksperimental'noi Biologii I Meditsiny* 77, 44–47.

Bencheikh, N. (1995) Effet de la fréquence de collecte de la semence sur les caractéristiques du sperme. *Annales Zootechnie* 44, 263–279.

Besenfelder, U., Haas, C. and Brem, G. (2000) Reproduction technology and gene transfer in rabbits. *World Rabbit Science* 8, 37–59.

Besenfelder, U., Theau-Clément, M., Sabbioni, E., Castellini, C., Renieri, T. *et al.* (2004) Effects of different light intensities on quality of spermatozoa in rabbits. *World Rabbit Science* 12, 227–234.

Boiti, C., Canali, C., Zerani, M. and Gobbetti, A. (1998) Changes in refractoriness of rabbit corpora lutea to a prostaglandin F2 alpha analogue, alfaprostol, during pseudopregnancy. *Prostaglandins Other Lipid Mediators* 56, 255–264.

Boiti, C., Castellini, C., Besenfelder, U., Theau-Clément, M., Liguori, L. *et al.* (2005). Guidelines for the handling of rabbit bucks and semen. *World Rabbit Science* 13, 71–91.

Boiti, C., Maranesi, M., Dall'Aglio, C., Pascucci, L., Brecchia, G. and Zerani, M. (2007) Vasoactive peptides in the luteolytic process activated by PGF2α in pseudopregnant rabbits at different luteal stages. *Biology of Reproduction* 77, 156–164.

Brambell, F.W.R. (1969) The transmission of immune globulins from the mother to the foetal and newborn young. *Proceedings of the Nutrition Society* 28, 35–41.

Bruce, N.W. and Abdul-Karim, R.W. (1973) Relationship between fetal weight, placental weight and maternal placental circulation in the rabbit at different stages of gestation. *Journal of Reproduction and Fertility* 32, 15–24.

Busch, L.C., Winterhager, E. and Fischer, B. (1986) Regeneration of the uterine epithelium in later stages of pseudopregnancy in the rabbit: an ultrastructural study. *Anatomy and Embryology* 174, 97–104.

Cardinali, R., Dal Bosco, A., Bonanno, A., Di Grigoli, A., Rebollar, P.G. *et al.* (2008) Connection between body condition score, chemical characteristics of body and reproductive traits of rabbit does. *Livestock Science* 116, 209–215.

Castellini, C. (1996) Recent advances in artificial insemination. *6th World Rabbit Congress*, Toulouse, France, 2, pp. 13–26.

Castellini, C., Lattaioli, P., Dal Bosco, A., Minelli, A. and Mugnai, C. (2003) Oxidative status and semen characteristics of rabbit buck as affected by dietary vitamin E, C and n-3 fatty acids. *Reproduction Nutrition Development* 43, 91–103.

Castellini, C., Cardinali, R., Dal Bosco, A., Minelli, A. and Camici, O. (2006) Lipid composition of the main fractions of rabbit semen. *Theriogenology* 65, 703–712.

Castellini, C., Mourvaki, E., Cardinali, R., Collodel, G., Renieri, T. *et al.* (2012) Secretion patterns and effect of prostate-derived granules on the sperm acrosome reaction of rabbit buck. *Theriogenology* 78, 715–723.

Castellini, C., Mattioli, S., Ruggeri, S., Dal Bosco, A. and Collodel, G. (2013) The time-dependent effects of prostate granules and seminal plasma on the capacitation acrosome reaction and motility of rabbit sperm. *Animal Reproduction Science* 140, 97–102.

Castellini, C., Mattioli, S., Signorini, C., Cotozzolo, E., Noto, D. *et al.* (2019) Effect of dietary source on rabbit male reproduction. *Oxidative Medicine and Cellular Longevity* 2019, 3279670.

Chavatte-Palmer, P., Laigre, P., Simonoff, E., Chesne, P., Challah-Jacques, M. and Renard, J.P. (2008) In utero characterisation of fetal growth by ultrasound scanning in the rabbit. *Theriogenology* 69, 859–869.

Chrenek, P., Grossmann, R. and Sirotkin, A.V. (2010) The cAMP analogue, dbcAMP affects release of steroid hormones by cultured rabbit ovarian cells and their response to FSH, IGF-I and ghrelin. *European Journal of Pharmacology* 640, 202–205.

Chrenek, P., Turanová, Z., Slamečka, J. and Makarevich, A. (2011) Quality of rabbit vitrified/thawed transgenic embryos. *Zygote* 21, 53–58.

Cordier, A-G., Léveillé, P., Dupont, C., Tarrade, A., Picone, O. *et al.* (2013) Dietary lipid and cholesterol induce ovarian dysfunction and abnormal LH response to stimulation in rabbits. *PLOS ONE* 8, e63101.

Dal Bosco, A., Rebollar, P.G., Boiti, C., Zerani, M. and Castellini, C. (2011) Ovulation induction in rabbit does: current knowledge and perspectives. *Animal Reproduction Science* 129, 106–117.

Daniel-Carlier, N., Harscoet, E., Thepot, D., Auguste, A., Pailhoux, E. and Jolivet, G. (2013) Gonad differentiation in the rabbit: evidence of species-specific features. *PLOS ONE* 8, e60451.

Darin-Bennett, A. and White, I.G. (1977) Influence of the cholesterol content of mammalian spermatozoa on susceptibility to cold-shock. *Cryobiology* 14, 466–470.

Denker, H.-W. (1977) Implantation: the role of proteinases, and blockage of implantation by proteinase inhibitors. *Advances in Anatomy, Embryology, and Cell Biology 53/5*. Springer, Berlin, Heidelberg, New York.

Denker, H.-W. and Gerdes, H.J. (1979) The dynamic structure of rabbit blastocyst coverings. I. Transformation during regular preimplantation development. *Anatomy and Embryology* 157, 15–34.

Duranthon, V., Beaujean, N., Brunner, M., Odening, K.E., Santos, A.N. *et al.* (2012) On the emerging role of rabbit as human disease model and the instrumental role of novel transgenic tools. *Transgenic Research* 21, 699–713.

Eixarch, E., Hernadez-Andrade, E., Crispi, F., Illa, M., Torre, I. *et al.* (2011) Impact on fetal mortality and cardiovascular Doppler of selective ligature of uteroplacental vessels compared with undernutrition in a rabbit model of intrauterine growth restriction. *Placenta* 32, 304–309.

El-Hajj Ghaoui, R., Thomson, P.C., Evans, G. and Maxwell, W.M. (2004) Characterization and localization of membrane vesicles in ejaculate fractions from the ram, boar and stallion. *Reproduction Domestic Animal* 39, 173–180.

Ertzeid, G. and Storeng, R. (1992) Adverse effects of gonadotrophin treatment on pre- and postimplantation development in mice. *Journal of Reproduction and Fertility* 96, 649–655.

Fair, T. and Lonergan, P. (2012) The role of progesterone in oocyte acquisition of developmental competence. *Reproduction in Domestic Animals* 47, 142–147.

Fischer, B., Winterhager, E. and Busch, L.C. (1986) Transformation of endometrium and fertility in late stages of pseudopregnancy in the rabbit. *Journal of Reproduction and Fertility* 78, 529–540.

Fischer, B., Mootz, U., Denker, H.-W., Lambertz, M. and Beier, H.M. (1991) The dynamic structure of coverings under non-physiological developmental conditions. III. Transformation of coverings under non-physiological developmental conditions. *Anatomy and Embryology* 183, 17–27.

Fischer, B., Chavatte-Palmer, P., Viebahn, C., Navarrete Santos, A. and Duranthon, V. (2012) Rabbit as a reproductive model for human health. *Reproduction* 144, 1–10.

Fortun-Lamothe, L., Prunier, A., Bolet, G. and Lebas, F. (1999) Physiological mechanisms involved in the effects of concurrent pregnancy and lactation on foetal growth and mortality in the rabbit. *Livestock Production Science* 60, 229–241.

Fujimoto, S., Pahlavan, N., Woody, H.D. and Dukelow, W.R. (1975) Cell numbers in rabbit pre-implantation blastocysts. *Cytologia* 40, 307–311.

Gogol, P., Bochenek, M. and Smora, Z. (2002) Effect of rabbit age on sperm chromatin structure. *Reproduction Domestic Animal* 37, 92–95.

Goodman, S.B., Kugu, K., Chen, S.H., Preuthipan, S., Tilly, K.I. *et al.* (1998) Estradiol-mediated suppression of apoptosis in the rabbit corpus luteum with a shift in expression of Bcl-2 family members favoring cellular survival. *Biology of Reproduction* 59, 820–827.

Guelfi, G., Zerani, M., Brecchia, G., Parillo, F., Dall'Aglio, C. *et al.* (2011) Direct actions of ACTH on ovarian function of pseudopregnant rabbits. *Molecular and Cellular Endocrinology* 339, 63–71.

Hutt, K., McLaughlin, E. and Holland, M. (2006) Primordial follicle activation and follicular development in the juvenile rabbit ovary. *Cell and Tissue Research* 326, 809–822.

Johansson, M., Bromfield, J.J., Jasper, M.J. and Robertson, S.A. (2004) Semen activates the female immune response during early pregnancy in mice. *Immunology* 112, 290–300.

Jost, A., Perlman, S. and Magre, S. (1985) The initial stages of testicular differentiation in the rabbit fetus. *Archives d'anatomie microscopique et de morphologie expérimentale* 74, 69–75.

Kadasi, A., Sirotkin, A.V., Maruniakova, N., Nath, S. and Kolesarova, A. (2012) The effect of rapamycin on secretory activity of the rabbit ovarian fragments. *Journal of Microbiology, Biotechnology and Food Sciences* 1, 488–497.

Kermabon, A.Y., Belair, L., Theau-Clément, M., Salesse, R. and Djiane, J. (1994) Effect of anoestrus and bromocryptine treatment on the expression of prolactin and LH receptors in the rabbit ovary during lactation. *Journal Reproduction Fertility* 102, 131–138.

Kovorkova, O., Ethier-Chiasson, M. and Lafond, J. (2007) Differential expression of glucose transporters in rabbit placenta: effect of hypercholesterolemia in dams. *Biology of Reproduction* 76, 487–495.

Kolesarova, A., Roychoudhury, S., Slivkova, J., Sirotkin, A., Capcarova, M. and Massanyi, P. (2010) In vitro study on the effect of lead and mercury on porcine ovarian granulosa cells. *Journal of Environmental Science and Health, Part A* 45, 320–331.

Kolesarova, A., Capcarova, M., Maruniakova, N., Lukac, N., Ciereszko, R.E. and Sirotkin, A.V. (2012) Resveratrol inhibits reproductive toxicity induced by deoxynivalenol. *Journal of Environmental Science and Health, Part A* 47, 1329–1334.

Kulíková, B., Oravcová, M., Baláži, A., Supuka, P. and Chrenek, P. (2017) Factors affecting storage of Slovak native rabbit semen in the gene bank. *Zygote* 25, 592–600.

Laukova, M. and Sirotkin, A.V. (2007) Follicle-stimulating hormone: effects and possible mechanisms of action in rabbit ovarian cells. *Slovak Journal of Animal Science* 40, 66–71.

Lavara, R., Mocé, E., Lavara, F., Viudes de Castro, M.P. and Vicente, J.S. (2005) Do parameters of semen quality correlate with the results of on-farm inseminations in rabbits? *Theriogenology* 64, 1130–1141.

Lavara, R., David, I., Mocé, E., Baselga, M. and Vicente, J.S. (2013) Environmental and male variation factors of freezability in rabbit semen. *Theriogenology* 79, 582e9.

Lavara, R., Mocé, E., Baselga, M. and Vicente, J.S. (2017) Freezability genetics in rabbit semen. *Theriogenology* 102, 54–58.

Lecarpentier, E., Morel, O., Tarrade, A., Dahirel, M., Bonneau, M. *et al.* (2012) Quantification of utero-placental vascularization in a rabbit model of IUGR with three-dimensional power Doppler angiography. *Placenta* 33, 769–775.

Leiser, R. and Denker, H.-W. (1988) The dynamic structure of rabbit blastocysts coverings. II. Ultrastructural evidence for a role of the trophoblast in neozona formation. *Anatomy and Embryology* 179, 129–134.

Lorenzo, P.L., Illera, J.C., Silvan, G., Munro, C.J., Illera, M.J. and Illera, M. (1997) Steroid-level response to insulin-like growth factor-1 in oocytes matured in vitro. *Journal of Reproductive Immunology* 35, 11–29.

Maranesi, M., Zerani, M., Lilli, L., Dall'Aglio, C., Brecchia, G. *et al.* (2010) Expression of luteal estrogen receptor, interleukin-1, and apoptosis-associated genes after PGF2alpha administration in rabbits at different stages of pseudopregnancy. *Domestic Animal Endocrinology* 39, 116–130.

Marseille-Tremblay, C., Gravel, A., Lafond, J. and Mounier, C. (2007) Effect of an enriched cholesterol diet during gestation on fatty acid synthase, HMG-CoA reductase and SREBP-1/2 expressions in rabbits. *Life Sciences* 81, 772–778.

McArdle, A.M., Denton, K.M., Maduwegedera, D., Moritz, K., Flower, R.L. and Roberts, C.T. (2009) Ontogeny of placental structural development and expression of the renin-angiotensin system and 11[beta]-HSD2 genes in the rabbit. *Placenta* 30, 590–598.

McCracken, J.A., Custer, E.E. and Lamsa, J.C. (1999) Luteolysis: a neuroendocrine-mediated event. *Physiological Reviews* 79, 263–323.

Mehaisen, G.M., Vicente, J.S. and Lavara, R. (2004) In vivo embryo recovery rate by laparoscopic technique from rabbit does selected for growth rate. *Reproduction in Domestic Animals* 39, 347–351.

Mocé, E. and Vicente, J.S. (2009) Rabbit sperm cryopreservation: a review. *Animal Reproduction Science* 110, 1e24.

Mocé, E., Vicente, J.S. and Lavara, R. (2003) Effect of freezing-thawing protocols on the performance of semen from three rabbit lines after artificial insemination. *Theriogenology* 60, 115e23.

Mocé, E., Vicente, J.S., Lavara, R., Viudes De Castro, M.P., Lopez, M. and Bolet, G. (2005) Characteristics of fresh semen from eight rabbit breeds. *Reproduction Domestic Animal* 40, 388–395.

Montoudis, A., Simoneau, L., Brissette, L., Forest, J.-C., Savard, R. and Lafond, J. (2004) Influence of a maternal cholesterol-enriched diet on [1-14C]-linoleic acid and L-[4, 5-3H]-leucine entry in plasma of rabbit offspring. *Life Sciences* 74, 1751–1762.

Mourvaki, E., Collodel, G., Moretti, E., Cosci, I. and Castellini, C. (2008) Distribution of alpha-, gamma (+beta)- and delta-tocopherol in the seminal plasma spermatozoa and seminal vesicles of rabbit. *Andrologia* 40, 282–285.

Napoli, C., Witztum, J.L., Calara, F., de Nigris, F. and Palinski, W. (2000) Maternal hypercholesterolemia enhances atherogenesis in normocholesterolemic rabbits, which is inhibited by antioxidant or lipid-lowering intervention during pregnancy: an experimental model of atherogenic mechanisms in human foetuses. *Circulation Research* 87, 946–952.

Nitray, J., Sirotkin, A. and Rafay J. (1992) Effect of FSH, LH, LH-RH and arginine-vasotocin on the production of steroids, nonapeptide hormones and cGMP by rabbits granulosa cells isolated at different stages of reproductive cycle. *Endocrine Regulations* 26, 95–99.

Parigi-Bini, R. and Xiccato, G. (1993) Recherches sur l'interaction reproduction et lactation chez la lapine. Une revue. *World Rabbit Science* 1, 155–161.

Parillo, F., Dall'Aglio, C., Brecchia, G., Maranesi, M., Polisca, A. *et al.* (2013) Aglepristone (RU534) effects on luteal function of pseudopregnant rabbits: steroid receptors, enzymatic activities, and hormone productions in corpus luteum and uterus. *Animal Reproduction Science* 138, 118–132.

Pendzialek, S.M., Schindler, M., Plösch, T., Gürke, J., Haucke, E. *et al.* (2017) Cholesterol metabolism in rabbit blastocysts under maternal diabetes. *Reproduction, Fertility and Development* 29, 1921–1931.

Picone, O., Laigre, P., Fortun-Lamothe, L., Archilla, C., Peynot, N. *et al.* (2011) Hyperlipidic hypercholesterolemic diet in prepubertal rabbits affects gene expression in the embryo, restricts fetal growth and increases offspring susceptibility to obesity. *Theriogenology* 75, 287–299.

Ramin, N., Thieme, R., Fischer, S., Schindler, M., Schmidt, T. *et al.* (2010) Maternal diabetes impairs gastrulation and insulin and IGF1 receptor expression in rabbit blastocysts. *Endocrinology* 151, 4158–4167.

Rebollar, P.G., Ubilla, E. and Rodriguez, J.M. (1992) Influence of the parturition–insemination interval on the conception rate in rabbits artificially inseminated with fresh semen. *Journal Applied Rabbit Research* 15, 407–411.

Rebollar, P.G., Milanès, A., Pereda, N., Millán, P., Cano, P. *et al.* (2006) Oestrus synchronisation of rabbit does at early post-partum by doe litter separation or eCG injection: reproductive parameters and endocrine profiles. *Animal Reproduction Science* 93, 218–230.

Rebollar, P.G., Bonanno, A., Di Grigoli, A., Tornambe, G. and Lorenzo P.L. (2008) Endocrine and ovarian response after a 2-day controlled suckling and eCG treatment in lactating rabbit does. *Animal Reproduction Science* 104, 316–328.

Rebollar, P.G., Dal Bosco, A., Millán, P., Cardinali, R., Brecchia, G. *et al.* (2012) Ovulating induction methods in rabbit does: the pituitary and ovarian responses. *Theriogenology* 77, 292–298.

Schindler, M., Pendzialek, S.M., Grybel, K., Seeling, T. and Navarrete Santos, A. (2020) Metabolic profiling in blastocoel fluid and blood plasma of diabetic rabbits. *International Journal of Molecular Sciences* 21, 919.

Setty, S.L. and Mills, T.M. (1987) The effects of progesterone on follicular growth in the rabbit ovary. *Biology of Reproduction* 36, 1247–1252.

Sirotkin, A.V. (2011) *Regulators of Ovarian Functions*. Nova Science Publishers, Inc., New York.

Sirotkin, A.V., Rafay, J., Kotwica, J., Darlak, K. and Valenzuela, F. (2009) Role of ghrelin in regulating rabbit ovarian function and the response to LH and IGF-I. *Domestic Animal Endocrinology* 36, 162–172.

Sirotkin, A.V., Chadio, S., Chrenek, P., Xylouri, E. and Fotopouloy, H. (2010) Phosphodiesterase inhibitor 3-isobutyl-methyl-xanthine stimulates reproduction in rabbit females. *Theriogenology* 74, 1321–1326.

Stephenson, T.J., Stammers, J.P. and Hull, D. (1990) Maternal to fetal transfer of free fatty-acids in the *in situ* perfused rabbit placenta. *Journal of Developmental Physiology* 13, 117–123.

Stocco, C., Telleria, C. and Gibori, G. (2007) The molecular control of corpus luteum formation, function, and regression. *Endocrine Review* 28, 117–149.

Teplán, I. (1989) Role, mechanism of action and application of gonadoliberins in reproductive processes. *Acta Biologica Hungarica* 40, 3–36.

Theau-Clément, M. (2007) Preparation of the rabbit doe to insemination: a review. *World Rabbit Science* 15, 61–80.

Theau-Clément, M., Lebas, F. and Falières, J. (2008) Influence of different eCG doses on the rabbit doe ovary response, fertilising aptitude and embryo development. *World Rabbit Science* 16, 73–79.

Thieme, R., Schindler, M., Ramin, N., Fischer, S., Muhleck, B. *et al.* (2012) Insulin growth factor adjustment in preimplantation rabbit blastocysts and uterine tissues in response to maternal type 1 diabetes. *Molecular and Cellular Endocrinology* 358, 96–103.

Tyl, R.W. (2010) In honor of the Teratology Society's 50th anniversary: the role of Teratology Society members in the development and evolution of *in vivo* developmental toxicity test guidelines. *Birth Defects Research Part C: Embryo Today: Reviews* 90, 99–102.

Webb, R., Woad, K.J. and Armstrong, D.G. (2002) Corpus luteum (CL) function: local control mechanism. *Domestic Animal Endocrinology* 23, 277–285.

Yamamoto, Y., Shimamamoto, K., Sofukitis, N. and Miyagawa, I. (1999) Effect of hypercholesterolemia on Leydig and Sertoli cell secretory function and the overall sperm fertilizing capacity in the rabbit. *Human Reproduction* 14, 1516–1521.

Zerani, M., Dall'Aglio, C., Maranesi, M., Gobbetti, A., Brecchia, G. *et al.* (2007) Intraluteal regulation of PG-F2α-induced prostaglandin biosynthesis in pseudopregnant rabbits. *Reproduction* 133, 1005–1016.

Zerani, M., Parillo, F., Brecchia, G., Guelfi, G., Dall'Aglio, C. *et al.* (2010) Expression of type I GNRH receptor and in vivo and in vitro effects in corpora lutea of pseudopregnant rabbits. *Journal of Endocrinology* 207, 289–300.

12 Genetics of Reproduction in the Rabbit

Miriam Piles[1]*, Maria Antonia Santacreu[2], Agustin Blasco[2] and Jun Pablo Sanchez[1]

[1]Institute for Food and Agriculture Research and Technology, Caldes de Montbui, Barcelona, Spain; [2]Institut for Animal Science and Technology, Universitàt Politecnica de Valencia, Spain

12.1 Genetics of Reproduction Traits

In this chapter, the genetic determinism of all traits involved in male and female reproductive performances are reviewed. All traits related to the underlying biological processes leading to the mating outcome, as well as some of its general features, such as the homogeneity of the reproductive performances, are considered: semen and ejaculate characteristics, ovulation rate, fertility, embryo survival, foetal survival and litter size. Different parameters of fertility and litter size are discussed, including the contribution of both sexes to each phase of the reproductive cycle. The female reproductive cycle is considered here as the period from mating to the end of lactation. Doe performance during lactation has also been included due to its contribution to litter size at weaning which is one of the most economically important traits in rabbit meat production. Homogeneity of litter size and kit weight at birth is reviewed because of its close relationship to prolificacy and kit survival during lactation. Finally, researches on doe longevity are also reviewed due to the impact of this trait on all the other traits and on farm management as well as on does' robustness.

12.2 Fertility

Because of the major economic importance of litter size (Cartuche *et al.*, 2014) most of the studies on reproductive performances have been focused on the female contribution to this trait and little attention has been paid to fertility. However, prolificacy depends on the rate of fertilized ova, which could be also improved by selection. The rate of ova that are not fertilized or that do not initiate the first stages of embryo development until day 7 after mating (i.e. just in the first quarter of gestation period) was about 16% in a maternal line selected for ovulation rate (Laborda *et al.*, 2012a). This rate would be expected to be even higher in paternal lines or in the final cross between a crossbred female and a sire from a paternal line, considering that male reproduction performances are usually worse in paternal than in maternal lines (Vicente *et al.*, 2000).

The outcome of a mating or of an artificial insemination is the result of the contribution of the male, the female and the interaction between the two. Although the doe has a major role on fertility at birth (Piles *et al.*, 2005) because of its effects during the whole gestation, failures of

Contact e-mail address: miriam.piles@irta.cat

DOI: 10.1079/9781780643342.0012

fertilization or embryogenesis due to male effects play important roles before implantation. In recent years, the use of artificial insemination (AI) in intensive rabbit-meat production systems has become a common practice in commercial farms (Piles *et al.*, 2013a). As a consequence, the contribution of the buck to fertility, as well as to seminal characteristics conditioning the production of potentially fertile AI doses, is achieving increased interest (Alvariño, 2000).

12.2.1 Between-breed/line variation

Differences between lines have been evaluated just for fertility at birth (Piles *et al.*, 2013a). Vicente (2000) compared male reproductive performances of a paternal line selected for growth and three maternal lines. In this study, fertility rate did not differ among lines whereas prolificacy differed, probably due to the selection process of the maternal lines. Garcia-Tomás *et al.* (2006a) found differences in fertility between two rabbit lines highly selected for growth rate. No relevant differences, however, were observed either for number of kits born alive or stillborn. Theau-Clément *et al.* (2007) did not evidence any differences in fertilizing ability between two lines that have been divergently selected for body weight at 63 days. Brun *et al.* (2002a) reported that crossbred males and females from two rabbit maternal lines had a better conception rate than purebred animals of the same lines used in the cross. However, it was not possible to ascertain whether those differences were due to the effect of the crossbred males or crossbred females. Using paternal lines, García-Tomás *et al.* (2006a) found unfavourable individual heterosis effects for male fertility at birth. However, fertility at birth could not be a good marker of male fertilizing capability, because it is masked by the female and environmental effects through gestation. Thus, this result cannot be considered conclusive regarding the potential usefulness on the use of a crossbred male.

12.2.2 Within-line variation

Few works have been carried out to investigate within-line sources of variation of rabbit fertility. The first studies showed that fertility after natural mating had an almost null male contribution (Piles *et al.*, 2005, 2013a), whereas female contribution was higher but still low (heritability of about 6%). Results from subsequent studies confirmed a similar effect when AI was performed at high sperm dosage (Tusell *et al.*, 2010). This study indicated that these conditions of AI were not optimal for the detection of individual variation among males, probably because the number and quality of sperms at mating time for most of the males exceeded the threshold needed to reach fertility (Amann and Hammerstedt, 2002). Thus, although differences among males that are independent of sperm dosage are maintained, differences among males that can, at least in part, be overcome by increasing the amount of sperm are not detected.

By using the character state model, Tusell *et al.* (2010) showed that male contribution to fertility and litter size after AI was low but higher in magnitude than the contribution obtained after natural mating. This result may indicate a potential interaction between the male genotype and mating conditions (Piles *et al.*, 2013a). On the other hand, Piles *et al.* (2013b) suggested that male contribution to fertility at birth is probably masked by female effects through gestation, because males are not involved in any biological processes regarding foetal development at late gestation. As a consequence, male contribution to fertility is better observed at early stages of gestation (i.e. around implantation time) when the male is involved in biological processes leading to fertilization and to first stages of embryo development. The genetic correlation between male and female contributions to fertility has been found to be moderate/high and positive in a maternal and in a paternal line of rabbits (Piles *et al.*, 2005; Tusell *et al.*, 2010; David *et al.*, 2011; Piles and Tusell, 2012), indicating that selection for the contribution to fertility of one of the sexes, if successful, could have a favourable correlated response in the contribution to the same trait of the other sex (Piles *et al.*, 2013a). This correlated response could be responsible, at least in part, for the observed differences in semen quality traits among bucks from maternal and paternal lines (Vicente *et al.*, 2000).

12.2.3 Relationship with other traits

Average daily gain (ADG) seems not to be genetically correlated with male contribution to fertility at birth after natural mating or after AI under common commercial conditions of use of the bucks (Piles and Tusell, 2012). ADG was negatively correlated with female contribution to fertility. The magnitude of this correlation was, however, not high enough to produce a relevant impairment of the reproductive performances of paternal lines of rabbits selected for ADG (since the probability of a genetic correlation lower than –0.5 was almost null). Moreover, several studies involving rabbit maternal lines indicated that the genetic correlation between growth and female contribution to litter size was negative, null or positive, but always of low magnitude (Camacho and Baselga, 1990; Gómez *et al.*, 1998; Garreau *et al.*, 2000; García and Baselga 2002c). Therefore, from these results it is possible to summarize that rabbit growth is not (or is poorly) genetically correlated with fertility.

12.3 Semen Production and Quality

Since the implementation of AI as a regular reproductive technique, the impact of reproductive performances of individual males is enormous. For this reason, there has been a growing interest in different traits involved in the efficient production of potentially fertile doses, which is still suboptimal from an animal production perspective (Piles *et al.*, 2012).

Several studies have been carried out to explore the sources of variation of semen production and quality traits. The genetic studies of traits that are only expressed in active adult bucks have special difficulties in achieving the dimension needed to obtain accurate-enough estimates of genetic parameters, such as heritabilities or genetic correlations. To obtain meaningful results, a strict collaboration between selection nuclei and AI centres is usually needed. However, most AI centres do not record the pedigrees of their bucks and only traits included in the classical spermiogram tests are usually evaluated, because of the high cost of other analyses. Table 12.1 summarizes information on heritability and repeatability of semen production and quality parameters reported by several studies.

12.3.1 Between-breed/line variation

It is usual to find relevant differences between lines in semen production and quality traits that are not necessarily related to their specialization as maternal or paternal lines (Piles *et al.*, 2013a).

Table 12.1. Heritability (h^2) and repeatability (r) for some semen traits from single ejaculates (ejac.) or combined pools of ejaculates. Data are from Bencheikh (1995), Garcia-Tomás *et al.* (2006b), Khalil *et al.* (2007), Lavara *et al.* (2008, 2012), Tusell *et al.* (2012) and Brun *et al.* (2016).

Trait	Range of h^2	Range of r	Range of n.[1]
Ejaculate volume			
Single ejaculate	0.06–0.13	0.29–0.38	172–412
pool	0.23	0.46	541
Ejaculate concentration			
Single ejaculate	0.08–0.10	0.35–0.39	172–412
pool	0.27	0.48	541
Total sperm amount			
Single ejaculate	0.07–0.12	0.17–0.18	172–412
pool	0.23	0.38	541
Semen pH	0.06–0.18	0.07–0.38	172–883
Sperm motility traits	0.05–0.18	0.24–0.37	172–883
Sperm abnormalities	0.19	0.41	412
Acrosome status	0.18	0.32	412
Sperm morphology traits	0.11–0.35	0.26–0.46	283

[1]Number of animals included in the estimations.

In addition, some of the differences between lines in semen characteristics do not correspond to differences in fertility or prolificacy at birth, suggesting the uselessness of these measurements to evaluate male contribution to reproductive performance in the current AI conditions. For example, lower sperm production, less motility and more acrosome defects in a paternal line selected for growth than in three maternal lines were reported by Vicente *et al.* (2000). The same study indicated that fertility rate did not differ among lines whereas prolificacy did, probably due to the selection process of the maternal lines. On the other hand, Brun *et al.* (2006) did not find any differences in male libido between two lines divergently selected for body weight at 63 days. However, males from the lighter line had higher ejaculate volume, sperm motility and number of ejaculates suitable for AI but lower sperm concentration than males from the heavier line (Brun *et al.*, 2006). In another study, no differences were found when the same lines were compared for fertilizing ability (Theau-Clément *et al.*, 2007). Finally, Garcia-Tomás *et al.* (2006c) investigated two rabbit lines highly selected for growth rate and reported differences in direct genetic effects for some semen traits that did not correspond to differences in male fertility and prolificacy at birth (Garcia-Tomás *et al.*, 2006a). Other studies, also reported differences between lines for sperm production and quality traits such as collection rate, ejaculate volume, sperm concentration, pH, sperm motility traits, morphological characteristics of the sperm and seminal plasma proteoma (Brun *et al.*, 2002b; Theau-Clément *et al.*, 2003; El-Tarabany *et al.*, 2015; Casares-Crespo *et al.*, 2016, 2018) and favourable maternal effects for ejaculate and sperm quality and production traits (Garcia-Tomás *et al.*, 2006c). These observed differences in semen traits could be related to different sexual development. Males from rabbit maternal and paternal lines have been found to have different percentages of seminiferous tubules with spermatozoa at different ages (García-Tomás *et al.*, 2009).

Non-additive genetic variability across lines (heterosis estimates), investigated by several studies for several semen parameters, reported high variability in the estimates of direct heterosis, largely depending on the lines involved in the crossbreds (Brun *et al.*, 2002b; García-Tomás *et al.*, 2006c; Khalil *et al.*, 2007; El-Tarabany *et al.*, 2015). Brun *et al.* (2002b) analysed semen characteristics in two maternal lines and their reciprocal crosses and reported positive direct heterosis for sperm concentration (37.5%), total number of sperm per ejaculate (37.6%), mass motility (6.8%) and percentage of motile spermatozoa (4.1%). Khalil *et al.* (2007) found favourable direct heterosis effect for ejaculate volume (10.6%), sperm concentration (13.6%), sperm motility (10.5%) and for percentage of spermatozoa with abnormal form and dead spermatozoa (–21.5% and –20.3%, respectively) in the cross scheme between a Spanish maternal line and a Saudi breed designed to obtain two new synthetic maternal lines. The same authors found favourable maternal heterosis for the same traits (24.0% for ejaculate volume, 10.3% for sperm concentration, 21.8% for sperm motility and –9.6% and –14.7% for percentage of spermatozoa with abnormal form and dead spermatozoa, respectively). El-Tarabany *et al.* (2015) found even opposite heterotic effects for some traits depending on the breeds involved in the crossbreed. However, the heterotic effects for semen traits obtained in crosses between two paternal lines of rabbits were of low relevance and only favourable for the presence of sperm with cytoplasmic droplets (García-Tomás *et al.*, 2006c), which do not have a clear relationship with fertility. Therefore, the superiority of crossbred bucks was not proven for those lines and traits.

12.3.2 Within-line variation

In general, a wide range of heritability (h^2) and repeatability estimates for semen traits can be found in the literature (Garcia-Tomás *et al.*, 2006b). The large variation in the magnitude of this parameter is due to several factors such as: (i) different genetic composition of populations of bucks among experiments; (ii) variation in the definition of the traits, which in some cases are means of observations, whereas in other cases corresponds to individual ejaculates (Ducrocq and Humblot, 1995; Wolf, 2009); and (iii) the possible effect of collection frequency on the individual variation of semen traits.

On the other hand, the heritability estimates are imprecise in most of the reviewed studies. This is commonly due to small experimental data sets. Moreover, a large amount of environmental variation is originated during semen manipulation and time to evaluation. The subjective methods used to evaluate several semen traits can contribute to explain this wide variability of results.

Only three studies evaluated the genetic determinism of male libido, based on a categorical classification with different classes. Panella *et al.* (1994) and Khalil *et al.* (2007) reported moderate h^2 values (0.30 and 0.17, respectively) that seem to be the result of inaccurate modelling or the result of genetically heterogeneous populations, which included 11 different genetic types of bucks. The study of Tusell *et al.* (2012) reported a very low h^2 (0.06) in a population selected for post-weaning growth rate.

Ejaculate parameters that are used to discard ejaculates in AI centres (i.e. presence of urine, calcium carbonates deposits and gel plugs) have low heritability (Tusell *et al.*, 2012). Factors involved in semen collection such as variation in the temperature of the artificial vagina (that could lead to a higher presence of urine and calcium carbonate deposits in the ejaculate or unsuccessful mountings) play major roles in affecting these parameters (Morrell, 1995). The suitability of the ejaculate for AI, which involves the subjective combination of several quality traits, as expected, had low heritability (h^2 = 0.06; Tusell *et al.*, 2012). Therefore, genetic selection for increasing semen production by improving male libido and reducing the number of rejected ejaculates may not be effective.

Heritability estimates of other semen production parameters, i.e. ejaculate volume and sperm concentration for single ejaculates, ranged from 0.06 to 0.13 and from 0.08 to 0.10, respectively (Lavara *et al.*, 2011; Brun *et al.*, 2016). For pools of two consecutive ejaculates, heritability for these parameters was estimated to be 0.23 and 0.27, respectively (Tusell *et al.*, 2012). Tusell *et al.* (2012) reported a moderate and negative genetic correlation between sperm concentration and ejaculate volume (mean of the marginal posterior distribution of this parameter (PM)) = –0.53; highest posterior density interval at 95% (HPD95%): –0.76, –0.27). Both traits determine the total amount of sperm produced per ejaculate, which was found to be moderately heritable in a pool of two consecutive ejaculates (PM: 0.23; HPD95%: 0.14, 0.31; Tusell *et al.*, 2012). Thus, this trait could be used as a selection criterion in a breeding programme to improve semen production.

Among semen quality characteristics, pH has been considered in several studies because it is directly related to the degree of activation of the sperm cells. The h^2 for pH has been reported to be low (range: 0.06–0.18; Khalil *et al.*, 2007; Tusell *et al.*, 2011, 2012; Brun *et al.*, 2016). The h^2 of sperm mass motility was estimated to be 0.05 (Brun *et al.*, 2016). Estimates of heritability for individual sperm motility, evaluated in a subjective manner, were similar to the values obtained for mass motility (0.08; Tusell *et al.*, 2012). During the last decade, computer-assisted sperm analysis (CASA) systems have been used to improve the accuracy of sperm motility records in domestic animals. In general, the sperm motility and movement traits evaluated in this way have shown low heritability. However, the rate of motile sperms per ejaculate is a trait that has been considered as convenient for selection (Brun *et al.*, 2016, h^2 = 0.18). Three studies until now (Napier, 1961; Lavara *et al.*, 2008, 2012), have actually examined quantitative variation of other semen-quality traits, such as sperm abnormalities, acrosome status or sperm morphology traits (length, width, area and perimeter of sperm head). The available data suggest that h^2 are medium to high and depend on the method of estimation.

Summarizing, genetic selection for increased semen production by improving male libido and reducing the number of rejected ejaculates may not be effective because of the low heritability of both male libido and ejaculate characteristics. Selection for a set of semen characteristics may have no correlated response in male fertility and prolificacy, at least for the semen parameters that are usually evaluated. It is necessary to find new effective, cheap and easy-to-measure fertility markers useful to improve semen quality and, indirectly, male reproductive performances. On the other hand, it seems possible to improve semen production by selecting for increased total amount of sperms in the ejaculate. However, the relationship between semen production and semen quality (which, in turn, requires a clear definition) is not yet established.

12.3.3 Relationship with other traits

Tusell *et al.* (2011) studied the genetic relationship between male fertility and one of the semen-quality traits, i.e. pH of the ejaculate, which is considered an indicator of the ejaculate capability to fertilize. The pH of the ejaculate is, in a major part, a consequence of the number and activity of the spermatozoa present in the ejaculate. This study reported a negative and linear relationship between pH and liability of fertility at both phenotypic and environmental levels and a negative genetic correlation between pH and fertility at birth.

Although estimates are generally imprecise, there is some evidence of the existence of a genetic relationship between semen production and quality with ADG (Piles *et al.*, 2013a). Tusell *et al.* (2012) found that ADG had a slightly favourable genetic correlation with sperm concentration (0.21, HPD95% = –0.03, 0.48) and a slightly unfavourable genetic correlation with ejaculate volume (–0.19, HPD95% = –0.47, 0.08). Moreover, ADG was genetically uncorrelated with all libido and semen traits that are usually included in the criteria for ejaculate rejection in AI centres (i.e. pH, individual motility and presence of urine, blood and other elements that preclude the use and/or evaluation of the ejaculate). Lavara *et al.* (2011) also obtained estimates of the genetic correlation among semen production and quality with ADG. Some of their estimates were in contrast to what was reported by Tusell *et al.* (2011), even not very precise, making it difficult to draw reliable conclusions. Lavara *et al.* (2012) obtained moderate and negative genetic correlation between sperm motility and ADG (–0.53, HPD95% = 0.95, 0.02). Sperm movement characteristics measured with CASA systems, such as average path velocity, straight-line velocity, curvilinear velocity and straightness, seem to be genetically uncorrelated with growth (the PM of the genetic correlations ranged between 0.03 and –0.14; and the HPD95% intervals were around –0.50, 0.50). For sperm morphology and acrosome membrane functionality, Lavara *et al.* (2012) reported a moderate genetic correlation between daily gain and normal sperms per ejaculate, but the magnitude of the genetic correlations does not seem to be high (–0.40, HPD9% = –0.78, –0.02 for normal acrosome status; 0.25, HPD95% = –0.18, 0.66 for sperm abnormalities).

Brun *et al.* (2006) did not find any differences in male libido between two lines divergently selected for body weight at 63 days. These authors, however, reported that males from the line with the lowest body weight had higher ejaculate volume, sperm motility and number of ejaculates suitable for AI, but lower sperm concentration than males from the line with the highest body weight. In a later study, the same lines were compared by their fertilizing ability, and no differences were encountered between them (Theau-Clement *et al.*, 2007). Because of the antagonism between volume and sperm concentration of the ejaculate, the genetic correlation between ADG and total number of sperms in the ejaculate was almost null.

In summary, it seems that semen traits and growth rate are not genetically correlated. Two practical consequences derive from this conclusion: (i) selection to improve ADG is not expected to impair semen traits involved in AI dose production; and (ii) multi-trait selection, including ADG and other semen traits directly related to efficient dose production (even if they have usually low heritability), is potentially feasible. Nevertheless, the decision of defining the objectives of a programme of selection has genetic and economic components. Consequently, the economic weights of growth rate and the semen traits need to be carefully evaluated to correctly define selection criteria.

12.4 Prolificacy

Litter size is considered one of the most important economic traits on intensive meat rabbit production (Eady and Garreau, 2008; Cartuche *et al.*, 2014). Accordingly, commercial maternal lines are usually selected on litter size at birth or litter size at weaning.

12.4.1 Between-breed/line variation

Mean value for litter size in selected maternal lines is about 8–10 young rabbits with a standard deviation of about three young rabbits. Differences between breeds in litter size have been reported by Bolet *et al.* (2000) in the first evaluation for zootechnical traits of original breeds or local populations in Europe. These authors evaluated prolificacy in eight breeds and in a control un-selected line (INRA9077 strain based on New Zealand breed). The breeds studied showed smaller litter size than the control line, either at birth or at weaning. Litter size at weaning ranged between 3.03 and 6.26 young rabbits. The difference between the studied breeds and the control line was about two rabbits at weaning and ranged from 1.5 to 3 young rabbits.

Relevant differences in litter size between lines or breeds have also been reported by other authors, often in the context of crossbreeding studies (Poujardieu and Vrillon, 1973; Hulot and Matheron, 1979; Brun *et al.*, 1992, 1998; Cifre *et al.*, 1998; Baselga *et al.*, 2003; Orengo *et al.*, 2003; Ragab and Baselga, 2011; Kabir *et al.*, 2012; Khalil and Al-Saef, 2012). Selected lines from Californian and New Zealand White breeds are usually involved in these studies. Some of these works found that differences between lines in litter size are due to differences for ovulation rate and prenatal survival (Hulot and Matheron, 1979; Bolet *et al.*, 1988; Brun *et al.*, 1992). In summary, relevant differences between breeds and lines show evidences of genetic variation in litter size and its components, ovulation rate and prenatal survival.

12.4.2 Within-line variation

Estimates of heritability for litter size indicated a general low value of about 0.10 (reviewed in: Rochambeau, 1988; Blasco, 1996). Table 12.2 summarizes information on the heritability of two parameters affecting litter size in rabbits: number born alive and number of weaned kits. It is relevant to note that estimates from selection experiments with high sample size are usually much lower (around 0.07 and 0.04 for number born alive and weaned; García and Baselga, 2002a; Bolet and Saleil, 2012a, 2012b, 2012c). Response estimates using a control population (no selected population or contemporary population from frozen embryos) are in agreement with these very low estimated heritability values (Rochambeau, 1998; García and Baselga, 2002a; Tudela *et al.*, 2003).

Estimated heritability decreases from birth to weaning (Table 12.2). Estimated correlations between litter size at different stages (birth, weaning and slaughtering age) are high and positive (García and Baselga, 2002a, 2002b; Piles *et al.*, 2006c; Ragab and Baselga, 2011). Differences between heritability and repeatability are usually small and permanent environmental effect shows similar values to heritability (Table 12.2). On the other hand, genetic maternal effects seem to be irrelevant. The contribution of the maternal effects is lower than 10% of the phenotypic variation (Ferraz *et al.*, 1992; Gómez *et al.*, 1994).

Litter size is a trait essentially attributed to the dam (Mocé *et al.*, 2004). Male contribution (i.e. male genetic and permanent environmental effects) to litter size at birth is almost null (Piles *et al.*, 2006c). Male contribution is higher for the number of implanted embryos than for litter size at birth (Piles *et al.*, 2013b) in agreement with results from Bolet *et al.* (1990). These authors reported an effect of sire line on embryo survival, but this effect may be due to line differences in fertilization rate. Thus, the male seems to be involved in biological processes leading to fertilization and embryo development at early stages of gestation and its contribution is lower at the end of gestation.

12.4.3 Relationship with other traits

Very few estimates of genetic correlations between litter size and growth traits have been published, indicating, in general, low correlation with a high standard error (Gómez *et al.*, 1998; Garreau *et al.*, 2000; García and Baselga, 2002c; Mínguez *et al.*, 2012).

García and Baselga (2002a) used a line selected for litter size at weaning for 20 generations and reported no correlated responses on growth traits. However, Brun and Ouhayoun (1994) observed lower weaning and adult weight in the line A1077, selected for litter size at weaning during 13 generations, versus its control line. In agreement with these results, Rochambeau (1998) reported that the individual weight at weaning decreased after 18 generations of selection for litter size in lines A1077 and A2026 (–3.4 g and –4.4 g per generation, respectively). On the other hand, it seems that litter size can affect growth traits of young rabbits from birth to weaning (Drummond *et al.*, 2000; Poigner *et al.*, 2000) as well as the productive performances of the future reproductive females (Rommers *et al.*, 2001). Thus, larger litters show a lower average birth weight, lower

Table 12.2. Estimates of heritability (h^2) and repeatability (r) for number born alive (NBA) and number weaned (NW).

h^2_{NBA}	r_{NBA}	h^2_{NW}	r_{NW}	Breed/line	Reference
0.08	0.15	0.06	0.12	H	Cifre *et al.* (1998)
0.03		0.02		Caldes	Gómez *et al.* (2000)
0.12	0.32	0.09	0.25	NZW	Rastogi *et al.* (2000)
0.05		0.03		Botucatu	Moura *et al.* (2001)
0.07	0.17	0.05	0.13	V	García and Baselga (2002a)
0.13	0.21	0.11	0.17	A	García and Baselga (2002b)
0.07		0.04		A1077	Bolet and Saleil (2002a)
0.06		0.04		A2066	Bolet and Saleil (2002b)
0.07		0.04		A9077	Bolet and Saleil (2002c)
0.12		0.11		A	Piles *et al.* (2006c)
0.08		0.06		Prat	Piles *et al.* (2006c)
0.07		0.04		V	Piles *et al.* (2006c)
0.15		0.08		LP	Sánchez *et al.* (2008)
0.05	0.11			CPC-Italy	Mantovani *et al.* (2008)
0.10		0.09		ITELV2006	Bolet *et al.* (2012)

birth-to-weaning growth rate, and a higher mortality than smaller litters. These results led to the creation of a new line, A1777. This new line is selected for number born alive at birth and direct and maternal effects on weaning weight. After 12 generations of selection, estimated genetic response by using mixed models was 1.8 kits at weaning (0.14 kits per generation), although phenotypic increment in litter size at weaning was 1 kit (Garreau *et al.*, 2015). The genetic progress of weaning weight was 246 g, mostly due to the progress in direct effects.

12.5 Components of Prolificacy

Fertilization rate is not considered a limiting factor to increase litter size as it generally exceeds 90% in selected maternal lines of rabbits (Adams, 1960; Torres *et al.*, 1984; Santacreu *et al.*, 1990; Theau-Clement *et al.*, 2009). An approach to increase litter size is to select for its main components, i.e. ovulation rate and prenatal survival. Results from three selection experiments for litter size support this approach since the observed improvement in litter size was due to increased ovulation rate (Brun *et al.*, 1992; García and Baselga, 2002a), or in foetal survival (line A; García and Baselga 2002b).

In rabbits, ovulation rate (i.e. number of ova shed per estrus) is usually estimated as the number of corpora lutea in both ovaries counted by laparoscopy or *post mortem* after the dissection of the ovary. Mean value for ovulation rate in selected maternal lines is 14–16 ± 2.5 ova (Brun *et al.*, 1992, 1998; García and Baselga, 2002a; Laborda *et al.*, 2011; Ragab *et al.*, 2012).

Prenatal survival comprises two periods, the embryonic period and the foetal period. In rabbits, it is common to consider the embryonic period, from ovulation to implantation (d 7) and the foetal period from the implantation to the birth (d 30) (Mocé and Santacreu, 2010). Embryonic survival is calculated as the ratio between the number of implanted embryos and the number of corpora lutea. Laparoscopy is usually used to measure ovulation rate and implanted embryos which allows to estimate embryonic and foetal survival in the same female (Santacreu *et al.*, 1990). Prenatal survival, calculated as a percentage, is around 70%. In rabbits, approximately 30% of ova shed are lost during prenatal development (Adams 1960; García and Baselga, 2002a; Laborda *et al.*, 2012a). About one third of the total losses corresponds to the embryonic period whereas foetal mortality comprises the remaining two thirds of the total losses. Thus, prenatal losses are considered the major limiting factor of litter size in rabbits, as in pigs and mice.

12.5.1 Within-line variation

Few estimates of genetic parameters for ovulation rate and survival rates in rabbits under natural mating have been published (Blasco *et al.*, 1993a; Laborda *et al.*, 2011; Laborda *et al.*, 2012a; Ziadi *et al.*, 2013).

Ovulation rate has higher heritability than litter size, but this heritability is lower in rabbits than in pigs and mice (see review by Blasco *et al.*, 1993b; Ruiz-Flores and Johnson, 2001; Rosendo *et al.*, 2007; Laborda *et al.*, 2011). These differences might be due to diversity in ovulation processes among species. For example, in rabbits, ovulation occurs in response to coitus.

The number of implanted embryos in rabbits had a low heritability (Bolet *et al.*, 1994; Argente *et al.*, 2000; Laborda *et al.*, 2011; Ziadi *et al.*, 2013); this was of the same magnitude as heritability of the number of foetuses at different moments of gestation in rabbits (d 12), pigs (d 50) and mice (d 17) (Clutter *et al.*, 1990; Blasco *et al.*, 1993; Johnson *et al.*, 1999, respectively).

Prenatal survival has also a low heritability, similar to that of litter size. The low estimate of heritability is close to the values published in pigs (0.11; Rosendo *et al.*, 2007) and mice (Clutter *et al.*, 1990). Heritability was also low for embryo survival and moderate for foetal survival (Laborda *et al.*, 2012a; Ziadi *et al.*, 2013).

12.5.1.1 Relationship between litter size, ovulation rate and prenatal survival

The efficiency for improving litter size by selection of its components depends on the genetic parameters of ovulation rate and prenatal survival. Estimated correlations between litter size and its components are scarce in the literature. Phenotypic correlation between ovulation rate and litter size is low and positive. The estimated genetic correlations between these two parameters is also usually low, ranging from –0.20 to 0.36, with a high standard error (Blasco *et al.*, 1993b, Laborda *et al.*, 2011; Ziadi *et al.*, 2013), mainly derived by the low number of animals investigated, considering the technical difficulties derived by the collection of information using laparoscopies, laparotomies or at slaughtering. However, the lack of correlated response on litter size in rabbits (Laborda *et al.*, 2012b estimated with control populations), pigs (Cunningham *et al.*, 1979; Leymaster and Christenson, 2000; Rosendo *et al.*, 2007) and mice (Bradford, 1969; Land and Falconer, 1969), corroborates that the correlation between ovulation rate and litter size should be close to zero.

Litter size is phenotypically and genetically correlated with embryo, foetal and prenatal survival. Genetic correlations were moderate with embryo survival and foetal survival and high with prenatal survival (Blasco *et al.*, 1993a, 1993b; Laborda *et al.*, 2012a; Ziadi *et al.*, 2013). The positive correlations between litter size and survival rates agree with the estimates found in pigs (Johnson *et al.*, 1999; Rosendo *et al.*, 2007). However, these results should be taken with caution due to high standard errors.

12.6 Selection Experiments

Few selection experiments for litter-size components have been reported in rabbits: two divergent selection experiments for uterine capacity, one selection experiment for ovulation rate and another experiment combining selection for ovulation rate and litter size.

12.6.1 Selection for uterine capacity

Estimated genetic parameters in rabbits suggested that prenatal survival could be a good proxy to improve litter size more efficiently than direct selection (Blasco *et al.*, 1993a, 1993b). Selection for increased uterine capacity was proposed as a means to change prenatal survival (Bennett and Leymaster, 1989). Uterine capacity was defined by Christenson *et al.* (1987) as the maximum number of foetuses that a female is able to support at birth when ovulation rate is not a limiting factor. To measure uterine capacity in does, Blasco *et al.* (1994) proposed to use unilateral ovariectomy. By removing one ovary, a duplication of the ovulation rate is produced in the remaining ovary, leading to an overcrowding of embryos in the adjacent uterine horn. The two uterine horns of rabbit does have separate cervical canals, thus transmigration of embryos does not take place (Fleming *et al.*, 1984).

Two divergent selection experiments for uterine capacity (based on different criteria of selection) were performed in rabbits. In the first experiment (Bolet *et al.*, 1994), selection was performed on a number of dead foetuses from implantation to birth. Estimation of the implantation sites can be obtained by laparoscopy without affecting litter size (Santacreu *et al.*, 1990). After four generations of selection, the number of dead foetuses did not change and no significant response was obtained in litter size and its components.

The second experiment based the selection strategy on litter size in unilateral ovariectomized females, which includes both embryo and foetal survival (Argente *et al.*, 1997). Selection for uterine capacity based on this approach through 10 generations was successful, although it did not appear to be more effective than direct selection for litter size. Similar results were obtained in pigs (Gion *et al.*, 1990; Leymaster and Christenson, 2000) and mice (Kirby and Nielsen, 1993). Analyses based on genetic trends showed that divergence in uterine capacity between lines was 1.5 rabbits born (Blasco *et al.*, 2005). Using a control population, it was evident that the response was asymmetric: no differences on uterine capacity were found between high and cryopreserved control lines, whereas the low line and the control line differed by 1.08 kits (Mocé *et al.*, 2005). The correlated response on litter size was also asymmetric and divergence between both lines was of 2.35 kits. The low uterine capacity line had 1.88 kits less than the control line, while the high uterine capacity line differed with the control line in 0.5 kits (Santacreu *et al.*, 2005). The correlated response on litter size in the low line was associated with a lower prenatal survival (difference between low and control lines around 7%). Most of the embryo mortality was produced before 72 hours of gestation. Moreover, embryos from the low line showed a less advanced stage of development at 72 hours of gestation than embryos from the high line (Mocé *et al.*, 2004). As approximately half of the response in uterine capacity was obtained in the first two generations, the presence of a major locus with moderate effect on uterine capacity and large effect on number of implanted embryos in this population was suggested (Argente *et al.*, 2003). This hypothesis was experimentally tested using an F2 population created by reciprocal mating between the two lines divergently selected for uterine capacity. Single nucleotide polymorphisms (SNPs) in several candidate genes involved in the reproductive functions were then used in the association analysis and combined with expression analysis between the two uterine capacity lines (Estellé *et al.*, 2006; Peiró *et al.*, 2008; Merchán *et al.*, 2009; Argente *et al.*, 2010; García *et al.*, 2010; Ballester *et al.*, 2013). Some of these genes (progesterone receptor, *PGR*; hydroxysteroid (17-beta) dehydrogenase 4, *HSD17B4*; and endoplasmic reticulum oxidoreductase 1, *ERO1*) showed different expression levels in the oviduct of the does of two lines, but these results could not be confirmed by the identification of any causative mutations. An interesting polymorphism was, however, identified in the promoter region of the *PGR* gene associated with increased litter size and expression level of this gene (Peiró *et al.*, 2008, 2010).

A subsequent genome-wide association study was carried out using the commercial high-density OrcunSNP array by analysing does from a control population and from the high and low uterine capacity lines (Sosa-Madrid *et al.*, 2020). Several reproduction traits were used in the association analyses. SNP regions on chromosome 17 (OCU17) were associated with total number born, number born alive and number of implanted embryos. Ovulation rate was associated with other regions on OCU9, OCU10 and OCU14. Regions on OCU11 were associated with number of implanted embryos.

12.6.2 Selection for ovulation rate

Only one selection experiment for ovulation rate has been carried out in rabbits. After ten generations of selection, ovulation rate increased (1% per generation), but no correlated response on litter size was found (Laborda *et al.*, 2011, 2012b). Similar results were obtained after selection for ovulation rate in pigs (Johnson *et al.*, 1984; Rosendo *et al.*, 2007) and mice (Bradford, 1969; Land and Falconer, 1969). The lack of correlated response on litter size was due to a correlated response in foetal mortality of about 1% per generation (Laborda *et al.*, 2012a, 2012b), whereas embryo mortality was not modified. An ovarian transcriptomic analysis showed differential expression genes between females of the line selected for ovulation rate (OR) and the control line that could help to elucidate the processes involved in oocyte maturation and ovulatory events (Serna-García *et al.*, 2020). Further studies are needed to know the possible causes of higher prenatal mortality in females selected for high ovulation rate.

12.6.3 Selection for ovulation rate and litter size

Several authors predicted greater response in litter size by selecting using an index based on its components rather than using direct selection for litter size itself, not only in rabbits (Blasco *et al.*, 1993b), but also in pigs (Johnson *et al.*, 1984; Bennett and Leymaster, 1989) and mice (Clutter *et al.*, 1990; Ribeiro *et al.*, 1997a, 1997b). An alternative to an index could be a two-stage selection approach (Ruíz-Flores and Johnson, 2001).

A two-stage selection experiment for ovulation rate and litter size has been carried out in rabbits. Results from six generations showed a change in litter size: the estimated overall response was approximately 0.9 kits, 0.13 kits/generation (Ziadi *et al.*, 2013). Similar results for litter size were obtained after 11 generations (Badawy *et al.*, 2019). In pigs, a two-stage experiment was also successful in increasing litter size (Ruíz-Flores and Johnson, 2001). The litter size increment in rabbit experiment was due to the direct response in ovulation rate (0.14 ova/generation) and the correlated response in prenatal survival. It seems that both embryonic and foetal survival has contributed to the increment of prenatal survival in the same amount. The two-stage selection also improved litter size at birth, weaning and slaughter. An improvement of uterine capacity has been suggested by Badawy *et al.* (2019) to explain why survival at birth was not reduced. Survival at weaning and slaughter did not change either. Moreover, selection by litter size and ovulation rate led to a small weaning and slaughter weight increase (Badawy *et al.*, 2019) in agreement with the uncorrelated response in survival at weaning and at slaughter. Therefore, a two-stage selection for ovulation rate and litter size could be effective to increase both litter size and prenatal survival rate without negative consequences on post-natal survival and growth traits.

12.7 Homogeneity and Residual Variance

A recent interest in several livestock species for the possibility of increasing the character homogeneity by selection has been expressed for several traits. In traits selected by their economic interest, selection for homogeneity would decrease the genetic variance of the trait, reducing the direct response to selection for the trait, which is economically undesirable. However, if only residual variance is diminished, this reduction in environmental variability will not affect the genetic variance of the trait when leading to more homogeneous products. Moreover, it will increase the heritability of the trait, since genetic variance is not affected, and residual variance is reduced.

12.7.1 Residual variance of litter size

Reducing environmental variability for litter size can be an interesting purpose for selection. Litter-size heritability is very low and its actual response to selection in several experiments has been low as well. Thus, reducing environmental variability would increase litter-size heritability, also improving management by reducing the number of adoptions. A divergent selection experiment for litter-size environmental variability has been performed (Blasco *et al.*, 2017). Assuming that all parities (i=1, 2, ...n) of a doe have the same genetic value (a) and permanent environmental effects (p), litter size only depends on systematic environmental effects (F) and random environmental effects e_i (residuals).

Litter size of a doe in parity i:

$$LS_i = m + F_i + a + p + e_i$$

Pre-correcting data for systematic environmental effects, within doe litter size phenotypic variance is a measure of the residual variance, which is the main component of environmental variability, although results without pre-correcting data were almost the same, indicating that the pre-correction had little effect in the variance. Response to selection was high, approximately 4.5% of the mean of the residual variance per generation (Fig. 12.1). A negative correlated response to selection with litter size, due to a lower early embryo development (García *et al.*, 2016) and implantation rate (Argente *et al.*, 2017), was also found, which means that the most homogeneous line had also the highest litter size; therefore, selection for homogeneity does not damage litter size.

Improving homogeneity by selection has shown a more appealing result: variability of

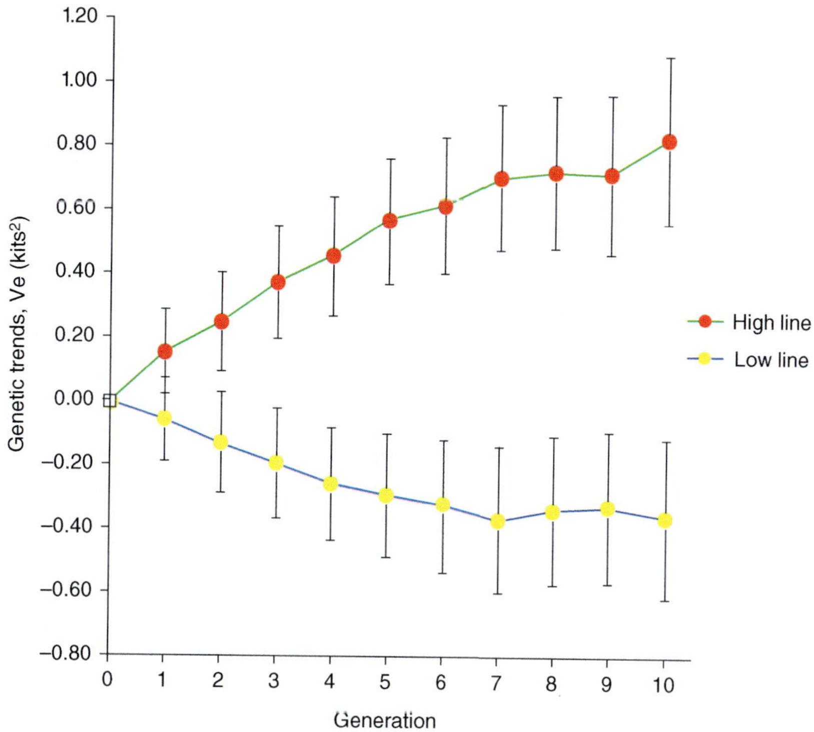

Fig. 12.1. Genetic trends for litter-size variability. (Adapted from Blasco *et al.*, 2017)

litter size is associated with stress sensitivity and disease resistance, as shown by Argente *et al.* (2019) and Beloumi *et al.* (2020). A doe that is more constant in its litter size (more homogeneous) shows a better response to stress, and lower inflammatory response and susceptibility to infectious disorders. Moreover, the line selected for litter-size homogeneity has shown a greater capacity to mobilize energy reserves at delivery, which implies better adaptation to adverse conditions when high-energy demand is required (Garcia *et al.*, 2019). In summary, the more homogeneous line is more robust and resilient. A recent study combining genome-wide association and a whole-genome sequencing of both lines found several candidate genes and functional mutations related to the immune system, stress response, and the nervous system (Casto-Rebollo *et al.*, 2020), contributing to explain the differences between lines in robustness and resilience. As both response to stress and disease resistance are traits related to welfare, selection for homogeneity is also a way of improving rabbit welfare, which is one of the new objectives in animal production.

12.7.2 Residual variance of birth weight

San Cristobal-Gaudy *et al.* (1998) proposed a model assuming that the environmental variability of residual variance is also partially controlled by genes. Using this model, an experiment of selection for residual variance of birth weight has been performed (Garreau *et al.*, 2008a, 2008b; Bodin *et al.*, 2010). After 10 generations of selection a response of about 1.7% per generation of litter weight standard deviation was achieved in each direction (Figure 12.2). Correlated response on birth weight was not significantly different from zero. It seems that mortality at birth and at weaning was lower in the more homogeneous line, which may be partially related to the greater length capacity for elongation of the uterine horn in the homogeneous line (Garreau *et al.*, 2008b).

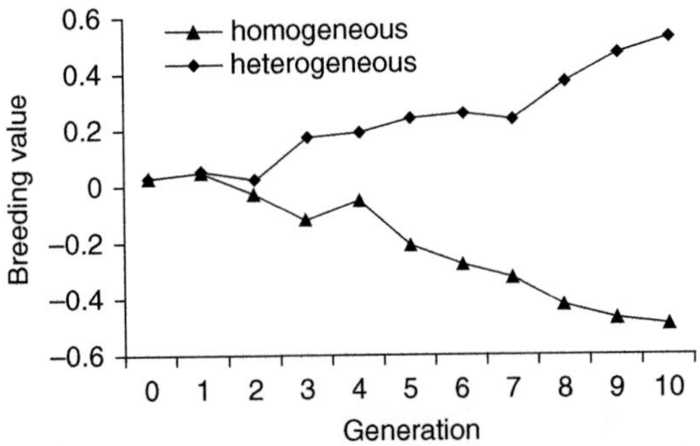

Fig. 12.2. Genetic trends for birth weight variability. (From Bodin *et al.*, 2010.)

12.8 Length of Productive Life

Length of productive life (LPL) records are time-to-event data, the event being death or culling for non-voluntary reasons. There is a body of statistical techniques specifically devoted to this type of record, called survival analysis. These techniques consider the information contained in the part of the records that could not have shown the final event. These types of records are known as right-censored records. With the availability of the Survival Kit software package, specifically designed for the genetic analysis of longevity data (Ducrocq and Sölkner, 1994), studies considering this trait in animal breeding have been produced in nearly all livestock species.

Different measurements have been used for studying the length of productive life in rabbits. The first references considering the survival ability of the females in farms did not directly use a temporal measurement of the life of the animals. Instead, considering survival rates, genetic variability across populations in survival rates as well as in the distribution of 'reasons for culling' was demonstrated (Torres *et al.*, 1986, 1987). In these studies and in Rosell (2003), a peak of mortality was systematically observed at first parity.

Other definitions of length of productive life, derived from studies in other livestock species, have also been applied to the rabbit, namely: (i) the number of days from first mating; (ii) first positive pregnancy test to death or culling (Sanchez *et al.*, 2004; Piles *et al.*, 2006a, b); and (iii) the number of inseminations for a female during its life (Piles *et al.*, 2006a). Length of productive life should be measured independently from the production level and management practices, i.e. length of productive life should be defined by involuntary culling reasons. Longevity records usually come from selection nuclei in which females are usually culled after they are genetically evaluated, irrespective of their sanitary status. Therefore, only a limited fraction of the culled animals might have true length of productive life measurements.

12.8.1 Between-breed/line variation

Piles *et al.* (2006b) analysed longevity, defined as the number of days between the first positive pregnancy test and the death or involuntary culling in three genetic types (A, V and Prat) developed from complete diallel crosses involving three maternal lines. Differences in longevity between types was explained by differences on direct genetic effects and by the existence of favourable heterotic effects in the crosses. More recently, Ragab *et al.* (2020) reported other results on longevity from a complete diallel cross involving four maternal lines (A, V, H and LP), some of which are included in the aforementioned study. Here the observed variability on the risk of death or involuntary culling across the genetic types involved in the crosses might be explained by differences in direct and maternal genetic effects in addition to heterotic effects. Based on these studies, additive and non-additive genetic variability between maternal lines of rabbits might exist. However, variability did not seem constant for the entire life of the animals, but it might depend on the age of the females, since the interaction between genetic type and parity order was significant (Ragab *et al.*, 2020).

12.8.2 Within-line variation

Studies have also been published on the genetic variability within population. Using as LPL trait the number of days between the first presentation to male and the date of death or culling, Sanchez *et al.* (2004) reported for V line a heritability of 0.05, when a sire-maternal grand sire proportional hazard model was used. The same dataset was analysed by using an animal proportional hazard model with residual term (Sanchez *et al.*, 2006b). In this case a slightly different definition of the trait was applied (i.e. the number of days between the first positive pregnancy test and the date of death or culling) and a higher heritability estimate was obtained (0.09). These two estimates were on the logarithmic scale, and these figures were not directly comparable to heritability value estimates from a linear model. Yazdi *et al.* (2002) defined the effective and equivalent heritabilities, with the objective of getting a heritability value to be used in the formula for approximating the accuracy of breeding value predictions. The equivalent heritability was obtained relying on index theory for a progeny test that can be interpreted similarly to the heritability of a Gaussian trait obtained using linear mixed models. In the two aforementioned studies (Sanchez *et al.*, 2004, 2006b) the equivalent heritabilities were found to be 0.04 and 0.08, respectively; these figures are directly comparable to heritability estimates of linear Gaussian traits.

Piles *et al.* (2006a) studied the length of productive life in two maternal populations (Prat and A1077). In the first population the trait was defined as the number of days between the first positive pregnancy test and the date of death or culling, whereas in the second line the trait was the number of artificial inseminations after the first fertile AI, defined as that conducting to the first parturition. In the two lines the estimated effective heritability was similar, being around 0.16. Sánchez *et al.* (2005) estimated dominance variance in the V line. They concluded that the magnitude of this parameter was low and hard to be properly estimated given the available data.

Recently, El Nagar *et al.* (2020) estimated genetic parameters and genetic trends for lines A, V, H, LP and R. The heritabilities of the five lines are low, as high as 0.14 for the H line. The magnitude of these heritabilities and the corresponding additive genetic variances could be linked with the criteria considered when the lines were created. For example, one of the lowest additive variances was estimated for the LP line, founded by selecting females with extremely large longevity (see the next section), this implies that the average of this trait is expected to be high but with reduced variability. On the contrary, for the line A, which was founded without attending any production criteria, the whole variation in the population from which founder animals were sampled was kept. El Nagar *et al.* (2020) also reported significant genetic trends for some of the lines under study, particularly for those with the lowest average longevity and with the largest genetic variation. As the reported genetic correlations between length of productive life and prolificacy traits seems to be null (see below) the observed trends must be a consequence of favourable but unintended selection for length of productive life. Only those females that successfully reach a certain number of parities can leave offspring for the next generation; genes in those dying in their firsts parities are not transmitted to the next generation and purged from the population.

From these studies it can be concluded that length of productive life has a low heritability and, as it could be expected, the magnitude pends on the population under study, the trait definition and the model of analysis. A very large generational interval would be required to have accurate-enough genetic predictions, preventing the implementation of this trait in a commercial selection programme.

12.8.3 Selection experiments

Two selection processes have been performed hitherto. The first one was the constitution of a robust line (LP) derived from a large population in which high selection pressure was applied, and the second one was based on a divergent selection experiment on longevity. The process of foundation of the robust line was conducted in the Polytechnic

University of Valencia and it was inspired by the hyper-prolificacy selection experiments previously proposed and used in pigs and rabbits (Sánchez *et al.*, 2006c, 2008). The length of productive life in line LP showed a better survival across farms than the V line, which is widely diffused in Spain as a component of crossbred females. The mean life was 31 days higher in the LP line than in the V line; this figure corresponds to nearly 60% of the interval between parities in the experiment. This difference was generated late in the life of the animals, and in the farm showing the worst environmental conditions. This is an evidence of the genetic determinism of longevity as a function of the age. Survival ability at young ages could be seen as a different but correlated trait. The divergent selection experiment for longevity was conducted in INRA, France (Garreau et *al.*, 2007, 2008b; Larzul *et al.*, 2014), comprising a single generation of selection exclusively conducted in males after a progeny test for its evaluation. The differences between lines were mostly established in the farm with the best sanitary conditions, the specific pathogen-free (SPF) farm. In this case, both differences on mortality and culling rate were observed between lines, while in other farms – non-SPF farms – only differences on culling rate were observed. The authors explained this result as a consequence of an antibiotic treatment in the non-SPF farm which attenuated the differences between lines regarding survival ability. In spite of the observed differences favouring high line, results must be considered with caution since response could be asymmetric and it might happen that only the action on the low longevity line was successful. This possibility cannot be discarded since no control unselected population was considered in the experiment.

12.8.4 Relationships with other traits

A few studies evaluated relationships between length of productive life and other traits in several lines (Tudela *et al.*, 2003; Sánchez *et al.*, 2006a, 2008; Theilgaard *et al.*, 2006; Garreau *et al.*, 2008b; Larzul *et al.*, 2014) but none of them reported any consistent genetic correlation with reproduction performances.

On the contrary, what seems to be clear is the favourable relationship between LPL and traits or physiological indicators of robustness. In this regard, several experiments have been conducted in which length of productive life in line LP as well as other populations have been subjected to different challenging factors. Theilgaard *et al.* (2007) reported a better tolerance to strong feed restriction of LP females than females from the V line. This effect could be partially mediated by a differential body reserves management across lines. Ferrian *et al.* (2013) observed a higher survival in LP females than in V females when they faced an immunological challenge (lipopolysaccharide injection), which could be explained by a differential immunological activity between lines (Ferrian *et al.*, 2012). A lower prevalence of digestive disorders was reported on LP kits than in kits from other lines selected for prolificacy or post-weaning growth (García-Quiros *et al.*, 2014). The observed differences in mortality were also linked to differential leukocytes populations across lines. All these results evidence that the selection for hyper-longevity has generated a more robust population that seems to be better prepared to face the challenging conditions that might occur in the course of the life of females in commercial rabbit farms.

References

Adams, C.E. (1960) Studies on prenatal mortality in the rabbit, *Oryctolagus cuniculus*: the amount and distribution of loss before and after implantation. *Journal of Endocrinology* 19, 325–344.

Alvariño, J.M.R. (2000) Reproductive performance of male rabbits. *In Proceedings of 7th World Rabbit Congress, 2000*, July, Valencia, Spain, A, 13–35.

Amann, R.P. and Hammerstedt, R.H. (2002) Detection of differences in fertility – andrology lab corner. *Journal of Andrology* 23, 317–325.

Argente, M.J., Santacreu, M.A., Climent, A. and Blasco, A. (1997) Divergent selection for uterine capacity in rabbits. *Journal of Animal Science* 75, 2350–2355.

Argente, M.J., Santacreu, M.A., Climent, A. and Blasco, A. (2000) Genetic correlation between litter size and uterine capacity. *Proceedings of 7th World Rabbit Congress*, Valencia, Spain, B:333–338.

Argente, M.J., Blasco, A., Ortega J.A., Haley, C.S. and Visscher, P.M. (2003) Analyses for the presence of a major gene affecting uterine capacity in unilaterally ovariectomized rabbits. *Genetics* 163, 1061–1068.

Argente, M.J., Merchán, M., Peiró, R., García, M.L., Santacreu, M.A. *et al.* (2010) Candidate gene analysis for reproductive traits in two lines of rabbits divergently selected for uterine capacity. *Journal of Animal Science* 88, 828–836.

Argente, M.J., García, M.L., Calle, E.W. and Blasco, A. (2017) Correlated response in litter size components in rabbits selected for litter size variability. *Journal of Animal Breeding and Genetics* 134, 505–511.

Argente, M.J., García, M.L., Zbynovska, M., Petruska, P., Capcarova, M. and Blasco, A. (2019) Correlated response to selection for litter size environmental variability in rabbits' resilience. *Animal* 13, 2348–2355.

Badawy, A.Y., Peiró, R., Blasco, A. and Santacreu, M.A. (2019) Correlated responses on litter size traits and survival traits after two-stage selection for ovulation rate and litter size in rabbits. *Animal* 13, 453–459.

Ballester, M., Castelló, A., Peiró, R., Argente, M.J., Santacreu, M.A. and Folch, J.M. (2013) Identification of differentially expressed genes in the oviduct of two rabbit lines divergently selected for uterine capacity using suppression subtractive hybridization. *Animal Genetics* 44, 296–304.

Baselga, M., García, M.L., Sánchez, J.P., Vicente, J.S. and Lavara, R. (2003) Analysis of reproductive traits in crosses among maternal lines of rabbits. *Animal Research* 52, 473–479.

Beloumi, D., Blasco, A., Muelas, R., Santacreu, M.A., García, M.L. and Argente, M.J. (2020) Inflammatory correlated response in two lines of rabbit selected divergently for litter size environmental variability. *Animals* 10, 1540.

Bencheikh, N. (1995) The effect of frequency of ejaculation on semen characteristics and sperm output in the rabbit. *Annales de Zootechnie* 44, 263–279.

Bennett, G.L. and Leymaster, K.A. (1989) Integration of ovulation rate, potential embryonic viability and uterine capacity into a model of litter size in swine. *Journal of Animal Science* 67, 1230–1241.

Blasco, A. (1996) Genetics of litter size and does fertility in the rabbit. In: *Proceedings of 6th World Rabbit Congress*, Toulouse, France, 2, 219–227.

Blasco, A., Santacreu, M.A., Thompson, R. and Haley, C. (1993a) Estimates of genetic parameters for ovulation rate, prenatal survival and litter size in rabbits from an elliptical selection experiment. *Livestock Production Science* 34, 163–174.

Blasco, A., Bidanel, J.P., Bolet, G., Haley, C.S. and Santacreu, M.A. (1993b) The genetics of prenatal survival of pigs and rabbits: a review. *Livestock Production Science* 37, 1–21.

Blasco, A., Argente, M.J., Haley, C.S. and Santacreu, M.A. (1994) Relationships between components of litter size in unilaterally ovariectomized and intact rabbit does. *Journal of Animal Science* 72, 3066–3072.

Blasco, A., Ortega, J.A., Santacreu, M.A. and Climent, A. (2005) Divergent selection for uterine capacity in rabbits. I. Genetic parameters and response to selection. *Journal of Animal Science* 83, 2297–2302.

Blasco, A., Martínez-Álvaro, M., García, M.L., Ibáñez-Escriche, N. and Argente, M.J. (2017) Selection for environmental variance of litter size in rabbits. *Genetics Selection Evolution* 49, 48.

Bodin, L., García, M., Saleil, G., Bolet, G. and Garreau, H. (2010) Results of 10 generations of canalising selection for rabbit birth weight. *In Proceedings of 9th World Congress on Genetics Applied to Livestock Production*. Leipzig, Germany, 1–6 August, 0391.

Bolet, G. and Saleil, G. (2002a) Strain INRA 1077. In: Rabbit genetic resources in Mediterranean countries. *Options Méditerranéennes, Serie B: Études et recherches*. CIHEAM, Zaragoza, Spain, 38, 109–116.

Bolet, G. and Saleil, G. (2002b) Strain INRA 2066. In: Rabbit genetic resources in Mediterranean countries. *Options Méditerranéennes, Serie B: Études et recherches*. CIHEAM, Zaragoza, Spain, 38, 117–124.

Bolet, G. and Saleil, G. (2002c) Strain INRA 9077. In: Rabbit genetic resources in Mediterranean countries. *Options Méditerranéennes, Serie B: Études et recherches*. CIHEAM, Zaragoza, Spain, 38, 124–132.

Bolet, G., Brun, J.M. and Hulot, F. (1988) Relationships between ovulation rate and embryonic survival in various strains of rabbit. *In Proceedings of 4th World Rabbit Congress,* Budapest, pp. 149–157.

Bolet, G., Brun, J.M., Hulot, F., Poujardieu, B., de Rochambeau, H. and Rouvier, R. (1990) Estrategies pour ameliorer la taille de portée chez le lapin: resultats et perspectives. In: *41st Annual Meeting of the European Association for Animal Production,* Toulouse, France.

Bolet, G., Santacreu, M.A., Argente, M.J., Climent, A. and Blasco, A. (1994) Divergent selection for uterine efficiency in unilaterally ovariectomized rabbits. I. Phenotypic and genetic parameters. In: *Proceedings of 5th World Congress on Genetics Applied to Livestock Production,* Guelph, Ontario, Canada 19, 261–264.

Bolet, G., Brun, J.M., Monnerot, M., Abeni, F. and Arnal, C. (2000) Evaluation and conservation of European rabbit (*Oryctolagus cuniculus*) genetic resources. First results and inferences. In: *Proceedings of 7th World Rabbit Congress,* Valencia, Spain, A, 281–323.

Bolet, G., Zerrouki, N., Gacem, M., Brun, J.M. and Lebas, F. (2012) Genetic parameters and trends for litter and growth traits in a synthetic line of rabbits created in Algeria. In: *Proceedings of 10th World Rabbit Congress, 3–6 September*, Sharm El-Sheikh, Egypt, pp. 195–199.

Bradford, G.E. (1969) Genetic control of ovulation rate and embryo survival in mice. I. Response to selection. *Genetics* 61, 907–918.

Brun, J.M. and Ouhayoun, J. (1994) Qualités bouchères de laperaux issus d'un croisement diallèle de 3 souches: interaction du type génétique et de la taille de portée d'origine. *Annales de Zootechnie* 43, 173–183.

Brun, J.M., Bolet, G. and Ouhayoun, J. (1992) The effects of crossbreeding and selection on productive and reproductive traits in a triallel experiment between three strains of rabbits. *Journal of Applied Rabbit Research* 15, 181–189.

Brun, J.M., Bolet, G., Baselga, M., Esparbie, J. and Falieres, J. (1998) Comparison de deux souches européennes de lapins sélectionnées sur la taille de portée: intérêt de leur croisemenet. In: *Proceedings de 7èmes Journées de la Recherche Cunicole,* Lyon, France, 1, 21–23.

Brun, J.M., Theau-Clement, M. and Bolet, G. (2002a) The relationship between rabbit semen characteristics and reproductive performance after artificial insemination. *Animal Reproduction Science* 70, 139–149.

Brun, J.M., Theau-Clement, M. and Bolet, G. (2002b) Evidence for heterosis and maternal effects on rabbit semen characteristics. *Animal Research* 51, 433–442.

Brun, J.M., Theau-Clément, M., Esparbié, J., Falières, J., Saleil, G. and Larzul, C. (2006) Semen production in two rabbit lines divergently selected for 63-d body weight. *Theriogenology* 66, 2165–2172.

Brun, J.M., Sanchez, A., Ailloud, E., Saleil, G. and Theau-Clement, M. (2016) Genetic parameters of rabbit semen traits and male fertilising ability. *Animal Reproduction Science*, 166, 15–21.

Camacho, J. and Baselga, M. (1990) Genetic correlation between reproductive and growth traits in rabbits. In: *Proceedings of 4th World Congress on Genetics Applied to the Livestock Production,* July, Edinburgh, 16, 366–369.

Cartuche, L., Pascual, M., Gómez, E.A. and Blasco, A. (2014) Economic weights in rabbits. *World Rabbit Science* 22, 165–177.

Casares-Crespo, L., Talavan, A.M. and Viudes-de-Castro, M.P. (2016) Can the genetic origin affect rabbit seminal plasma protein profile along the Year? *Reproduction in Domestic Animals* 51, 294–300.

Casares-Crespo, L., Fernandez-Serrano, P., Vicente, J.S., Marco-Jimenez, F. and Viudes-de-Castro, M.P. (2018) Rabbit seminal plasma proteome: the importance of the genetic origin. *Animal Reproduction Science* 189, 30–42

Casto-Rebollo, C., Argente, M.J., García, M.L., Pena, R. and Ibáñez-Escriche, N. (2020) Identification of functional mutations associated with environmental variance of litter size in rabbits. *Genetics Selection Evolution* 52, 22.

Christenson, R.K., Leymaster, K.A. and Young, L.D. (1987) Justification of unilateral hysterectomy-ovariectomy as a model to evaluate uterine capacity in swine. *Journal of Animal Science* 67, 738–744.

Cifre, P., Baselga, M., García-Ximénez, F. and Vicente, J.S. (1998) Performance of hyperprolific rabbit line. I. Litter size traits. *Journal of Animal Breeding and Genetics* 115, 131–138.

Clutter, A.C., Nielsen, M.K. and Johnson, R.K. (1990) Alternative methods of selection for litter size in mice. I. Characterization of base population and development of methods. *Journal of Animal Science* 68, 3536–3542.

Cunningham, P.J., England, M.E., Young, L.D. and Zimmerman, D.R. (1979). Selection for ovulation rate in swine: correlated response in litter size and weight. *Journal of Animal Science* 48, 509–516.

David, I., Carabano, M.J., Tusell, L., Diaz, C., Gonzalez-Recio, O. *et al.* (2011) Product versus additive model for studying artificial insemination results in several livestock populations. *Journal of Animal Science* 89, 321–328.

Drummond, H., Vázquez, E., Sánchez-Colón, S., Martínez-Gómez, M. and Hudson, R. (2000) Competition for milk in the domestic rabbits: survivors benefit from littermate deaths. *Ethology* 106, 511–526.

Ducrocq, V. and Humblot, P. (1995) Genetic characteristics and evolution of semen production of young normande bulls. *Livestock Production Science* 41, 1–10.

Ducrocq, V. and Sölkner, J. (1994) The Survival Kit. A fortran package for the analysis of survival data. In: *Proceedings of 5th World Congress on Genetics Applied to Livestock Production,* University of Guelph, Canada.

Eady, S.J. and Garreau, H. (2008) An enterprise gross margin model to explore the influence of selection criteria for breeding programs and changes to management systems. In: *Proceedings of 9th World Rabbit Congress*, A, 61–65.

El Nagar, A.G., Sánchez, J.P., Ragab, M., Mínguez, C. and Baselga M. (2020) Genetic variability of functional longevity in five rabbit lines. *Animal* 14, 1111–1119.

El-Tarabany, M.S., El-Bayomi, K. and Abdelhamid, T. (2015) Semen characteristics of purebred and crossbred male rabbits. *PLOS ONE* 10, e0128435.

Estellé, J., Sastre, Y., Merchán, M., Peiró, R., Santacreu, M.A. and Folch, J.M. (2006) TIMP-1 as candidate gene for embryo survival in two divergent lines selected for uterine capacity in rabbits. *Molecular Reproduction and Devevopment* 73, 678–684.

Ferraz, J.B.S., Johnson, R.K. and Van Vleck, L.D. (1992) Estimation of genetic trends and genetic parameters for reproductive and growth traits of rabbits raised in subtropics with animal model. In: *Proceedings of 5th World Rabbit Congress*, Corvallis, Oregon, A, 132–142.

Ferrian, S., Guerrero, I., Blas, E., García-Diego, F.J., Viana, D. *et al.* (2012) How selection for reproduction or foundation for longevity could have affected blood lymphocyte populations of rabbit does under conventional and heat stress conditions. *Veterinary Immunology and Immunopathology* 150, 53–60.

Ferrian, S., Blas, E., Larsen, T., Sánchez, J.P., Friggens, N.C. *et al.* (2013) Comparison of immune response to lipopolysaccharide of rabbit does selected for litter size at weaning or founded for reproductive longevity. *Research in Veterinary Science* 94, 518–525.

Fleming, M.W., Rhodes, R.C. and Dailey, R.A. (1984) Compensatory responses after unilateral ovariectomy in rabbits. *Biology of Reproduction* 30, 82–86.

García, M.L. and Baselga, M. (2002a) Estimation of genetic response to selection in litter size of rabbits using a cryopreserved control population. *Livestock Production Science* 74, 45–53.

García, M.L. and Baselga, M. (2002b) Genetic response to selection for reproductive performance in a maternal line of rabbits. *World Rabbit Science* 10, 71–76.

García, M.L. and Baselga, M. (2002c) Estimation of correlated response on growth traits to selection in litter size of rabbits using a cryopreserved control population and genetic trends. *Livestock Production Science* 78, 91–98.

García, M.L., Peiró, R., Argente, M.J., Merchán, M., Folch, J.M. *et al.* (2010) Investigation of the oviductal glycoprotein 1 (OVGP1) gene associated with embryo survival and development in the rabbit. *Journal of Animal Science* 88, 1597–1602.

García, M.L., Blasco, A. and Argente, M.J. (2016) Embryological changes in rabbit lines selected for litter size variability. *Theriogenology* 56, 1247–1250.

García, M.L., Blasco, A., García, M.E. and Argente, M.J. (2019) Body condition and energy mobilisation in rabbits selected for litter size variability. *Animal* 13, 784–789

García-Quirós, A., Arnau-Bonachera, A., Penadés, M., Cervera Fras, M.C., Martinez-Paredes, E. *et al.* (2014) A robust rabbit line increases leucocyte counts at weaning and reduces mortality by digestive disorder during fattening. *Veterinary Immunology and Immunopathology* 161, 123–131.

Garcia-Tomás, M., Sánchez, J., Rafel, O., Ramon, J. and Piles, M. (2006a) Reproductive performance of crossbred and purebred male rabbits. *Livestock Science* 104, 233–243.

Garcia-Tomás, M., Sánchez, J., Rafel, O., Ramon, J. and Piles, M. (2006b) Variability, repeatability and phenotypic relationships of several characteristics of production and semen quality in rabbit. *Animal Reproduction Science* 93, 88–100.

Garcia-Tomás, M., Sánchez, J., Rafel, O., Ramon, J. and Piles, M. (2006c) Heterosis, direct and maternal genetic effects on semen quality traits of rabbits. *Livestock Science* 100, 111–120.

García-Tomás, M., Sánchez, J. and Piles, M. (2009) Post-natal sexual development of testis and epididymis in the rabbit: variability and relationships among macroscopic and microscopic markers. *Animal Reproduction Science* 110, 347–355.

Garreau, H., Szendro, Z.S., Larzul, C. and Rochambeau, H. de. (2000) Genetic parameters and genetic trends of growth and litter size traits in the White Pannon breed. In: *Proceedings of 7th World Rabbit Congress*. Valencia, *Spain*, vol. 1, 403–408.

Garreau, H., Ducrocq, V., Tudela, F., Saleil, G., Esparbie, J. *et al.* (2007) Sélection divergente pour la longévité de la lapine en reproduction. In: *Proceedings de 12èmes Journées de la Recherche Cunicole*, Le Mans, France.

Garreau, H., Bolet, G., Larzul, C., Robert-Granié, C., Saleil, G. *et al.* (2008a) Results of four generations of a canalising selection for rabbit birth weight. *Livestock Production Science* 119, 55–62.

Garreau, H., Ducrocq, V., Tudela, F., Saleil, G., Juin, H. and Larzul, C. (2008b) Divergent selection for longevity in breeding does. In: *Proceedings of 9th World Rabbit Conference*, Verona, Italy.

Garreau, H., Balmisse, E., Ruesche, J., Hurtaud, J., Maupin, M. *et al.* (2015) Relation entre les critères de sélection de la lignée INRA 1777 et la survie des lapereaux. *Proceedings de 16èmes Journées de la Recherche Cunicole,* Le Mans, France, 77–80.

Gion, J.M., Clutter, A.C. and Nielsen, M.K. (1990) Alternative methods of selection for litter size in mice. II. Response to thirteen generations of selection. *Journal of Animal Science* 68, 3543–3556.

Gómez, E.A., Baselga, M. and Cifre, J. (1994) Influence des effets maternels sur la selection de la taille de portée chez le lapin. *Proceedings de 6èmes Journées de la Recherche Cunicole,* La Rochelle, France, 1, 217–224.

Gómez, E.A., Rafel, O. and Ramón, J. (1998) Genetic relationships between growth and litter size traits at first parity in a specialized dam line. *Proceedings of 6th World Congress Genetics Applied Livestock Production,* Armidale, Australia, 25, 552–555.

Gómez, E.A., Rafel, O. and Ramón, J. (2000) Preliminary genetic analysis of Caldes line: a selection experiment for a global objective. In: *Proceedings of 7th World Rabbit Congress,* Valencia, Spain, A: 417–424.

Hulot, F. and Matheron, G. (1979) Analyses des variations génétiques entre trois races de lapins de la taille de portée et de ses composantes biologiques en saillie post-partum. *Annales de Génétique et de Sélection Animale* 11, 53–77.

Johnson, R.K., Zimmerman, D.R. and Kittok, R.J. (1984) Selection for components of reproduction in swine. *Livestock Production Science* 11, 541–558.

Johnson, R.K., Nielsen, M.K. and Casey, D.S. (1999) Responses in ovulation rate, embryonal survival, and litter traits in swine to 14 generations of selection to increase litter size. *Journal of Animal Science* 77, 541–557.

Kabir, M., Akpa, G.N., Nwagu, B.I. and Adeyinka, I.A. (2012) Estimating additive and dominance variance for litter traits in purebred California White kits using different models. *Nigerian Veterinary Journal* 33(2), 448–454.

Khalil, M.H. and Al-Saef, A.M. (2012) Genetic groups comparisons for litter and lactational traits and feeding parameters in program of synthesizing new lines of rabbits. In: *Proceedings of 10th World Rabbit Congress,* Sharm El-Sheikh, Egypt, 235–239.

Khalil, M.H., Al-Sobayil, K.A., Al-Saef, A.M., García, M.L. and Baselga, M. (2007) Genetic evaluation for semen characteristics in a crossbreeding project involving Saudi and Spanish V-line rabbits. *Animal* 1, 923–928.

Kirby, Y.K. and Nielsen, M.K. (1993) Alternative methods of selection for litter size in mice: III. Response to 21 generations of selection. *Journal of Animal Science* 71, 571–578.

Laborda, P., Mocé, M.L., Santacreu, M.A. and Blasco, A. (2011) Selection for ovulation rate in rabbits: genetic parameters, direct response, and correlated response on litter size. *Journal of Animal Science* 89, 2981–2987.

Laborda, P., Mocé, M.L., Blasco, A. and Santacreu, M.A. (2012a) Selection for ovulation rate in rabbits: genetic parameters and correlated responses on survival rates. *Journal of Animal Science* 90, 439–446.

Laborda, P., Santacreu, M.A., Blasco, A. and Mocé, M.L. (2012b) Selection for ovulation rate in rabbits: direct and correlated responses estimated with a cryopreserved control population. *Journal of Animal Science* 90, 3392–3397.

Land, R.B., and Falconer, D.S. (1969) Genetic studies of ovulation rate in the mouse. *Genetical Research* 13, 25–46.

Larzul, C., Ducrocq, V., Tudela, F., Juin, H. and Garreau, H. (2014) The length of productive life can be modified through selection: an experimental demonstration in the rabbit. *Journal of Animal Science* 92, 2395–2401.

Lavara, R., García, M.L., Torres, C., Vicente, J.S. and Baselga, M. (2008) Genetic parameters for semen traits of rabbit males. I. Production, morphology, and sperm head morphometry. In: *Proceedings of 9th World Rabbit Congress,* Verona, Italy.

Lavara, R., Vicente, J.S. and Baselga, M. (2011) Genetic parameter estimates for semen production traits and growth rate of a paternal rabbit line. *Journal of Animal Breeding and Genetics* 128, 44–51.

Lavara, R., Vicente, J.S. and Baselga, M. (2012) Estimation of genetic parameters for sperm quality traits and growth rate in a paternal rabbit line. *Theriogenology* 78, 567–575.

Leymaster, K.A. and Christenson, R.K. (2000) Direct and correlated responses to selection for ovulation rate or uterine capacity in swine. *Journal of Animal Science* 78 (Suppl. 1), 68. (Abstr.).

Mantovani, R., Sartori, A., Mezzadri, M. and Lenarduzzi, M. (2008) Genetics of maternal traits in a new synthetic rabbit line under selection. In: *Proceedings of 9th World Rabbit Congress,* Verona, Italy, 169–174.

Merchán, M., Peiró, R., Argente, M.J., Santacreu, M.A., García, M.L. *et al.* (2009) Analysis of the oviductal glycoprotein 1 polymorphisms and their effects on components of litter size in rabbits. *Animal Genetics* 40, 756–758.

Mínguez, C., Sánchez, J.P., Ragab, M., El Nagar, A.G. and Baselga, M. (2012) Growth traits in four maternal rabbit lines. In: *Proceedings of 10th World Rabbit Congress,* Sharm El-Sheikh, Egypt, 55–59.

Mocé, M.L. and Santacreu, M.A. (2010) Genetic improvement of litter size in rabbits: a review. In: *Proceedings of the 9th World Congress on Genetics Applied to Livestock Production*, 1–6 August 2010, Leipzig, Germany, communication 25.

Mocé, M.L., Santacreu, M.A., Climent, A. and Blasco, A. (2004) The effect of divergent selection for uterine capacity on prenatal survival in rabbits: maternal and embryonic genetic effects. *Journal of Animal Science* 82, 68–73.

Mocé, M.L., Santacreu, M.A., Climent, A. and Blasco, A. (2005) Divergent selection for uterine capacity in rabbits. III. Responses in uterine capacity and its components estimated with a cryopreserved control population. *Journal of Animal Science* 83, 2308–2312.

Morrell, J.M. (1995) Artificial insemination in rabbits. *British Veterinary Journal* 151, 477–488.

Moura, A.S.A.M.T., Costa, A.R.C. and Polastre, R. (2001) Variance components and response to selection for reproductive litter and growth traits through a multi-purpose index. *World Rabbit Science* 9, 77–86.

Napier, R.A.N. (1961) Fertility in the male rabbit. III. Estimation of spermatozoan quality by mixed insemination, and the inheritance of spermatozoa characters. *Journal of Reproduction and Fertility* 2, 273–289.

Orengo, J., Gómez, E.A., Piles, M., Ramón, J. and Rafel, O. (2003) Étude des caracteres de reproduction en croisement entre trois lignées femelles espagnoles. In: *Proceedings 10èmes Journées de la Recherche Cunicole.* Paris, 57–60.

Panella, F., Castellini, C. and Facchin, E. (1994) Heritability of some male reproductive traits in rabbit. *Cah. Options Méditerranéennes* 8, 279–283.

Peiró, R., Merchán, M., Santacreu, M.A., Argente, M.J., García, M.L. *et al.* (2008) Progesterone receptor gene as candidate gene for reproductive traits in rabbits. *Genetics* 180, 1699–1705.

Peiró, R., Herrler, A., Santacreu, M.A., Merchán, M., Argente, M.J. *et al.* (2010) Expression of progesterone receptor related to the polymorphismin: the PGR gene in the rabbit reproductive tract. *Journal of Animal Science* 88, 421–427.

Piles, M. and Tusell, L. (2012) Genetic correlation between growth and female and male contributions to fertility in rabbit. *Journal of Animal Breeding and Genetics* 129, 298–305.

Piles, M., Rafel, O., Ramon, J. and Varona, L. (2005) Genetic parameters of fertility in two lines of rabbits with different reproductive potential. *Journal of Animal Science* 83, 340–343.

Piles, M., Garreau, H., Rafel, O., Larzul, C, Ramon, J. and Ducrocq, V. (2006a) Survival analysis in two lines of rabbits selected for reproductive traits. *Journal of Animal Science* 84, 1658–1665.

Piles, M., Sánchez, J.P., Orengo, J., Rafel, O., Ramon, J. and Baselga, M. (2006b) Crossbreeding parameter estimation for functional longevity in rabbits using survival analysis methodology. *Journal of Animal Science* 84, 58–62.

Piles, M., Garcia, M.L., Rafel, O., Ramon, J. and Baselga, M. (2006c) Genetics of litter size in three maternal lines of rabbits: repeatability versus multiple-trait models. *Journal of Animal Science* 84, 2309–2315.

Piles, M., Díez, J., del Coz, J.J., Montañés, E., Quevedo, J.R. *et al.* (2012) Predicting male fertility from semen quality traits: performance of several parametric and non-parametric procedures. *Livestock Sciences* 155, 137–147.

Piles, M., Tusell, L., Lavara García, R. and Baselga M. (2013a) Breeding programmes to improve male reproductive performance and efficiency of insemination dose production in paternal lines: feasibility and limitations. *World Rabbit Science* 21, 61–75.

Piles, M., Mocé, M.L., Laborda, P. and Santacreu, M.A. (2013b) Feasibility of selection for male contribution to embryo survival as a way of improving male reproductive performance and semen quality in rabbits. *Journal of Animal Science* 91, 4654–4658.

Poigner, J., Szendro, Z.S., Lévai, A. Radnai, I. and Biro-Németh, E. (2000) Effect of birth weight and litter size on growth and mortality in rabbits. *World Rabbit Science* 8, 17–22.

Poujardieu, B. and Vrillon, J.L. (1973) Variation de la productivité numérique au sevrage et de ses composantes entre génotypes de lapines croisées et de race pure. *Proceedings of the Journées de Recherche Avicule et Cunicole* 4, 89–93.

Ragab, M. and Baselga, M. (2011) A comparison of reproductive traits of four maternal lines of rabbits selected for litter size at weaning and founded on different criteria. *Livestock Science* 136, 201–206.

Ragab, M., Lavara, R., Vicente, J.S., Mínguez, C. and Baselga, M. (2012) Effect of lactation stage on litter size components in rabbits. In: *Proceedings of 10th World Rabbit Congress,* Sharm El-Sheikh, Egypt, 373–377.

Ragab, M., Baselga, M., El Nagar, A., Mínguez, C. and Sánchez, J.P. (2020) Genetic analysis of functional longevity in a complete diallel crossing experiment involving four maternal rabbit lines. *Journal of Animal Breeding and Genetics.* DOI: 10.1111/jbg.12520

Rastogi, R.K., Lukefahr, S.D. and Lauckner, F.B. (2000) Maternal heritability and repeatability for litter traits in rabbits in a humid tropical environment. *Livestock Production Science* 67, 123–128.

Ribeiro, E.L., Nielsen, M.K., Bennett, G.L. and Leymaster, K.A. (1997a) A simulation model including ovulation rate, potential embryonic viability, and uterine capacity to explain litter size in mice. *I.* Model development and implementation. *Journal of Animal Science* 75, 641–651.

Ribeiro, E.L., Nielsen, M.K., Leymaster, K.A. and Bennett, G.L. (1997b) A simulation model including ovulation rate, potential embryonic viability, and uterine capacity to explain litter size in mice. II. Responses to alternative criteria of selection. *Journal of Animal Science* 75, 652–656.

Rochambeau, H. (1988) Genetics of the rabbit for wool and meat production. In: *Proceedings of 4th World Rabbit Congress,* Budapest.

Rochambeau, H. (1998) La femelle parentale issue des souches expèrimentales de l'INRA evolutions génétiques et perspectives. *Proceedings de 7èmes Journées de la Recherche Cunicole,* Lyon, France, 3–14.

Rommers, J.M., Kemp, B., Meijerhoff, R. and Noordhuisen, J.P.T.M. (2001) The effect of litter size before weaning on subsequent body development, feed intake, and reproductive performance of young rabbits does. *Journal of Animal Science* 79, 1973–1982.

Rosell, J.M. (2003) Health status of commercial rabbitries in the Iberian Peninsula: a practitioner's study. *World Rabbit Science* 11, 157–169.

Rosendo, A., Druet, T., Gogué, J. and Bidanel, J.P. (2007) Direct responses to six generations of selection for ovulation rate or prenatal survival in Large White pigs. *Journal of Animal Science* 85, 356–364.

Ruiz-Flores, A. and Johnson, R.K. (2001) Direct and correlated responses to two-stage selection for ovulation rate and number of fully formed pigs at birth in swine. *Journal of Animal Science* 79, 2286–2299.

Sánchez, J.P., Baselga, M., Peiro, R. and Silvestre, M.A. (2004) Analysis of factors influencing longevity of rabbit does. *Livestock Production Science* 90, 227–234.

Sánchez, J.P., Baselga, M. and Korsgaard, I. (2005) Estudio de los efectos de dominancia en la determinación de la longevidad del conejo de carne. In: *XI Jornadas sobre Producción Animal*, Zaragoza, Spain.

Sánchez, J.P., Baselga, M. and Ducrocq, V. (2006a) Genetic and environmental correlations between longevity and litter size in rabbits. *Journal of Animal Breeding and Genetics* 123, 180–185.

Sánchez, J.P., Korsgaard, I.R., Damgaard, L.H. and Baselga, M. (2006b) Analysis of rabbit doe longevity using a semiparametric log-Normal animal frailty model with time-dependent covariates. *Genetics Selection Evolution* 38, 281–295.

Sánchez, J.P., Theilgaard, P., Mínguez, C. and Baselga, M. (2006c) Constitution and evaluation of a long-lived productive maternal rabbit line. In: *8th World Congress on Genetics Applied to Livestock Production,* Belo Horizonte, Brazil.

Sánchez, J.P., Theilgaard, P., Mínguez, C. and Baselga, M. (2008). Constitution and evaluation of a long-lived productive rabbit line. *Journal of Animal Science* 86, 515–525.

San Cristobal-Gaudy, M., Elsen, J.M., Bodin, L. and Chevalet, C. (1998) Prediction of the response to a selection for canalisation of a continuous trait in animal breeding. *Genetics Selection Evolution* 30, 423–451.

Santacreu, M.A., Viudes, M.P. and Blasco, A. (1990) Evaluation par coelioscopie des corps jaunes et des embryons. Influence sur la taille de portée chez la lapine. *Reproduction Nutrition Development* 30, 583–588.

Santacreu, M.A., Mocé, M.L., Climent, A. and Blasco, A. (2005) Divergent selection for uterine capacity in rabbits. II. Correlated response in litter size and its components estimated with a cryopreserved control population. *Journal of Animal Science* 83, 2303–2307.

Serna-García, M., Peiró, R., Serna, E. and Santacreu, M.A. (2020) Ovarian transcriptomic analysis reveals differential expression genes associated with cell death process after selection for ovulation rate in rabbits. *Animals* 10, 1924.

Sosa-Madrid, B.S., Santacreu, M.A., Blasco, A., Fontanesi, L., Pena, R.N. and Ibáñez-Escriche, N. (2020) A genome-wide association study in divergently selected lines in rabbits reveals novel genomic regions associated with litter size traits. *Journal of Animal Breeding and Genetics* 137, 123–138.

Theau-Clément, M., Brun, J.M., Sabbioni, E., Castellini, C., Renieri, T. *et al.* (2003) Comparaison de la production spermatique de trois souches de lapins: moyennes et variabilités. *Proceedings de 10èmes Journées de la Recherche Cunicole,* Paris, 81–84.

Theau-Clément, M., Brun, J.M., Esparbié, J., Falières, J., Garvanèse, J. *et al.* (2007) Une selection pour la vitesse de croissance influence-t-elle la fécondance de la semence de lapin? *Proceedings de12émes Journées de la Recherche Cunicole,* Le Mans, France, 53–56.

Theau-Clement, M., Salvetti, P., Bolet, G., Saleil, G. and Joly, T. (2009) Influence de l'intervalle entre le sevrage et l'insemination sur la production d'embryons et leur qualité chez la lapine. *Proceedings de 13émes Journées de la Recherche Cunicole,* INRA-ITAVI, Le Mans, France, 125–128.

Theilgaard, P., Sanchez, J.P., Pascual, J.J., Friggens, N.C. and Baselga, M. (2006) Effect of body fatness and selection for prolificacy on survival of rabbit does assessed using a cryopreserved control population. *Livestock Science* 103, 65–73.

Theilgaard, P., Sanchez, J.P., Pascual, J.J., Baselga, M., Friggens, N.C. and Berg, P. (2007) Late reproductive senescence in a rabbit line hyper selected for reproductive longevity, and its association with body reserves. *Genetics Selection Evolution* 39, 207–223.

Torres, C., Pla, M. and García, F. (1986) Nivel de respuesta en el tiempo a un control de seguimiento sanitario de conejos. *XI Symposium de Cunicultura,* Teruel, Spain, pp. 145–152.

Torres, C., Garcés, M., Fabado, F. and Pla, M. (1987) Causas de eliminación de reproductores en función de la línea y época. *XII Symposium de Cunicultura,* Guadalajara, Spain.

Torres, S., Hulot, F. and Meunier, M. (1984) Étude comparée du développment et de la mortalité embryonnaire chez deux genotypes de lapines. *Proceedings of 3th World Rabbit Congress, Rome,* 273–277.

Tudela, F., Hurtaud, J., Garreau, H. and de Rochambeau, H. (2003) Comparaison des performances zootechniques de femelles parentales issues d'une souche témoin et d'une souche sélectionnée sur la productivité numérique. *Proceedings de 10èmes Journées de la Recherche Cunicole,* Paris.

Tusell, L., García-Tomás, M., Baselga, M., Rekaya, R., Rafel, O. *et al.* (2010) Genotype x AI conditions interaction for male effect on fertility and prolificacy. *Journal of Animal Science* 88, 3475–3485.

Tusell, L., Legarra, A., García-Tomás, M., Rafel, O., Ramon, J. and Piles M. (2011) Different ways to model biological relationships between fertility and the pH of the semen in rabbits. *Journal of Animal Science* 89, 1294–1303.

Tusell, L., Legarra, A., García-Tomás, M., Rafel, O., Ramon, J. and Piles, M. (2012) Genetic basis of semen traits and their relationship with growth rate in rabbits. *Journal of Animal Science* 90, 1385–1397.

Vicente, J.S., Viudes-de-Castro, M.P., Lavara, R. and Lavara, F. (2000) Effect of male line on prolificacy from does inseminated with low sperm doses. In: *Proceedings of 7th World Rabbit Congress,* Valencia, Spain, A, 273–277.

Wolf, J. (2009) Genetic correlations between production and semen traits in pig. *Animal* 3, 1094–1099.

Yazdi, M.H., Visscher, P.M., Ducrocq, V. and Thompson, R. (2002) Heritability, reliability of genetic evaluations and response to selection in proportional hazard models. *Journal of Dairy Science* 85, 1563–1577.

Ziadi, C., Mocé, M.L., Laborda, P., Blasco, A. and Santacreu, M.A. (2013) Genetic selection for ovulation rate and litter size in rabbits: estimation of genetic parameters and direct and correlated responses. *Journal of Animal Science* 91, 3113–3120.

13 Genetic Improvement in the Meat Rabbit

Manuel Baselga[1], István Nagy[2], Miriam Piles[3]*, Hervé Garreau[4], Luca Buttazzoni[5], Zsolt Szendrő[2] and María-Luz García[6]

[1]Institut for Animal Science and Technology, Universitàt Politecnica de Valencia, Spain; [2]Department of Animal Genetics and Biotechnology, Kaposvár University, Hungary; [3]Genetica i Millora Animal, Institut de Recerca i Tecnologia Agroalimentàries (IRTA), Caldes de Montbui, Barcelona, Spain; [4]INRAE, Station d'Amélioration Génétique des Animaux, Castanet-Tolosan, France; [5]Centro di ricerca di Zootecnia e Acquacoltura, Consiglio per la ricerca in agricoltura e l'analisi dell'economia agraria (CREA), Monterotondo (Roma), Italy; [6]Department of Agrofood Technology, Escuela Politécnica Superior de Orihuela, Universidad Miguel Hernández de Elche, Orihuela, Spain

13.1 Introduction

The genetic improvement is directed towards the competitiveness of rabbit farming. Classical quantitative traits, such as litter size in maternal lines and growth in paternal lines, remain the most common selection criteria, but functional traits, such as longevity, health and maternal traits, are emerging targets in breeding schemes. Selection of meat rabbits is based on the well-established concepts of quantitative genetics but could also benefit from molecular marker information. Recently, it is common to find association studies between genetic markers and production or health traits aiming to find causative genes of them. The interest of the genomic selection in rabbit needs also to be discussed. This chapter aims to present the organization of rabbit-breeding schemes, the development of lines and the diffusion of the genetic improvement for meat rabbits. The genetics of rabbit wool and fur, and the genetic improvement of these traits are the subject of Chapter 7.

13.2 Rabbit Production Schemes

From a production point of view, the two most important types of animals in a rabbit farm are the does and the young rabbits (Baselga, 2004). Does should be efficient producers of weaned rabbits and these rabbits should grow fast, have a good feed conversion rate and acceptable carcass quality. Therefore, a selection programme in rabbits must be driven by two consequent main aims: improving the doe performances and the efficiency of the growing rabbits. These aims could be achieved using crossbred does and young, mating specialized lines. The programme requires high levels of organization and appropriate facilities (Baselga, 2004). It could also be possible to use only one line or breed with acceptable reproduction and growth performances. The implementation of the last option is less demanding in terms of resources and organization than the former and could be the most convenient for less-developed countries.

Contact e-mail address: miriam.piles@irta.cat

 The Genetics and Genomics of the Rabbit (ed Luca Fontanesi)
DOI: 10.1079/9781780643342.0013

13.2.1 Crossbreeding schemes

Intensive meat production in rabbits is based on a three- or four-way crossbreeding scheme. The first cross involves two maternal lines generating crossbred does in order to take advantage of the expected positive heterosis in reproductive traits, the eventual complementarity among the lines and the dissipation of the inbreeding accumulated within the lines. In the second cross, the former crossbred does are mated to males from a third line (three-way crossbreeding) or crossbred males from a third and a fourth paternal line (four-way crossbreeding) to produce meat rabbits.

The first and oldest selection programme, still active, is the French programme carried out by INRAE, which in 1969 began the development of specialized rabbit lines (Rouvier, 1981). This programme was followed by the Spanish selection programme that started in 1976. The Spanish programme is carried out jointly by the Polytechnic University of Valencia and IRTA (Baselga, 2004). Other countries, Egypt, Saudi Arabia (Youssef *et al.*, 2008), Hungary and Italy, now have their selection programmes based on the use of crossbreeding schemes (Baselga, 2004).

13.2.2 Schemes based on multi-purpose lines

In countries where the rabbit industry has not yet reached a high level of organization, it may not be possible to select sire and dam lines for a subsequent crossbreeding programme. An alternative could be the development of a multi-purpose line, through simultaneous selection for litter size and growth traits. This alternate strategy was successfully tested in Brazil, Spain and France. Moura *et al.* (2001) initiated the selection of a multi-purpose line in 1992. There were four selection criteria: (i) litter size at weaning; (ii) individual weaning weight; (iii) litter weaning weight; and (iv) individual body weight at 70 days. These traits were combined in an index. Estimated annual genetic trends for litter size at weaning (0.04 young rabbits per litter and per year) and litter weaning weight were positive and significant. A positive direct linear trend was observed for weaning weight, but no maternal linear trend was detected. Direct and maternal trends were favourable for time-to-market individual weight. Moura *et al.* (2001) observed a more consistent response for this trait, which received a higher relative economic weight in the selection index. Gómez *et al.* (2000) chose two selection criteria for their multi-purpose strain: (i) litter size at weaning; and (ii) individual daily weight gain. An independent level culling selection method was used. Renewal bucks (and does) were selected from the 50% (and 80%) best-ranked dams and from the 15% (and 25%) fast-growing young rabbits. A genetic trend in litter weight at weaning (30.9 g per litter and per year) was the result of a positive genetic response in individual weight at weaning (11 g per young rabbit and per year) and litter size at weaning (0.03 young rabbits per litter and per year). Direct response on daily gain was around 1.06 g per day, with an indirect response on individual 60 days weight (38 g per young rabbit and per year). The ability of does to raise litters was also improved since genetic response on litter size at weaning was higher than the response on litter size at birth. Salaün *et al.* (2001) selected a Rex line on a single criterion: total litter weight at weaning per doe and per year. The heritability of the trait was low (0.06) and the genetic correlation with the total litter weight at weaning (0.98) and the litter size at weaning (0.87) were high. An annual genetic trend was observed for the selected trait (342 g), which was equal to 1.7% of the phenotypic mean or 3.22% of the phenotypic standard deviation.

A multi-purpose line has been selected by ITELV in co-operation with INRA in Algeria. This line was founded from local rabbits and the INRA strain A2666 (Gacem *et al.*, 2008). After four generations of the foundational mating, the new line has improved its prolificacy (1.8 young rabbits born alive), the live weight of the females at mating (more than 500 g) and the rate of stillbirth (13.5%) with regard to the local breed, without losing productivity in summer. The litter size at weaning and the slaughter weight are the objectives of selection of the line, considered too low when the synthesis of the line was finished. The simplest way to diffuse the improvement achieved in the line along the generations of selection is regularly providing males from the nucleus of selection to the farms to replace the bucks. The replacement of the does would be done with the female progeny obtained in the farm sired by the bucks of the nucleus.

A synthetic line, obtained by crossing New Zealand, Californian and Flemish Giant, was set up in Australia to establish a national breeding programme (Eady, 2003). This multi-purpose line is selected for number of kits weaned per litter, average daily gain (ADG) and bacterial disease resistance. The breeding programme, named Crusader, was designed to provide improved bucks directly to farmers. Genetic links have been established between Crusader and some commercial breeding programmes by the reciprocal use of bucks. A data-management system has been set up for a national rabbit evaluation programme.

13.3 Development of Lines

When the use of crossbreeding is possible, the development of maternal and paternal lines is a crucial activity (Baselga, 2004). Chapter 2 of this book is devoted to the description of the existing rabbit genetic resources that, in principle, could be the base for a crossbreeding programme. In general, the breeds are entities that are numerically too large and genetically very variable to obtain repeatable results, particularly if the most cosmopolitan breeds are considered as they could be constituted by many sub-populations of very different performances (Baselga, 2004). Thus it is preferable to rely on smaller groups of animals, the lines, to which a definite programme of selection can be applied, depending on the specialization of the lines, with an appropriate crossbreeding plan designed to maximize the heterotic effects.

13.3.1 Criteria for the development of new lines

In order to create a new population of small size that, after two or three generations of inter-se mating, gives rise to a new line (Khalil and Baselga, 2002), the most common practice in the past was to rely on the existing breeds to get samples of one or several of them (Lukefahr *et al.*, 1996; Baselga, 2004). The final size of the line can range between 100 and 200 does and 20–40 bucks (Baselga, 2004). This procedure is simple and easy to perform but some caveats must be taken into account. The first one is that given the enormous diversity that can exist within a breed, one may sample the founder stock for the new line from populations of the breed that are genetically poor for the traits of interest, and consequently the initial performance of the new line will be low and possibly non-competitive. Another one is that health problems could appear when all the founders coming from a relatively large number of different farms are put together in the same herd. This problem can be overcome by using hysterectomy or other techniques such as freezing and embryo transfer to obtain animals from the farms (García-Ximénez *et al.*, 1996).

There are alternative procedures to the one described above to develop a new line; two of them are described below. In both cases, the first point is to realistically define the desired specialization of the line. The first alternative attempts to find two or three populations, no matter their genetic origin (pure breed, synthetics or crossbred), that are clearly outstanding on the important traits regarding the desired specialization of the line (Baselga, 2002). The next step is to obtain animals from these populations and mate them without selection for two or three generations. The second alternative relies on applying very high intensities of selection for the traits of interest in very large populations (for example, commercial populations). We illustrate this procedure describing the foundation of line H (Universitàt Politecnica de Valencia, UPV; Spain) which was based on the detection of hyperprolific does, screening a large group of does, spread over many Spanish farms. A doe was classified as hyperprolific if it had a parity with 17 or more kits born alive or if it was within the group of best 1% of does based on its accumulated number of kits born alive across parities. In a first step of the foundation process, male progeny was obtained from matings between 20 hyperprolific does and 9 bucks pertaining to line V. In a second step, those young males were mated to a new and larger batch of hyperprolific does (backcross) to accumulate the genes for prolificacy in the progeny, which constituted generation 0 of the new line H. To avoid health problems, hysterectomies were performed in the first step and embryo vitrification in the second step. After thawing and transferring these embryos, a set of 474 rabbits belonging to generation 0 were allowable for maintaining the line and for studies of comparison of the line H with the line V and crossbred does AxV which demonstrated the superiority of line H (Cifre *et al.*, 1998a, 1998b). The same procedure has been applied to create the line LP for which the screening in commercial farms was based on doe hyperlongevity and prolificacy over defined thresholds (mean prolificacy higher than 7.5 rabbits born alive in more than 28 parities) (Baselga, 2004). In this case, three steps of backcrosses were performed. Details regarding the foundation of this line as well as results on the comparison with line V have been reported by Sánchez *et al.* (2008). Those results confirm that applying very high intensities of selection is a successful strategy for the creation of new lines. In addition, these authors showed that line LP resulted peculiarly robust in challenging environmental conditions regarding management, temperature, feeding or immune response (Theilgaard *et al.*, 2009; Savietto *et al.*, 2012, 2013; Ferrian *et al.*, 2013).

Using a mixed animal model (additive and non-additive permanent effects), the litter size data from June 1980 to February 2009 of the lines A, V, H and LP, and their full pedigree to take into account the process of selection, Ragab and Baselga (2011) estimated the differences in litter size between the lines at their respective times of foundation. The line A showed the lowest litter sizes with respect to the others with differences of 1.39 rabbits/litter for total born, 1.20 for number born alive, 0.84 for number weaned and 1.06 for number marketed. These important differences were attributed to the different criteria used for founding the lines, revealing that the criterion of pertaining to a breed, used for creating the line A, was the worst.

13.3.2 Criteria and methods of selection – paternal lines

13.3.2.1 The current approaches

Paternal lines are commonly selected for post-weaning daily gain (Rochambeau *et al.*, 1989; Estany *et al.*, 1992; Gómez *et al.*, 2002) or for body weight at a time close to the market age (Lukefahr *et al.*, 1996; Larzul *et al.*, 2003a; Drouilhet *et al.*, 2013). These parameters are very easy to record and have a moderate and favourable genetic correlation with the conversion index (Moura *et al.*, 1997; Piles *et al.*, 2004; Drouilhet *et al.*, 2013), which is very important for an efficient production, because feeding accounts for the highest proportion of total costs (60–80% in rabbits; Baselga and Blasco, 1989; Drouilhet *et al.*, 2013). Thus, economic weights for a standard Spanish farm of intensive rabbit-meat production have been estimated, being the conversion index, the trait with the highest value of this parameter, among several production and reproduction traits (Cartuche *et al.*, 2014). In meat-production farms, feed efficiency can be improved by limiting the amount of food to 75–80% of *ad libitum* feeding during part of the fattening period. This is mainly performed limiting the period in which feed is supplied daily. Although growth is reduced during restriction, feed restriction reduces the risk of digestive diseases and improves feed efficiency both during the period of feed restriction and especially after it, when the amount of food is gradually increased (reviewed by Gidenne *et al.*, 2012). Therefore, feeding costs and the use of antibiotics are reduced, which result in economic profit and a reduced environmental impact of rabbit-meat production. As a consequence, feed restriction has become a common management technique on commercial farms in Europe.

Despite its importance, the cost and technical difficulties in measuring individual feed intake makes exceptional the direct genetic selection for feed efficiency. When animals are housed in collective cages, electronic devices are required to enable recording of the individual feed intake. If animals are housed in individual cages, feed consumption could be measured manually. However, it would still be expensive and, in addition, there could exist a genotype by type of housing interaction, which could make selection under individual housing not advisable when housing conditions at fattening consist of group cages (Piles *et al.*, 2017). Selection for feed efficiency has commonly been performed by indirect selection for growth traits (i.e. ADG or live weight at a fixed age) of animals fed *ad libitum* and housed in groups (Piles *et al.*, 2004). The methodology used to select growth traits has been, in general, individual selection. It is the simplest procedure of selection because the traits are expressed in both sexes and the heritability is medium. In this way it is possible to save time, labour and resources. The interval between generations could be around six months. However, the correlated response on feed efficiency is small

(Sánchez *et al.*, 2004; Larzul *et al.*, 2005) because the genetic correlation of this trait with the selection criteria could not be high enough (Piles *et al.*, 2004). In France, the rabbit breeder Hypharm in collaboration with INRA has selected two paternal lines directly for feed efficiency using as selection criterion the residual feed intake (RFI; i.e. the amount of food consumed above or below the expected needs) and ADG under restricted feeding, respectively, measured on a limited number of selection candidates kept in individual cages. Results that compare the production performance of young rabbits selected for ADG and RFI and bred under different feeding regimens suggest that there is an interaction effect between genotype and feeding regimen on ADG but not on other traits such as body weight, feed intake or feed efficiency when animals are in individual cages (Drouilhet *et al.*, 2016). However, this interaction effect is relevant when animals are bred in collective cages, which elicits competition for feed intake between cage mates that is particularly important when feed restriction is applied at fattening (Dalmau *et al.*, 2015). Competition effects are a kind of social effect (also known as indirect genetic effects; Bijma, 2011) which have a genetic determinism and contribute largely to the total heritable variance of ADG when growing rabbits are raised on restricted feeding but not when they are fed *ad libitum*. Ignoring those effects in a breeding programme for increasing rabbit growth is likely to have negative consequences on the productive performance of young rabbits and eventually on animal well-being when the amount of food is limited. The interaction between genotype and feeding regimen will lead to a substantial re-ranking of the selection candidates under different conditions because of the null correlation between total heritable variance for animals on different feeding regimens. This is mainly due to the null and negative genetic correlations between direct and social genetic effects on full and restricted feeding regimens, respectively. Therefore, selection of rabbits for ADG under *ad libitum* feeding may completely fail to improve ADG in rabbits on restricted feeding. It is recommended to select animals under the same conditions of feeding and housing as those applied on production farms for rabbit-meat production, by accounting for social effects if the amount of food is limited (Piles *et al.*, 2017). In Spain, since 2018, three paternal lines belonging to IRTA are being directly selected for feed efficiency of growing rabbits housed in groups following different strategies which account for social effects when needed. An electronic feeding station has been developed and produced to this end by IRTA in collaboration with a private company.

In Hungary, the Pannon White breed is being selected for daily gain and carcass traits, using computerized tomography (Szendrő *et al.*, 1996; Garreau *et al.*, 2000; Nagy *et al.*, 2010, 2013a; Gyovai *et al.*, 2012; Ács *et al.*, 2019). The development of this breed was initiated at the Kaposvár University in the late 1980s. During the first stage of its development (1988–1990), the population consisted of roughly 100 New Zealand White does selected for average daily gain and dressing-out percentage by progeny test. During the second stage, in 1991, New Zealand White rabbits were reciprocally crossed with Californian rabbits and New Zealand-type rabbits of various origins. The progeny of the crosses showing the best body-weight gain and dressing-out percentage served as the basis of the Pannon White rabbits, a new synthetic population. The third stage of the programme (since 1992) involves a mass selection for ADG and carcass traits (applying computerized tomography) using a closed herd book. Until 2004, the trait evaluated by computer tomography was the so-called L-value (the average surface of muscle *Longissimus dorsi* between the 2nd and 3rd, and 4th and 5th lumbar vertebrae), which has a favourable genetic correlation with dressing-out percentage (Nagy *et al.*, 2006). Since 2004, this trait has been replaced with thigh muscle volume (determined by 10–11 computer tomography scans) as it determines the weight and the ratio of the hind part of the carcass (Szendrő *et al.*, 2010). Besides the Pannon White breed, the foundation of the Pannon paternal line (Nagy *et al.*, 2013b) started in 2004 using sperm from some Hungarian and foreign rabbits showing high ADG. The sperm of these animals were used to inseminate Pannon White does showing high weight gain, being the resulting progeny of the founders of the paternal line. Since 2006, this line has been kept as a closed population and selected for ADG and thigh muscle volume applying computer tomography. This line was officially recognized as a new synthetic rabbit breed in 2013.

In France, the economic weight of carcass yield becomes more and more important because of the demand of slaughterhouses. Breeding for disease resistance has also become a major challenge. The AGP39 and AGP59 lines of the breeding company Hypharm are currently selected for carcass yield and against digestive disorders incidence, in addition to the weight at slaughter age (Garreau *et al.*, 2008b). Carcass yield is measured on four offspring of the first litter of each dam of the line. Each growing rabbit is inspected for general health and well-being. A binary disease score is defined according to the presence or the absence of digestive disorders. The genetic parameters of these traits were estimated using a Bayesian methodology. Estimated heritabilities were equal to 0.08 and 0.24 for the disease trait and carcass yield, respectively (Garreau *et al.*, 2008b).

A number of attempts were performed in Italy over the 1990s in order to establish a national breeding programme for rabbits. Finally, it started in 1999 in a central herd with three breeds: Californian, New Zealand and Argenté de Champagne. The latter was chosen after different attempts with Burgundy Fawn, Vienna Blue and Giant Chinchilla. The Argenté de Champagne breed has been selected for body weight at 60 ± 2 d of age, using an animal mixed model (BLUP) to evaluate the animals. Besides usual environmental factors, the model also accounts for the number of weaned kits in the litter and for individual inbreeding. The main challenge to overcome in selecting Argenté de Champagne for body weight at 60 d was the scarce adaptation of the breed to intensive breeding: it took over seven years to obtain robust rabbits by keeping those capable of surviving, reproducing and growing in cages. Due to these difficulties, from 1999 through 2012, only 759 females and 160 males could be weighed and evaluated. During this period a sharp reduction in rabbit adult body size has been observed, and selection response for daily gain up to 60 days was negative till 2007; later on it turned positive, with an average response of 11 grams per year up to 2012. Nowadays, the breed, quite different from the original stock, is known as Italian Silver.

13.3.2.2 *Alternative approaches*

In the future, for lines highly selected for growth, traits related to the production of potentially fertile artificial insemination doses (such as semen production and quality traits) or to the male contribution to fertility and prolificacy (such as the rate of implanted embryos) (Piles *et al.*, 2013), could be incorporated to the objectives of selection of the paternal lines. In all cases a close connection between the nucleus and the artificial insemination centres will be required. Detailed information regarding the genetic determinism of such traits is provided in Chapter 12.

13.3.3 Criteria and methods of selection – maternal lines

13.3.3.1 *The current approaches*

The most common criteria for selection of maternal lines have been related to litter size at birth or at weaning (Estany *et al.*, 1989; Rochambeau *et al.*, 1994; Gómez *et al.*, 1996; Nagy *et al.*, 2011). Only in one case the selection criteria included litter size at birth and kit weight at nine weeks of life to prevent negative responses in adult body weight (Bolet and Saleil, 2002).

Selection methods in maternal lines are more complicated than in sire lines. This complexity is due to the fact that males do not express the prolificacy traits themselves, and to the low values of the heritabilities of reproduction traits, which lead to longer generation intervals than in selection of sire lines. Family indexes were proposed to integrate the does' own information and the information from relatives to carry out the genetic evaluation (Matheron and Rouvier, 1977; Baselga *et al.*, 1984). This methodology is still applied for the selection of line A (UPV, Spain) where the family index has a maximum of four items: average litter sizes at weaning of the doe, its dam, full sisters and half- (paternal or maternal) sisters.

Currently, mixed-model methodology (BLUP) is the most used procedure for evaluation. One of the biggest differences with the family index is that some environmental and physiological effects are considered in the model (Estany *et al.*, 1989; Gómez *et al.*, 1996; Rochambeau *et al.*, 1998). Simulation studies with real data have shown similar efficiencies of both methodologies for selecting litter size. The loss of response due to selection on a family index instead of a BLUP is around 8% (Armero *et al.*, 1995) when the generations of selection do not overlap. All these methods of selection increase the inbreeding of the lines, generation after generation, and it is necessary to minimize the losses of genetic variability by managing the mating (Ragab *et al.*, 2015). This aspect is analysed by Kerdiles and Rochambeau (2002) along 20 generations of selection of the INRA maternal lines.

In the Italian breeding scheme, Californian and New Zealand breeds have been selected for both maternal and growth traits. They are intended to produce an F_1 doe to be mated to an Italian Silver buck. A two-trait animal model is used to jointly analyse body weight at 60 ± 2 d (like in the paternal line Italian Silver) and litter weight at 19 ± 2 d. Only the maternal effect (the effect of the littering doe) on litter weight was considered and no correction for litter size was used in order to achieve a positive indirect selection response for that trait. Heritability for body weight was 0.31 and for litter weight 0.22 while the genetic correlation between them was 0.61. From 1999 through 2012, realized genetic trend for New Zealand breed was 43.8 g/y for litter weight and 30.9 g/y for body weight at 60 d, while for Californian breed the trends were 27.4 g/y and 17.4 g/y, respectively, for the two traits. From 1999 through 2012, 22,927 females and 791 New Zealand males, 4210 females and 275 Californian males were weighed and indexed. In order to reduce generation interval, both male and female replacements are produced by primiparous does only. Selection indexes are computed from data on body weight at 60 days of age and weight of litters at 19 days of age produced by first-littering does. Genetic evaluations are run after collecting litter weights on each batch of primiparous does. From each of the best litters (rank 95) according to their pedigree index two young bucks are selected, while from each of the next-best ranking litters (rank 50) up to three young females are kept for replacement in the nucleus. Selection index weights were computed by setting as constraints the desirable correlations between the breeding values for the two traits and the selection index to be built. Expected correlations of the selection index with breeding values for litter weight at 19 days and individual body weight at 60 days were 0.99 and 0.70, respectively.

13.3.3.2 Alternative approaches

Different selection criteria have been proposed, as alternative methods, in order to improve response to selection for litter size or the ability of the doe to nourish the lactating progeny and avoid kit mortality.

Some experiments have explored the genetic improvement of litter size selecting for uterine capacity (Santacreu *et al.*, 2005) or for ovulation rate (Laborda *et al.*, 2012) but their responses are not higher than the ones obtained directly selecting for number of kits born alive or litter size at weaning, responses that will be commented on later in this chapter. However, an experiment that has reached the nine generations of selection with the same objective of improving litter size is achieving promising results selecting jointly for litter size and ovulation rate (Badawy *et al.*, 2013, Ziadi *et al.*, 2013).

Selection criteria for increasing kit survival during lactation include direct and maternal genetic effects of kit weight at weaning (Garreau and Rochambeau, 2003; Garreau *et al.*, 2004), or direct genetic effects of doe litter weight at weaning or total milk production (Garreau *et al.*, 2004; Youssef *et al.*, 2008). Particularly, Garreau *et al.* (2008a) proposed a selection for the within-litter homogeneity of kit birth weight. This involved a new model (canalizing selection) incorporating a genotypic value for the mean and a genotypic value for the environmental variance. This method is currently applied by the French breeding companies.

A divergent selection for functional longevity in breeding does was carried out in an INRA experimental farm, using the survival analysis methodology (Garreau *et al.*, 2008c; Larzul *et al.*, 2014). After only one generation of selection there was a significant difference of longevity between the two lines (39 days), demonstrating the efficiency of selection for this trait. Furthermore, except for the total number born, which was higher in the low line, there was no difference between the two lines for reproduction traits. This selection criterion is now included in the breeding objective of the AGP77 line of the French breeding company Hypharm.

A divergent selection experiment of environmental variance of litter size at birth over 13 generations is being carried out in Miguel Hérnandez University (Spain). Successful response to selection of this trait as well as differences in resilience between the divergent lines has been obtained (Blasco *et al.*, 2017; Argente *et al.*, 2019). A low variability in litter size in the homogenous line is related to better adaptation to environment with less response to stress and diseases (Beloumi *et al.*, 2020). Moreover, a reduction in litter-size variability shows a higher litter size at birth, as a consequence of higher embryonic development at early gestation and number of implanted embryos in the homogenous line (García *et al.*, 2016; Argente *et al.*, 2017).

In an attempt to improve the maternal ability of does, in the Italian scheme, great importance was given to the number of functional teats. These are counted on does a few days after their first littering and does are selected by independent culling levels. Dams of males should have at least ten functional teats and dams of females at least nine. In 2004, New Zealand females with 8, 9, 10, 11 and 12 functional teats, respectively, were 65.6, 19.7, 14.7, 0.0 and 0.0%, while in 2012 the same distribution was 20.9, 33.1, 45.7, 0.2 and 0.1%. Studies on the identification of genes associated with teat number in Italian breeds by using genome-wide association studies have been started in 2019 by the University of Bologna in collaboration with the Italian Rabbit Breeders Association (ANCI). The final aim is to identify markers that could be used in marker-assisted selection for this trait (Bovo *et al.*, 2021).

13.3.4 Responses to selection

13.3.4.1 Within-line responses

The standard way of estimating the response in selection experiments has been the use of a control population that is developed contemporarily to the selected population by random mating and no selection. Another common method is the analysis of the differences between two lines selected contemporarily, one to increase a trait and another to decrease it, which is called divergent selection (Santacreu *et al.*, 2000). The use of frozen-thawed embryos allows the contemporary comparison of two different generations of the same line (Santacreu *et al.*, 2000). All these methods are neither model-dependent nor dependent on the genetic parameters of the traits.

On the other hand, statistical methods, such as mixed-model methodology (BLUP and REML) and Bayesian Inference, are frequently used to estimate genetic trends or responses to selection that are dependent on the models, the parameters and the information used *a priori*. These statistical methods have been applied in different selection experiments in rabbits. In many cases there was a good agreement between the responses estimated by both types of approaches, but not always. The implication is that the statistical methods and models used to estimate response to selection require a previous validation in the populations and traits under analysis before they are implemented.

The responses reported in experiments of selection for weight at market time, 63–70 d, range between 18 g and 35 g per generation (Rochambeau *et al.*, 1994; Lukefahr *et al.*, 1996; Garreau *et al.*, 2000; Larzul *et al.*, 2003a) and there is a good agreement between the estimates of the responses obtained using control lines or mixed-model methods (Lukefahr *et al.*, 1996; Larzul *et al.*, 2003a). When selection was for growth rate, the responses were between 0.45 and 1.50 g/d (Rochambeau *et al.*, 1989; Estany *et al.*, 1992; Piles and Blasco, 2003; Gyovai *et al.*, 2008; Drouilhet *et al.*, 2013; Nagy *et al.*, 2013b). There are correlated responses to selection for growth as an increase in the adult weight (Blasco *et al.*, 2003). At a fixed slaughter weight, the feed conversion decreases and feed consumption increases (Feki *et al.*, 1996); intestinal content increases, and the dressing percentage is reduced (Gómez *et al.*, 1998; Pla *et al.*, 1998) because of the lower maturity. Consequences of this lower maturity are also reduced fat deposits, diminished water-holding capacity of the meat (Piles *et al.*, 2000) and lower ultimate pH in muscle (Gondret *et al.*, 2003). Some of the negative consequences of selection for growth rate are not quantitatively important and can be reduced by increasing the market weight and imposing a light fasting before slaughter. In an analysis at a constant age, the improvement in the conversion index can disappear as well as the negative effects on dressing percentage and maturity (Garreau *et al.*, 2000; Larzul *et al.*, 2003b; Hernández *et al.*, 2004). The efficiency of the computer tomography-aided selection is observed in the Pannon White and Pannon terminal rabbit breeds where the estimated annual genetic trend for the thigh muscle volume was 4 cm^3 and 5.8 cm^3, respectively (Gyovai *et al.* 2008; Nagy *et al.*, 2013b).

The analysis of the responses to selection in maternal lines by comparison to a control population (Rochambeau *et al.*, 1998; Tudela *et al.*,

2003) or by the use of frozen embryos (García and Baselga, 2002a, b), estimate responses between 0.08 and 0.09 total kits born, born alive or weaned per litter and generation. In the same lines, the responses estimated, as genetic trends, by mixed-model methods (BLUP and REML) completely agree in the first three experiments (Rochambeau *et al.*, 1998; García and Baselga, 2002a; Tudela *et al.*, 2003), but the genetic trend estimated in the fourth is 0.175 weaned rabbits per litter and generation, which is approximately double the response estimated using frozen embryos. There are also reports of responses estimated exclusively by mixed-model methods, the estimates ranging between 0.05 and 0.13 live-born or weaned rabbits per litter and generation (Estany *et al.*, 1989; Rochambeau *et al.*, 1994; Gómez *et al.*, 1996). In some of these lines with significant responses in litter size, studies have been carried out to assess which component of litter size had been modified. García and Baselga (2002a) found that for one of the lines, the improvement in litter size was probably due to an increase in ovulation rate but for another line an increase in foetal survival was probably the trait actually improved that could explain the observed response in litter size (García and Baselga, 2002b). The correlated responses in growth traits when selection is for litter size have also been investigated. García and Baselga (2002c) did not find significant responses for weight at weaning, weight at market time, post-weaning daily gain, daily feed intake and conversion index, when the comparisons were done at a constant litter size at birth. However, Rochambeau *et al.* (1994) reported that selection for increased litter size resulted in a decrease of individual weight at weaning but the total weight of the litter at weaning increased in two different maternal lines. This consideration has meant the modification of the selection objective in the INRA 1077 maternal line, including in the objective the weight at 63 days, in addition to litter size performances, attempting to increase litter size and individual weight at the same time (Rochambeau, 1998). In 2003, the new INRA 1777 maternal line was created with the aim to improve directly weaning weight in addition to prolificacy. This line is selected for direct and maternal genetic effects for weaning weight and for number of kits born alive per litter (Garreau and Rochambeau, 2003; Garreau *et al.*, 2005). The same selection objective is applied to the AGP77 line of the breeding company Hypharm.

13.3.4.2 Crossbred responses

It has been mentioned before that the final aim of the genetic improvement of the lines is the improvement of the performance of the crossbred doe and crossbred kit. However, no matter which lines are used in crossbreeding, their selection, as it has been presented, is on a within-line basis expecting that the response will also be expressed in the crosses. In this sense, it is crucial to evaluate the response of the selection programmes on the crossbred doe and kits.

Contemporary comparisons among animals (pure or crossbred), representing different stages of the programmes, have been made by Tudela *et al.* (2003), Costa *et al.* (2004) and Quevedo *et al.* (2005). The maternal lines involved in the first experiment were the INRA A1077 at the 30th generation of selection and the control INRA A9077 both crossed to a second French maternal line at its current generation. The difference in total litter size between both types of crossbred does was 1.43 kits, a little higher than the expected values from selection of the line INRA A1077 (1.12). In the second experiment the maternal lines were A (at two different generations of selection) and V. In this experiment the crossbred does came from mating does of the V line to bucks of the A line. The young rabbits were the progeny of the cross between the crossbred does and bucks of the paternal line R. Two types of crossbred animals, hereafter called H1 and H2, were compared. The H1 (H2) does came from the cross of bucks of generation 16 (29) of line A with females of generation 26 of line V. The young rabbits H1 (H2) were obtained by mating does H1 (H2) to bucks of generation 6 (18) of line R. Animals from previous generations were conserved as frozen embryos and thawed and transferred to produce adults contemporary to the current generations. Litter size and daily body gain were all favourable to current crossbreeds. Concerning litter-size traits, the differences were 0.83, 1.16 and 0.74, respectively, for number of total born, born alive and weaned kits. These responses were higher than expected from the responses evaluated in the pure lines. For example, the expected response in litter size at weaning was 0.55. In contrast, the responses in the crossbred young kits were lower than expected. Thus, the response for post-weaning daily gain was 0.6 g/d although the expected response was 2.4 g/d. No responses were obtained for daily feed intake and conversion index, but the consumption was expected to increase and the conversion index decrease. The authors had no explanation for these results, and suggest that they could be a consequence of the type of feed currently used to control a new disease, spread all over Europe, called epizootic enterocolitis, which can have different effects on the growth of the animals depending on the genetic type source. The results of Costa *et al.* (2004) were confirmed in a third experiment (Quevedo *et al.*, 2005, 2006a, 2006b) using the same lines. The difference between both experiments was that different generations of the V line were used (generation 15th for the dams of H1 does, and generation 26th for the dams of H2 does). Thus, the H1 does represented old crossbred because their sires and dams belonged to old generations and H2 does represented current crossbred does. Analysing the data of primiparous does, Quevedo *et al.* (2005) showed that the H2 does had a number of born alive significantly higher (2.06) than the H1 does and higher than expected from the response within line (1.06). Also a higher foetal growth in the H2 does was observed, consuming the same amount of energy that was interpreted as an increase in the efficiency of the use of energy as a consequence of selection. The results for the multiparous does confirmed the superiority of the does H2 in number born alive (1.1) and showed that these does consumed more feed (3%) and produced more milk (6%) during the first 21 days of lactation than the H1 does (Quevedo *et al.*, 2006a). In this experiment, as occurred in the second, the superiority in post-weaning daily gain of the current young rabbits with respect to the ones from previous generations was only one third of the expected value, apparently, because of the same circumstances.

Ragab (2012) has reported results regarding reproduction traits of the 16 genetic types of does coming from a complete diallel cross between the lines A (generation 41), V (generation 38), H (generation 20) and LP (generation 8). The raw means obtained for 34,546 litters in four farms were: 10.54 total born/litter, 9.79 born alive/litter, 7.97 weaned kits/litter and 49.9 d of kindling interval. The crossbred does were compared to those from line V and showed better performances than this line: 0.46 total born/litter, 0.56 born alive/litter, 0.75 weaned/litter and −2.2 d of kindling interval.

No comparison between pure and crossbred lines is available for the Italian scheme. However, some interesting results for the three-way cross have recently been provided: from 37 d to 86 d of age the rabbits grew 38.9 g/d or 39.8 g/d with a density of 48 kg/m^2 or 36 kg/m^2, respectively.

13.4 Dissemination of the Genetic Improvement

The final step of a programme is to disseminate the achieved improvement from the nucleus to the farms. The standard way is to follow a pyramidal organization, including the multiplication of the nucleus stock, as an intermediate step, before the commercial farms. At the multiplication stage, the cross between maternal lines is finally done and the crossbred does are supplied to the farmers, but this step can be preceded by a multiplication of the pure lines. In this way, the costs of selection are spread to a higher number of animals. The genetic lag between the nucleus and the farms increases as the stages of multiplication increase. In order to minimize the lag, the French and Spanish programmes have modified the standard of multiplying the stock converting the intermediate stage of multiplication of the pure lines in a true stage of selection. The French programme does it through what is called 'demultiplication'. On the demultiplication farms, the replacement does come from the progeny of the does evaluated as in the nucleus, but the replacement bucks are supplied regularly to the farms directly from the nucleus. The companies that demultiply can be big companies aiming to sell pure or crossbred stock throughout France and other countries (Rochambeau, 1998). In France, three private companies are involved in the meat-rabbit genetic improvement. They have created and selected their own male and female lines. The Spanish approach is slightly different, and completely integrates the multiplication of the maternal lines and their selection. The chosen way is to create what could be named secondary nuclei of selection, owned by farmers, co-operatives or small companies. One maternal line is replicated in each secondary nucleus and it is selected, under the responsibility of the geneticists of the primary nucleus, in the same way as in the latter. The line selected in the secondary nucleus is used as the dam to produce crossbred does that are commonly distributed to, or produced by, the farmers of the area near the secondary nucleus. The sires required to get the crossbred doe pertain to another maternal line and are supplied by the primary nucleus. The main feature of this approach is its simplicity, allowing the production of pure line stock and crossbred does at very low prices with a minimum lag between breeders and producers. This approach reduces the health and adaptation problems originating from the introduction of foreign animals into the farms and improves the communication between breeding companies and farmers. The main disadvantage is the difficulty of operating on a large scale. The dissemination of the paternal-line bucks for the terminal cross can have different modalities. The paternal lines can be owned and selected by the demultipliers, by the secondary nucleus, by the primary nucleus or by breeders not related to the programme, associated or not to artificial insemination centres. It is very common in the Spanish programme that the secondary nuclei have their own artificial insemination centre, facilitating in this way the complete dissemination of the genetic gain achieved in all lines involved in the double cross. In France, breeding companies are also involved in artificial insemination centres.

In Hungary, the majority of the production is accomplished through two companies from which the first one applies the French system using one of the French crossbred types. There are one-one stations in France and in Hungary, being animals from both stations connected by artificial insemination and jointly genetically evaluated to ensure a homogeneous genetic progress. The other company follows a conventional pyramidal organization in which the nucleus is located at the Kaposvár University, while the company owns both the multiplier and the commercial rabbit farms.

In the Italian scheme, the unique nucleus herd produces pure breeding stock for contracted farms where the cross between maternal lines is performed, while Italian Silver bucks are sold directly to commercial farms.

13.5 Perspectives of the Genomic Selection

Genomic selection (Meuwissen *et al.*, 2001) has not been implemented in rabbits. In general, genomic selection could lead to an increased annual response to selection, compared with the traditional schemes, by means of a reduction of the generation interval, an increase in the selection intensity and a higher accuracy of the predicted breeding values. However, current rabbit breeding programmes are characterized by high selection intensities and short generation intervals, reducing the potential impact of genomic selection; therefore, major gain should come from a gain in the accuracy of prediction of breeding values. The gain in accuracy also depends on: (i) the linkage disequilibrium between the markers and the causative genes, which in turn depends on the marker density and the effective population size; (ii) the genetic determinism of the traits; and (iii) the size of the reference population (phenotyped and genotyped) and its proximity to the population to be evaluated as candidates to selection. Given that the rabbit populations under selection are very small, the size of the reference population could probably not achieve the minimum size to obtain accurate predictions. In addition, this reference population should be frequently updated because the short generation interval in this species will make to break down very quickly the linkage disequilibrium pattern reducing the efficiency of the prediction equation obtained at a given time.

No summaries have been published thus far on simulation studies about the potential efficiency of this kind of selection in rabbits but they exist in pigs (reviewed by Samoré and Fontanesi, 2016). Some of them compare several schemes of genomic selection to conventional schemes of selection of maternal breeds, when the objective of selection includes only maternal traits (Lillehammer *et al.*, 2011) and when production and maternal traits are considered (Lillehammer *et al.*, 2013). In these studies, the reference population was continuously updated with the individuals selected per round of selection that are genotyped and phenotyped, and no decline of the accuracy was observed. The cost-benefit analysis was also made and the number of slaughter pigs necessary to pay back, with the extra gains achieved by the genomic selection, the extra cost of this selection was computed. This number varied between 110,000 and 169,000 slaughter pigs per round of selection. Taking into account that the cost of genotyping might be similar for one pig and for one rabbit, but the benefits that could be obtained from one slaughter pig will be much higher than from one rabbit, it seems very unlikely that the genomic selection in rabbits could be currently affordable. Further development of genomic tools should also be needed to obtain cheaper SNP genotyping arrays. However, future changes in the conditions of production of the rabbits and in the cost of the SNP chips could make feasible the genomic selection, but research is needed in order to evaluate the practical and logistical applications, benefits and strategies of selection before any implementation is made.

References

Ács, V., Nagy, I. and Donkó, T. (2019) A selection index for improving the carcass traits in the Pannon Large rabbit breed. *Acta Universitatis Agriculturae et Silviculturae Mendelianae Brunensis* 67, 1125–1129.

Argente, M.J., Calle, E.W., García, M.L. and Blasco, A. (2017) Correlated response in litter size components in rabbits selected for litter size variability. *Journal of Animal Breeding and Genetic* 34, 505–511. DOI: 10.1111/jbg.12283

Argente, M.J., García, M.L., Zbyňovská, K., Petruška, P., Capcarová, M. and Blasco, A. (2019) Correlated response to selection for litter size environmental variability in rabbits' resilience. *Animal* 13, 2348–2355.

Armero, E., Baselga, M. and Cifre, J. (1995) Selecting litter size in rabbits: analysis of different strategies. *World Rabbit Science* 3, 179–186.

Badawy, A.Y., Peiró, R., Mocé, M.L., Blasco, A. and Santacreu, M.A. (2013) Selección por tasa de ovulación y tamaño de camada en conejo. Estimación de la respuesta a la selección. In: *Proceedings of XV Jornadas de Producción Animal,* AIDA, Zaragoza, Spain, pp. 457–459.

Baselga, M. (2002) Line V (Spain). In: Khalil, M.H. and Baselga, M. (eds) *Rabbit Genetic Resources in Mediterranean Countries*. CIHEAM, Zaragoza, Spain, pp. 231–241.

Baselga, M. (2004) Genetic improvement of meat rabbits: programmes and diffusion. In: *Proceedings of 8th World Rabbit Congress,* Puebla, Mexico, pp. 1–13.

Baselga, M. and Blasco, A. (1989) *Mejora Genética del Conejo de producción de carne*. Mundi-Prensa.

Baselga, M., Blasco, A. and Estany, J. (1984) Indice de selección de caracteres reproductivos con información variable. In: *Proceedings of 3rd World Rabbit Congress, Rome, pp.* 1, 62–65.

Beloumi, D., Blasco, A., Muelas, R., Santacreu, M.A., García, M.L. and Argente, M.J. (2020) Inflammatory correlated response in two lines of rabbit selected divergently for litter size environmental variability. *Animals* 10, 1540.

Bijma, P. (2011) A general definition of the heritable variation that determines the potential of a population to respond to selection. *Genetics* 189, 1347–1359.

Blasco, A., Piles, M. and Varona, L. (2003) A Bayesian analysis of the effect of selection for growth rate on growth curves in rabbits. *Genetics Selection Evolution* 35, 21–41.

Blasco, A., Martínez-Álvaro, M., García, M.L., Ibáñez-Escriche, N. and Argente, M.J. (2017) Selection for environmental variance of litter size in rabbit. *Genetics Selection Evolution* 49, 48.

Bolet, G. and Saleil, G. (2002) Strain INRA1077 (France). In: Khalil, M.H. and Baselga, M. (eds) *Rabbit Genetic Resources in Mediterranean Countries*. CIHEAM, Zaragoza, Spain, pp. 109–116

Bovo, S., Schiavo, G., Utzeri, V.J., Ribani, A., Schiavitto, M., Buttazzoni, L., Negrini, R. and Fontanesi, L. (2021) A genome-wide association study for the number of teats in European rabbits (*Oryctolagus cuniculus*) identifies several candidate genes affecting this trait. *Animal Genetics* 52, 237–243.

Cartuche, L., Pascual, M., Gómez, E.A. and Blasco, A. (2014) Economic weights in rabbit meat production. *World Rabbit Science* 22, 165–177.

Cifre, J., Baselga, M., Garcia-Jimenez, F. and Vicente, J.S. (1998a) Performance of a hyperprolofic rabbit line. I. Litter size traits. *Journal of Animal Breeding and Genetics* 115, 131–138.

Cifre, J., Baselga, M., Garcia-Jimenez, F. and Vicente, J.S. (1998b) Performance of a hyperprolofic rabbit line. II. Maternal and growth performances. *Journal of Animal Breeding and Genetics* 115, 139–147.

Costa, C., Baselga, M., Lobera, J., Cervera, C. and Pascual J.J. (2004) Evaluating response to selection and nutritional needs in a three-way cross of rabbits. *Journal of Animal Breeding and Genetics* 121, 186–196.

Dalmau, A., Abdel-Khalek, A.M., Ramon, J., Piles, M., Sanchez, J.P. *et al.* (2015) Comparison of behaviour, performance and mortality in restricted and ad libitum-fed growing rabbits. *Animal* 9, 1172–1180.

Drouilhet, L., Gilbert, H., Balmisse, E., Ruesche, J., Tircazes, A. *et al.* (2013) Genetic parameters for two selection criteria for feed efficiency in rabbits. *Journal of Animal Science* 91, 3121–3128.

Drouilhet, L., Achard, C.S., Zemb, O., Molette, C., Gidenne, T. *et al.* (2016) Direct and correlated responses to selection in two lines of rabbits selected for feed efficiency under ad libitum and restricted feeding. I. Production traits and gut microbiota characteristics. *Journal of Animal Science* 94, 38–48.

Eady, S.J. (2003) *Farmed Rabbits in Australia*. RIRDC Publication No. 02/144. Rural Industries Research and Development Corporation, Canberra.

Estany, J., Baselga, M., Blasco, A. and Camacho, J. (1989) Mixed model methodology for the estimation of genetic response to selection in litter size of rabbits. *Livestock Production Science* 21, 67–76.

Estany, J., Camacho, J., Baselga, M. and Blasco, A. (1992) Selection response of growth rate in rabbits for meat production. *Genetics Selection Evolution* 24, 527–537.

Feki, S., Baselga, M., Blas, E., Cervera, C. and Gómez, E.A. (1996) Comparison of growth and feed efficiency among rabbit lines selected for different objectives. *Livestock Production Science* 45, 87–92.

Ferrian, S., Blas, E., Larsen, T., Sánchez, J.P., Friggens, N.C. *et al.* (2013) Comparison of immune response to lipopolysaccharide of rabbit does selected for litter size at weaning or founded for reproductive longevity. *Research in Veterinary Science* 94, 518–525.

Gacem, M., Zerrouki, N., Lebas, F. and Bolet, G. (2008) Strategy for developing rabbit meat production in Algeria: creation and selection of a synthetic strain. In: Xiccato, G., Trocino, A. and Lukefahr, S. (eds) *Proceedings of 9th World Rabbit Congress*, Verona, Italy, pp. 85–89.

García, M.L. and Baselga, M. (2002a) Genetic response to selection for reproductive performance in a maternal line of rabbit. *World Rabbit Science* 10, 71–76.

García, M.L. and Baselga, M. (2002b) Estimation of genetic response in litter size of rabbits using a cryopreserved control population. *Livestock Production Science* 74, 45–53.

García, M.L. and Baselga, M. (2002c) Estimation of correlated response on growth traits to selection in litter size of rabbits using a cryopreserved control population and genetic trends. *Livestock Production Science* 78, 91–98.

García, M.L., Blasco, A. and Argente, M.J. (2016) Embryologic changes in rabbit lines selected for litter size variability. *Theriogenology* 86, 1247–1250.

García-Ximénez, F., Vicente, J.S., Cifre, P. and Baselga, M. (1996) Foundation of a maternal rabbit line using hysterectomy and embryo cryopreservation. In: *Proceedings of 6th World Rabbit Congress*, Toulouse, France, pp. 285–288.

Garreau, H. and Rochambeau, H. de (2003) La sélection des qualités maternelles pour la croissance du lapereau. In: *Proceedings of 10ème Journées de la recherche Cunicole*, Paris, pp. 61–64.

Garreau, H., Szendrő, Z.S., Larzul, C. and Rochambeau, H. de (2000) Genetic parameters and genetic trends of growth and litter size traits in the White Pannon breed. In: *Proceedings of 7th World Rabbit Congress*, Valencia, Spain, pp. 403–408.

Garreau, H., Piles, M., Larzul, C., Baselga, M. and Rochambeau, H. de (2004) Selection of maternal lines: last results and prospects In: *Procedings of 8th World Rabbit Congress*, Puebla, Mexico, pp. 14–25.

Garreau, H., Duzert, R., Tudela, F., Baillot C., Ruesche J. *et al.* (2005) Gestion et sélection de la souche INRA 1777: résultats de trois générations de sélection. In: *Proceedings of 11ème Journées de la recherche Cunicole*, Paris, pp. 19–22.

Garreau, H., Bolet, G., Larzul, C., Robert-Granié, C., Saleil, G. *et al.* (2008a) Results of four generations of a canalising selection for rabbit birth weight. *Livestock Science* 119(1–3), 55–62.

Garreau, H., Eady, S.J., Hurtaud, J. and Legarra, A. (2008b) Genetic parameters of production traits and resistance to digestive disorders in a commercial rabbit population. In: Xiccato, G., Trocino, A. and Lukefahr, S. (eds) *Proceedings of 9th World Rabbit Congress*, Verona, Italy, pp. 103–107.

Garreau, H., Ducrocq, V., Tudela, F., Saleil, G., Juin, H. and Larzul, C. (2008c) Divergent selection for longevity in breeding does. In: Xiccato, G., Trocino, A. and Lukefahr, S. (eds) *Proceedings of 9th World Rabbit Congress*, Verona, Italy, pp. 97–101.

Gidenne, T., Combes, S. and Fortun-Lamothe, L. (2012) Feed intake limitation strategies for the growing rabbit: effect on feeding behaviour, welfare, performance, digestive physiology and health: a review. *Animal* 6, 1407–1419.
Gómez, E., Rafel, O., Ramon, J. and Baselga, M. (1996) A genetic study of a line selected on litter size at weaning. In: *Proceedings of 6th World Rabbit Congress*, Toulouse, France, pp. 289–292.
Gómez, E.A., Baselga, M., Rafel, O. and Ramon, J. (1998) Comparison of carcass characteristics in five strains of meat rabbit selected on different traits. *Livestock Production Science* 55, 53–64.
Gómez, E.A., Rafel, O. and Ramon, J. (2000) Preliminary genetic analysis of Caldes line: a selection experiment for a global objective. In: *Proceedings of 7th World Rabbit Congress*, Valencia, Spain, pp. 417–424.
Gómez, E.A., Rafel, O. and Ramon, J. (2002) The Caldes strain. In: Khalil, M.H. and Baselga, M. (eds). *Rabbit Genetic Resources in Mediterranean Countries*, CIHEAM. Zaragoza, Spain, pp. 187–198.
Gondret, F., Combes, S. and Larzul, C. (2003) Sélection divergente sur le poids a 63 jours: conséquences sur les caractéristiques musculaires à même âge ou à même poids. In: *Proceedings of 10ème Journées de la recherche Cunicole*, Paris, pp. 153–156.
Gyovai, P., Nagy, I., Gerencsér, Zs., Metzger, Sz., Radnai, I. *et al.* (2008) Genetic parameters and trends of the thigh muscle volume in Pannon White rabbits. In: Xiccato, G., Trocino, A. and Lukefahr, S. (eds) *Proceedings of the 9th World Rabbit Congress*, Verona, Italy, pp. 115–119
Gyovai, P., Nagy, I., Gerencsér, Z.S., Matics, Z.S., Radnai, I. *et al.* (2012) Genetic parameters for litter weight, average daily gain and thigh muscle volume measured by in vivo Computer Tomography technique in Pannon White rabbits. *Livestock Science* 144, 119–123.
Hernández, P., Aliaga, S., Pla, M. and Blasco, A. (2004) The effect of selection for growth rate and slaughter age on carcass composition and meat quality traits in rabbits. *Journal of Animal Science* 82, 3138–3143.
Kerdiles, V. and Rochambeau, H. de (2002) A genetic description of two strains of rabbits. *Journal of Animal Breeding and Genetics* 119, 25–33.
Khalil, M.H. and Baselga, M. (eds) (2002) *Rabbit Genetic Resources in Mediterranean Countries*. CIHEAM. Zaragoza, Spain.
Laborda, P., Mocé, M.L., Santacreu, M.A. and Blasco, A. (2012) Selection for ovulation rate in rabbits. I. Genetic parameters, direct response and correlated response on litter size. *Journal of Animal Science* 90, 439–446.
Larzul, C., Gondret, F., Combes, S., Garreau, H. and Rochambeau, H. de (2003a) Analyse d´une expérience de sélection sur le poids à 63 jours. I – Déterminisme génétique de la croissance. In: *Proceedings of 10ème Journées de la recherche Cunicole*, Paris, pp. 149–152.
Larzul, C., Gondret, F., Combes, S., Garreau, H. and Rochambeau H. de (2003b) Analyse d´une expérience de sélection sur le poids à 63 jours. II – Déterminisme génétique de la composition corporelle. In: *Proceedings of 10ème Journées de la recherche Cunicole*, Paris, pp. 145–148.
Larzul, C., Gondret F., Combes, S. and Rochambeau, H. de (2005). Divergent selection on 63-day body weight in the rabbit: response on growth, carcass and muscle traits. *Genetics Selection Evolution* 37, 105–122.
Larzul, C., Ducrocq, V., Tudela, F., Juin, H. and Garreau, H. (2014) The length of productive life can be modified through selection: an experimental demonstration in the rabbit. *Journal of Animal Science* 92, 2395–2401.
Lillehammer, M., Meuwissen, T.H.E. and Sonesson, A.K. (2011) Genomic selection for maternal traits in pigs. *Journal of Animal Science* 89, 3908–3916.
Lillehammer, M., Meuwissen, T.H.E. and Sonesson, A.K. (2013) Genomic selection for two traits in a maternal pig breeding scheme. *Journal of Animal Science* 91, 3079–3087.
Lukefahr, S.D., Odi, H.B. and Atakora, J.K.A. (1996) Mass selection for 70 body weight in rabbits. *Journal of Animal Science* 74, 1481–1489.
Matheron, G. and Rouvier, R. (1977) Optimisation du progrés génétique sur la prolificité chez le lapin. *Annales de Génétique et Sélection Animal* 9(3), 393–405.
Meuwissen, T.H.E., Hayes, B.J. and Goddard, M.E. (2001) Prediction of total genetic value using genome-wide dense marker maps. *Genetics* 157, 1819–1829.
Moura, A.S.A.M.T., Kaps, M., Vogt, D.W. and Lamberson, M. (1997) Two-way selection for daily gain and feed conversion in a composite rabbit population. *Journal of Animal Science* 75, 2344–2349.
Moura, A.S.A.M.T., Costa, A.R.C. and Polastre, R. (2001) Variance components and response to selection for reproductive, litter and growth traits through a multi-purpose index. *World Rabbit Science* 9, 77–86.

Nagy, I., Ibáñez, N., Romvári, R., Mekkawy, W., Metzger, S. *et al.* (2006) Genetic parameters of growth and in vivo computerized tomography based carcass traits in Pannon White Rabbits. *Livestock Science* 104, 46–52.

Nagy, I., Curik, I., Radnai, I., Cervantes, I., Gyovai, P. *et al.* (2010) Genetic diversity and population structure of the synthetic Pannon White rabbit revealed by pedigree analyses. *Journal of Animal Science* 88, 1267–1275.

Nagy, I., Radnai, I., Nagyné-Kiszlinger, H., Farkas, J. and Szendrő, Zs. (2011) Genetic parameters and genetic trends of reproduction traits in synthetic Pannon rabbits using repeatability and multi-trait animal models. *Archiv Tierzucht* 54, 297–307.

Nagy, I., Gorjanc, G., Curik, I., Farkas, J., Kiszlinger, H. and Szendrő, Z. (2013a) The contribution of dominance and inbreeding depression in estimating variance components for litter size in Pannon White rabbits. *Journal of Animal Breeding and Genetics* 130, 303–311.

Nagy, I., Gyovai, P., Radnai, I., Nagyné-Kiszlinger, H., Farkas, J. and Szendrő, Z. (2013b) Genetic parameters, genetic trends and inbreeding depression of growth and carcass traits in Pannon terminal line rabbits. *Archiv Tierzucht* 56,191–199.

Piles, M. and Blasco, A. (2003) Response to selection for growth rate in rabbits estimated by using a control cryopreserved population. *World Rabbit Science* 11, 53–62.

Piles, M., Blasco, A. and Pla, M. (2000) The effect of selection for growth rate on carcass composition and meat characteristics of rabbits. *Meat Science* 54, 347–355.

Piles, M., Gómez, E.A., Rafel, O., Ramon, J. and Blasco, A. (2004) Elliptical selection experiment for the estimation of genetic parameters of the growth rate and feed conversion ratio in rabbits. *Journal of Animal Science* 82, 654–660.

Piles, M., Moce, M.L., Laborda, P. and Santacreu, M.A. (2013) Feasibility of selection for male contribution to embryo survival as a way of improving male reproductive performance and semen quality in rabbits. *Journal of Animal Science* 91, 4654–4658.

Piles, M., David, I., Ramon, J., Canario, L., Rafel, O. *et al.* (2017) Interaction of direct and social genetic effects with feeding regime in growing rabbits. *Genetics Selection Evolution* 49, 58.

Pla, M., Guerrero, L., Guardia, D., Oliver, M.A. and Blasco, A. (1998) Carcass characteristics and meat quality of rabbit lines selected for different objectives. I. Between lines comparison. *Livestock Production Science* 54, 115–123.

Quevedo, F., Cervera, C., Blas, E., Baselga, M., Costa, C. and Pascual J.J. (2005) Effect of selection for litter size and feeding programme on the performance of young rabbit females during rearing and first pregnancy. *Animal Science* 80, 161–168.

Quevedo, F., Cervera, C., Blas, E., Baselga, M. and Pascual, J.J. (2006a) Long-term effect of selection for litter size and feeding programme on the performance of reproductive rabbit does. 1. Pregnancy of multiparous does. *Animal Science* 82, 739–750.

Quevedo, F., Cervera, C., Blas, E., Baselga, M. and Pascual, J.J. (2006b) Long-term effect of selection for litter size and feeding programme on the performance of reproductive rabbit does. 2. Lactation and growing period. *Animal Science* 82, 751–762.

Ragab, M. (2012) Genetic analyses of reproductive traits in maternal lines of rabbits and in their diallel cross. PhD thesis, Polytechnic University of Valencia, Spain.

Ragab, M. and Baselga, M. (2011) A comparison of reproductive traits of four maternal lines of rabbits selected for litter size at weaning and founded on different criteria. *Livestock Science* 136, 201–206.

Ragab, M., Sánchez, J.P. and Baselga, M. (2015) Effective population size and inbreeding depression on litter size in rabbits: a case study. *Journal of Animal Breeding and Genetics* 132, 68–73.

Rochambeau, H. de (1998) La femelle parentale issue des souches expérimentales de l'INRA: évolutions génétiques et perspectives. In: *Proceedings of 7èmes Journées de la recherche Cunicole*, Lyon, France, pp. 3–14.

Rochambeau, H. de, Fuente, L.F. de la and Rouvier, R. (1989) Sélection sur la vitesse de croissance post-sevrage chez le lapin. *Genetics Selection Evolution* 21, 527–546.

Rochambeau, H. de, Bolet, G. and Tudela, F. (1994) Long term selection: comparison of two rabbit strains. In: *Proceedings of 5th World Congress on Genetics Applied to Livestock Production*, Guelph, Canada, pp. 257–260.

Rochambeau, H. de, Duzert, R. and Tudela, F. (1998) Long term selection experiments in rabbit: estimation of genetic progress on litter size at weaning. In: *Proceedings of 6th World Congress on Genetics Applied to Livestock Production*, Armidale, Australia, pp. 112–115.

Rouvier, R. (1981) Les travaux de recherche français sur la sélection du lapin au cours des 10 dernières années (1970–1980). *Comptes Rendus de l'Académie d'Agriculture de France* 61, 151–159.

Salaün, J.M., Garreau, H., Thebault, R.G., Allain, D. and Rochambeau, H. de (2001) Sélection pour le poids total de lapereaux par femelle et par an dans une souche de lapins Castor Rex. In: *Proceedings of 9èmes Journées de la recherche Cunicole*, Paris, pp. 221–224.

Samoré, A.B. and Fontanesi, L. (2016) Genomic selection in pigs: state of the art and perspectives. *Italian Journal of Animal Science* 15, 211–232.

Sánchez, J.P., Baselga, M., Silvestre, M.A. and Sahuquillo, J. (2004) Direct and correlated responses to selection for daily gain in rabbits. In: *Proceedings of 8th World Rabbit Congress*, Puebla, Mexico, pp. 169–174.

Sánchez, J.P., Theilgaard, P., Mínguez, C. and Baselga, M. (2008) Constitution and evaluation of a long-lived productive rabbit line. *Journal of Animal Science* 86, 515–525.

Santacreu. M.A., Argente, M.J., Mocé, M.L. and Blasco, A. (2000) Selection for uterine capacity. II. Response to selection estimated with a cryopreserved control population. In: *Proceedings of 7th World Rabbit Congress*, Valencia, Spain, pp. 491–496.

Santacreu, M.A., Mocé, M.L., Climent, A. and Blasco, A. (2005) Divergent selection for uterine capacity in rabbits. II. Correlated response on litter size and its components estimated with a cryopreserved control population. *Journal of Animal Science* 83, 2303–2307.

Savietto, D., Blas Ferrer, E., Cervera Fras, M., Baselga Izquierdo, M., Friggens, N.C. *et al.* (2012) Digestive efficiency in rabbit does according to environment and genetic type. *World Rabbit Science* 20, 131–140.

Savietto, D., Cervera, C., Blas, E., Baselga, M., Larsen, T. *et al.* (2013) Environmental sensitivity differs between rabbit lines selected for reproductive intensity and longevity. *Animal* 7, 1969–1977.

Szendrő, Zs., Romváry, R., Horn, P., Biró-Nemeth, E. and Milisits, G. (1996) Two-way selection for carcass traits by computerised tomography. In: *Proceedings of the 6th World Rabbit Congress*, Toulouse, France, pp. 371–375.

Szendrő, Zs., Matics, Z., Gerencsér, Z., Nagy, I., Lengyel, M. *et al.* (2010) Effect of dam and sire genotypes on productive and carcass traits of rabbits. *Journal of Animal Science* 88, 533–543.

Theilgaard, P., Baselga, M., Blas, E., Friggens, N.C., Cervera, C. and Pascual, J.J. (2009) Differences in productive robustness in rabbits selected for reproductive longevity or litter size. *Animal* 3, 637–646.

Tudela, F., Hurtaud, J., Garreau, H. and Rochambeau, H. de (2003) Comparaison des performances zootechniques de femelles parentales issues d'une souche témoin et d'une souche sélectionnée sur la productivité numérique. In: *Proceedings of 10èmes Journées de la recherche Cunicole*, Paris, pp. 53–56.

Youssef, Y.K., Iraqi, M.M., El-Raffa, A.M., Afifi, E.A., Khalil, M.H. *et al.* (2008) A joint project to synthesize new lines of rabbits in Egypt and Saudi Arabia: emphasis for results and prospects. In: Xiccato, G., Trocino, A. and Lukefahr, S. (eds) *Proceedings of 9th World Rabbit Congress*, Verona, Italy, pp. 1637–1641.

Ziadi, J.C., Mocé, M.L., Laborda, P., Blasco, A. and Santacreu, M.A. (2013) Genetic selection for ovulation rate and litter size in rabbits: estimation of genetic parameters and direct and correlated responses. *Animal Science* 91, 3113–3120.

14 Rabbit Research in the Post-genomic Era: Transcriptome, Proteome and Metabolome Analyses

Veronique Duranthon[1], Susana S. Araújo[2], Mariana Palma[2,3], Andrea Rau[4], Manolis Matzapetakis[2] and André M. Almeida[2,5,6*]

[1]Université Paris-Saclay, INRAE, ENVA, UVSQ, BREED, Jouy-en-Josas, France; [2]Instituto de Tecnologia Química e Biológica António Xavier, Universidade Nova de Lisboa, Portugal; [3]Centre for Funtional Ecology, Departmento de Ciências da Vida, Universidade de Coimbra, Coimbra, Portugal; [4]Université Paris-Saclay, INRAE, AgroParisTech, GABI, Jouy-en-Josas, France; [5]IBET – Instituto de Biologia Experimental e Tecnológica, Oeiras, Portugal and CIISA – Centro Interdisciplinar de Investigação em Sanidade Animal, Faculdade de Medicina Veterinária da Universidade de Lisboa, Portugal; [6]LEAF, Instituto Superior de Agronomia, Universidade de Lisboa, Portugal

14.1 Introduction

During the last 40 years, developments in the fields of genetics and molecular biology have led to important science breakthroughs clarifying gene functions and metabolic pathways involved in numerous diseases or physiological states of relevance to human health. During the last 20 years, researchers have been gradually focusing on obtaining a complete picture of biochemical mechanisms underlying their research subjects. To do so, they have gradually increased the scale of their analysis. In fact, from a handful of genes or proteins studied, the depth of the biochemical pathway has moved to an analysis of hundreds of genes and gene products, allowing a fully integrated overview of the biological issue. Such an upscaling was made possible by growing developments (and corresponding price decreases) in several technologies: protein electrophoresis, mass spectrometry-based protein identification or Nucleic Magnetic Resonance (NMR). Nevertheless, high throughput capacity for gene sequencing that led to complete genome sequencing of several animal and plant species of major use in agriculture and biomedical research, as well as a growing capacity to analyse such heavy and complex datasets through bioinformatics, were probably the most determinant factors. Such a large-scale analysis of genes, transcripts, proteins or metabolites is known collectively as the 'Omics' technologies, respectively referring to genomics, transcriptomics, proteomics and metabolomics, as represented in Fig. 14.1. Similarly, transcripts, proteins or metabolites in a given tissue, organ, organism or cell culture are respectively termed as the transcriptome, the proteome and the metabolome.

Numerous applications for genomics, transcriptomics, proteomics and metabolomics may be found in the literature for all fields of science and their importance has been continuously increasing. Accordingly, the rabbit model has been of extensive use in Omics-driven research.

Contact e-mail address: aalmeida@isa.ulisboa.pt

DOI: 10.1079/9781780643342.0014

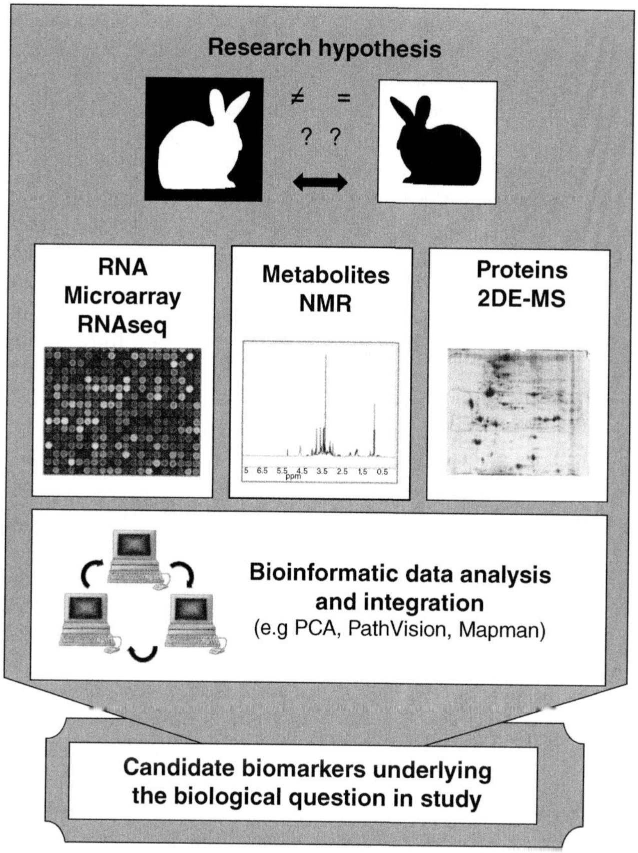

Fig. 14.1. Schematic overview of a common systems biology approach in the rabbit.

In this chapter we aim to introduce readers interested in rabbit genetics but not familiar with Omics principles and applications to the general principles of each of the most used technologies of importance in the post-genomic era, i.e. the technologies that underlie gene function: transcriptomics, proteomics and metabolomics. In each section, specific examples of the use of the technology and achievements with the *Oryctolagus cuniculus* species will be provided.

14.2 Transcriptomics

14.2.1 Microarrays: general principles, applications and case studies in the rabbit

DNA molecules constitute the complete genomic information defining the structure and function of living organisms. The flow of this genomic information within a biological system, which depends on the cell type, environmental conditions, etc., occurs through the processes of transcription into messenger RNA (mRNA) and eventually translation into proteins. As such, it is often of great interest to obtain a comprehensive understanding of the behaviour of the transcriptome, i.e. the full set of mRNA molecules present in a cell, tissue or organ, under different sets of experimental conditions (for example, treatment versus placebo) or in different groups or tissue types (for example, in liver versus brain or across time).

With the introduction of high-throughput biological assays in the mid-1990s, it has become possible to measure the abundance of mRNA molecules in different cells for thousands of genes simultaneously. In particular, the use of microarray technology, and more recently next-generation sequencing technology (see para.

14.2.2), has become a prominent way to study gene expression. Microarrays are a multiplex technology based on the affinity of single-stranded DNA sequences to bind to complementary sequences of nucleotides. Particularly, thousands of features, made up of specific probe sequences, are fixed or synthesized to a physical surface, such as a glass slide (the so-called chip), allowing for a genome-wide assay of gene expression. In a typical experiment, RNA molecules are extracted from a biological sample, converted into complementary DNA (cDNA), amplified, and labelled with fluorescent dyes. The cDNA molecules (or targets) are then hybridized on the chip, where cDNA molecules bind to their complementary template. The relative level of hybridization of each sequence is subsequently calculated as a function of the fluorescence level recovered after the hybridization process. We note that a variety of commercial and custom-made microarrays are currently available, with the most substantial distinction among them lying in the process by which probes are created and affixed to the slide. In this chapter, we focus on the use of the single-channel Agilent 60-mer gene expression microarray for the rabbit, for which the comparison of two conditions for the same gene requires two separate single-dye hybridizations. Note that the same array may also be used for two-dye hybridizations.

As with any scientific experiment, a full microarray experiment may be broadly broken down into three steps: (i) the experimental design phase; (ii) performing the experiment itself under the conditions defined in the previous step; and (iii) the statistical analysis of experimental data.

Although it is often neglected, the experimental design phase is an indispensable part of the experimental pipeline in order to clearly specify the scientific objectives, identify available resources, and define appropriate methods for the subsequent data analysis. Summarizing, an experimental design should enable the data analysis and interpretation of results to be as simple and efficient as possible, while accounting for the specific research question, as well as experimental and financial constraints. Practical considerations during this stage include identifying the quantity of biological samples that may be collected, the type and number of microarrays to be used, and the number and type of replicates to be included in the experiment. For the latter, we may distinguish between technical and biological replicates. Technical replicates correspond to the use of multiple microarrays for the same individual, or several probes for the same genes; these replicates enable an estimation of technical effects (e.g. spatial effects on the microarray), but allow inference only about a single, particular biological sample. On the other hand, biological replicates correspond to mRNA extractions from separate organisms or cell lines, followed by separate labelling and hybridization, under the same experimental conditions. Biological replicates tend to be more variable than technical replicates, and they are essential to estimate the variability among individuals and to perform inference on the biological populations represented in the experiment.

Often in practice, microarray experiments are performed using measurements taken on a mixture, or pool, of biological material collected from different biological replicates. There may be several reasons to use pooled measurements, including difficulties in recovering sufficient biological material per individual, to reduce the experimental cost, and to reduce biological variability (as expression measures are averaged across the pool). However, it is worth noting that pooling can introduce biases in expression measurements, for example in terms of the difference between the mean raw individual signals and the pooled mean signal.

Another important aspect of the experimental design phase is to identify and anticipate the potential sources of technical and biological variability, in particular with respect to potential confounding factors. Indeed, an often-cited remark by R.A. Fisher (1938) is the following: 'To consult the statistician after an experiment is finished is often merely to ask him to conduct a *post mortem* examination. He can perhaps say what the experiment died of.' To prevent confounding from adversely affecting an experiment, it is essential to make use of the principle of randomization. Randomization refers to the random ordering of treatments on experimental units. In some cases, a portion (or block) of experiments cannot be completely randomized (i.e. the entire experiment cannot be performed at one time); in such cases, it is important to balance samples within each block.

14.2.1.1 Performing a microarray experiment

During the experiment itself, it is important to consider the principles of repeatability and reproducibility. By performing experiments under the planned homogeneous conditions (e.g. RNA extractions performed on the same day by the same technician, etc.), the subsequent measurements of gene expression represent to the greatest extent possible the biological phenomenon of interest, rather than artifacts.

14.2.1.2 Arrays for rabbit analysis

To our knowledge, a single chip is commercially available for rabbit transcriptome analysis. It is produced by Agilent (commercial reference G2519F-020908) and is designed as one glass slide formatted with four 'high definition' 44k arrays. It displays 43,803 rabbit probes sourced from different databases (their early versions): RefSeq (Release 29), May 2008, UniGene (Build 11), March 2008 and Ensembl (Release 49), February 2008. This array has been successfully used for the transcriptome analysis of several rabbit tissues, including spleen (Dkhil *et al.*, 2013), retina (Zhao *et al.*, 2012), infected rabbit appendices (Dkhil *et al.*, 2012), liver (Al-Quraishy *et al.*, 2012) and early embryos (Naturil-Alfonso *et al.*, 2012; Saenz-de-Juano *et al.*, 2012). A re-annotation of this chip has been performed using the sigReannot pipeline (Casel *et al.*, 2009) and is available in the biomaRt package of Bioconductor (https://bioconductor.org/packages/release/bioc/html/biomaRt.html, accessed 18 January 2021) and at http://www.sigenae.org/fileadmin/_temp_/annotation/ens61/rabbit/agilent, accessed 18 January 2021. This re-annotation can be done again at the request of users and establishes that oligos that may be linked to a gene from among the 13,798 different known genes. However, when performed in 2009, such a re-annotation showed that some oligos have either multiple hits (about 1950 oligos) or no hits (about 12,000 oligos) among rabbit Ensembl transcripts or Unigene clusters.

This commercial array can be 'easily' customized. For example, we have chosen to design a 60 k array by removing from the commercial array all probes with either multiple hits or no hits among Ensembl transcripts or Unigene clusters and by adding probes corresponding to genes previously identified as expressed during early embryo development, or involved in particular biological functions of interest (gonad differentiation, embryo metabolism, X chromosome inactivation, pluripotency and epigenetic modifications of the genome). This 60 k array was successfully used to analyse the transcriptome of rabbit Embryonic Stem Cells and Induced Pluripotent Stem cells (GO accession number GPL16482) (Osteil *et al.*, 2013) and to compare these *in vitro* derived pluripotent cells to pluripotent cells present in the embryo (inner cell mass and epiblast cells) (Schmaltz-Panneau *et al.*, 2014). It was also used to analyse the effects of rabbit embryo vitrification on foetal placenta transcriptome (Saenz de Juano *et al.*, 2014). A second generation of 60 k customized array has been further developed that includes more candidate genes involved in pluripotency and embryo development or metabolism, epigenetics, immunology and mammary gland development (GO accession number GPL16709). This second-generation customized array has been successfully used to analyse transcriptome variations in peripheral blood mononuclear cells after *in vitro* stimulation by LPS or PMA-Ionomycin (Jacquier *et al.*, 2015), and to further characterize *in vitro* derived rabbit pluripotent cells (Osteil *et al.*, 2016; Tapponnier *et al.*, 2017). We then developed a third-generation customized rabbit array displaying two probes per rabbit transcript identified in Ensembl database after the annotation of the rabbit genome (Carneiro *et al.*, 2014). This array contains probes corresponding to about 24,000 rabbit Ensembl transcripts. Interestingly, it also contains embryo-specific probes and probes corresponding to about 1600 transcripts still not annotated in Ensembl database but identified in rabbit spleen, adrenal, PBMC, mammary gland and placenta (Carneiro *et al.*, 2014). This last array is being used for further analyses of rabbit embryo and brain transcriptomes.

14.2.1.3 Target preparation

Starting from total RNA (from 20 to 200 ng of total RNA), targets are synthesized using the 'RNA amplification and labeling kit' (Agilent). This kit makes it possible to both amplify and label RNA of each target sample. Briefly, starting from total RNA, spike RNA are added to each RNA sample in order to serve as a quality control after hybridization. A first cDNA strand is reverse transcribed using oligo dT primers coupled to a T7 RNA polymerase promoter at their 5' end. The oligo dT part of the primer selects messenger RNAs to be reverse transcribed, while the T7 RNA polymerase promoter is used for target amplification and labelling by *in vitro* transcription. The second cDNA strand is then synthesized, and *in vitro* transcription using T7 RNA polymerase is carried out. The addition of Cyanine3 labelled CTP in the reaction results in the synthesis of a labelled cRNA that can be hybridized to the array. Because several cRNA molecules are transcribed from one single double-stranded cDNA matrix, the reaction results in a linear amplification of the initial target RNAs. Both the yield and specific activity of the labelled cRNA must be controlled according to manufacturer's instructions. Specific activity should be at least 6 pmole Cy3 per microgram of cRNA.

14.2.1.4 Hybridization to the array

Labelled cRNA is then fragmented into about 200 bp fragments and hybridized to the array (600 ng labelled cRNA diluted into 40 µl hybridization buffer Agilent). Hybridization is performed for 17 hours in a rotating oven (10 rpm) at 65°C. After two washes in Gene expression wash buffers 1 and 2 (Agilent), respectively, slides are analysed by the Agilent array scanner, and fluorescent signal intensities associated to each probe are quantified using feature extraction software (Agilent). This software also extracts signals corresponding to spike RNAs, from which the quality of the whole hybridization procedure can be deduced.

14.2.1.5 Data analysis for a microarray experiment

Finally, during the analysis itself, the model defined during the initial design phase may be used to determine the influence of different factors (e.g. the effect of a treatment as compared to a placebo) by estimating their effects as well as their variance. One of the most common statistical analyses performed for microarray experiments is a differential analysis, where the objective is to identify a subset of differentially expressed (DE) genes. A gene is declared DE if an observed difference or change in expression between two (or more) experimental conditions is statistically significant, i.e. greater than expected simply due to natural random variation. Such an analysis can be broadly broken into four steps: (i) data exploration and pre-processing; (ii) normalization; (iii) fitting a statistical model; and (iv) performing a statistical test.

In the data exploration and pre-processing step, probe intensities are quantified (typically directly via software provided by the microarray manufacturer). Subsequently, descriptive statistics and graphical summaries may be used (e.g. calculating data quantiles, plotting linear associations between two samples, visualizing spatial effects on the array, plotting histograms or boxplots of probe intensities prior to and following normalization) to explore the data prior to the analysis. Such data exploration enables an appreciation of potential biases or technical artifacts that may be present in the data. In addition, differences between arrays in the shape or centre of the distributions of probe intensities can help highlight an eventual need for data normalization. Finally, prior to normalizing and analysing the data, a log2 transformation is typically applied to the raw probe intensities; this transformation leads to a stabilization of the variance as well as more easily interpretable results in the following steps.

The objective of the normalization step is to make measurements from different arrays comparable, i.e. to remove background noise from meaningful signal and ensure that observed differences between samples are indeed due to differential expression and not simply technical biases. We note that an appropriate statistical design, as described above, can help avoid confusion between biological and technical effects. A large variety of methods have been proposed for the normalization of microarray data, including scale, non-linear, quantile, and background-correction normalization methods; a common hypothesis to most such methods is that a majority of genes are not DE among samples. The specific type of normalization to be used depends on the specific platform used for the experiment. For Agilent single-channel arrays, following log2 transformation of the raw data, the probe intensities are typically centred per slide and, if present, within-array replicate probes are averaged across technical replicates.

A statistical model (previously defined in the experimental design phase) is subsequently fitted to the pre-processed and normalized data to identify DE genes. The statistical model includes both the construction of a test statistic and an appropriate modelling of per-gene variances. Typically, the per-gene statistical significance is calculated via a two-sample t-statistic:

$$t_i = \frac{\bar{x}_{i1} - \bar{x}_{i2}}{S_i\sqrt{\frac{1}{n_1}+\frac{1}{n_2}}} = \frac{M_i}{SE(M_i)}$$

Where $\bar{x}_{ij}$ represents the mean of gene i in condition j, s_i the standard deviation of gene i, n_j the number of biological replicates for condition j, M_i the average log-fold change of gene i, and SE the standard error. However, the per-gene fold change M_i and standard error $SE(M_i)$ tend to be very unstable when a small number of replicates are present (a typical situation in practice for microarray data), leading to a lack of power to detect DE. If, instead, a common estimate of this variance is estimated for all genes, $SE(M_i) = SE(M)$, a more stable estimate may be obtained at the expense of a high rate of false positives (i.e. genes incorrectly identified as DE). For this reason, a more appropriate estimation of variances is required, such as the significance analysis of microarrays (SAM) *t*-test (Tusher *et al.*, 2001), where a constant is added to per-gene variance estimates to stabilize small variances, or limma (Smyth, 2004), an empirical Bayesian method used to obtain per-gene estimates of variances that borrow information across all genes to stabilize estimates.

Once the statistical model has been fitted to the data and per-gene test statistics calculated, a statistical test is performed to assess DE. Under the hypothesis that a gene i is non-DE (i.e. the null hypothesis), the p-value for that gene represents the probability of observing a test statistic at least as extreme as the observed value t_i. Because tests are performed individually for a large number of genes, a large number of false positives will be identified using the raw per-gene p-values. For this reason, an adjustment for multiple testing is necessary to justify statistical significance. This multiple testing correction controls the global risk of having a false positive among the full set of genes. One commonly used approach to this end is that of Benjamini and Hochberg (1995), which is used to control the false discovery rate (i.e. the proportion of false positives expected among all genes identified as DE). After applying this correction, differentially expressed genes are subsequently identified using these adjusted p-values.

In addition to differential analyses, a variety of additional analyses may be useful to aid in the biological interpretation of microarray data. In particular, it may be useful to consider the functional enrichment among the set of DE genes (also referred to as a gene category enrichment analysis), co-expression analyses of genes or samples, inference of gene regulatory networks, and integration with other sources of data (e.g. epigenomic, metabolomic, proteomic). For most of these analyses, including differential analyses, open-source software packages are available within the Bioconductor project (http://www.bioconductor.org, accessed 18 January 2021; Gentleman *et al.*, 2004).

14.2.2 RNA sequencing application in rabbits

Next-Generation Sequencing (NGS) refers to emerging high-throughput DNA sequencing methods that are available over the last decade, after earlier capillary sequencing methods that relied upon 'Sanger sequencing' (Soon *et al.*, 2013). Commonly, NGS techniques are often named as second- and third-generation sequencing. NGS was selected as the method of the year in 2007 (Schuster, 2008) and it is opening fascinating opportunities in life sciences (Ansorge, 2009). The second-generation and main commercially available high-throughput sequencing NGS platforms that emerged were the 454 Genome Sequencer FLX Instrument (Roche Applied), the Ilumina/HiSeq Genome Analyzer (Solexa) and the ABI SOLID system (Applied Biosystems). However, the pace of change in this area is rapid and three major new sequencing platforms have been released in 2011: Ion Torrent's PGM, Pacific Biosciences' S and the Illumina MiSeq (Quail *et al.*, 2012). Third-generation sequencing, characterized by the single-molecule sequencing, includes platforms such as the Helicos™ Genetic Analysis System (SeqLL, LLC), the single-molecule real-time sequencing approach

(SMRT, Pacific Biosciences), Nanopore sequencing (Oxford Nanopore's), Complete Genomics (Beijing Genomics Institute) and GnuBIO (Bio-Rad), among others (Levy and Myers, 2016). Comprehensive principles description, as well as a comparison of their applications, advantages and limitations, may be found in Ansorge (2009), Marguerat and Bähler (2010), Metzker (2010), Liu *et al.* (2012), Quail *et al.* (2012) and Levy and Myers (2016).

NGS is not restricted to the analysis of DNA sequences and is now routinely used to analyse other biological components such as RNA and protein libraries, as well as how they interact in complex networks (Soon *et al.*, 2013). RNA-sequencing (RNA-seq) detects all the transcripts present in an experimental sample by direct sequencing. This allows transcriptome analysis at a maximal resolution and dynamic range, independently of the transcript size and above the knowledge of the genome they are derived from (Marguerat and Bähler, 2010). The short-sequenced fragments that are usually produced (reads) are countable and the digital data requires, afterwards, huge computing resources for *de novo* assembling and quantifying the number of reads. If a reference genome is available, reads are mapped back to reference genome being the number of mapped reads a measure of expression level for that gene (Malone and Oliver, 2011).

RNA-sequencing has largely surpassed hybridization-based approaches by overcoming some limitation associated with probe microarray hybridization by overcoming hybridization background or probe cross-hybridization issues and, importantly, the restriction on transcript detection imposed by the array design (Costa *et al.*, 2010). As RNA-Seq is quantitative, it can be used to determine RNA expression levels more accurately than microarrays (Wang *et al.*, 2009), allowing the detection of low abundant or novel transcripts (Roberts *et al.*, 2011) or small RNA molecules (small RNAs) without requiring a prior knowledge of the genome sequence of the target organism (Mortazavi *et al.*, 2008). Additionally, RNA–Seq is also well suited for detecting new RNA editing events or sites (Picardi and Pesole, 2013), novel transcripts (Sánchez-Pla *et al.*, 2012), splice variants (Bryant *et al.*, 2012) or allelic specific differences in gene expression resulting from Genome Wide Association Studies (Quinn *et al.*, 2013).

Whole-transcriptome characterization or differential gene expression studies using RNA-seq have become a cost-effective option for many experimental approaches providing valuable and complete insights underlying the complexity of many biological processes (Wilhelm and Landry, 2009). The main goal of transcriptome analysis is to identify, characterize, catalogue and quantify all the transcripts expressed within a specific cell or tissue at a specific time point. An enormous number of bibliographic reports is available, describing the use such type of approach in studies targeting a broad range of mammals, including human (Sultan *et al.*, 2008) or model animals like mouse (Mortazavi *et al.*, 2008) or farm animals like sheep (Jäger *et al.*, 2011). An exponential amount of sequence data resulting from multiple studies using NGS technologies is available and can be mined in public sequence data repository, such as the NCBI Sequence Read Archive (Wheeler *et al.*, 2008). With the publication of the first and then updated version of the rabbit reference genome (Lindblad-Toh *et al.*, 2011; Carneiro *et al.*, 2014), numerous studies on rabbit transcriptome successfully using RNA-seq technology have been published.

The discovery of small non-coding RNA (sRNA) genes, which produce functional RNA molecules rather than encoding proteins, has undoubtedly changed the way to address gene expression regulation in living organisms (Eddy, 2001). MicroRNAs (miRNAs) are small non-coding RNAs about 22 nucleotides long that modulate the expression of complementary messenger RNAs; thus the precise amounts of proteins expressed in a cell (Libri *et al.*, 2013). Hundreds of miRNA genes have been found in diverse animals, and many of these are phylogenetically conserved (Ambros, 2004). The identification of the pathways by which animal miRNAs are produced, matured and turned over has revealed many aspects of their biogenesis that are subject to regulation (for review see Libri *et al.*, 2013). MicroRNA actions have been shown to modulate several developmental and physiological processes in animals including stem cell differentiation, haematopoiesis, cardiac and skeletal muscle development, neurogenesis, insulin secretion, cholesterol metabolism and the immune response (for review, see Williams, 2008). Importantly, miRNA aberrant expression has been implicated in a number of diseases including cancer (O'Day and Lal, 2010), neurodegenerative (Shioya *et al.*, 2010) and heart disease (Cheng *et al.*, 2007). The understanding of miRNA function is still unclear relying on the existence mRNA, miRNA sequence and expression data to be integrated with other comparative genomic data (Rajewsky, 2006). Consequently, it can be anticipated that the application of RNA-seq for miRNA identification and profiling may provide valuable insights into the function and evolution of this layer of post-transcriptional control of gene expression in the future as recently achieved by Zhang *et al.* (2018).

A total of 579 mature miRNAs (306 precursors) have been annotated on the *Oryctolagus cuniculus* genome (oryCun2.0) and included in the miRNA repertoire existing in miRBase (Kozomara and Griffiths-Jones, 2011). This highlights the amount of research, done as well as the wide range of biomedical research applications. Li *et al.* (2011) adopted the SOLID platform for rabbit miRNA expression profiling and identification from 11 different tissues and organs. In this study, the authors have confidently identified 886 rabbit mature miRNAs and 464 pre-miRNA harpins mining NGS datasets in their own miRNA discovery pipeline. Despite the limitations imposed by the incomplete rabbit genome annotation at the time, this study was one of the most comprehensive rabbit miRNA resources available with advantages in both biomedical research and insight into mammals' miRNA gene evolution. More recently, Zhao *et al.* (2019) investigated the role of non-coding RNAs (ncRNAs) in the regulation of the hair follicle development in Angora rabbits. Beside the already mentioned miRNAs, the authors searched for long non-coding RNAs (lncRNAs, with more than 200 base pairs) and circular RNAs (circRNAs, continuous loop structures more stable than linear mRNAs). These classes of ncRNAs play regulatory roles in translation by interacting with mRNAs, miRNAs or mRNA-binding proteins (Chekulaeva and Rajewsky, 2019). A multiple high-throughput sequencing approach (Illumina HiSeq) was used to investigate changes in the abundance of coding and non-coding mRNA species during three hair-cycle stages. A total of 111 lncRNAs, 247 circRNAs, 97 miRNAs, and 1168 mRNAs were found differentially expressed during the hair-cycle stages. To integrate the data, lncRNA-miRNA-mRNA and circRNA-miRNA-mRNA ceRNA networks were built to extract underlying relationships between ncRNAs and mRNAs they might be regulating. As one example, LNC_002919 and novel_circ_0026326 were found to act as competing endogenous RNAs (ceRNAs) for miR-320-3p, which targets HIV-1 TAT interactive protein 2 (HTATIP2) transcripts, implicated in hair follicle development. The authors suggested that these two ncRNAs participate in the regulation of the HF cycle as miR-320-3p sponges. Also, this study highlights the complexity of the mechanisms implicated in the regulation of gene expression underlying the hair follicle cycle. Kuang *et al.* (2019) performed a comprehensive analysis of lncRNA and circRNA profiles in rabbit embryos at different stages by whole-transcriptome sequencing and identified several candidate lncRNAs and circRNAs that may be indispensable for the morphogenesis and development of rabbit embryos.

Rabbits and their wild relative counterparts (e.g. *O. cuniculus algirus*) have been extensively used as models to study questions related to domestication (Carneiro *et al.*, 2011, 2012, 2014; Albert *et al.*, 2012; Campos *et al.*, 2012). Domesticated animals differ from their wild relatives in appearance, physiology and behaviour and it is widely accepted that these differences are mostly genetically encoded (Carneiro *et al.*, 2011; Albert *et al.*, 2012). However, despite the work of Carneiro *et al.* (2014), it is still unclear which genes contribute to these domestication traits and how changes in expression levels of a set of genes might lead to phenotypic changes observed during the domestication process. RNA-seq has been used to perform a differential expression study where in frontal cortex brain gene expression was compared in four pairs of domesticated and wild animals, including rabbits (Albert *et al.*, 2012). The overall results pointed out that gene expression differences observed that could correlate with behavioural differences when domestic are compared to wild animals within each species and unique to each domestication event, suggesting that domestication has proceeded through different genetic routes in different species. Moreover, this study raised an interesting open question aiming to understand if the genetic variants that cause similar traits in different domestication events are, similarly, species-specific.

Induced pluripotent stem cells (iPSCs) and embryonic stem cells (ESCs) are two types of pluripotent stem cells that have the capacity to divide indefinitely and to differentiate into all somatic cells and tissue lines (Saha and Jaenisch, 2009). Pluripotent stem cells hold promises for biomedical research and medical applications including stem cell biology, regenerative medicine and therapeutics (Saha and Jaenisch, 2009; Bai *et al.*, 2013). Several putative rabbit ESCs and iPSCs have been established which constitute a promise for translational biomedical research, due to phylogenetic proximity to primates (Tancos *et al.*, 2012; Osteil *et al.*, 2016). Rabbit iPSC technology offers a unique tool for deriving specific stem cells to study disease, developing possible treatments for degenerative disorders, drug screening, toxicology testing, and for generating knockout or transgenic animal models for some human diseases (Tancos *et al.*, 2012). RNA-seq approaches still encloses myriad unexploited opportunities to support the above-mentioned aspects but also to go deeper into the molecular and functional features of iPSC in rabbits (Osteil *et al.*, 2013; Tapponier *et al.*, 2018). One challenging way might be related to the post-transcriptional regulation cell reprogramming by miRNA mediated modulation.

This section aimed to provide one overview about new opportunities opened by NGS, especially the application of RNA-Seq to take advantage of rabbit genomic and genetic resources. Some applications and putative challenges in translational biomedical research were emphasized. Microarray have dominated rabbit transcriptome profiling and differential gene expression-based studies till the last years. Nevertheless, the increasing public information available in the NCBI Sequence Read Archive database for *Oryctolagus cuniculus* allow us to anticipate that NGS based studies addressing topics like rabbit embryogenesis and other fundamental biological aspects in this species will be the matter of intense analyses and exploitation.

14.3 Proteomics

Every biological study will have one or more of the following objectives: (i) to characterize a biological system (or set of systems); (ii) to understand how the biological system works; (iii) to understand how it is controlled; and (iv) to manipulate it for a variety of purposes. To achieve these objectives, it is vital to understand that from DNA strands and genes to actual protein synthesis and phenotypes, numerous levels of biological information may be present: from DNA to RNA to proteins to metabolisms to tissues, organs, individuals and populations. If we consider a mammalian organism as having a total of 3×10^4 genes, we realize that the number of transcripts and proteins it generates is much higher ($1\text{-}2 \times 10^5$ mRNAs and proteins). The number is further increased if we consider the number of post-translational modifications (PTMs) such proteins may present: 2×10^6. As such, it is quite clear that, in order to understand the mechanisms behind a biological phenomenon, it is vital to understand the proteins that are involved in it. The proteome may be defined as the proteins present in a given cell, fluid, tissue, organ, organism or population. Therefore, the study of the proteome can be defined as the description and explanation of quantitative and qualitative changes that occur in it as a consequence of a certain stimulus. The science that studies the proteome is termed proteomics. In this section we will present the major features of proteomics methodologies and techniques, in an introductory fashion. We will subsequently focus on the application of proteomics to a specific topic: skeletal muscle research.

14.3.1 Principles and methodologies in proteomics: a brief introduction

Two major trends are considered in proteomics: gel-based proteomics (GBP) and gel-free proteomics. A broader view on these trends is presented in Brewis and Brennan (2010). GBP is the oldest and most widely used method. It is based on the separation of protein extracts (e.g. from a certain tissue or fluid) in an acrylamide gel. Generally speaking, three major types of gels may be used in proteomics experiments: (i) one-dimensional electrophoresis (1DE); (ii) two-dimensional electrophoresis (2DE); and (iii) Blue-Native (BN) poly-acrylamide gel-electrophoresis (PAGE). In 1DE, denatured protein extracts are separated according to their molecular mass on an acrylamide gel, whereas in 2DE, denatured proteins are first separated according to their isoelectric point in a process termed isoelectric focusing and, secondly, separated according to their molecular weight in a process very similar to 1DE. In the third type, BN-PAGE, proteins are first extracted in their native form (no denaturing) and the complexes they form are then separated according to their molecular mass. This lane may then be cut and stripped from the gel. After a 90° rotation, it may fit in a standard SDS-PAGE gel where the separation of the proteins that make up a complex occurs.

On the initial stages of proteomic studies, 1DE was the method of choice for protein separation. This resulted from the easy access to this technique by researchers that were familiar with it, particularly through the use of Western Blot. Nevertheless, and given the large amount of proteins present in one proteome and that could overlap in the same gel band, 1DE was soon considered to be insufficient and was replaced by 2DE. Two-dimensional electrophoresis is currently the most widely used gel-based method for protein separation and has become very standardized over the last 20 years as equipment and consumables to conduct the first-dimension electrophoresis have become available from several commercial companies. Initially, 2DE gels were single-stained with silver nitrate (SS). This compound has severe problems of compatibility with mass spectrometry and also a low dynamic range that renders spot quantification extremely difficult. For these reasons, SS was quickly replaced by Coomassie blue staining, in particular by the so-called Colloidal Coomassie that avoided the above-mentioned problems. Presently, there are numerous commercial single-staining procedures available for 2DE that may include the detection of phosphorylated proteins as well as other PTMs. Nevertheless, single-stained 2DE has one important drawback as the electrophoresis patterns change considerably from one electrophoretic run to the other, among numerous other factors (Miller, 2012). To circumvent this problem, researchers have turned to Differential In Gel Electrophoresis or DIGE (Timms and Cramer, 2008). In DIGE, two samples can be loaded in one 2DE gel, each of them stained with a different dye that can only be read at a particular wavelength. A third dye, read at a third different wavelength, is then used to stain a pool of all the samples available in that study. The gel pattern obtained for the first two dyes is then compared with the latter. Albeit the fact that DIGE has the advantage of bringing an extraordinary robustness and reproducibility to 2DE that is not possible with single staining, it has one major drawback: it is very expensive in consumables and apparatus needs and is not frequently available in most animal research institutes.

After protein separation, gels are then analysed using specific software that compares the gel patterns of the different experimental groups. For further information on gel analysis software and their comparison, readers should refer for instance to the reviews by Marengo *et al.* (2008). These software tools typically align the gels with each other, so that protein spots in the different gels may have the same coordinates of spatial distribution. Software will then attribute a reference number to each spot and will calculate the relative intensity (volume of the spot divided by the sum of volumes of all spots) for each spot. Most contemporary software includes statistical packages that allow users to compare, for each spot, the average relative intensity for each spot in the experimental groups of the analysis. The previously described process is generally used for single-staining gels, whereas in DIGE, as referred, the relative intensities of each spot are compared to the pool pattern. In general, vendor companies have a different version of the software, according to the use of DIGE or single stain. After gel analysis, researchers will have a list of spots over-expressed or under-expressed that can be of interest to the studied biological condition.

Proteins are then excised from the gel, digested with a protease, usually trypsin, and identified using Mass Spectrometry (MS), in which Matrix Assisted Lased Desorption Ionization-Time Of Flight/Time Of Flight (MALDI-TOF/TOF) is frequently the method of choice. In MALDI-TOF, the peptide mixtures resulting from trypsin digestion are co-crystallized with a matrix (e.g. α-Cyano-4-hydroxycinnamic acid) in a MALDI target or plate. A laser beam then ionizes the mixture. Resulting ions are then transported across the time of flight (TOF) tube until they reach a detector. The time ions take to travel through the TOF tube is dependent on their mass. When ions reach the detector, a mass spectrum is created that is composed of several peaks with a relation of the masses of the peptides obtained through the tryptic digestion.

The obtained spectrum creates a list of masses, each characteristic of a certain peptide, which may then be compared to mass lists available in databases and resulting from the theoretical digestion of proteins with known sequences available in public databases, creating a putative identification. This analysis method is termed Peptide Mass Fingerprinting or PMF (Thiede *et al.*, 2005). Over the last 10 years, advances in mass spectrometry have also allowed researchers to sequence some of the peptides, hence determining their amino acid sequencing, bringing higher confidence levels to the identification. Search programs finally provide, in the form of a score and sequence coverage, a measure of the reliability of the identification.

Gel-based proteomics and, in particular, 2DE, is considered to be the workhorse of proteomics studies in farm animals, representing the vast majority of proteomics research in farm-animal subjects. In fact, their relatively cheap prices and general availability of electrophoresis equipment (with the exception of the mass spectrometer, that tends to be only present in specialized platforms), makes this set of techniques highly preferred by animal researchers. 2DE has therefore been considered to still have an important role to play in proteomics (Rogowska-Wrzesinska *et al.*, 2013). Nevertheless, gel-based proteomics still has one major limitation, related to the fact that it is very difficult to locate in a gel low abundant, and membrane proteins. To get around this problem, Gel-free proteomics (also known as bottom-up proteomics or shotgun proteomics) has been gaining wide recognition during the last decade. In shotgun proteomics, the whole proteome is first digested with a protease such a trypsin that is subsequently separated, using Liquid Chromatography for instance. Finally, each peptide is quantified and identified using mass spectrometry. Shotgun proteomics relies chiefly on the identification of unique peptides that are characteristic of a particular protein. It is therefore heavily dependent on the accuracy and the annotation level of the database where searches are conducted. Shotgun proteomics has therefore a limited use in studies involving farm animals, rabbit included, as only cattle (*Bos taurus*), pig (*Sus scrofa*) and chicken (*Gallus gallus*) are considered to be sequenced organisms with rather complete annotated protein databases (Soares *et al.*, 2012). Nevertheless, the research using this tool is increasing and will likely gain importance and become quite common in the years to come. In the following section, we will present a specific case study where proteomics has been successfully applied: the rabbit skeletal muscle research.

14.3.2 Proteomics in rabbit skeletal muscle research

Similar to other species such as the pig (Almeida and Bendixen, 2012) that is a widely used animal in biological research as well as for meat production, the rabbit is one of the most commonly studied species using proteomics. In fact, examples of the application of proteomics to rabbit research include, for instance, topics related to the respiratory system, eye research, body fluids and disease detection. Interestingly, it is in the context of muscle research (skeletal and cardiac) that proteomics has known its higher development level in research using rabbits. For a complete detailed review on the uses of proteomics in the rabbit species, readers are advised to refer to Miller *et al.* (2014). In this section, we will focus on the use of proteomics in skeletal muscle research using the rabbit as a model.

To know the distribution of protein spots within a two-dimensional electrophoresis gel is of key importance, as it enables fellow researchers to assess and compare extraction protocols, as well as differences established as a consequence of experimental procedures. This is generally referred to as mapping a proteome. Likewise, the characterization of full proteomes using novel gel-free, high-throughput mass spectrometry equipment or *de novo* sequencing allows researchers to obtain a full grasp of the proteome of a particular tissue or fluid. The bovine muscle proteome, for instance, has been extensively studied (Bouley *et al.*, 2004; Chaze *et al.*, 2006). Nevertheless, for the rabbit, little information was available. The first reported 2DE mapping of the rabbit skeletal (*gastrocnemius*) muscle proteome was conducted by Almeida *et al.* (2009). Authors identified a total of 45 different proteins and were further classified into major functional classes: contractile apparatus, cell structure, cell defence and metabolic. Structural proteins such as myosin heavy and light chains or actin, and different isoforms represented the majority of the mapped proteins. None the less, other relevant metabolic proteins, particularly those of the glycolytic pathway (Almeida *et al.*, 2009), were equally mapped. Then, the sarcoplasmic reticulum of rabbit skeletal muscle was also characterized (Staunton and Ohlendieck, 2012) using one-dimensional gradient gels and on-membrane digestion. Authors identified over 30 different protein species, not frequently found in 2DE-based studies, such as key calcium-handling proteins like ryanodine receptor, Ca^{2+}-ATPase, calsequestrin and sarcalumenin. As already described by Almeida *et al.* (2009), aldolase and phosphofructokinase were present in the purified sarcoplasmic reticulum, supporting the hypothesis of a close physical coupling between the glycolytic pathway and the energy-dependent metabolism (Staunton and Ohlendieck, 2012). More recently, Liu *et al.* (2013) used shotgun proteomics to characterize the proteins from the sarcoplasmic reticulum membrane fractions of New Zealand white rabbit skeletal muscle. The authors identified nearly 500 proteins in this characterization effort, which was considered a major breakthrough (Almeida, 2013) in skeletal muscle proteome characterization and muscle contractibility, highlighting the promise of shotgun proteomics by comparison to standard 2DE-based techniques.

The rabbit has also been used to determine breed differences and nutritional status based on protein expression in the *gastrocnemius* muscle. In a study on the effect of weight loss on the proteomics profiles of the *gastrocnemius* muscles (Almeida *et al.*, 2010), a 20% weight reduction was induced in New Zealand White, a selected meat producer (*Oryctolagus cuniculus cuniculus*), and in the Iberian wild rabbit (*Oryctolagus cuniculus algirus*). Using 2DE/MS, differential protein expression between control (*ad libitum*) and restricted diet experimental rabbit groups was related to spots belonging to several energy metabolism enzymes, as well as structural proteins. The authors determined several putative candidates of tolerance to weight loss markers and proposed their use as a selection tool in rabbit, as well as for other species breeding. Additionally, a spot identified as mitochondrial import stimulation factor seems of capital interest as a marker of undernutrition and a possible object of further studies aiming to better understand its physiological role.

Proteomics was also used in the functional characterization of specific states, such as the study of muscle contractility as inhibited by oxidation using hydrogen peroxide on purified myosin from the psoas muscle fibres (Prochniewicz *et al.*, 2008). These authors established that 5 mM hydrogen peroxide decreases fibre contractility without changing the enzymatic activity of myofibrils and myosin, while a 50 mM concentration had deep and irreversible inhibitory effects on fibre contractility. Mass spectrometric analysis established that the oxidative modifications had an effect on multiple methionine residues in the myosin light and heavy chains, suggesting an unknown role for methionines in muscle contractility (Prochiniewicz *et al.*, 2008). Muscle contractility is affected by specific conditions and particularly neuromuscular activity, disorders or ageing. Donoghue *et al.* (2007) used DIGE to study the effects of chronic low-frequency stimulation of the rabbit tibialis anterior muscle, a model for the study of the response of fast fibres to enhanced neuromuscular activity under conditions of maximum activation. The authors compared un-stimulated control specimens to 14- and 60-day conditioned muscles. Different expression was recorded for a total of 41 proteins. Differential expression was particularly noteworthy on contractile, ion homeostasis, excitation-contraction coupling, metabolism and stress-response proteins. Results established two markers of muscle stimulation: cofilin-2 and transgelin (Donoghue *et al.*, 2007). More recently, the same research group used rabbit muscle proteomics to study dystrophinopathy, a form of muscle dystrophy (Lewis and Ohlendiek, 2010). In this study, proteins under-represented in classic two-dimensional electrophoresis due to their low molecular weight and membrane association and a 427 kDa cytoskeletal dystrophin, a protein associated to a glycoprotein complex and known to be affected by muscular dystrophy was highlighted. Using on-membrane digestion of one-dimensional blots of the sarcolemma enriched fraction, dystrophin isoform Dp427 and associated glycoproteins as well as sarcolemmal dysferlin were successfully identified, demonstrating the potential of this technique to study membrane associated proteins (Lewis and Ohlendiek, 2010).

The above-mentioned case study clearly shows the importance and potential of proteomics techniques to understand the biochemical mechanisms behind any biological system function. It also reveals that, concerning the rabbit species, the use of proteomics tools is still very incipient. Two keys point in addressing this issue seems to be, on the one hand, the finalization of the rabbit genome sequencing and, on the other hand – for rabbit researchers, both from the productive or the biomedical perspective – to have access to proteomics and mass spectrometry platforms, services and means that will enable and increase the amount of knowledge generated from this species.

14.4 Metabolomics

The term metabolomics is widely used in the literature to describe the study of the low molecular weight compounds (MW < 1.500 Da) in a biological system (German *et al.*, 2005; Bernini *et al.*, 2011). It started appearing in literature in the late 1990s (Oliver *et al.*, 1998; Hunter, 2009), along with transcriptomics and proteomics; however, studies of metabolites and metabolism were being reported from the 1960s (Hunter, 2009). The number of metabolomics-based studies continues to increase, either as methodology development or in other specific applications, rendering it a useful approach to study the metabolome of several organisms (Nagana Gowda and Raftery, 2015). The metabolome, defined as the total set of metabolites present in a given cell or tissue of an organism, is a complex mixture whose composition is directly related to both intrinsic and extrinsic factors. Metabolome directly reflects not only the underlying biochemical processes at the transcriptome, proteome and genome levels, but also environmental factors (Nicholson and Wilson, 2003). The metabolite profile present in a biological sample provides a snapshot of its physiological state at the sampling (or sample quenching) moment (Nagana Gowda and Raftery, 2015; Markley *et al.*, 2017).

The most widely used techniques in metabolomics are Nuclear Magnetic Resonance (NMR) and Mass Spectrometry (MS) coupled to Gas Chromatography (GC) or Liquid Chromatography (LC) (Griffin, 2004). Methodology selection depends on the purpose and the requirements of the intended studies, which can demand either quantitative or qualitative information. Given the complexity and the large amount of compounds in some samples, the use of both methods is required (Griffin, 2004). The concentration and chemical properties of the metabolites of interest can also determine the technique selection.

Regardless of the chosen technique, sample preparation and handling is a key step in the entire process. The metabolome of diverse samples of an organism, either a tissue or biofluid, require specific procedures for collection, handling and preparation. It is known that different protocols for sample collection, preparation (such as filtration, tissue extraction and centrifugation) and storage (freezing) can significantly affect the sample composition and lead to different results, affecting the reproducibility, resolution and further correlation with proteomics and transcriptomics data (Macomber, 1998; Martinez *et al.*, 2005; Bernini *et al.*, 2011). Thus, strict reproducible and robust protocols should be established. Furthermore, the subsequent statistical analysis requires that all spectra are obtained and processed by identical protocols (Moestue *et al.*, 2011).

MS determines the mass/charge ratio of ions in a sample and plots it as a mass spectrum. The output expresses the intensity as a function of the mass/charge ratio. MS is more sensitive than NMR, although its reproducibility is in some way lower. It detects the molecular mass of compounds often with isotopic resolution that can reveal their elemental composition. However, it is not ideally suited for the direct analysis of mixtures and thus it is commonly coupled with chromatographic techniques, such as LC and GC for the pre-separation of analytes. These chromatographic techniques are, in general, only suitable for specific classes of compounds and thus cannot always provide a complete description of the mixture under analysis. In addition, MS analysis requires sample preparation and thus could be time-consuming and often has issues in direct quantification of compounds (Griffin, 2004). MS is also sample-destructive, preventing further analyses by other methods.

NMR technique exploits the magnetic properties of specific atom nuclei to determine the physical and chemical attributes of atoms or molecules. NMR is less sensitive than MS but is more versatile, and due to its high spectral resolution it allows, simultaneously, to identify and quantify several small molecules in a mixture. Its low sensitivity has been countered at the instrumental level with the development of spectrometers with higher magnetic fields and cryoprobes, which can enhance the NMR dynamic range, increasing the detection limits to the nanomolar range. NMR has the advantages of requiring little sample preparation and spectra acquisition can be fully automated with a high degree of reproducibility (Griffin, 2004).

NMR can be applied for tissues extracts (Serkova *et al.*, 2007; Mannina *et al.*, 2008; Qwele *et al.*, 2013), either liquid, solid (Macomber, 1998) or viscous samples (Jiménez *et al.*, 2013), intact tissues (Tilgner *et al.*, 2019) or to the whole living organism (Blaise, 2009). Liquid samples, including urine, plasma (Bernini *et al.*, 2011), cerebrospinal fluid (Wevers *et al.*, 1995) and saliva (Santone *et al.*, 2014) are the most common types of samples since they require little processing and are commonly collected for medical/analytical purposes. Intact tissues, such as muscle (Nestor *et al.*, 2010), fat (Millis *et al.*, 1997), skin (Lucas *et al.*, 2005) and other organs (Barba *et al.*, 2007; Merrifield *et al.*, 2011) can be analysed without any pre-treatment resorting to High Resolution-Magic Angle Spinning (HR-MAS) NMR (Moestue *et al.*, 2011). Biopsy samples can be analysed within a few minutes after collection and contribute to quicker and reliable diagnoses, and to adequate medical treatments assignment (Ben Sellem *et al.*, 2011; Kumar *et al.*, 2015; Rezig *et al.*, 2018). Besides this advantage, when compared with other NMR tools that require previous tissue extractions (Cheng *et al.*, 1998; Griffin, 2004), HR-MAS also allows the identification of molecules with different physical properties and changes in cytosolic composition, whereas the extraction procedures imply a separation of the sample constituents (Griffin, 2004). These methodologies, that detect simultaneously a large number of metabolites, are useful to establish molecular biomarkers for specific physiological conditions, such as diseases (Zhang *et al.*, 2013; Cui *et al.*, 2017; Yilmaz *et al.*, 2017; Serkova *et al.*, 2019). In particular, metabolomic studies have been shown to be more effective in distinguishing different samples than the genomic and proteomic studies (Griffin, 2004; Serkova *et al.*, 2007). This is explained by the fact that metabolic fluxes in given pathways are the result of multiple steps of upstream regulation both at the proteomic and the transcriptomic level and the resulting metabolite composition is characteristic of the integral biological system at a given condition (Griffin, 2004; Moesture *et al.*, 2011).

Metabolomics studies have some constraints to study the inherent dynamic nature of the systems that leads to constant variability. These fluctuations can derive from fluxes in response to stimuli, variations of environmental factors including diet and drug administration, and in the case of multicellular systems, cells' specific composition, among others. Such information can be accessed by more specific methodologies, albeit with reduced sensitivity, than the methods that directly probe metabolite concentrations (Griffin, 2004). Metabolic fluxes can be studied by deuterium (2^{H}) NMR and carbon (13^{C}) NMR techniques that allow following labelled compound and assess changes in specific metabolic pathways (de Graaf *et al.*, 2003; Pichumani *et al.*, 2016; Silva *et al.*, 2019).

Nevertheless, it should be noted that the complete assignment of a metabolome is virtually impossible, especially for less abundant compounds and unstable or transient metabolites (reaction intermediates). In such cases, specially designed experiments have to be devised for their detection (Serkova *et al.*, 2007; Mannina *et al.*, 2008; Bernini *et al.*, 2011; Qwele *et al.*, 2013). Ultimately it is the sensitivity and the resolving capacity of the combined methods that define the number of observable molecules (Griffin, 2004).

Regardless of the selected technique, metabolomic studies provide information on the concentration and structure of multiple metabolites that require additional processing and interpretation (Nicholson and Wilson, 2003; Beckonert *et al.*, 2007). Acquired spectra need processing to extract the workable information to be further interpreted. Data analysis can be performed following two main approaches that can be applied alone or in a complementary way.

The Untargeted or Chemometric Analysis bypasses the need for the previous identification of the metabolites and instead attempts to analyse differences in a spectra set. The spectra are divided into a number of sections called 'bins' or 'buckets' and the area under the spectrum line is determined in each bin, for all the spectra under analysis. The area of each bin is then compared with the corresponding ones from all the spectra set. The sample clustering can then be assessed by Multivariate Statistical Analysis, from which the most common models are Principal Component Analysis (PCA), Partial Least-Squares Discriminant Analysis (PLS-DA), and Orthogonal Projections to Latent Structures (OPLS). PCA is an unsupervised analysis and gives information about the general sample clustering and helps identification of outliers. PLS and OPLS are supervised analyses and could be used to identify the features (variables important in projection) responsible for the groups' separation (Moestue *et al.*, 2011).

The Targeted or Profiling Approach resorts to the identification and quantification of the metabolites present in a sample. Metabolite identification is based on a series of NMR experiments traditionally used for compound structure identification in organic chemistry (Beckonert *et al.*, 2007). The most common metabolites are already extensively listed in literature and databases such as the the Human Metabolome Database (HMDB) that can be queried to assist compound identification. Metabolite identification can also be semi-automatically addressed using specific commercial software or open-access online platforms. The list of identified compounds with their respective concentration can then be analysed using Univariate Analysis and/or Multivariate Analysis.

NMR-Metabolomics constitutes a powerful approach to identify the small molecules present in most of the biological system and organisms. Few studies on rabbit metabolomics have been developed, and in all of them rabbit is used as a model organism for specific pathologies (Tessem *et al.*, 2005, 2006; van Vliet *et al.*, 2013; Ibarra *et al.*, 2014; Andreadou *et al.*, 2015; Shi *et al.*, 2015; Zelentsova *et al.*, 2016; Locci *et al.*, 2019). These studies highlighted the importance of the rabbit as a model, especially for studies on human diseases (Mapara *et al.*, 2012). However, these investigations pointed out the potential of this approach to investigate rabbit metabolome *per se*. Rabbits had also been frequently used as a model in Magnetic Resonance Imaging (MRI) studies (Li *et al.*, 2012; Illa *et al.*, 2013; Wei *et al.*, 2013; Yan *et al.*, 2013; Zhang *et al.*, 2013) and Magnetic Resonance Angiography (MRA) (Damianou *et al.*, 2014), which are not covered in this chapter.

Metabolomics offers a very powerful tool for the study of various species, and when combined with proteomics and transcriptomics, it has the potential to answer many fundamental questions on animal physiology (Griffin and Shockcor, 2004).

14.5 Conclusion

The so-called post-genomic tools (i.e. transcriptomics, proteomics and metabolomics) are becoming increasingly popular and disseminated across various areas of research. They are generating knowledge on a growing number of genes, proteins and metabolites that are relevant to understanding pathways of considerable importance to the development of sciences such as nutrition, reproduction, embryology, physiology, vaccine development, as well as animal production and food security. The integration of data from these three different omics with previous knowledge, integrating the information at the whole organism level, will necessarily have strong implications on health, disease and production, affecting all domestic animal species. The rabbit, being a species with a multi-importance, e.g. for production and as experimental model, will necessarily be in the forefront of the development of these sciences, with ultimate benefits to the society itself. To achieve such an objective, it will be necessary for rabbit researchers to reinforce or gain access to state-of-the-art, high-throughput omics platforms and means, which are seldom available. With this chapter we have transmitted the basics of the principles behind these techniques, as well as their advantages and drawbacks when specifically applied to this species. We hope therefore that it will become an incentive for animal scientists to become increasingly aware of the advantages and possibilities of omics in the context of rabbit research to complement genetics in this species.

Acknowledgements

Funding from *Fundação para a Ciência e a Tecnologia* is acknowledged in the form of grants SFRH/BD/85391/2012 (M. Palma), SFRH/BPD/90916/2012 (A.M. Almeida) and research contracts from the Ciência (A.M. Almeida, S.S. Araújo and M. Matzapetakis) and Investigador FCT (M. Matzapetakis) programmes, current PhD holder contract of S.S. Araújo (DL57) and through research unit GREEN-it 'Bioresources for Sustainability' (UID/Multi/04551/2019). A.M. Almeida and V. Duranthon were members of the TD1101 COST action RGB-net, collaborative European network on rabbit genome biology, funded by the European Science Foundation to which networking funding is acknowledged.

References

Albert, F.W., Somel, M., Carneiro, M., Aximu-Petri, A., Halbwax, M. *et al.* (2012) A comparison of brain gene expression levels in domesticated and wild animals. *PLOS Genetics* 8, e1002962.

Almeida, A.M. (2013) Rabbit muscle proteomics: a great leap forward. *Proteomics* 13, 2225–2226.

Almeida, A.M. and Bendixen, E. (2012) Pig proteomics: a review of a species in the crossroad between biomedical and food sciences. *Journal of Proteomics* 75, 4296–4314.

Almeida, A.M., Campos, A., Van Harten, S., Cardoso, L.A. and Coelho, A.V. (2009) Establishment of a proteomic reference map for the gastrocnemius muscle in the rabbit (*Oryctolagus cuniculus*). *Research in Veterinary Science* 87, 196–199.

Almeida, A.M., Campos, A., Francisco, R., Van Harten, S., Cardoso, L.A. and Coelho, A.V. (2010) Proteomic investigation of the effects of weight loss in the gastrocnemius muscle of wild and NZW rabbits via 2D-electrophoresis and MALDI-TOF MS. *Animal Genetics* 41, 260–272.

Al-Quraishy, S., Metwaly, M.S., Dkhil, M.A., Abdel-Baki, A.A. and Wunderlich, F. (2012) Liver response of rabbits to *Eimeria coecicola* infections. *Parasitology Research* 110, 901–911.

Ambros, V. (2004) The functions of animal microRNAs. *Nature* 431, 350–355.

Andreadou, I., Benaki, D., Efentakis, P., Bibli, S.I., Milioni, A.I. *et al.* (2015) The natural olive constituent oleuropein induces nutritional cardioprotection in normal and cholesterol-fed rabbits: comparison with preconditioning. *Planta Medica* 81, 655–663.

Ansorge, W.J. (2009) Next-generation DNA sequencing techniques. *New Biotechnology* 25, 195–203.

Bai, Q., Desprat, R., Klein, B., Lemaitre, J.M. and De Vos, J. (2013) Embryonic stem cells or induced pluripotent stem cells? A DNA integrity perspective. *Current Gene Therapy* 13, 93–98.

Barba, I., Jaimez-Auguets, E., Rodriguez-Sinovas, A. and Garcia-Dorado, D. (2007) 1H NMR-based metabolomic identification of at-risk areas after myocardial infarction in swine. *Magnetic Resonance Materials in Physics* 20, 265–271.

Beckonert, O., Keun, H.C., Ebbels, T.M.D., Bundy, J., Holmes, E. *et al.* (2007) Metabolic profiling, metabolomic and metabonomic procedures for NMR spectroscopy of urine, plasma, serum and tissue extracts. *Nature Protocols* 2, 2692–2703.

Benjamini, Y. and Hochberg, Y. (1995) Controlling the false discovery rate: a practical and powerful approach for multiple testing. *Journal of the Royal Statistics Society: Series B* 57, 289–300.

Ben Sellem, D., Elbayed, K., Neuville, A., Moussallieh, F.M., Lang-Averous, G. *et al.* (2011) Metabolomic characterization of ovarian epithelial carcinomas by HRMAS-NMR spectroscopy. *Journal of Oncology* 2011, 174019.

Bernini, P., Bertini, I., Luchinat, C., Nincheri, P., Staderini, S. and Turano, P. (2011) Standard operating procedures for pre-analytical handling of blood and urine for metabolomic studies and biobanks. *Journal of Biomolecular NMR* 49, 231–243.

Blaise, B.J., Giacomotto, J., Triba, M.N., Toulhoat, P., Piotto, M. *et al.* (2009) Metabolic profiling strategy of *Caenorhabditis elegans* by whole-organism nuclear magnetic resonance. *Journal of Proteome Research* 8, 2542–2550.

Bouley, J., Chambon, C. and Picard, B. (2004) Mapping of bovine skeletal muscle proteins using two-dimensional gel electrophoresis and mass spectrometry. *Proteomics* 4, 1811–1824.

Brewis, I.A. and Brennan, P. (2010) Proteomics technologies for the global identification and quantification of proteins. *Advances in Protein Chemistry and Structural Biology* 80, 1–44.

Bryant, D.W., Priest, H.D. and Mockler, T.C. (2012) Detection and quantification of alternative splicing variants using RNA-seq. *Methods in Molecular Biology* 883, 97–110.

Campos, R., Storz, J.F. and Ferrand, N. (2012) Copy number polymorphism in the α-globin gene cluster of European rabbit (*Oryctolagus cuniculus*). *Heredity* 108, 531–536.

Carneiro, M., Afonso, S., Geraldes, A., Garreau, H., Bolet, G. *et al.* (2011) The genetic structure of domestic rabbits. *Molecular Biology and Evolution* 28, 1801–1816.

Carneiro, M., Albert, F.W., Melo-Ferreira, J., Galtier, N., Gayral, P. *et al.* (2012) Evidence for widespread positive and purifying selection across the European rabbit (*Oryctolagus cuniculus*) genome. *Molecular Biology and Evolution* 29, 1837–1849.

Carneiro, M., Rubin, C.J., Di Palma, F., Albert, F.W., Alföldi, J. *et al.* (2014) Rabbit genome analysis reveals a polygenic basis for phenotypic change during domestication. *Science* 345, 1074–1079.

Casel, P., Moreews, F., Lagarrigue, S. and Klopp, C. (2009) sigReannot: an oligo-set re-annotation pipeline based on similarities with the Ensembl transcripts and Unigene clusters. *BMC Proceedings* 16, 3 Suppl. 4, S3.

Chaze, T., Bouley, J., Chambon, C., Barboiron, C. and Picard, B. (2006) Mapping of alkaline proteins in bovine skeletal muscle. *Proteomics* 6, 2571–2575.

Chekulaeva, M. and Rajewsky, N. (2019) Roles of long noncoding RNAs and circular RNAs in translation. In: Mathews, M.B., Sonenberg, N. and Hershey, J.W.B. (eds) *Translation Mechanisms and Control*. Cold Spring Harbor Perspectives in Biology. Cold Spring Harbor Press, Cold Spring Harbor, New York, 3;11(6), ii: a032680.

Cheng, L.L, Chang, I.W., Smith, B.L. and Gonzalez, R.G. (1998) Evaluating human breast ductal carcinomas with high-resolution magic-angle spinning proton magnetic resonance spectroscopy. *Journal of Magnetic Resonance* 135, 194–202.

Cheng, Y., Ji, R., Yue, J., Yang, J., Liu, X. *et al.* (2007) MicroRNAs are aberrantly expressed in hypertrophic heart: do they play a role in cardiac hypertrophy? *American Journal of Pathology* 170, 1831–1840.

Costa, V., Angelini, C., De Feis, I. and Ciccodicola, A. (2010) Uncovering the complexity of transcriptomes with RNA-Seq. *Journal of Biomedicine & Biotechnology*, 853916.

Cui, J., Liu, Y., Hu, Y., Tong, J., Li, A. *et al.* (2017) NMR-based metabonomics and correlation analysis reveal potential biomarkers associated with chronic atrophic gastritis. *Journal of Pharmaceutical and Biomedical Analysis* 132, 77–86.

Damianou, C., Hadjisavvas, V., Mylonas, N., Couppis, A. and Ioannides, K. (2014) MRI-Guided sonothrombolysis of rabbit carotid artery. *Journal of Stroke and Cerebrovascular Diseases* 23, e113–121.

de Graaf, R.A., Mason, G.F., Patel, A.B., Behar, K.L. and Rothman, D.L. (2003) In vivo 1H-[13C]-NMR spectroscopy of cerebral metabolism. *NMR in Biomedicine* 16, 339–357.

Dkhil, M.A., Abdel-Maksoud, M.A., Al-Quraishy, S., Abdel-Baki, A.A. and Wunderlich, F. (2012) Gene expression in rabbit appendices infected with *Eimeria coecicola*. *Veterinary Parasitology* 186, 222–228.

Dkhil, M.A., Al-Quraishy, S., Abdel-Baki, A.A., Delic, D. and Wunderlich, F. (2013) *Eimeria coecicola*: spleen response of *Oryctolagus cuniculus*. *Experimental Parasitology* 133, 137–143.

Donoghue, P., Doran, P., Wynne, K., Pedersen, K., Dunn, M.J. and Ohlendieck, K. (2007) Proteomic profiling of chronic low-frequency stimulated fast muscle. *Proteomics* 7, 3417–3430.

Eddy, S.R. (2001) Non-coding RNA genes and the modern RNA world. *Nature Reviews. Genetics* 2, 919–929.

Fisher, R.A. (1938) Presidential address by Professor R.A. Fisher, Sc.D., FRS Sankhyā: *The Indian Journal of Statistics* (1933–1960) 4, 14–17.

Gentleman, R.C., Carey, V.J., Bates, D.M., Bolstad, B., Dettling, M. *et al.* (2004) Bioconductor: open software development for computational biology and bioinformatics. *Genome Biology* 5, 10, R80.

German, J.B., Hammock, B.D. and Watkins, S.M. (2005) Metabolomics: building on a century of biochemistry to guide human health. *Metabolomics* 1, 3–9.

Griffin, J.L. (2004) Metabolic profiles to define the genome: can we hear the phenotypes? *Philosophical Transactions of the Royal Society of London Series B-Biological Sciences* 359, 857–871.

Griffin, J.L. and Shockcor, J.P. (2004) Metabolic profiles of cancer cells. *Nature Reviews Cancer* 4, 551–561.

Hunter, P. (2009) Reading the metabolic fine print: the application of metabolomics to diagnostics, drug research and nutrition might be integral to improved health and personalized medicine. *EMBO Reports* 10, 20–23.

Ibarra, R., Dazard, J.E., Sandlers, Y., Rehman, F., Abbas, R. *et al.* (2014) Metabolomic analysis of liver tissue from the VX2 rabbit model of secondary liver tumors. *HBP Surgery* 2014, 310372.

Illa, M., Eixarch, E., Batalle, D., Arbat-Plana, A., Muñoz-Moreno, E. *et al.* (2013) Long-term functional outcomes and correlation with regional brain connectivity by MRI diffusion tractography metrics in a near-term rabbit model of intrauterine growth restriction. *PLOS ONE* 8, e76453.

Jacquier, V., Estellé, J., Schmaltz-Panneau, B., Lecardonnel, J., Moroldo, M. *et al.* (2015) Genome-wide immunity studies in the rabbit: transcriptome variations in peripheral blood mononuclear cells after in vitro stimulation by LPS or PMA-Ionomycin. *BMC Genomics* 23, 16–26.

Jäger, M., Ott, C.E., Grunhagen, J., Hecht, J., Schell, H. *et al.* (2011) Composite transcriptome assembly of RNA-seq data in a sheep model for delayed bone healing. *BMC Genomics* 12, 158.

Jiménez, B., Mirnezami, R., Kinross, J., Cloarec, O., Keun, H.C. *et al.* (2013) 1H HR-MAS NMR Spectroscopy of tumor-induced local metabolic 'field-effects' enables colorectal cancer staging and prognostication. *Journal of Protcome Research* 12, 959–968.

Kozomara, A. and Griffiths-Jones, S. (2011) miRBase: integrating microRNA annotation and deep-sequencing data. *Nucleic Acids Research* 39, D152–157.

Kuang, L., Lei, M., Li, C., Guo, Z., Ren, Y. *et al.* (2019) Whole transcriptome sequencing reveals that non-coding RNAs are related to embryo morphogenesis and development in rabbits. *Genomics*, ii: S0888–7543(19)30551-30558.

Kumar, S., Kumar, S., Roy, R., Rathore, A.S., Goel, M.M. *et al.* (2015) High Resolution Magic Angle Proton Magnetic Resonance Spectroscopy (HRMAS) in intact sentinel node biopsy from breast cancer patients: a new diagnostic tool. *Journal of Surgery* 11, 227–231.

Levy, S.E. and Myers, R.M (2016) Advancements in next-generation sequencing. *Annual Review of Genomics and Human Genetics* 17, 95–115.

Lewis, C. and Ohlendieck, K. (2010) Mass spectrometric identification of dystrophin isoform Dp427 by on-membrane digestion of sarcolemma from skeletal muscle. *Analytical Biochemistry* 404, 197–203.

Li, S.C., Liao, Y.L., Chan, W.C., Ho, M.R., Tsai, K.W. *et al.* (2011) Interrogation of rabbit miRNAs and their isomiRs. *Genomics* 98, 453–459.

Li, Z., Shen, H., Zhang, Y., Lu, M., Qiao, X. *et al.* (2012) Metabolomic study of serum from rabbits with acute acalculous cholecystitis. *Inflammation Research* 61, 987–995.

Libri, V., Miesen, P., van Rij, R.P. and Buck, A.H. (2013) Regulation of microRNA biogenesis and turnover by animals and their viruses. *Cell and Molecular Life Sciences* 70, 3525–3544.

Lindblad-Toh, K., Garber, M., Zuk, O., Lin, M.F., Parker, B.J. *et al.* (2011) A high-resolution map of human evolutionary constraint using 29 mammals. *Nature* 478, 476–482.

Liu, L., Li, Y., Li, S., Hu, N., He, Y. *et al.* (2012) Comparison of next-generation sequencing systems. *Journal of Biomedicine & Biotechnology* 2012, 251364.

Liu, Z., Du, X., Yin, C. and Chang, Z. (2013) Shotgun proteomic analysis of sarcoplasmic reticulum preparations from rabbit skeletal muscle. *Proteomics* 13, 2335–2338.

Locci, E., Stocchero, M., Noto, A., Chighine, A., Natali, L. *et al.* (2019) A 1H NMR metabolomic approach for the estimation of the time since death using aqueous humour: an animal model. *Metabolomics* 15, 76.

Lucas, L.H., Wilson, S.F., Lunte, C.E. and Larive, C.K. (2005) Concentration profiling in rat tissue by high-resolution magic-angle spinning NMR spectroscopy: investigation of a model drug. *Analytical Chemistry* 77, 2978–2984.

Macomber, R.S. (1998) *A Complete Introduction to Modern NMR Spectroscopy*. Wiley-Interscience, New York.

Malone, J.H. and Oliver, B. (2011) Microarrays, deep sequencing and the true measure of the transcriptome. *BMC Biology* 9, 34.

Mannina, L., Sobolev, A.P., Capitani, D., Iaffaldano, N., Rosato, M.P. *et al.* (2008) NMR metabolic profiling of organic and aqueous sea bass extracts: implications in the discrimination of wild and cultured sea bass. *Talanta* 77, 433–444.

Mapara, M., Thomas, B.S. and Bhat, K.M. (2012) Rabbit as an animal model for experimental research. *Dental Research Journal* 9, 111–118.

Marengo, E., Robotti, E. and Bobba, M. (2008) 2D-PAGE maps analysis. *Methods in Molecular Biology* 428, 291–325.

Marguerat, S. and Bähler, J. (2010) RNA-seq: from technology to biology. *Cellular and Molecular Life Sciences* 67, 569–579.

Markley, J.L., Brüschweiler, R., Edison, A.S., Eghbalnia, H.R., Powers, R. *et al.* (2017) The future of NMR-based metabolomics. *Current Opinion in Biotechnology* 43, 34–40.

Martinez, I., Bathen, T., Standal, I.B., Halvorsen J., Aursand, M. *et al.* (2005) Bioactive compounds in cod (*Gadus morhua*) products and suitability of 1H NMR metabolite profiling for classification of the products using multivariate data analyses. *Journal of Agricultural and Food Chemistry* 53, 6889–6895.

Merrifield, C.A., Lewis, M., Claus, S.P., Beckonert, O.P., Dumas, M.E. *et al.* (2011) A metabolic system-wide characterisation of the pig: a model for human physiology. *Molecular BioSystems* 7, 2577–2588.

Metzker, M.L. (2010) Sequencing technologies – the next generation. *Nature Reviews Genetics* 11, 31–46.

Miller, I. (2012) Application of 2D DIGE in animal proteomics. *Methods in Molecular Biology* 854, 373–396.

Miller, I., Rogel-Gaillard, C., Spina, D., Fontanesi, L. and de Almeida, A.M. (2014) The rabbit as an experimental and production animal: from genomics to proteomics. *Current Protein and Peptide Science* 15, 134–145.

Millis, K.K., Maas, W.E., Cory, D.G. and Singer, S. (1997) Gradient, high-resolution, magic-angle spinning nuclear magnetic resonance spectroscopy of human adipocyte tissue. *Magnetic Resonance in Medicine* 38, 399–403.

Moestue, S., Sitter, B., Bathen, T.F., Tessem, M.B. and Gribbestad, I.S. (2011) HR MAS MR spectroscopy in metabolic characterization of cancer. *Current Topics in Medicinal Chemistry* 11, 2–26.

Mortazavi, A., Williams, B.A., McCue, K., Schaeffer, L. and Wold, B. (2008) Mapping and quantifying mammalian transcriptomes by RNA-Seq. *Nature Methods* 5, 621–628.

Nagana Gowda, G.A. and Raftery, D. (2015) Can NMR solve some significant challenges in metabolomics? *Journal of Magnetic Resonance* 260, 144–160.

Naturil-Alfonso, C., Saenz-de-Juano, M.D., Peñaranda, D.S., Vicente, J.S. and Marco-Jiménez, F. (2012) Transcriptome profiling of rabbit parthenogenetic blastocysts developed under in vivo conditions. *PLOS ONE* 7, 12.

Nestor, G., Bankefors, J., Schlechtriem, C., Brännäs, E., Pickova, J. and Sandström, C. (2010) High-resolution 1H magic angle spinning NMR spectroscopy of intact Arctic char (*Salvelinus Alpinus*) muscle: quantitative analysis of n-3 fatty acids, EPA and DHA. *Journal of Agricultural and Food Chemistry* 58, 10799–10803.

Nicholson, J.K. and Wilson, I.D. (2003) Understanding 'global' systems biology: metabonomics and the continuum of metabolism. *Nature Reviews Drug Discovery* 2, 668–676.

O'Day, E. and Lal, A. (2010) MicroRNAs and their target gene networks in breast cancer. *Breast Cancer Research* 12, 201.

Oliver, S.G., Winson, M.K., Kell, D.B. and Baganz, F. (1998) Systematic functional analysis of the yeast genome. *Trends in Biotechnology* 16, 373–378.

Osteil, P., Tapponnier, Y., Markossian, S., Godet, M., Schmaltz-Panneau, B. *et al.* (2013) Induced pluripotent stem cells derived from rabbits exhibit some characteristics of naïve pluripotency. *Biology Open* 2, 613–628.

Osteil, P., Moulin, A., Santamaria, C., Joly, T., Jouneau, L. *et al.* (2016). A panel of embryonic stem cell lines reveals the variety and dynamic of pluripotent states in rabbits. *Stem Cell Reports* 7, 383–398.

Picardi, E. and Pesole, G. (2013) REDItools: high-throughput RNA editing detection made easy. *Bioinformatics* 29, 1813–1814.

Pichumani, K., Mashimo, T., Vemireddy, V., Kovacs, Z., Ratnakar, J. *et al.* (2016) Hepatic gluconeogenesis influences 13C enrichment in lactate in human brain tumors during metabolism of [1,2-13C]acetate. *Neurochemistry International* 97, 133–136.

Prochniewicz, E., Lowe, D.A., Spakowicz, D.J., Higgins, L., O'Conor, K. *et al.* (2008) Functional, structural, and chemical changes in myosin associated with hydrogen peroxide treatment of skeletal muscle fibers. *American Journal of Physiology – Cell Physiology* 294, 613–626.

Quail, M.A., Smith, M., Coupland, P., Otto, T.D., Harris, S.R. *et al.* (2012) A tale of three next generation sequencing platforms: comparison of Ion Torrent, Pacific Biosciences and Illumina MiSeq sequencers. *BMC Genomics* 13, 341.

Quinn, E.M., Cormican, P., Kenny, E.M., Hill, M., Anney, R. *et al.* (2013) Development of strategies for SNP detection in RNA-seq data: application to lymphoblastoid cell lines and evaluation using 1000 genomes data. *PLOS ONE* 8, e58815.

Qwele, K., Hugo, A., Oyedemi, S.O., Moyo, B., Masika, P.J. and Muchenje, V. (2013) Chemical composition, fatty acid content and antioxidant potential of meat from goats supplemented with Moringa (*Moringa oleifera*) leaves, sunflower cake and grass hay. *Meat Science* 93, 455–462.

Rajewsky, N. (2006) microRNA target predictions in animals. *Nature Genetics* 38 Suppl, S8–13.

Rezig, L., Servadio, A., Torregrossa, L., Miccoli, P., Basolo, F. *et al.* (2018) Diagnosis of post-surgical fine-needle aspiration biopsies of thyroid lesions with indeterminate cytology using HRMAS NMR-based metabolomics. *Metabolomics* 14, 141.

Roberts, A., Pimentel, H., Trapnell, C. and Pachter, L. (2011) Identification of novel transcripts in annotated genomes using RNA-Seq. *Bioinformatics* 27, 2325–2329.

Rogowska-Wrzesinska, A., Le Bihan, M.C., Thaysen-Andersen, M. and Roepstorff, P.J. (2013) 2D gels still have a niche in proteomics. *Proteomics* 88, 4–13.

Saenz-de-Juano, M.D., Marco-Jiménez, F., Peñaranda, D.S., Joly, T. and Vicente, J.S. (2012) Effects of slow freezing procedure on late blastocyst gene expression and survival rate in rabbit. *Biology of Reproduction* 87, 91.

Saenz-de-Juano, M.D., Marco-Jimenez, F., Schmaltz-Panneau, B., Jimenez-Trigos, E., Viudes-de-Castro, M.P. *et al.* (2014) Vitrification alters rabbit foetal placenta at transcriptomic and proteomic level. *Reproduction* 147, 789–801.

Saha, K. and Jaenisch, R. (2009) Technical challenges in using human induced pluripotent stem cells to model disease. *Cell Stem Cell* 5, 584–595.

Sánchez-Pla, A., Reverter, F., Ruíz de Villa, M.C. and Comabella, M. (2012) Transcriptomics: mRNA and alternative splicing. *Journal of Neuroimmunology* 248, 23–31.

Santone, C., Dinallod, V., Paci, M., D'Ottavio, S., Barbato, G. and Bernardini, S. (2014) Saliva metabolomics by NMR for the evaluation of sport performance. *Journal of Pharmaceutical and Biomedical Analysis* 88, 441–446.

Schmaltz-Panneau, B., Jouneau, L., Osteil, P., Tapponnier, Y., Afanassieff, M. *et al.* (2014) Contrasting transcriptome landscapes of rabbit pluripotent stem cells in vitro and in vivo. *Animal Reproduction Science* 149, 67–79.

Schuster, S.C. (2008) Next-generation sequencing transforms today's biology. *Nature Methods* 5, 16–18.

Serkova, N.J., Rose J.C., Epperson, L.E., Carey, H.V. and Martin, S.L. (2007) Quantitative analysis of liver metabolites in three stages of the circannual hibernation cycle in 13-lined ground squirrels by NMR. *Physiological Genomics* 31, 15–24.

Serkova, N.J., Davis, D.M., Steiner, J. and Agarwal, R. (2019) Quantitative NMR-based metabolomics on tissue biomarkers and its translation into in vivo magnetic resonance spectroscopy. *Methods in Molecular Biology* 1978, 369–387.

Shi, Q., Yanhua, K., He, B., Chen, X., Yana, Y. and Li, Y. (2015) Metabolomics study on serum of allergic bronchial asthma rabbits treated by recuperating lung decoction. *RSC Advances* 5, 13768–13776.

Shioya, M., Obayashi, S., Tabunoki, H., Arima, K., Saito, Y. *et al.* (2010) Aberrant microRNA expression in the brains of neurodegenerative diseases: miR-29a decreased in Alzheimer disease brains targets neurone navigator 3. *Neuropathology and Applied Neurobiology* 36, 320–330.

Silva, J.C., Marques, C., Martins, F.O., Viegas, I., Tavares, L. *et al.* (2019) Determining contributions of exogenous glucose and fructose to de novo fatty acid and glycerol synthesis in liver and adipose tissue. *Metabolic Engineering* 56, 69–76.

Smyth, G.K. (2004) Linear models and empirical Bayes methods for assessing differential expression in microarray experiments. *Statistical Applications in Genetics and Molecular Biology* 3, Article 3.

Soares, R., Franco, C., Pires, E., Ventosa, M., Palhinhas, R. *et al.* (2012) Mass spectrometry and animal science: protein identification strategies and particularities of farm animal species. *Journal of Proteomics* 75, 4190–4206.

Soon, W.W., Hariharan, M. and Snyder, M.P. (2013) High-throughput sequencing for biology and medicine. *Molecular Systems Biology* 9, 640.

Staunton, L. and Ohlendieck, K. (2012) Mass spectrometric characterization of the sarcoplasmic reticulum from rabbit skeletal muscle by on-membrane digestion. *Protein and Peptide Letters* 19, 252–263.

Sultan, M., Schulz, M.H., Richard, H., Magen, A., Klingenhoff, A. *et al.* (2008) A global view of gene activity and alternative splicing by deep sequencing of the human transcriptome. *Science* 321, 956–960.

Tancos, Z., Nemes, C., Polgar, Z., Gocza, E., Daniel, N. *et al.* (2012) Generation of rabbit pluripotent stem cell lines. *Theriogenology* 78, 1774–1786.

Tapponnier, Y., Afanassieff, M., Aksoy, I., Aubry, M., Moulin, A. *et al.* (2017) Reprogramming of rabbit induced pluripotent stem cells toward epiblast and chimeric competency using Krüppel-like factors. *Stem Cell Reseach* 24, 106–117.

Tessem, M.B., Bathen, T.F., Cejkova, J. and Midelfart, A. (2005) Effect of UV-A and UV-B irradiation on the metabolic profile of aqueous humor in rabbits analyzed by 1H NMR Spectroscopy. *Investigative Ophthalmology & Visual Science* 46, 776–781.

Tessem, M.B., Midelfart, A., Cejkova, J. and Bathen, T.F. (2006) Effect of UVA and UVB irradiation on the metabolic profile of rabbit cornea and lens analysed by HR-MAS 1H NMR Spectroscopy. *Ophthalmic Research* 38, 105–114.

Thiede, B., Höhenwarter, W., Krah, A., Mattow, J., Schmid, M. *et al.* (2005) Peptide mass fingerprinting. *Methods* 35, 237–247.

Tilgner, M., Vater, T.S., Habbel, P. and Cheng, L.L. (2019) High-Resolution Magic Angle Spinning (HRMAS) NMR methods in metabolomics. *Methods in Molecular Biology* 2037, 49–67.
Timms, J.F. and Cramer, R. (2008) Difference gel electrophoresis. *Proteomics* 8, 4886–4897.
Tusher, V.G., Tibshirani, R. and Chu, G. (2001) Significance analysis of microarrays applied to the ionizing radiation response. *Proceedings of the National Academy of Sciences* 98, 5116–5121.
van Vliet, E., Eixarch, E., Illa, M., Arbat-Plana, A., Gonzalez-Tendero, A. *et al.* (2013) Metabolomics reveals metabolic alterations by intrauterine growth restriction in the fetal rabbit brain. *PLOS ONE* 8, e64545.
Wang, Z., Gerstein, M. and Snyder, M. (2009) RNA-Seq: a revolutionary tool for transcriptomics. *Nature Reviews Genetics* 10, 57–63.
Wei, X.E., Li, Y.H., Zhao, H., Li, M.H., Fu, M. and Li, W.B. (2013) Quantitative evaluation of hyperbaric oxygen efficacy in experimental traumatic brain injury: an MRI study. *Neurological Sciences* 2013, 18.
Wevers, R.A., Engelke, U., Wendel, U., de Jong, J.G., Gabreels, F.J. and Heerschap, A. (1995) Standardized method for high-resolution 1H-NMR of cerebrospinal fluid. *Clinical Chemistry* 41, 744–751.
Wheeler, D.L., Barrett, T., Benson, D.A., Bryant, S.H., Canese, K. *et al.* (2008) Database resources of the National Center for Biotechnology Information. *Nucleic Acids Research* 36, D13–D21.
Wilhelm, B.T. and Landry, J.R. (2009) RNA-Seq-quantitative measurement of expression through massively parallel RNA-sequencing. *Methods* 48, 249–257.
Williams, A.E. (2008) Functional aspects of animal microRNAs. *Cellular and Molecular Life Sciences* 65, 545–562.
Yan, D., Zhang, J., Liang, W., Sun, J., Liu, B.Y. *et al.* (2013) Magnetic resonance imaging and histopathological analysis of experimental muscle injuries in a rabbit. *Biomedical and Environmental Sciences* 26, 841–848.
Yilmaz, A., Geddes, T., Han, B., Bahado-Singh, R.O., Wilson, G.D. *et al.* (2017) Diagnostic biomarkers of Alzheimer's disease as identified in saliva using 1H NMR-based metabolomics. *Journal of Alzheimer's Disease* 58, 355–359.
Zelentsova, E.A., Yanshole, L.V., Snytnikova, O.A., Yanshole, V.V., Tsentalovich, Y.P. and Sagdeev, R.Z. (2016) Post-mortem changes in the metabolomic compositions of rabbit blood, aqueous and vitreous humors. *Metabolomics* 12, 172.
Zhang, A.H., Sun, H., Qiu, S. and Wang, X.J. (2013) NMR-based metabolomics coupled with pattern recognition methods in biomarker discovery and disease diagnosis. *Magnetic Resonance in Chemistry* 51, 549–556.
Zhang, F., Zhang, R., Zhang, X., Wu, Y., Li, X. *et al.* (2018) Comprehensive analysis of circRNA expression pattern and circRNA-miRNA-mRNA network in the pathogenesis of atherosclerosis in rabbits. *Aging (Albany NY)* 10, 2266–2283.
Zhao, B., Chen, Y., Hu, S., Yang, N., Wang, M. *et al.* (2019) Systematic analysis of non-coding RNAs involved in the Angora rabbit (*Oryctolagus cuniculus*) hair follicle cycle by RNA sequencing. *Frontiers in Genetics* 10, 407.
Zhao, H.X., Niu, C.M. and Guan, W.Y. (2012) Microarray-based gene expression profiles in rabbit retina due to negative pressure suction. *Genetics and Molecular Research* 11, 2277–2287.

15 Methods to Create Transgenic and Genome-edited Rabbits

Zsuzsanna Bősze[1]*, Peter Chrenek[2, 3], Andreas Ritsch[4] and László Hiripi[1]

[1]National Agricultural Research and Innovation Centre, Agricultural Biotechnology Institute, Gödöllő, Hungary; [2]Department of Biochemistry and Biotechnology, Faculty of Biotechnology and Food Sciences, Slovak University of Agriculture in Nitra, Slovak Republic; [3]NPPC—Research Institute for Animal Production Nitra, Lužianky, Slovak Republic; [4]Department of Internal Medicine I, Innsbruck Medical University, Austria

15.1 Introduction

A transgenic/gene-targeted rabbit is one animal whose genome has been altered artificially (i) to introduce a mutation in a given gene, (ii) to alter one or more gene's expression pattern, or (iii) to modify the abundance of the target gene. The ultimate aim of these genetic modifications is to produce a desired phenotype, which does not exist in the naturally occurring breeds or populations. Prior to the current revolution in applied molecular genetics, the only practical method for studying the regulation and function of mammalian genes was to use spontaneous mutants. To reveal the genetic basis of the mutation, the animals had to transmit an observable trait to offspring. Successful attempts were made to isolate genetic defects on a different genomic background by performing congenic breeding experiments. The major problem with this method of genetic evaluation is that, along with the gene(s) of interest, a large amount of DNA flanking the mutant genetic locus is invariably transferred from animal to animal during meiotic recombination.

The first and, for a long time, the only available method to introduce exogenous recombinant DNA in rabbits was the pronuclear microinjection into the one-cell embryo. The efficiency of the pronucleus microinjection was and stayed very low. However, our ability to produce transgenic rabbits with increased efficiency has been improved dramatically during the last five years, through transfer of emerging new technologies established in laboratory mice. However, transfer to rabbit in some cases was not without difficulties. Numerous milestones have been achieved in the development of methods capable of producing transgenic rabbits (Table 15.1). Here we describe in detail the different approaches that have been used to produce transgenic and gene targeted rabbits and analyse their potential for future applications.

15.2 Pronucleus Microinjection

Artificial gene transfer is a form of genetic engineering. Genetic engineering, recombinant DNA technology, genetic modification/manipulation (GM) and gene splicing are terms that are applied to the manipulation of genes, generally implying that the process is outside the organism's natural reproductive process. It involves the isolation, manipulation and reintroduction of DNA into cells or model organisms, usually to express, as well as to enhance or inhibit, expression of a protein. Gene transfer involves two distinctly

Contact e-mail address: bosze.zsuzsanna@abc.naik.hu

DOI: 10.1079/9781780643342.0015

Table 15.1. A timeline of the significant achievements leading to the ability to produce transgenic and gene-targeted rabbits.

Milestones	References
Transgenic rabbit via pronuclear microinjection	Hammer *et al.* (1985)
Cloned rabbit from embryonic cells	Stice and Robl (1988)
Live rabbit from intracytoplasmic injection	Deng and Yang (2001)
Cloned rabbit from adult cumulus/fibroblast cells	Chesne *et al.* (2002); Li *et al.* (2006)
Artificial chromosome-type transgenesis in rabbit	Brem *et al.* (1996)
Transgenic rabbit with lentiviral transduction	Hiripi *et al.* (2010)
Gene-deleted rabbit with ZFN nuclease	Flisikowska *et al.* (2011)
Transposon-mediated transgenesis in rabbit	Katter *et al.* (2013)
Targeted gene deletion and knock-in with TALEN nuclease in rabbit	Song *et al.* (2013); Song *et al.* (2016)
Gene deletion and targeted mutations in rabbits by CRISPR/Cas9	Yang *et al.* (2014); Song *et al.* (2018)

separate processes. The first step in the process must provide a mechanism by which genetic information can be transported from extracellular space, across biological membranes, and into the nucleus so that the incoming genetic information comingles with the genome of the target organism. The second step in the process affords a means for the new genetic information to become part of the target genome.

The mutual contributions of developmental biology and genetic engineering permitted the developments of techniques for the creation of transgenic animals. DNA microinjection was the first successful technique in laboratory mouse (Gordon and Ruddle, 1981) and then to various other species such as rats, rabbits, sheep, pigs, birds and fish. Since 1981, when the term transgenic was introduced, there have been rapid developments in the use of genetically engineered animals derived by the increasing number of applications of this technology.

Gene transfer can be realized through:

(a) the fertilized eggs and embryos;
(b) the germ cells (sperms and oocytes);
(c) the gonads (ovary and testis); and
(d) the somatic cells.

This method involves the direct microinjection of a chosen gene construct (plasmid-based or artificial chromosome-type transgene) from another member of the same species or from a different species, into the pronucleus of a fertilized ovum. It is one of the first methods that proved to be effective in mammals. Microinjection in mice and rabbit one-cell embryos is relatively simple, thanks to the visible pronucleus (Fig. 15.1A). The total volume of the microinjected DNA is about 1–2 pL, which at a concentration of 1–2 ng/μL represents around 200–400 copies of DNA, depending on the type of gene construct. The fertilized oocytes are fixed on one side with the help of a holder pipette and are microinjected from the other end with the help of the microinjection capillary filled with the transgene solution using automatic or manual microinjector (Fig. 15.1B). The microinjection of the transgene solution results in a visible increase in the volume of the injected pronucleus (Fig. 15.1C). After microinjection and *in vitro* embryo culture, the viable embryos are transferred to the oviduct of recipient females,

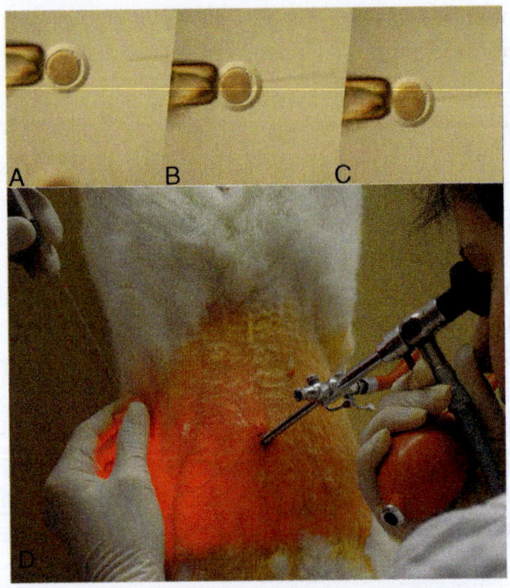

Fig. 15.1. Pronucleus microinjection and endoscopic embryo transfer: A–C: phases of pronucleus microinjection; D: embryo transfer with endoscopic method.

which were made pseudo-pregnant with hormonal treatment or by mating with vasectomized males. The integration of DNA is, however, a random process: the transgene copies integrate in tandem and neither the transgene copy number nor the integration site could be planned. Therefore, the plasmid-based transgenes with some probability became either silenced or show ectopic, unwanted expression patterns.

The first transgenic rabbit line was established in 1985 (Hammer *et al.*, 1985). In this pioneering experiment a well-characterized transgene (the human growth hormone gene under the mouse metallothionein I promoter) was microinjected into rabbit zygotes. Successful integration and expression were reported in rabbits, sheep and pigs. The pronucleus microinjection technology in rabbits is very similar to that in mouse with slight modifications. Rabbit zygotes are different in size so holder capillaries should also be different. Rabbit embryos produce a special mucoprotein coat on the zona pellucida (Fig. 15.1A–C). This layer is very important for implantation of rabbit embryos in the uterus. Although successful transfer of *in vitro* cultured blastocyst stage embryos was reported (Jin *et al.*, 2000), only a short culturing period is recommended for transgenic rabbit programmes. Most laboratories use a surgical method (Yang and Foote, 1990), but the endoscopic system is also an effective fast way for rabbit embryo transfer (Fig. 15.1D) (Besenfelder and Brem, 1993). The microinjection method has been improved and refined, but its efficiency is still about 1–3 (up to 10)%. It means that 100 microinjected zygotes result in average 1–3 founder rabbits. It is possible to double the efficiency by microinjection into both nuclei (Chrenek *et al.*, 2005), but this technique requires very well trained professional micromanipulators.

The specific feature of rabbit transgenesis is the high proportion of mosaic founders. Mosaicism is common in transgenic animals when foreign DNA is integrated into the chromosomal environment after the first division of the zygote. Rabbit embryos undergo a rapid series of cell division between two-cell stage and morula stage (Sultana *et al.*, 2009), when integration occurred in later stages, it resulted 30% mosaic founders (Castro *et al.*, 1999). Fortunately, this phenomenon did not cause any problem in germline transmission of the transgene.

Rabbit was also used for different sperm-mediated gene transfer trials in the past. The first live young from intracytoplasmic sperm injections were born in 2001 (Deng and Yang, 2001). In another study, rabbit sperm cells were treated by freezing without cryoprotectants, and their ability for DNA uptakc was determined, resulting in genetically modified rabbit blastocysts; however, transgenic rabbits and the inheritance of transgene expression has not been reported (Li *et al.*, 2010). Beyond that, the intracytoplasmic sperm injection requires a highly sophisticated microinjection system, which restricts its widespread application.

Transgenic rabbits produced by pronuclear microinjection are valuable animal models for human diseases (see Chapter 18).

Monoclonal antibodies from rabbits could not be generated, because a plasmacytoma cell line as a fusion partner was not available. To obtain this rabbit cell line, transgenic rabbits were generated carrying two transgenes, c-myc and v-abl (Spieker-Polet *et al.*, 1995). Since then, producing rabbit monoclonal antibodies became a successful business.

Transgenic animals can be used to produce pharmaceuticals in large amounts. Mostly mammary gland serves as bioreactor. Transgenic rabbit is a good alternative to larger livestock animals to produce protein in their milk. Detailed description of transgenic rabbits as bioreactors was published in 2013 (Wang *et al.*, 2013). The most successful story on this field is Ruconest® (Europe), Rhucin® (USA), which is a human recombinant protein produced by the Pharming Group. This recombinant human C1 inhibitor is on the market in Europe and in the USA. It is the second milk-derived recombinant protein, which reached the market. This product is manufactured in a transgenic rabbit platform for treatment of hereditary angioedema (www.pharming.com, accessed 6 March 2021). The most recent transgenic technologies like transposon-mediated transgenesis and DNA nuclease technologies are also based on pronuclear microinjection, so the renaissance of the first successful transgenic method has come through.

15.3 Artificial Chromosome-type Transgenesis

Traditional pronuclear microinjection is mostly used for the analysis of dominant gain-of-function mutations. Usually a transgene expression cassette harbours the gene of interest (in most cases cDNA) under the control of an artificial or tissue-specific promoter. To ensure the desired expression level, other regulatory regions are commonly used (homologous or heterologous introns, insulator sequences etc.). There are limitations in this traditional transgenic strategy. Plasmid vector has about 20 kb capacity for DNA cargo. For important gene regulatory elements such as enhancers, suppressors cannot be always fitted into the transgenic cassettes perfectly. Due to positional effects of transgene integrations and the lack of some regulatory elements, transgene expression may not be accurately simulated in a spatial and temporal manner of the endogenous gene. On the other hand, alternative splicing events are not appearing with traditional cDNA transgenic methods.

To prevent the limitations of the transgene cassette method, artificial chromosome-mediated transgenesis was developed and applied in transgenic animals. The most commonly used vectors are BAC- (bacterial artificial chromosome) or YAC-based (yeast artificial chromosome) but artificial mammalian minichromosomes are also possible to deliver (Co *et al.*, 2000). The first YAC transgenic mice were produced in 1992 (Schedl *et al.*, 1992). BAC transgenic mouse lines were established in 1997 (Yang *et al.*, 1997). Compared to the traditional method, BACs have up to 400 kb capacity while YACs can carry up to 1 Mb inserts. This capacity is usually enough for most mammalian genes to contain the most important 5' and 3' regulatory elements which can warrant the precise transgene expression levels in a copy number-dependent way. Latest results show that gene expression levels are also regulated by very far elements (ENCODE Project Consortium, 2004) but usually these elements play important roles in fine tuning of gene expression. Expression from an artificial chromosome-type transgene recapitulates the endogenous gene, including tissue specificity, alternative splicing, cell cycle and mRNA regulation by miRNAs. Unfortunately, on average, 100–300 kb genomic regions may contain other genes, which could interfere with the planned experiment.

BAC and YAC libraries are opened and offered widely. A BAC library from the rabbit is publicly available for the academic scientific community on a cost-recovery basis from the BACPAC Resources Center at the Children's Hospital of Oakland Research Institute (http://bacpac.chori.org/rabbit, accessed 19 January 2021). BAC recombineering kits are also available to make any modifications (insertions, deletions) easily and rapidly. Commercially available kits are based on λ-mediated Red/ET recombination (http://www.genebridges.com/, accessed 19 January 2021). The first artificial chromosome-based transgenic rabbit was produced in 1996 (Brem *et al.*, 1996). A YAC clone was microinjected and resulted in transgenic rabbits with a 250 kb long mouse tyrosinase gene to rescue the albino phenotype. It was the first demonstration of a large DNA insertion in farm animals, where the expression level of the transgene was as expected.

Transgenic rabbits expressing a high level of apolipoprotein (a) and apolipoprotein B were also produced with a YAC vector (Rouy *et al.*, 1998). These authors used a 270-kb YAC clone with the human apo (a) gene and a 90-kb P1 phagemid clone containing the apolipoprotein B gene to express either one or both transgenes in rabbits as a model of atherosclerosis and restenosis. Expression of both transgenes was tissue-specific in the liver.

A rabbit model of progressive retinal degeneration, a disease affecting more than one million people worldwide, was developed by Kondo *et al.* (2009). As a rabbit has large eyes and its eye anatomy and physiology is close to that of humans, it is an ideal animal model for studying eye diseases. Autosomal dominant mutations of rhodopsin gene are responsible for ~25–40% of progressive retinal degeneration cases. A BAC clone harbouring the rabbit rhodopsin gene was engineered to exchange Pro347 to Leu by Red/ET recombination (a C-to-T transition in exon 5). The mutated BAC clone was microinjected and resulted in the first rabbit model of progressive retinal degeneration. Those models are important in preclinical trials of treatments including surgical procedures such as intraocular devices, subretinal injection of genes for gene therapy, and implantation of retinal prostheses.

BAC transgenic rabbits were created with augmented humoral immune response (Catunda Lemos *et al.*, 2012). A transgenic rabbit with a 110 kb rabbit BAC clone was generated to overexpress the neonatal Fc receptor, which is the receptor for IgG molecules. The complex phenotype with the desired temporal and spatial expression levels could be achieved only by the use of the artificial type transgenesis.

Interestingly, in contrast to the conventional speculations, which assume that transgenes are integrating as intact molecules into chromosomes, two recent experiments sequenced the genome of transgenic mice and sheep with traditional transgenes (Chiang *et al.*, 2012) and a BAC transgenic mouse (Dubose *et al.*, 2013). These studies showed that DNA integration can be a very complex phenomenon with complex rearrangements in the chromosomes of the transgenic animals even using artificial chromosomes as transgenes. Therefore, it should be emphasized that microinjection of transgenes maintained in artificial chromosomes is technically demanding due to their large size and fragility. Deletions at both ends of the injected DNA or fragmentation might occur; therefore every individual transgene integration should be analysed rigorously.

15.4 Lentiviral Transgenesis

At the beginning of the 21st century, the ultimate transgenic technique was the lentiviral transgenesis. Lentiviral vectors have a capability to infect dividing and non-dividing cells with stable integration into the host genome. The most commonly used lentivirus vectors are derived from human immunodeficiency virus (HIV). Besides HIV, other lentiviral vectors can be used to produce transgenic animals. Equine infectious anaemia virus derived vector (Whitelaw *et al.*, 2004) was used to produce transgenic swine, while transgenic rhesus monkeys and rodents (Whitelaw *et al.*, 2004; Niu *et al.*, 2010; Bender *et al.*, 2013) were created with Simian Immunodeficiency Virus (SIV). The HIV genome contains 9 open reading frames encoding 15 proteins and further cis-acting elements required for the viral life cycle. For biosafety reasons, third-generation lentiviral vectors harbour only genes and cis-acting sequences which are required for particle production infection and integration. Generally, the lentiviral system consists of three or four vectors. A transfer vector with the gene of interest driven by a promoter of choice in a lentiviral backbone is required for genomic RNA production and packaging. This vector contains a woodchuck hepatitis virus posttranslational regulatory element (WPRE) to enhance transgene expression and a central polypurine tract which is important for nuclear transfer. The packaging vector system provides trans-acting factors like Gag-pol, Rev (integrase, reverse transcriptase) and envelope protein usually originated from Vesicular Stomatitis Virus (VSV-G). To further increase the safety, a deleted 3' long terminal repeat (LTR) is used resulting in replication defective particles. The vector system mentioned before is being transfected into 293T human kidney cell line, which produces lentiviral particles. For successful transgenesis at least 10^8 infectious units per mL should be used. Lentiviral vectors are usually delivered into the perivitelline space of zygotes or oocytes (Pfeifer, 2006).

Lentiviral transgenesis has some practical disadvantages. Its maximal cargo capacity is about 10 kb, which is not necessarily enough when special expression profiles are required or when only genomic constructs are planned to be used. Of note, the lentiviral transgenesis, especially if the concentration of the injected viral stock is high, might result in multiple integration events (Ritchie *et al.*, 2009). Since lentiviral transduction results are of one transgene copy integrated in each integration site, in case of more than one integrated transgene copy, they are inherited independently from each other on the different chromosomes. This phenomenon is aggravating to establish well-characterized transgenic lines and probably can cause unwanted effects like oncogene activation or insertional mutagenesis. An additional problem is gene silencing of the transgene which is caused by DNA methylation due to the presence of sequences with viral origin (Hofmann *et al.*, 2006). High frequency of mosaicism in founder animals was also reported.

Contrary to the successful application of lentiviral transgenesis in different livestock species (Hofmann *et al.*, 2004; Whitelaw *et al.*, 2004; Cornetta *et al.*, 2013), this method turned out to be inefficient in rabbit. Only one published paper described the production of transgenic rabbits using the Simian Immunodeficiency Virus (Hiripi *et al.*, 2010). Green fluorescent protein as a reporter gene under the CAG promoter was delivered by perivitelline space microinjection into one-cell-stage rabbit zygotes (lentivirus integrates into the genome through its existing receptor on the zygote membrane; therefore it is enough to introduce the transgene solution under the zona pellucida, which is less harmful to embryos than pronucleus microinjection). Among the 87 offspring in the F0 generation, 28 were found to be transgenic by PCR and 18 expressed the transgene. Transgene expression was revealed in tissues derived from all three primary germ layers in all founders tested. This data seemed very promising; however, all founders showed mosaic expression of the transgene. Astonishingly, germ-line transmission of the transgene was very limited. Less than 1% of F1 generation expressed the transgene. Single-copy integration of the transgene was most likely due to the low titer of the SIV vector.

Similarities between human and rabbit embryonal development and placentation made rabbit a popular model in pregnancy-related toxicological studies (Pentsuk and van der Laan, 2009; Bourdon *et al.*, 2018) and in revealing the effect of maternal Diabetes mellitus on early embryogenesis (Ramin *et al.*, 2010). Another study showed that the GFP reporter gene expression in an HIV-derived lentiviral construct was found in the placenta and yolk sac but foetuses never expressed the transgene (Skoda *et al.*, 2017).

In conclusion, lentiviral transgenesis was demonstrated in rabbit and the direction of its future application might be an HIV vector-based extraembryonic specific gene expression to create human placentation defective models for translational studies.

15.5 Transposon-mediated Transgenesis

In the last few years a promising alternative transgenic method has been developed based on DNA transposons. DNA transposons are mobile genetic elements which can integrate into the genome of the host cell by a simple 'cut-and-paste' mechanism. That is why DNA transposons became an alternative non-infectious gene-delivery machine. The newly developed transposon vectors cut out the transgene of interest flanked by inverted terminal repeats from the plasmid and integrate it into the host genome (Fig. 15.2). The excisions from the plasmid vector and the integration to genomic region at special target sequences are catalysed by a transposase. The transposase can be provided on a plasmid DNA or even better as an *in vitro* synthetized transcript. Using transposase mRNA form ensures a shorter half-life, which helps to avoid side effects. To date, different transposon-based systems have been used to produce transgenic mammals. The *Sleeping Beauty* was applied in rodents (Dupuy *et al.*, 2002; Carlson *et al.*, 2011), pigs (Garrels *et al.*, 2011) and rabbits (Katter *et al.*, 2013) and *PiggyBac* (Ding *et al.*, 2005) and *Tol2* (Sumiyama *et al.*, 2010) in mice.

Sleeping Beauty belongs to the TC1/mariner family of DNA transposons and was reconstructed by molecular methods from a fish (Izsvak *et al.*, 2000). The hyperactive variant (SB 100X) was later developed to increase its efficacy in biotechnological processes (Mates *et al.*, 2009). The SB100X vector system can be used in non-dividing cells also with low silencing activity in mammalian cells (Grabundzija *et al.*, 2010). The *Sleeping Beauty* transposon directs integration into the genome quasi randomly (its target sequence is AT) avoiding coding region of genes ('gene-free zones' or introns). Contrary to this, the *PiggyBac* transposon targets transcriptionally active regions (Wilson *et al.*, 2007; Ammar *et al.*, 2012).

The first transposon-mediated transgenesis in rabbit was reported by Katter *et al.* (2013) using 0.4 ng/μl circular pT2/*Venus* plasmid (*Venus* is a reporter gene driven by CAG promoter) and 5 ng/μl *SleepingBeauty* 100X transposase mRNA. This cocktail was microinjected into the pronucleus of early rabbit zygotes. A total of 46 newborns from 10 successful embryo transfers were born. The reported transgenic rate was 15% of live founder rabbits. All F1 and later-generation transgenic rabbits expressed the *Venus* reporter gene ubiquitously. Interestingly, transgenic silencing was not registered in rabbits; however it was detected in some cases in transgenic mice. Based on the available information, the transposon-based method seems to be far more efficient than any other way of additive transgenesis in rabbit. The complete transposon-mediated transgenesis protocol in rabbit was published (Ivics *et al.*, 2014).

This method has only one limiting factor; over 7 kb transgene integration efficiency is reduced in the SB system (Izsvak *et al.*, 2000), but in *PiggyBac* transposase-mediated transgenesis, the BAC-sized transgenes were integrated at high frequency via pronuclear microinjection (Rostovskaya *et al.*, 2013). In summary, transposon-mediated transgenesis could advance not just rabbit but also farm-animal transgenesis in general through providing an efficient alternative

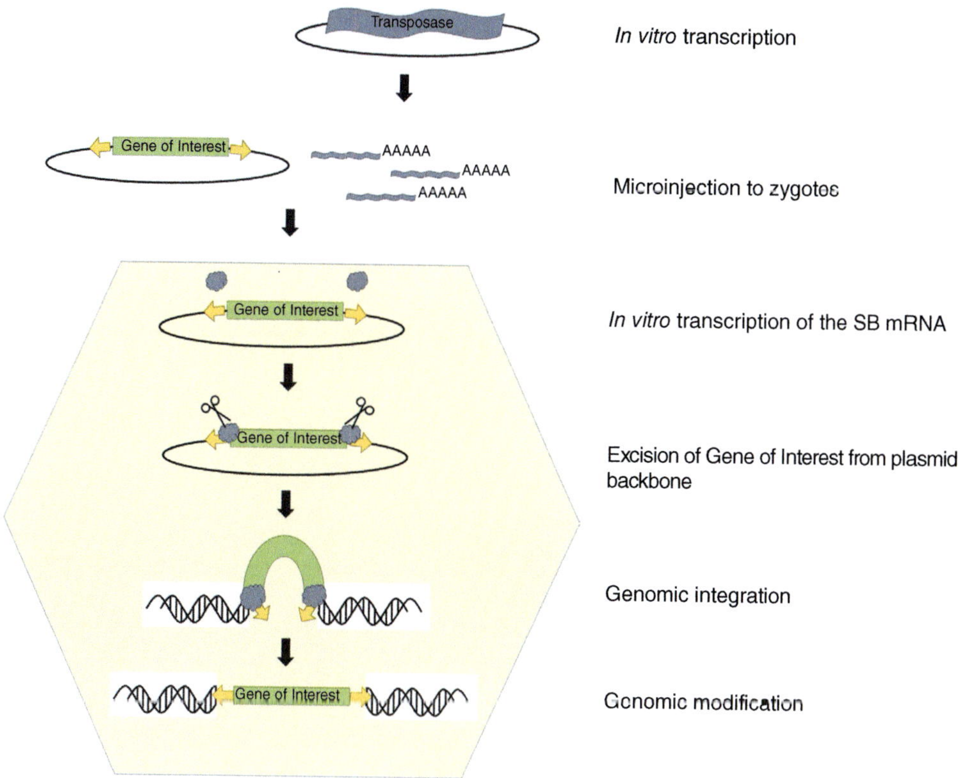

Fig. 15.2. The bicomponent vector system and mode of action in transposon-mediated transgenesis (Izsvák *et al.*, 2000).

and cost-effective method for agricultural, pharmaceutical and biomedical applications.

15.6 The Role of Somatic Nuclear Transfer in Rabbit Transgenesis

During the last decade, somatic cell nuclear transfer (SCNT) became an efficient and routinely used method to create genetically modified swine and ruminant models (for review: Kues and Niemann, 2011). Somatic cells, which were pre-selected *in vitro* for a targeted gene modification can be introduced into enucleated oocytes by SCNT to produce live transgenic animals. However, contrary to mice, pigs and ruminants, the efficiency of SCNT in rabbits, regardless of the type of nuclear donor cells, using standard methods, is very low and only a negligible percentage of offspring remain healthy and reach sexual maturity (Chesne *et al.*, 2002; Zakhartchenko *et al.*, 2011). The experimental strategy of rabbit nuclear transfer is similar to the methods used in other species: the main steps are enucleation of oocyte, selection of donor cells, activation of the reconstructed embryo and embryo transfer (Fig. 15.3). The first liveborn rabbits from nuclear transfer were produced 30 years ago, from eight cell-stage rabbit blastomeres (Stice and Robl, 1988). SCNT using adult rabbit cells as donors were not published until 2002, when a French research group used cumulus cells as nucleus donors and reported that healthy and fertile rabbits were born from the transfer of the reconstructed embryos (Chesne *et al.*, 2002). Among the difficulties of increasing the efficiency of rabbit somatic cloning are the rapid kinetics of the cell cycle of early rabbit embryonal development (Sultana *et al.*, 2009) and the narrow window of time for

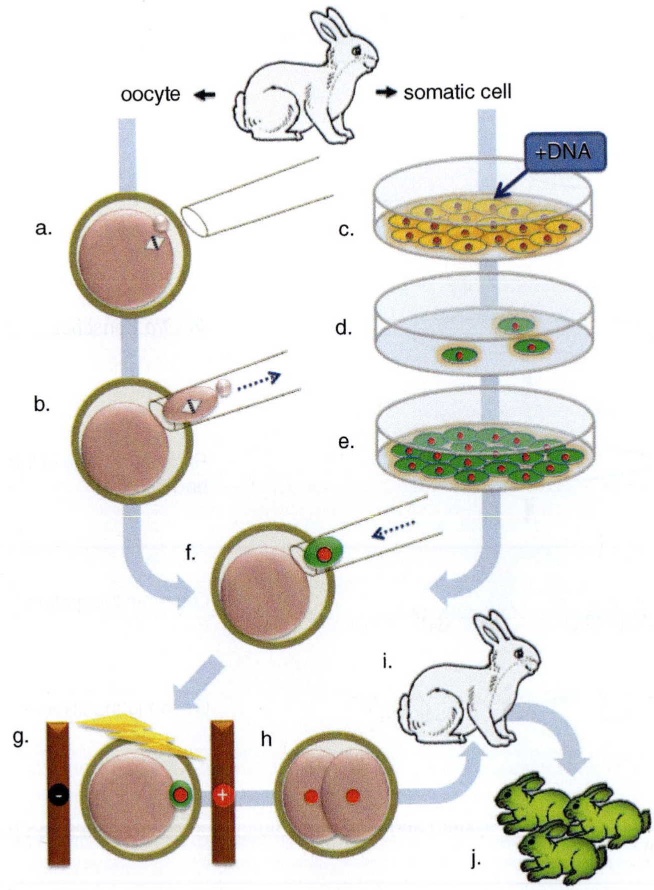

Fig. 15.3. Steps of rabbit somatic cell nuclear transfer: (a) matured oocyte; (b) enucleation; (c) *in vitro* transfection of cultured donor cells; (d) selection of transgenic cells; (e) *in vitro* culture of transgenic cells; (f) insertion of donor cell into enucleated oocyte; (g) fusion and activation; (h) developing reconstructed embryo; (i) embryo transfer into recipient foster does; (j) phenotypic analysis of transgenic offspring.

their implantation following embryo transfer. Recently, two independent methods were published to enhance the effectiveness of SCNT in rabbits. Optimization of parthenogenetic activation of rabbit oocytes and the use of oocytes selected by brilliant cresyl-blue staining both enhance cloned rabbit embryo development (Jia *et al.*, 2019; Zhang *et al.*, 2019). From the point of transgenesis, the somatic cell cloning would be attractive if gene-targeted embryonal stem (or induced pluripotent) cell lines would be available, which were not reported so far (Madeja *et al.*, 2019). Therefore this method has limited practical use in rabbit transgenesis; however, gene-targeted rabbits by cloning were produced using recombinant adeno-associated virus-mediated homologous recombination in fibroblast cells following SCNT that has as targeted locus the hypoxanthine phosphoribosyltransferase gene (Yin *et al.*, 2015).

15.7 Adenoviral Transduction into Rabbit Somatic Cells for Local *In Vivo* Gene Transfer

Recombinant adenoviral vectors provide an efficient, feasible, clinically applicable technique for arterial gene transfer to various cell types including liver, endothelial and smooth muscle cells. Durable, high-level expression of recombinant genes, using *in vivo* gene transfer is a potential powerful tool for understanding biological processes and for preventing or treating human diseases such as atherosclerosis and thrombosis by gene therapy. The first experiments in rabbits were performed by Steg *et al.* (1994) and since then, efforts to optimize the expression cassettes continued to ensure high expression levels and tissue specificity, as well as to avoid toxicity and prevent immune responses (Dronadula *et al.*, 2011). Fig. 15.4A shows the maps of adenovirus genome serotype 5 and different generations of recombinant adenoviral vectors. First-generation adenoviral vectors contain the entire genome of wild-type virus except for early genes E1 and E3. As E1 is essential for virus replication, these recombinant vectors infect target cells but are not replicated. However, these vectors can be easily produced by simple infection of packaging cell lines stably expressing the E1 gene (Fig. 15.4B). For use in animal systems, recombinant viral particles have to be purified from cell lysates of infected packaging cells. This can be done via density gradient ultracentrifugation. However, higher yields and ratios of infective particles can be achieved using anion exchange chromatography. Accordingly, 2×10^{13} adenoviral particles of high purity can be isolated from 60 mL crude lysate of infected packaging cells (Tancevski *et al.*, 2006). Second-generation vectors contain less viral sequences leading to decreased immunotoxicity. To further improve in this direction, vector development finally led to the generation of gutless or helper-dependent recombinant adenoviral vectors in which viral sequences are restricted to inverted terminal repeats and the packaging signal (Fig. 15.4A). However, production of this vector requires transfection of packaging cell lines with corresponding helper plasmids or helper viruses in addition to the recombinant viral vector (Alba *et al.*, 2005). Fig. 15.4B describes the main steps of adenoviral vector production. Adenoviral vectors have been successfully employed to study the function of certain genes in rabbit systems. Liver cells can be transfected simply by intravenous injection of recombinant adenoviral particles. Accordingly, intravenous injection of 1×10^{12} particles has been shown to result in enhanced expression of scavenger receptor BI in liver cells leading to major changes of plasma lipoprotein profiles in treated New Zealand White rabbits (Tancevski *et al.*, 2005). Additionally, *in situ* injection of adenoviral particles was used to express various transgenes in cardiomyocytes and skeletal muscle cells (Flores-Munoz *et al.*, 2012). In an atherosclerotic rabbit model, expression of ATP-binding cassette transporter G1 via vessel wall-directed local adenoviral gene transfer attenuated arteriosclerosis and endothelial dysfunction in New Zealand White rabbits (Munch *et al.*, 2012), which further demonstrated the practical applicability of this method.

15.8 Genome Editor/Designer Nucleases

The development of targeted nuclease technologies like Zinc-Finger Nucleases (ZFN), Transcriptional Activator-Like Effector Nucleases (TALEN) and RNA-Guided Nucleases (CRISPR/Cas9) has completely changed the potential of transgenic technology by opening new perspectives in livestock genetic modifications in the last years. Targeted gene deletions or modifications have expanded to new research fields in new species. All these technologies are based on programmable, sequence-specific DNA-binding followed by a double-stranded DNA break at the targeted site. The activity of these nucleases recruits non-homologous end-joining or using donor plasmids inducing homology-directed repair at the locus of interest. These technologies promise all types of targeted genetic modifications: knock-out, knock-in, allele exchange, introduction of point mutations, large insertions and deletions, down- and upregulation of genes. The resulting genetic modifications are not specific to any species as the systems work in all animals tested to the present. Genetic modification using 'designer nucleases' can be carried out without leaving any trace of the modification. Only the desired mutation will occur. For an overview of genome engineering see Gaj *et al.* (2013). These fresh technologies have already hit rabbit biotechnology as described in Duranthon *et al.* (2012).

In rabbit, the first application was a Zinc Finger Nuclease targeting event. ZFNs were introduced into fertilized rabbit oocytes, where

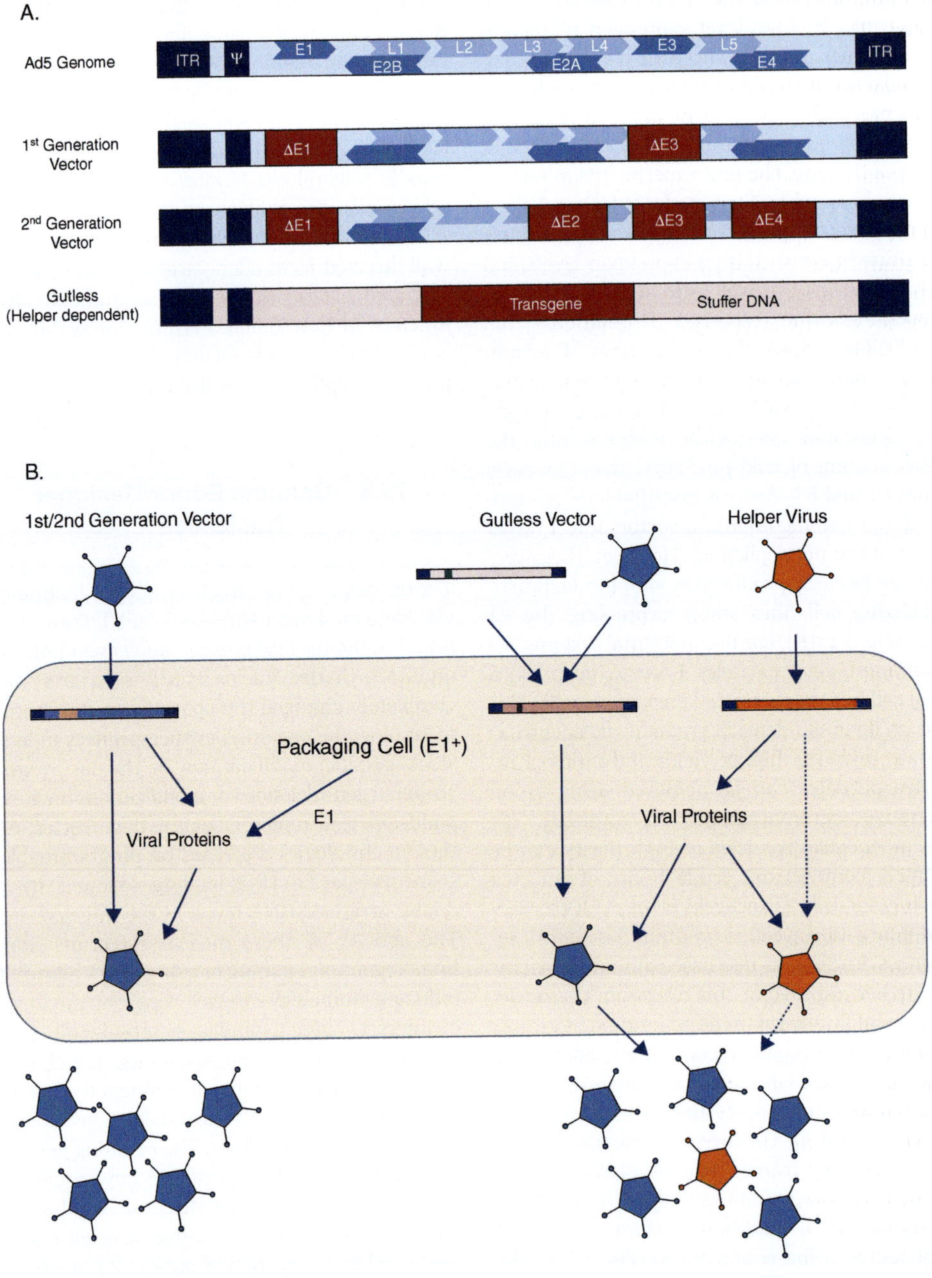

Fig. 15.4. A – map of adenovirus genome serotype 5 and different generations of adenoviral vectors. Positions and direction of early transcripts (E1-E4) and late transcripts (L1-L5) are indicated by arrows. First-generation vectors are containing the whole adenoviral genome with the exception of the E1 and E3 regions, which are replaced by transgene sequences. However, expression of viral genes is causing an immune response towards infected cells leading to the loss of transgene expression. Second-generation vectors were developed lacking different early transcript regions. This enhances cloning capacity but is not able to completely avoid associated immunogenicity due to residual expression of viral genes. Third-generation vectors, called gutless or helper-dependent vectors, are devoid of all viral coding sequences.

the immunoglobulin M (IgM) locus was disrupted. The functional knock-out phenotype was established with high efficiency in a heritable way (Flisikowska *et al.*, 2011). This should be the first step to create a transgenic rabbit platform for the production of therapeutic human polyclonal antibodies.

Another publication announced the first TALEN knock-out rabbits (Song *et al.*, 2013). In this study, RAG-1- and RAG-2-deficient rabbits without mature T and B cells were created. The efficiency of genetic modification was extremely high. These rabbits could be used as valuable animal models for drug discovery. TALEN efficiency to produce targeted events (knock-in events also) was enhanced in rabbit embryos *in vitro* and *in vivo* using homology-directed repair enhancer RS-1 (Song *et al.*, 2016). The first transgenic rabbits produced by RNA-Guided Nucleases (CRISPR/Cas9) were reported by Yang *et al.* (2014) who successfully knocked out nine rabbit genes in *in vitro* cultured rabbit embryos. Four gene-specific RNA mixtures were used to create founders and eventually establish novel knocked-out rabbit lines (CD36, LDLR, RyR2, and APOE). Some founder rabbits carried a biallelic mutation for the targeted gene. That means that it was possible to create a one-step homozygous knocked-out rabbit. Off-target effects, at least in the potential exons, were not identified. Honda *et al.* (2014) effectively disrupted the rabbit tyrosinase gene (*TYR*) and confirmed germline transmission by pronuclear injection of circular plasmid expressing humanized Cas9 and the single-guide RNA. Those pilot single-gene editing studies were followed with multi-gene knocked-out rabbit models in which three or five genes were simultaneously targeted in a single embryo, with high efficiency (Yan *et al.*, 2014). Myostatin knocked-out rabbits with CRISPR/Cas9 were produced by Lv *et al.* (2016). The rabbits showed the characteristic double-muscle phenotype and had no reproduction problems (Lv *et al.*, 2016). The same research group published two congenital cataract rabbit models, both of them created with CRISPR/Cas9 method targeting two different genes (Yuan *et al.*, 2016; Yuan *et al.*, 2017). Gene-mutation efficiency in the GJA8 locus reached 98.7% in embryos and 100% in pups. Mutations in all 19 F0-generation pups (100%) with indel mutations in the αA-crystallin gene, ranging from 3 bp to 52 bp, were reported. Off-target assay for the αA-crystallin gene revealed that none of the potential off-target sites exhibited mutations, demonstrating that off-target mutagenesis was not induced by cytoplasmic microinjection of *in vitro*-transcribed Cas9 mRNA. Recently, two muscular dystrophy model gene-targeted rabbits were published, underlining the efficiency of the CRISPR/Cas9 approach. The rabbit model of Duchenne muscular dystrophy (DMD) was created with a dual sgRNA-directed CRISPR/Cas9 (Sui *et al.*, 2018a). In the original publication the authors targeted the exon number 51 of dystrophin gene to create a duchenne muscular dystrophy model. Mutation in the top ten off-target sites was not detected in the DMD-knocked-out rabbits.

Limb-girdle muscular dystrophy type 2L (LGMD2L) and Miyoshi myopathy type 3 (MMD3) are autosomal recessive muscular dystrophy caused by mutations in the gene encoding anoctamin-5 (ANO5). The rabbit model was created with a pair of sgRNA targeting exon 12 and exon 13 of rabbit anoctamin-5 gene. Fifty per cent of newborn pups carried a mutation in the target gene (Sui *et al.*, 2018b).The top 12 potential off-target sites in those ANO5-KO rabbits were analysed but no mutations were detected.

Fig. 15.4. Continued
and contain only inverted terminal repeat (ITR) and packaging signal (ψ) regions. Packaging efficiency is dependent on a vector genome size close to that of the wild-type virus (36 kb). Therefore, stuffer DNA has to be included to ensure an efficient packaging of the vector genome. B – production of recombinant adenoviral particles. First- and second-generation adenoviral particles lack the E1 gene, which is needed for replication of the viral genome. Production of these particles is therefore enabled by the use of a transgenic packaging cell line expressing the adenoviral E1 gene. To produce gutless vectors, proteins needed for replication, packaging and capsid formation have to be supplied in trans. This can be achieved by co-transfection of the recombinant vector with a helper adenovirus. To avoid production of helper viruses the packaging efficiency of the helper virus has to be reduced by altering helper virus genome size, by mutating its packaging signal or by using strategies to eliminate the packaging signal of the helper virus during viral production.

Another rabbit model generated using the CRISPR/Cas9 system was produced to study severe combined immunodeficiency (X-SCID) (Hashikawa *et al.*, 2020).

CRISPR/Cas9 system was used to target not only autosomes or chromosome X but also the sex chromosome Y in rabbit (SSCY). SRY is a sex determination gene on chromosome Y. Interestingly, genome-edited but chimeric F0 rabbits were diagnosed with hermaphroditism, characterized by possessing ovotestis, testis, ovary and uterus simultaneously. This rabbit model would be optimal for understanding the pathogenesis of hermaphroditism (Song *et al.*, 2018).

Targeted knock-in rabbits have also been developed in the last few years. As an example, the VP6 gene of the rotavirus was site-specifically integrated into the rabbit β-casein locus by CRISPR/Cas9. The aim was to design rotavirus vaccine suitable for both mammary-gland-based production and milk-based administration (Li *et al.*, 2019).

Recently, an inactive Cas9 variant (dead Cas9, dCas9) has been fused to diverse functional domains for targeting genetic and epigenetic modifications, including base editing. Base-editing systems can produce the desired point mutations in the genome. Base editing was successfully tested and applied in rabbits. xCas9-derived base editors, exBE4 and exABE, can dramatically improve the base-editing efficiencies in rabbits (Liu *et al.*, 2018; Liu *et al.*, 2019a). Further optimization of base editors enabled efficient conversion of the C·G base pair to T·A in GC rich region of the genome in rabbits (Liu *et al.*, 2019b).

All these techniques seem promising although ZFN technology is less comfortable and flexible, while for the highly efficiency of CRISPR (50–100% mutant founder rabbits) more *in vivo* data will be needed to analyse the potential off-target effect compared to the competitor technologies.

15.9 The Influence of Transgene Expression on Productivity Traits

The need for a systematic study of biological properties in transgenic livestock animals is a consequence of their potential commercial utilization in various spheres of human activity. Possible interactions of the integrated transgenes with the given genotype of the livestock animal are one of the intensively discussed problems, which prevented the authorization of any kind of transgenic animals in the food chain. Another risk factor is the emigration of integrated foreign genes from the experimental population into production herds. Therefore, a detailed knowledge on the relations between performance indexes and the expression of the integrated gene is a prerequisite for the objective evaluation of the potential benefits in transgenesis of rabbit as a production animal.

15.9.1 Effect of transgene expression on milk quality and lactation

Milk is the sole source of nourishment of the newborns and remains so until the offspring become three weeks old. Rabbit milk yield may be affected by the breed of doe (Lukefahr *et al.*, 1983a), number of kits suckling and their age at weaning (Lukefahr *et al.*, 1983b; Taranto, 2003), a consecutive pregnancy while still lactating (Lukefahr *et al.*, 1983c). High-level recombinant human protein production by the mammary gland of a transgenic rabbit might influence the lactation curve and/or the milk composition.

Six generations of transgenic rabbits with stable integration and production of biologically active IGF-1 without any negative effect on their physiological or reproductive performance were reported (Zinovieva *et al.*, 1998). Since then the majority of transgenic rabbit lines producing recombinant proteins in their milk avoided any side effect. For an overview of recombinant milk production see Bosze and Houdebine (2006). However, side effects resulting from premature mammary gland involution were published in a few cases: e.g. in a transgenic rabbit line, which expressed very high level of enzymatically active alkaline phosphatase in its milk. It is important to note that lactating does belonging to another transgenic line, but expressing the same transgene at one magnitude lower of quantity, were totally healthy and fertile (Bodrogi *et al.*, 2006). Later on, transgenic rabbits were created, which expressed a glycosyltransferase, FucT1 in their milk, which interfered with the lactation and resulted in mammary gland involution in less than six days after parturition (Prieto, 2012). It is important to emphasize the need to examine, case by case, each transgenic line for unwanted side effects, influencing the welfare of the experimental animals and always to be in compliance with the *Guide for the Care and Use of Laboratory Animals* (USA NIH publication NO 85-23, revised 1996) and to conform to the Directive 2010/63/EU of the European Parliament.

15.9.2 Meat quality of transgenic rabbits

In the current practice, specialized rabbit lines are used for meat production that maximizes heterosis and complementary effects in the generation of commercial hybrids. Lines are created on the basis of multi-breed crossing and various types of selection. The outcomes are populations of animals with fixed genes for the complex of maternal properties (e.g. numerical productivity, homogeneity of pups, longevity of females and milk production) and for intensive growth and fattening capacity of the male lines (weight, meat yield, food conversion ratio, resistance to digestive disease). For more detailed information see Chapters 10 and 13. Since differences in meat-quality parameters are constant among genotypes, Chrenek *et al.* (2012) examined if a transgene with expression restricted to the mammary gland, but present in all cells of the organism as genomic information, could influence meat quality. After careful analysis, they concluded that the altered genome of the transgenic rabbits did not influence meat quality (Chrenek *et al.*, 2012). It is an important observation, if we foresee transgenic livestock animals in the food chain, in which transgene expression is restricted to non-consumed tissues.

15.10 Conclusion

Methods to modify specific genome sequences and, based on them, the opportunities to improve biomedicine using rabbit as model or producer of pharmaceutically important recombinant proteins became unlimited recently. The potential of precise directed genetic modifications paved the way to evaluate the effect of naturally occurring allelic variants on economically important traits, using *in vivo* models on the most appropriate genetic background, thereby contributing to an improved production animal.

Acknowledgements

The authors thank S. Bodó (Fig. 15.3) and O.I. Hoffmann (Fig. 15.2) for their artwork. This work was supported by OTKA grants NK 104397, NN108921 and by NVKP_16-.1-2016-0039.

References

Alba, R., Bosch, A. and Chillon, M. (2005) Gutless adenovirus: last-generation adenovirus for gene therapy. *Gene Therapy* 12, S18–S27.

Ammar, I., Gogol-Doring, A., Miskey, C., Chen, W., Cathomen, T. *et al.* (2012) Retargeting transposon insertions by the adeno-associated virus Rep protein. *Nucleic Acids Research* 40, 6693–6712.

Bender, B., Hoffmann, O.I., Negre, D., Kvell, K., Bosze, Z. and Hiripi, L. (2013. Low titer lentiviral transgenesis in rodents with simian immundeficiency virus vector. *BioTechniques* 55, 137–140.

Besenfelder, U. and Brem, G. (1993) Laparoscopic embryo transfer in rabbits. *Journal of Reproduction and Fertility* 99, 53–56.

Bodrogi, L., Brands, R., Raaben, W., Seinen, W., Baranyi, M. *et al.* (2006) High level expression of tissue-nonspecific alkaline phosphatase in the milk of transgenic rabbits. *Transgenic Research* 15, 627–636.

Bosze, Z. and Houdebine, L.M. (2006) Application of rabbits in biomedical research: a review. *World Rabbit Science* 14, 1–14.

Bourdon, M., Torres-Rovira, L., Monniaux, D., Faure, C., Levy, R. *et al.* (2018) Impact of a gestational exposure to diesel exhaust on offspring gonadal development: experimental study in the rabbit. *Journal of Developmental Origins of Health and Disease* 9, 519–529.

Brem, G., Besenfelder, U., Aigner, B., Muller, M., Liebl, I. *et al.* (1996) YAC transgenesis in farm animals: rescue of albinism in rabbits. *Molecular Reproduction and Development* 44, 56–62.

Carlson, D.F., Garbe, J.R., Tan, W., Martin, M.J., Dobrinsky, J.R. *et al.* (2011) Strategies for selection marker-free swine transgenesis using the Sleeping Beauty transposon system. *Transgenic Research* 20, 1125–1137.

Castro, F.O., Limonta, J., Rodriguez, A., Aguirre, A., de la Fuente, J. *et al.* (1999) Transgenic rabbits for the production of biologically-active recombinant proteins in the milk. *Genetic Analysis: Biomolecular Engineering* 15, 179–187.

Catunda Lemos, A.P., Cervenak, J., Bender, B., Hoffmann, O.I., Baranyi, M. *et al.* (2012) Characterization of the rabbit neonatal Fc receptor (FcRn) and analyzing the immunophenotype of the transgenic rabbits that overexpresses FcRn. *PLOS ONE* 7, e28869.

Chesne, P., Adenot, P.G., Viglietta, C., Baratte, M., Boulanger, L. and Renard, J.P. (2002) Cloned rabbits produced by nuclear transfer from adult somatic cells. *Nature Biotechnology* 20, 366–369.

Chiang, C., Jacobsen, J.C., Ernst, C., Hanscom, C., Heilbut, A. *et al.* (2012) Complex reorganization and predominant non-homologous repair following chromosomal breakage in karyotypically balanced germline rearrangements and transgenic integration. *Nature Genetics* 44, 390-397, S391.

Chrenek, P., Vasicek, D., Makarevich, A.V., Jurcik, R., Suvegova, K. *et al.* (2005) Increased transgene integration efficiency upon microinjection of DNA into both pronuclei of rabbit embryos. *Transgenic Research* 14, 417–428.

Chrenek, P., Makarevich, A.V., Kozelova, D., and Rafay, J. (2012) Meat quality of transgenic rabbit. *Slovak Journal of Animal Science* 45, 60–62.

Co, D.O., Borowski, A.H., Leung, J.D., van der Kaa, J., Hengst, S. *et al.* (2000) Generation of transgenic mice and germline transmission of a mammalian artificial chromosome introduced into embryos by pronuclear microinjection. *Chromosome Research* 8, 183–191.

Cornetta, K., Tessanne, K., Long, C., Yao, J., Satterfield, C. and Westhusin, M. (2013) Transgenic sheep generated by lentiviral vectors: safety and integration analysis of surrogates and their offspring. *Transgenic Research* 22, 737–745.

Deng, M. and Yang, X.J. (2001) Full term development of rabbit oocytes fertilized by intracytoplasmic sperm injection. *Molecular Reproduction and Development* 59, 38–43.

Ding, S., Wu, X., Li, G., Han, M., Zhuang, Y. and Xu, T. (2005) Efficient transposition of the piggyBac (PB) transposon in mammalian cells and mice. *Cell* 122, 473–483.

Dronadula, N., Du, L., Flynn, R., Buckler, J., Kho, J. *et al.* (2011) Construction of a novel expression cassette for increasing transgene expression in vivo in endothelial cells of large blood vessels. *Gene Therapy* 18, 501–508.

Dubose, A.J., Lichtenstein, S.T., Narisu, N., Bonnycastle, L.L., Swift, A.J. *et al.* (2013) Use of microarray hybrid capture and next-generation sequencing to identify the anatomy of a transgene. *Nucleic Acids Research* 41, e70.

Dupuy, A.J., Clark, K., Carlson, C.M., Fritz, S., Davidson, A.E. *et al.* (2002) Mammalian germ-line transgenesis by transposition. *Proceedings of the National Academy of Sciences of the USA* 99, 4495–4499.

Duranthon, V., Beaujean, N., Brunner, M., Odening, K.E., Santos, A.N. *et al.* (2012) On the emerging role of rabbit as human disease model and the instrumental role of novel transgenic tools. *Transgenic Research* 21, 699–713.

ENCODE Project Consortium (2004) The ENCODE (ENCyclopedia of DNA elements) project. *Science* 306, 636–640.

Flisikowska, T., Thorey, I.S., Offner, S., Ros, F., Lifke, V. *et al.* (2011) Efficient immunoglobulin gene disruption and targeted replacement in rabbit using zinc finger nucleases. *PLOS ONE* 6, e21045.

Flores-Munoz, M., Godinho, B.M., Almalik, A. and Nicklin, S.A. (2012) Adenoviral delivery of angiotensin-(1-7) or angiotensin-(1-9) inhibits cardiomyocyte hypertrophy via the mas or angiotensin type 2 receptor. *PLOS ONE* 7, e45564.

Gaj, T., Gersbach, C.A. and Barbas, C.F. (2013) ZFN, TALEN, and CRISPR/Cas-based methods for genome engineering. *Trends in Biotechnology* 31, 397–405.

Garrels, W., Mates, L., Holler, S., Dalda, A., Taylor, U. *et al.* (2011) Germline transgenic pigs by Sleeping Beauty transposition in porcine zygotes and targeted integration in the pig genome. *PLOS ONE* 6, e23573.

Gordon, J.W. and Ruddle, F.H. (1981) Integration and stable germ line transmission of genes injected into mouse pronuclei. *Science* 214, 1244–1246.

Grabundzija, I., Irgang, M., Mates, L., Belay, E., Matrai, J. *et al.* (2010) Comparative analysis of transposable element vector systems in human cells. *Molecular Therapy* 18, 1200–1209.

Hammer, R.E., Pursel, V.G., Rexroad, C.E., Jr., Wall, R.J., Bolt, D.J. *et al.* (1985) Production of transgenic rabbits, sheep and pigs by microinjection. *Nature* 315, 680–683.

Hashikawa, Y., Hayashi, R., Tajima, M., Okubo, T., Azuma, S. *et al.* (2020) Generation of knockout rabbits with X-linked severe combined immunodeficiency (X-SCID) using CRISPR/Cas9. *Scientific Reports* 10, 9957.

Hiripi, L., Negre, D., Cosset, F.L., Kvell, K., Czompoly, T. *et al.* (2010) Transgenic rabbit production with simian immunodeficiency virus-derived lentiviral vector. *Transgenic Research* 19, 799–808.

Hofmann, A., Zakhartchenko, V., Weppert, M., Sebald, H., Wenigerkind, H. *et al.* (2004) Generation of transgenic cattle by lentiviral gene transfer into oocytes. *Biology of Reproduction* 71, 405–409.

Hofmann, A., Kessler, B., Ewerling, S., Kabermann, A., Brem, G. *et al.* (2006) Epigenetic regulation of lentiviral transgene vectors in a large animal model. *Molecular Therapy* 13, 59–66.

Honda, A., Hirose, M., Sankai, T., Yasmin, L., Yuzawa, K. *et al.* (2014) Single-step generation of rabbits carrying a targeted allele of the tyrosinase gene using CRISPR/Cas9. *Experimental Animals* 64, 31–37.

Ivics, Z., Hiripi, L., Hoffmann, O.I., Mates, L., Yau, T.Y. *et al.* (2014) Germline transgenesis in rabbits by pronuclear microinjection of Sleeping Beauty transposons. *Nature Protocols* 9, 794–809.

Izsvak, Z., Ivics, Z. and Plasterk, R.H. (2000) Sleeping Beauty, a wide host-range transposon vector for genetic transformation in vertebrates. *Journal of Molecular Biology* 302, 93–102.

Jia, L.Y., Ding, B., Shen, C., Luo, S.W., Zhang, Y.R. *et al.* (2019) Use of oocytes selected by brilliant cresyl blue staining enhances rabbit cloned embryo development in vitro. *Zygote* 27, 166–172.

Jin, D.I., Kim, D.K., Im, K.S. and Choi, W.S. (2000) Successful pregnancy after transfer of rabbit blastocysts grown in vitro from single-cell zygotes. *Theriogenology* 54, 1109–1116.

Katter, K., Geurts, A.M., Hoffmann, O., Mates, L., Landa, V. *et al.* (2013) Transposon-mediated transgenesis, transgenic rescue, and tissue-specific gene expression in rodents and rabbits. *FASEB Journal* 27, 930–941.

Kondo, M., Sakai, T., Komeima, K., Kurimoto, Y., Ueno, S. *et al.* (2009) Generation of a transgenic rabbit model of retinal degeneration. *Investigative Ophthalmology & Visual Science* 50, 1371–1377.

Kues, W.A. and Niemann, H. (2011) Advances in farm animal transgenesis. *Preventive Veterinary Medicine* 102, 146–156.

Li, H., Li, Z., Xiao, N., Su, X., Zhao, S. *et al.* (2019). Site-specific integration of rotavirus VP6 gene in rabbit β-casein locus by CRISPR/Cas9 system. *In Vitro Cellular & Developmental Biology-Animal* 55, 586–597.

Li, Q., Hou, J., Wang, S., Chen, Y. and An, X.R. (2010) Production of transgenic rabbit embryos through intracytoplasmic sperm injection. *Zygote* 18, 301–307.

Li, S., Chen, X., Fang, Z., Shi, J. and Sheng, H.Z. (2006) Rabbits generated from fibroblasts through nuclear transfer. *Reproduction* 131, 1085–1090.

Liu, Z.Q., Chen, M., Chen, S.Y., Deng, J.C., Song, Y.N. *et al.* (2018) Highly efficient RNA-guided base editing in rabbit. *Nature Communication* 9, 2717.

Liu, Z.Q., Chen, M., Shan, H., Chen, S., Xu, Y. *et al.* (2019a) Expanded targeting scope and enhanced base editing efficiency in rabbit using optimized xCas9(3.7) *Cellular and Molecular Life Sciences* 76, 4155–4164.

Liu, Z.Q., Shan, H.H., Chen, S.Y., Chen, M., Zhang, Q.J. *et al.* (2019b) Improved base editor for efficient editing in GC contexts in rabbits with an optimized AID-Cas9 fusion. *FASEB Journal* 33, 9210–9219.

Lukefahr, S., Hohenboken, W.D., Cheeke, P.R. and Patton, N.M. (1983a) Characterization of straightbred and crossbred rabbits for milk production and associative traits. *Journal of Animal Science* 57, 1100–1107.

Lukefahr, S., Hohenboken, W.D., Cheeke, P.R. and Patton, N.M. (1983b) Doe reproduction and preweaning litter performance of straightbred and crossbred rabbits. *Journal of Animal Science* 57, 1090–1099.

Lukefahr, S., Hohenboken, W.D., Cheeke, P.R., and Patton, N.M. (1983c) Breed, heterotic and diet effects on postweaning litter growth and mortality in rabbits. *Journal of Animal Science* 57, 1108–1116.

Lv, Q.Y., Yuan, L., Deng, J.C., Chen, M., Wang, Y. *et al.* (2016) Efficient generation of myostatin gene mutated rabbit by CRISPR/Cas9. *Scientific Reports* 6, 25029.

Madeja, Z.E., Pawlak, P. and Piliszek, A. (2019) Beyond the mouse: non-rodent animal models for study of early mammalian development and biomedical research. *International Journal of Developmental Biology* 63, 187–201.

Mates, L., Chuah, M.K., Belay, E., Jerchow, B., Manoj, N. *et al.* (2009) Molecular evolution of a novel hyperactive Sleeping Beauty transposase enables robust stable gene transfer in vertebrates. *Nature Genetics* 41, 753–761.

Munch, G., Bultmann, A., Li, Z., Holthoff, H.P., Ullrich, J. *et al.* (2012) Overexpression of ABCG1 protein attenuates arteriosclerosis and endothelial dysfunction in atherosclerotic rabbits. *Heart International* 7, e12.

Niu, Y., Yu, Y., Bernat, A., Yang, S., He, X. *et al.* (2010) Transgenic rhesus monkeys produced by gene transfer into early-cleavage-stage embryos using a simian immunodeficiency virus-based vector. *Proceedings of the National Academy of Sciences of the USA* 107, 17663–17667.

Pentsuk, N. and van der Laan, J.W. (2009) An interspecies comparison of placental antibody transfer: new insights into developmental toxicity testing of monoclonal antibodies. *Birth Defects Research Part B: Developmental and Reproductive Toxicology* 86, 328–344.

Pfeifer, A. (2006) Lentiviral transgenesis: a versatile tool for basic research and gene therapy. *Current Gene Therapy* 6, 535–542.

Prieto, P.A. (2012) Profiles of human milk oligosaccharides and production of some human milk oligosaccharides in transgenic animals. *Advances in Nutrition* 3, 456S–464S.

Ramin, N., Thieme, R., Fischer, S., Schindler, M., Schmidt, T. *et al.* (2010) Maternal diabetes impairs gastrulation and insulin and IGF-I receptor expression in rabbit blastocysts. *Endocrinology* 151, 4158–4167.

Ritchie, W.A., King, T., Neil, C., Carlisle, A.J., Lillico, S. *et al.* (2009) Transgenic sheep designed for transplantation studies. *Molecular Reproduction and Development* 76, 61–64.

Rostovskaya, M., Naumann, R., Fu, J., Obst, M., Mueller, D. *et al.* (2013) Transposon mediated BAC transgenesis via pronuclear injection of mouse zygotes. *Genesis* 51, 135–141.

Rouy, D., Duverger, N., Lin, S.D., Emmanuel, F., Houdebine, L.M. *et al.* (1998) Apolipoprotein(a) yeast artificial chromosome transgenic rabbits. Lipoprotein(a) assembly with human and rabbit apolipoprotein B. *Journal of Biological Chemistry* 273, 1247–1251.

Schedl, A., Beermann, F., Thies, E., Montoliu, L., Kelsey, G. and Schutz, G. (1992) Transgenic mice generated by pronuclear injection of a yeast artificial chromosome. *Nucleic Acids Research* 20, 3073–3077.

Skoda, G., Hoffmann, O.I., Gocza, E., Bodrogi, L., Kerekes, A. *et al.* (2017) Placenta-specific gene manipulation in rabbits. *Journal of Biotechnology* 259, 86–90.

Song, J., Zhong, J., Guo, X., Chen, Y., Zou, Q. *et al.* (2013) Generation of RAG 1- and 2-deficient rabbits by embryo microinjection of TALENs. *Cell Research* 23, 1059–1062.

Song, J., Yang, D.S., Xu, J., Zhu, T.Q., Chen, Y.E. and Zhang, J.F. (2016) RS-1 enhances CRISPR/Cas9-and TALEN-mediated knock-in efficiency. *Nature Communication* 7, 10548.

Song, Y.N., Xu, Y.Y., Liang, M.M., Zhang, Y.X., Chen, M. *et al.* (2018) CRISPR/Cas9-mediated mosaic mutation of SRY gene induces hermaphroditism in rabbits. *Bioscience Reports* 38, BSR20171490

Spieker-Polet, H., Sethupathi, P., Yam, P.C. and Knight, K.L. (1995) Rabbit monoclonal antibodies: generating a fusion partner to produce rabbit-rabbit hybridomas. *Proceedings of the National Academy of Sciences of the USA* 92, 9348–9352.

Steg, P.G., Feldman, L.J., Scoazec, J.Y., Tahlil, O., Barry, J.J. *et al.* (1994) Arterial gene transfer to rabbit endothelial and smooth muscle cells using percutaneous delivery of an adenoviral vector. *Circulation* 90, 1648–1656.

Stice, S.L. and Robl, J.M. (1988) Nuclear reprogramming in nuclear transplant rabbit embryos. *Biology of Reproduction* 39, 657–664.

Sui, T.T., Lau, Y.S., Liu, D., Liu, T.J., Xu, L. *et al.* (2018a) A novel rabbit model of Duchenne muscular dystrophy generated by CRISPR/Cas9. *Disease Models & Mechanisms* 11, *dmm032201*.

Sui, T.T., Xu, L., Lau, Y.S., Liu, D., Liu, T.J. *et al.* (2018b) Development of muscular dystrophy in a CRISPR-engineered mutant rabbit model with frame-disrupting ANO5 mutations. *Cell Death & Disease* 9, 609.

Sultana, F., Hatori, M., Shimozawa, N., Ebisawa, T. and Sankai, T. (2009) Continuous observation of rabbit preimplantation embryos in vitro by using a culture device connected to a microscope. *Journal of the American Association for Laboratory Animal Science* 48, 52–56.

Sumiyama, K., Kawakami, K., and Yagita, K. (2010) A simple and highly efficient transgenesis method in mice with the Tol2 transposon system and cytoplasmic microinjection. *Genomics* 95, 306–311.

Tancevski, I., Frank, S., Massoner, P., Stanzl, U., Schgoer, W. *et al.* (2005) Increased plasma levels of LDL cholesterol in rabbits after adenoviral overexpression of human scavenger receptor class B type I. *Journal of Molecular Medicine* 83, 927–932.

Tancevski, I., Wehinger, A., Patsch, J.R. and Ritsch, A. (2006) In vivo application of adenoviral vectors purified by a Taqman Real Time PCR-supported chromatographic protocol. *International Journal of Biological Macromolecules* 39, 77–82.

Taranto, S., Meo, C. D., Stanco, G., Piccolo, G., Gazaneo, M.P. and Nizza, A. (2003) Influence of age at weaning on caecal content characteristics and post-weaning performance and health of rabbits. *Asian-Australasian Journal of Animal Science* 16, 1540–1544.

Wang, Y., Zhao, S., Bai, L., Fan, J. and Liu, E. (2013) Expression systems and species used for transgenic animal bioreactors. *BioMed Research International* 2013, 580463.

Whitelaw, C.B., Radcliffe, P.A., Ritchie, W.A., Carlisle, A., Ellard, F.M. *et al.* (2004) Efficient generation of transgenic pigs using equine infectious anaemia virus (EIAV) derived vector. *FEBS Letters* 571, 233–236.

Wilson, M.H., Coates, C.J. and George, A.L., Jr. (2007) PiggyBac transposon-mediated gene transfer in human cells. *Molecular Therapy* 15, 139–145.

Yan, Q., Zhang, Q., Yang, H., Zou, Q., Tang, C. *et al.* (2014) Generation of multi-gene knockout rabbits using the Cas9/gRNA system. *Cell Regeneration* 3, 12.

Yang, D., Xu, J., Zhu, T., Fan, J., Lai, L. *et al.* (2014) Effective gene targeting in rabbits using RNA-guided Cas9 nucleases. *Journal of Molecular and Cell Biology* 6, 97–99.

Yang, X.W., Model, P. and Heintz, N. (1997) Homologous recombination based modification in Escherichia coli and germline transmission in transgenic mice of a bacterial artificial chromosome. *Nature Biotechnology* 15, 859–865.

Yang, X.Z. and Foote, R.H. (1990) Survival of bisected rabbit morulae transferred to synchronous and asynchronous recipients. *Molecular Reproduction and Development* 26, 6–11.

Yin, M.R., Jiang, W.H., Fang, Z.F., Kong, P.C., Xing, F.Y. *et al.* (2015) Generation of hypoxanthine phosphoribosyltransferase gene knockout rabbits by homologous recombination and gene trapping through somatic cell nuclear transfer. *Scientific Reports* 5, 16023.

Yuan, L., Sui, T.T., Chen, M., Deng, J.C., Huang, Y.Y. *et al.* (2016) CRISPR/Cas9-mediated GJA8 knockout in rabbits recapitulates human congenital cataracts. *Scientific Reports* 6, 22024.

Yuan, L., Yao, H., Xu, Y., Chen, M., Deng, J. *et al.* (2017) CRISPR/Cas9-Mediated Mutation of αACrystallin Gene Induces Congenital Cataracts in Rabbits. *Investigative Ophthalmology and Visual Science* 58, BIO34–BIO41.

Zakhartchenko, V., Flisikowska, T., Li, S., Richter, T., Wieland, H. *et al.* (2011) Cell-mediated transgenesis in rabbits: chimeric and nuclear transfer animals. *Biology of Reproduction* 84, 229–237.

Zhang, S., Xiang, S., Yang, J.J., Shi, J.Y., Guan, X.M. *et al.* (2019) Optimization of parthenogenetic activation of rabbit oocytes and development of rabbit embryo by somatic cell nuclear transfer. *Reproduction of Domestic Animals* 54, 258–269.

Zinovieva, N., Lassnig, C., Schams, D., Besenfelder, U., Wolf, E. *et al.* (1998) Stable production of human insulin-like growth factor 1 (IGF-1) in the milk of hemi- and homozygous transgenic rabbits over several generations. *Transgenic Research* 7, 437–447.

16 Pluripotent Stem Cells in Rabbits

Marielle Afanassieff[1]*, Arata Honda[2], Elen Gócza[3], Darko Bosnakovski[4] and Pierre Savatier[1]

[1]Université Claude Bernard Lyon 1, Inserm, INRA, Stem Cell and Brain Research Institute, Bron, France; [2]Institute of Laboratory Animals, Kyoto University Graduate School of Medicine, Kyoto, Japan; [3]National Agricultural Research and Innovation Centre, Agricultural Biotechnology Institute, Gödöllő, Hungary; [4]Faculty of Medical Sciences, University Goce Delcev, Štip, Republic of North Macedonia

16.1 Introduction

Pluripotency defines the capacity of a single cell to differentiate into all cell types in a developing foetus, including germ cells. In mice, pluripotency can be captured and propagated *in vitro* from the early epiblasts of the blastocyst in the form of embryonic stem cells (ESCs) (Evans and Kaufman, 1981; Martin, 1981). Pluripotency can also be achieved *in vitro* by reprogramming somatic cells using a cocktail of transcription factors to produce induced pluripotent stem cells (iPSCs) (Takahashi and Yamanaka, 2006; Okita *et al.*, 2007). Both ESCs and iPSCs have the capacity to colonize the blastocyst and after embryo transfer into surrogate mothers, contribute to the development of all tissue types, including germ cells. This has led to the generation of mutant mice harbouring inactivated genes and, therefore, to numerous breakthroughs in the elucidation of the genetic pathways that regulate embryonic and foetal development in rodents (Saito *et al.*, 2004). The development of similar technologies in rabbits offers a potentially broad scope of applications, including modelling of human early development, human diseases, and the generation of bioreactors. This chapter will survey the values of genome modification in rabbits and will review the current state of research on ESCs and iPSCs in the prospect of generating germline chimaeras.

16.2 Genome Modifications in Rabbits

16.2.1 Modelling human diseases in rabbits

The rabbit is a relevant animal model for the study of a wide range of human physiopathologies such as cardiovascular diseases, hypertension, atherosclerosis and metabolic syndrome (Marian *et al.*, 1999; Palinski *et al.*, 2001; Brunner *et al.*, 2008; Fan *et al.*, 2018; Lozano *et al.*, 2019; Odening *et al.*, 2019). It is also used to study infectious diseases including bacterial, fungal and viral pathogens (reviewed in Fox, 1984). The rabbit is invaluable in investigations on the mechanisms of progression or regression of papillomavirus-induced tumours (Kreider and Bartlett, 1981; Cladel *et al.*, 2019), the analysis of tuberculosis latency and reactivation (Kesavan *et al.*, 2009), and the study of electric signal transmission in ocular degeneration diseases (Kominami *et al.*, 2019). Rabbits are of interest in cancer research,

Contact e-mail address: marielle.afanassieff@inserm.fr

 The Genetics and Genomics of the Rabbit (ed Luca Fontanesi)
DOI: 10.1079/9781780643342.0016

for instance, to study renal tumours induced by ethylnitrosourea, which resemble Wilm's tumour in humans (Sharpe and Franco, 1995). Transgenic rabbits expressing oncogenes have been developed as models for oncology studies and for comparisons of the therapeutic efficacy of various anti-tumour treatments (Knight *et al.*, 1988). These examples stress the biomedical benefits of producing genetically modified rabbit models. In this context, pluripotent stem cells are invaluable in studies involving targeting mutations in specific loci and in producing germline chimaeras that harbour various genotypes of research interest.

16.2.2 Modelling human early development in rabbits

Mouse ESCs (mESCs) have been extensively used in studies on early embryonic development and in the analysis of the determination and regulation of stem cell fate (Niwa *et al.*, 2005; Ralston and Rossant, 2005). However, the development of the mouse preimplantation embryo shows unique characteristics regarding the timing of embryonic genome activation (Hamatani *et al.*, 2004), the inactivation of the X chromosome (Gendrel and Heard, 2011), the nuclear organization of heterochromatin (Andrey *et al.*, 2010), the shape of the gastrulating embryo (Tam and Behringer, 1997), and the molecular regulation of early lineage differentiation (Berg *et al.*, 2011). Similar to humans, the embryonic genome activation in the rabbit spans over several cell cycles from the one-cell stage, and the embryonic genome is fully active from the 8- to 16-cell stage onwards (Christians *et al.*, 1994). Concomitant with zygote genome activation, rabbit embryos show a reorganization of pericentromeric heterochromatin (Yang *et al.*, 2013) and a specific pattern of DNA demethylation in trophectoderm cells similar to most mammalian species (Reis Silva *et al.*, 2011). Human and rabbit embryos utilize a similar mechanism in the regulation of X chromosome inactivation, which involves the activation of both X chromosomes in the inner cell mass (ICM) cells and the absence of *Xist* gene imprinting (Okamoto *et al.*, 2011). Rabbits also possess a hemochorial placental structure that is more similar to that in humans than to the placenta of rodents. This feature offers certain advantages to study the placental nutrient transport and foeto-maternal exchanges (Kevorkova *et al.*, 2007). In contrast to the 3D egg-cylinder shape of gastrulating rodent embryos, all other mammalian embryos are organized as a flattened disk at the surface of the conceptus at the time of gastrulation, showing consequently different cellular interactions. Based on this spatial organization and the partial time overlaps between gastrulation and implantation, the rabbit embryo is thus a powerful model for studying early gastrulation events (Duranthon *et al.*, 2012). Therefore, rabbit ESCs harbouring knock-in or mutated genes involved in the determination and regulation of stem cell fate could be useful tools in investigating early mammalian development.

16.2.3 Rabbits as bioreactors

Transgenic rabbits offer particularly attractive possibilities in the preparation of recombinant pharmaceutical proteins in milk. The advantages include low-cost production and high quality of the proteins. Among other theoretically possible fluids such as blood, egg white, seminal plasma, or urine, milk is presently the most mature system to produce high levels of stable recombinant proteins, without altering the health of the animals (Wang *et al.*, 2013). Although several mammalian species, including pig, sheep, goat and cow, can be used as animal bioreactors, rabbits present a number of advantages such as easier generation of transgenic founders and offspring, high fertility, relatively high milk production, insensitivity to prion diseases, and no transmission of severe diseases to humans (Houdebine, 2009). Rabbit milk naturally contains 2.5 times more protein than goat's milk and 4.8 times more than cow's milk. A lactating female rabbit can produce 170–220 g of milk per day or a yield of up to 10 kg of milk per year under semi-automatic hygienic milking conditions. Expression levels of transgenic protein can be as high as 20 g/L. For small and medium-sized facilities, the rabbit system can ideally produce up to 50 kg of protein per year (Wang *et al.*, 2013). For these reasons, the generation of germline chimaeras harbouring genetic modifications to produce recombinant proteins in milk would be invaluable. Several recombinant human proteins are currently produced through conventional methods of genome modification, and transgenesis vectors are still under improvement for exocrine secretion of pharmaceutical molecules (Kerekes *et al.*, 2017; Lu *et al.* 2019). Some of these proteins are being tested in preclinical trials, including acid alpha-glucosidase for the treatment of Pompe disease (Van den Hout *et al.*, 2001), coagulation factor VIIa for haemophilic patients (Chevreux *et al.*, 2013, 2017), tissue non-specific alkaline phosphatase as a therapeutic agent in Gram-negative bacterial lipolysacchraride-mediated acute and chronic diseases (Bodrogi *et al.*, 2006), and VP2/VP6 proteins of rotaviruses for the development of vaccines (Soler *et al.*, 2005). Ruconest®, a human C1 inhibitor produced in transgenic rabbits, has been approved and commercialized for the treatment of patients with hereditary angioedema (Koles *et al.*, 2004).

16.2.4 Transgenesis in rabbits

Despite the success in the use of rabbits as bioreactors, transgenesis in rabbits is still limited by various technical problems. Gene transfer remains poorly efficient, and transgene expression and interference with the host genome are not fully controlled (see Chapter 15 for more details). A commonly used method consists of injecting linear DNA into one of the two pronuclei of the fertilized oocyte. However, the yield of transgenic rabbits is low (1–3%) and foreign DNA is often randomly integrated into the host genome. Moreover, linear DNA is usually integrated as head-to-tail and tandem concatemers, showing DNA rearrangements and resulting in poor transgene expression (Houdebine, 2005). The use of lentiviral vectors has resulted in a 50-fold increase in the efficiency of transgenesis in farm animals compared to DNA microinjection (Lillico *et al.*, 2011). However, the only transgenic rabbits created with a lentiviral vector expressing the enhanced green fluorescent protein (GFP) under the transcriptional control of the ubiquitous CAG promoter showed very low frequency of germline transmission (1.3%) (Hiripi *et al.*, 2010). Transposon systems are efficient and reliable tools for generating transgenic animals, but their efficiency decreases with the size of transgenes (Ivics *et al.*, 2009). The *Sleeping Beauty* transposon was recently used to generate transgenic rabbits expressing the *Venus reporter* gene, driven by the CAG promoter with high efficiency (Bosze *et al.*, 2012; Kerekes *et al.*, 2017). This result is promising for the transgenesis in rabbits, particularly in the case of small transgenes, and for short hairpin RNA-mediated gene silencing. Accessibility of a rabbit bacterial artificial chromosome (BAC) library at the BACPAC Resources Center at the Children's Hospital of Oakland Research Institute (http://bacpac.chori.org/rabbit, accessed 20 January 2021) makes BAC transgenesis attractive (Duranthon *et al.*, 2012). BAC technology was used to create transgenic rabbit lines. A rabbit model of retinal degeneration was generated by recombination with a mutated rabbit rhodopsin BAC clone (Kondo *et al.*, 2009). A rabbit with increased humoral immune response as a result of FcRn BAC clone overexpression was generated (Catunda Lemos *et al.*, 2012). Recently, new genome-editing methods using specific DNA endonucleases, i.e. Zing Finger Nucleases (ZFNs) and Transcription Activator-Like Effector Nucleases (TALEN) or RNA-dependent DNA endonucleases (CRISPR/Cas9 system), have shown remarkably high efficiency in producing transgenic animals (Garrels *et al.*, 2012; Gasiunas and Siksnys, 2013; Song *et al.*, 2013; Yang *et al.* 2013; Bosze *et al.*, 2016; Honda and Ogura, 2017). However, the generation of healthy offspring needs to be confirmed as ZFNs, similar to other DNA endonucleases, may produce off-target effects (Perez *et al.*, 2008; Li *et al.*, 2013). Moreover, the CRISPR/Cas 9 system is effective to create gene knock-out models in rabbits (Yang *et al.* 2014; Honda *et al.*, 2015; Yuan *et al.*, 2016; Sui *et al.*, 2018; Wan *et al.*, 2019), but is still inefficient for gene knock-in models with the exception of the ROSA26 locus (Yang *et al.*, 2016). In this context, the generation of rabbit ESCs and iPSCs suitable for gene targeting by homologous recombination and the subsequent production of transgenic animals would represent a major breakthrough. ESCs and iPSCs could be used in the production of germline chimaeras. ESC-like cells have been used in generating somatic chimaeras albeit at a very low rate (Giles *et al.*, 1993; Schoonjans *et al.*, 1996; Bodo *et al.*, 2004; Chrenek *et al.*, 2008; Zakhartchenko *et al.*, 2011; Tapponnier *et al.*, 2017). The transfer of ESC and iPSC nuclei into enucleated oocytes is another option in generating transgenic rabbits (Chesne *et al.*, 2002; Matsuda *et al.*, 2002; Quan *et al.*, 2014; Yin *et al.*, 2015).

16.2.5 Rabbit iPSCs and the preservation of genetic resources

Mouse iPSCs (miPSCs) are generated by reprogramming somatic cells using a cocktail of genes involved in the control of pluripotency. miPSCs have the same capacity as mESCs in producing somatic and germline chimaeras after injection into preimplantation embryos (Okita *et al.*, 2007; Boland *et al.*, 2009). Although previous studies have identified differences between ESCs and iPSCs in terms of chromatin structure (Guenther *et al.*, 2010), epigenetic marks (Kim *et al.*, 2010), genes expression, and potential of differentiation (Miura *et al.*, 2009), iPSCs serve as alternatives to ESCs in species where ESCs have not yet been generated (Ezashi *et al.*, 2012). During the past 30 years, extensive breeding has focused on more economically competitive breeds, resulting in the erosion of genetic diversity (Blackburn, 2012). The creation of biological resources centres (BRCs) responds to this critical necessity of genetic resource conservation by cryobanking reproductive cells, embryos and genomic materials of domestic animals and endangered species (Blackburn, 2004). Cryobanking of iPSCs and somatic cells can facilitate the propagation of samples for BRCs (Afanassieff *et al.*, 2019).

16.3 Rabbit Embryonic Stem Cells (Rabbit ESCs)

16.3.1 Derivation of rabbit ESCs

Since the publication of mESCs in 1981 (Evans and Kaufman, 1981; Martin, 1981), several attempts have been made to derive rabbit ESCs using the same culture conditions as their rodent counterparts, but none has been successful in establishing long-term stable lines (Giles *et al.*, 1993; Graves and Moreadith, 1993; Schoonjans *et al.*, 1996). The first rabbit ESC lines were derived from ICMs of fertilized and parthenogenetic blastocysts using fibroblast growth factor (FGF)-2 and feeder layers of irradiated mouse embryonic fibroblasts (MEFs) (Fang *et al.*, 2006; Wang *et al.*, 2007). Although these lines were capable of differentiating into derivatives of all three embryonic germ layers *in vitro* through the formation of embryoid bodies and *in vivo* formation of teratomas, these showed heterogeneity in cell morphology, cell growth and cell-surface marker expression. A more robust method was then developed, in which the feeder cell density was exquisitely controlled to generate stable ESC lines (Honda *et al.*, 2008). In brief, two- to four-cell embryos are collected from expanded oviducts approximately 30 hours after mating and cultured *in vitro* until the blastocyst stage. Blastocysts are physically pushed out from the zona pellucida using a piezo micromanipulator to prevent the exposure of the embryos to the enzymes. These were cultured on mitomycin C-treated Mouse Embryonic Fibroblasts (MEFs) at a concentration of 36×10^3 cells/cm^2, in DMEM/F12 medium supplemented with 20% Knockout Serum Replacement (KOSR), 8 ng/mL of Fibroblast Growth Factor 2 (FGF2) and 10^3 U/mL of Leukemia Inhibitory Factor (LIF). Six to eight days after the initial plating, the ICM cells generally form colonies of very compact cells, and these outgrowths are dissected mechanically and transferred to fresh feeder cells. Rabbit ESC lines are established after five to eight passages performed every three to four days by disaggregating cells with trypsin and plating single cells at a density of 3×10^3 cells/cm^2 on fresh feeder cells and medium.

16.3.2 Characterization of rabbit ESCs

Rabbit ESCs form flattened colonies of compact cells resembling those of primate ESCs (Fig. 16.1). These are positive for alkaline phosphatase activity, show a low expression level for Stage-Specific Embryonic Antigen-1 (SSEA1), a heterogeneous expression for SSEA4 and Tra-1-60 cell-surface markers, and a strong expression for *Oct4* and *Nanog* pluripotency genes. They have a stable karyotype (44,XX or 44,XY). They are pluripotent because they are able to produce teratomas and embryoid bodies containing derivatives of the three germ layers, namely ectoderm, mesoderm and endoderm (Wang *et al.*, 2007; Honda *et al.*, 2008; Osteil *et al.*, 2013). One characteristic feature of ESCs in rodents and primates is their unusual cell-cycle distribution, which is characterized by a short G1 phase and a proportionally high number of cells in the S phase. Moreover, unlike somatic cells they do not undergo cell-cycle arrest in G1 in response to DNA damage (Aladjem *et al.*, 1998; Fluckiger

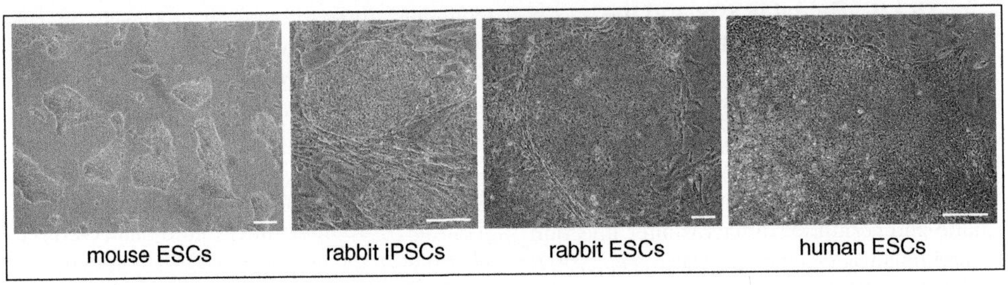

Fig. 16.1. Morphology of mouse, human and rabbit ESCs and iPSCs.

et al., 2006). Thus, rodent and primate ESCs lack a G1 checkpoint. Surprisingly, the cell-cycle characteristics of rabbit ESCs resemble those of somatic cells. Rabbit ESCs exhibit a longer G1 phase, and similar to somatic cells, undergo growth arrest in the G1 phase after DNA damage (Osteil *et al.*, 2013). All attempts to generate chimaeras with rabbit ESCs have been unsuccessful (Wang *et al.*, 2007; Honda *et al.*, 2008; Osteil *et al.*, 2013). Only one live-born ESC-derived rabbit chimaera has been reported after injection of albino Zika rabbit ESCs into eight-cell-stage embryos of the Black Alaska rabbit breed (Zakhartchenko *et al.*, 2011). However, the single live-born offspring, which showed evidence of low-level coat colour chimerism, died after two months, thus preventing investigations on its germline inheritance. Similarly, only one Caesarean newborn could be obtained after nuclear transfer of gray Chinchilla rabbit ESCs into enucleated oocytes of the white New Zealand rabbit breed (Fang *et al.*, 2006). These results show the low capacity of rabbit ESCs to colonize and reprogramme the host embryo through nuclear transfer.

16.3.3 Signalling pathways in rabbit ESCs

Rabbit ESCs are strictly dependent on FGF2 for self-renewal, as revealed by their altered morphology, decrease in cell growth, loss of alkaline phosphatase activity, and downregulation of pluripotency gene expression after FGF2 withdrawal (Honda *et al.*, 2009; Osteil *et al.*, 2013). In human ESCs, FGF2 activates two pathways, the phosphoinositide-3-kinase (PI3K)/AKT and the extracellular signal-regulated kinase (ERK) pathways (Thisse and Thisse, 2005; Okita and Yamanaka, 2006). The inhibition of either pathway by small molecules promotes differentiation. The situation in rabbit ESCs seems markedly different in that the pharmacological inhibition of PI3K activity does not induce growth retardation or differentiation (Honda *et al.*, 2009). FGF2 also induces activin secretion in MEF feeder cells, which results in the paracrine activation of the activin/Smad2/3 pathway in human ESCs to sustain self-renewal (Greber *et al.*, 2007; Greber *et al.*, 2010). Rabbit ESCs readily differentiate in the absence of feeder cells, and the inhibition of TGFβ receptor activity with SB431542 reduces rabbit ESC proliferation, *Oct4* expression, and Smad2/3 phosphorylation (Wang *et al.*, 2008; Honda *et al.*, 2009), suggesting that the same paracrine mechanism operates in rabbit ESCs. FGF2 has also been identified as an inhibitor of the BMP4/Smad1/5/8 pathway. BMP4 induces the differentiation of human ESCs, a process that can be prevented through the addition of the BMP antagonist, Noggin (Greber *et al.*, 2007). In rabbit ESCs, inhibition of FGF receptor 1 provokes Smad1 phosphorylation and differentiation, whereas the addition of Noggin to the cell-culture medium sustains self-renewal (Wang *et al.*, 2008). This finding suggests that a BMP4-dependent regulatory pathway similar to that described in human ESCs operates in rabbit ESCs.

The LIF/Stat3 signalling pathway is essential to self-renewal in mouse ESCs (Niwa *et al.*, 1998). In contrast, it fails to maintain the undifferentiated status of human ESCs (Daheron *et al.*, 2004). The role of LIF/Stat3 signalling in rabbit ESC derivation and culture is still a matter of debate. On the one hand, the addition of LIF to blastocyst outgrowths cultured on MEF feeders has been reported to increase the derivation efficiency of rabbit ESCs (Catunda *et al.*, 2008; Intawicha *et al.*, 2009; Xue *et al.*, 2012). Preliminary evidence has indicated that LIF and FGF2 could co-operatively sustain self-renewal of rabbit ESCs derived from parthenogenetically activated embryos (Hsieh *et al.*, 2011). On the other hand, the withdrawal of LIF from rabbit ESC cultures does not alter *Oct4* expression and does not induce differentiation. LIF signalling is therefore dispensable for the maintenance of rabbit ESC self-renewal (Honda *et al.*, 2009). It is possible to derive rabbit ESC lines on feeder cells without LIF and FGF2, suggesting that feeder cells are essential to rabbit ESC self-renewal, acting probably through the activin/Smad2/3 pathway (Osteil *et al.*, 2016).

16.4 Rabbit Induced Pluripotent Stem Cells (Rabbit iPSCs)

16.4.1 Reprogramming of somatic cells into rabbit iPSCs

In 2006, Shinya Yamanaka proved that the introduction of a small set of transcription factors into a differentiated cell was sufficient to revert the cell to a pluripotent state (Takahashi and Yamanaka, 2006). This technology was applied to rabbits to generate rabbit iPSCs using the human genes, *Oct4*, *Sox2*, *Klf4* and *c-Myc*. The first report made use of HIV-based lentiviral vectors expressing the reprogramming factors and GFP under the transcriptional control of an EF1 eukaryote promoter to infect adult stomach and liver cells (Honda *et al.*, 2010). Two days after in fection, GFP-positive cells were replated onto MEF feeders layers with DMEM/F12 medium supplemented with 20% KOSR, 4 ng/mL of FGF2, and 10^3 U/mL of LIF. Eight to 11 days after infection, ES-like colonies were harvested, and further propagated, later generating 11 independent rabbit iPSC lines. The second report made use of Moloney murine leukemia virus (MoMuLV)-based retroviral vectors that express the reprogramming factors under the transcriptional control of the retroviral 5' long terminal repeat (LTR) to infect adult ear skin fibroblasts (Osteil *et al.*, 2013). Three days after infection, infected fibroblasts were replated on MEF feeder layers with DMEM/F12 medium supplemented with 20% KOSR, and either 10 ng/mL of human FGF2 or 10^4 U/mL of human LIF. Two to four weeks after infection, 111 ES-like colonies were selected in the presence of FGF2, resulting in three stable rabbit iPSC lines. In parallel, 96 clones were selected from the infected fibroblasts cultured with LIF, although none of these was able to yield rabbit iPSC lines (unpublished data). The third report made use of a self-silencing lentiviral polycistronic vector expressing the four reprogramming factors driven by the spleen focus-forming virus (SFFV) retroviral promoter to infect embryonic fibroblasts (Tancos *et al.*, 2017). Two days after infection, infected fibroblasts were plated on feeder cells in DMEM/F12 medium supplemented with 20% KOSR, 4 ng/mL of human FGF2 and 10^3 U/mL of mouse LIF. After four weeks, colonies were manually picked and produced an iPSC line.

16.4.2 Characterization of rabbit iPSCs

Rabbit iPSCs form flattened colonies of compact cells that are similar to rabbit ESCs (Fig. 16.1). They show complete silencing of the four transgenes after 17–25 passages, indicating that they are fully reprogrammed. Most of the rabbit iPSC lines karyotyped using the G-banding technique exhibit normal chromosome complements. Rabbit iPSCs are positive for alkaline phosphatase activity, show heterogeneous expression of SSEA1, SSEA4 and Tra-1-60 cell-surface markers, and express *Oct4* and *Nanog* pluripotency genes. They are pluripotent because they are capable of forming teratomas and embryoid bodies that consist of three germ layers (Honda *et al.*, 2010; Osteil *et al.*, 2013; Tancos *et al.*, 2017). Rabbit iPSCs show a rapid and indefinite proliferation potential, with a doubling time of approximately 13 hours, and a high telomerase activity (Honda *et al.*, 2010). Contrary to rabbit ESCs, the rabbit iPSCs show the characteristic features of the pluripotent cell cycle such as a short G1 phase, a high proportion of cells in the S phase, and the lack of DNA damage checkpoint in G1 phase (Osteil *et al.*, 2013). Lastly, rabbit iPSCs are capable of colonizing the ICM after injection into blastocysts or aggregation with morula, albeit with low efficiency (Osteil *et al.*, 2013).

Rabbit iPSCs, similar to its embryonic counterpart, rely strictly on FGF2 and activin signalling pathways for its maintenance in the undifferentiated state. Withdrawal of FGF2 or treatment with a TGFβ receptor inhibitor induces dramatic morphological changes within 48 hours and a significant reduction in alkaline phosphatase activity, indicative of differentiation (Honda *et al.*, 2010; Osteil *et al.*, 2013). By contrast, LIF fails to sustain self-renewal in rabbit iPSCs (unpublished data).

16.5 Transcription Factors and miRNAs Involved in Rabbit Pluripotency

16.5.1 Naïve versus primed states of pluripotency

In mice, pluripotent stem cells exist in three forms: the ESCs derived from the early epiblast of the preimplantation embryo, the epiblast stem cells (EpiSCs) derived from the late epiblast of the post-implantation embryo, and the iPSCs generated by reprogramming adult cells with defined factors. Both mESCs and miPSCs exploit the LIF/STAT3 signalling pathway to fuel a complex network of transcription factors, known as the core pluripotency network, which is considered as the global activity that is essential for maintaining a robust cycle of self-renewal. In contrast, self-renewal of EpiSCs depends on FGF2 and activin/Smad signalling pathways, which activate a much less sophisticated network of transcription factors, leading to genetic instability and a high propensity to spontaneous differentiation. Moreover, only the mESCs and iPSCs, following embryo transfer into surrogate mothers, can colonize the ICM of the blastocyst and contribute to the development of all tissue types, including the germ cells. In contrast, EpiSCs fail to colonize the epiblast of the blastocyst and consequently are not suitable for the generation of knock-out mice. It is believed that these differences are largely due to differences in their developmental maturity. mESCs and miPSCs represent a pristine state known as the naïve or ground state of pluripotency, which characterizes the early epiblasts of the preimplantation embryo. EpiSCs represent an impaired state, known as the primed state of pluripotency, which is observed in the late epiblast of the post-implantation embryo (Nichols and Smith, 2009). As previously discussed, rabbit ESCs and iPSCs both rely on FGF2 and activin for self-renewal, which suggests that rabbit pluripotent stem cells have the characteristic features of primed pluripotency as defined in rodents. However, as discussed in the next section, rabbit ESCs and iPSCs also exhibit meaningful differences in their transcriptome.

16.5.2 Transcriptome of rabbit ESCs and rabbit iPSCs

A study on gene expression profiles using Agilent microarrays revealed that rabbit ESC and iPSC lines split into two separate groups, wherein the rabbit ESCs are grouped together and the rabbit iPSCs are clustered together in the other. Thus, despite the heterogeneity observed within and between cell lines, the variability is mainly associated with cell origin (rabbit ESC *vs.* iPSC; Osteil *et al.*, 2013). A detailed analysis of the expression of 22 genes (*Blimp1, Cdx2, Cdh1, Cdh2, Cldn6, Dax1, Dazl, Essrb, Fbxo15, Fgf4, Gbx2, Klf4, Lefty2, Nanog, Oct4, Otx2, Pecam1, Pitx2, Piwil2, Rex1, Tbx3,* and *Tcfcp2l1*), in which the mRNA levels were examined to determine stemness and to discriminate the naïve and primed pluripotent states in rodent cells, has confirmed and extended the microarray data. The expression of the rabbit homologs of these 22 genes was sufficient to distinguish rabbit ESCs and rabbit iPSCs. Moreover, compared to rabbit ESCs, rabbit iPSCs express naïve pluripotency markers at a higher level and primed pluripotency markers at a lower level. In particular, *Essrb, Klf4, Piwil2, Cdh1, Dazl,* and *Pecam1* were expressed at high levels in rabbit iPSCs, similar to that observed in mESCs. The expression levels of the 22 genes were measured in the ICM cells of 3.5-day rabbit blastocysts and were compared to the levels measured in ESCs and iPSCs. The levels of SSEA1 expression in rabbit iPSCs were the closest to that observed in the rabbit ICM, whereas rabbit ESC lines were the farthest (Osteil *et al.*, 2013). These findings indicate that rabbit iPSCs possess some features of mouse naïve pluripotency.

16.5.3 miRNAsome of rabbit ESCs and iPSCs

There is an increasing number of experimental evidence that suggests the important regulatory role of microRNAs (miRNAs) during early embryonic development (Yang *et al.*, 2005; Tang *et al.*, 2007) and in embryonic stem cell biology (Cao *et al.*, 2008; Gu *et al.*, 2008; Marson *et al.*, 2008). Notably, pluripotent stem cells display a distinct expression profile in relation to their state of pluripotency, in which naïve mouse ESCs mainly express the miR-290 cluster, whereas primed human ESCs and EpiSCs show a predominant expression level of the miR-302 cluster (Jouneau *et al.*, 2012). Moreover, apart from the function of miRNAs in ESC maintenance, recent reports describe the possible role of miRNAs in somatic cell reprogramming (Judson *et al.*, 2009; Anokye-Danso *et al.*, 2011; Miyoshi *et al.*, 2011). The expression profile of pluripotency-associated miRNAs in rabbit embryos and ESC cells has been analysed using SOLiD deep sequencing (Maraghechi *et al.*, 2013). A total of 1693 rabbit expressed miRNAs have been identified based on a comparison of sequence reads of human, mouse, and bovine miRNA databases. The rabbit-specific ocu-miR-302 and ocu-miR-290 clusters and three homologs of the human C19MC cluster (ocu-miR-512, ocu-miR-520e, and ocu-miR-498) are expressed in rabbit preimplantation embryos and ESC cells. The ocu-miR-302 cluster is highly similar to its human homolog, whereas the ocu-miR-290 cluster showed a low level of evolutionary conservation with its mouse homolog. The expression of the members of the ocu-miR-302 cluster begins at the 3.5-day blastocyst stage and remains high in rabbit ESC cells. In contrast, a high expression level of the members of the ocu-miR-290 cluster has been detected during preimplantation embryonic development, whereas a low level of expression was observed in rabbit ESC cells. These results show that the expression of the ocu-miR-302 cluster is characteristic of rabbit ESC cells, whereas the ocu-miR-290 cluster possibly plays a crucial role during early embryonic development (Maraghechi *et al.*, 2013).

16.6 *In Vitro* Differentiation of Rabbit ESCs and iPSCs

Rabbit ESCs spontaneously differentiate *in vitro* in the absence of growth factors to preferentially give rise to neural-type cells and beating cardiac cells (Wang *et al.*, 2007; Honda *et al.*, 2008). After culturing for three weeks, the presence of progesterone, estrogen and chorionic gonadotropin in harvested conditioned medium also indicates trophoblast differentiation of rabbit ESCs (Wang *et al.*, 2007). When cultured in handing drops, rabbit ESCs and iPSCs form 3D structures called embryoid bodies (EBs), which are composed of derivatives of the three germ layers (Fang *et al.*, 2006; Honda *et al.*, 2008, 2010; Hsieh *et al.*, 2011; Osteil *et al.*, 2013). A protocol for the analysis of the differentiation potential into neural lineage cells has been published (Honda *et al.*, 2013); EBs are formed after five days in suspension culture with rabbit ESC medium supplemented with 4 μM of retinoic acid (RA) and 10 μM of SB431542 (TGFβ receptor inhibitor), then incubated for 10 days in an adherent culture with neural differentiation medium supplemented with both RA and SB431542. This protocol induces rabbit ESC differentiation into nestin-positive neural stem cells (10%), TUJ1-positive neurons (20%), and GFAP-positive glial astrocytes (30%). To obtain O1- and CNPase-positive oligodendrocytes, RA and SB431542 are withdrawn from the culture medium after five days, and replaced by Noggin (100 ng/mL) for 20 days. Another study showed that stomach endoderm-derived rabbit iPSCs had a higher capacity to differentiate along the neural lineage compared to liver-derived rabbit iPSCs (Honda *et al.*, 2013). Thus, as previously observed in mouse and human iPSCs, the differentiation ability of rabbit iPSCs is influenced by the origin of the somatic tissues.

16.7 Current Bottlenecks

One key issue in the generation of transgenic rabbits is the availability of naïve ESCs and iPSCs that are capable of colonizing the preimplantation embryo to generate germline chimaeras. As discussed earlier, rabbit ESCs show all features of primed pluripotency, whereas rabbit iPSCs seem to self-renew in an intermediate state, which is closer to the naïve state of pluripotency as defined in rodents (Osteil *et al.*, 2013). Nevertheless, both types of cells fail to colonize efficiently the rabbit preimplantation embryo to produce chimaeras, which precludes the generation of somatic chimaeras. To resolve this issue, one needs to reprogramme rabbit ESCs and iPSCs to *bona fide* naïve pluripotency.

In mice, EpiSCs can spontaneously revert to LIF-dependent naïve ESCs, although the efficiency of this conversion is extremely low (Bao *et al.*, 2009). It can be dramatically increased by transiently overexpressing naïve state-specific transcription factors such as Klf2, Klf4 and NrOb1, in combination with LIF and pharmacological inhibitors of ERK and GSK3β signalling pathways (Guo *et al.*, 2009; Hall *et al.*, 2009). These reverted EpiSCs acquire the capacity to colonize the blastocyst and to generate germline chimaeras. Similarly, conversion of FGF2-dependent into LIF-dependent human ESCs and iPSCs was achieved by overexpressing *Klf2*, *Klf4*, and *Nanog*, in combination with ERK and GSK3β inhibitors. The resulting cells exhibited the genetic and epigenetic characteristics of naïve pluripotency (Hanna *et al.*, 2010). Recently, novel conditions in producing human ESCs and iPSCs were established, resulting in cells with molecular characteristics and functional properties that are highly similar to human preimplantation epiblast stem cells or naïve mESCs (Chan *et al.*, 2013; Gafni *et al.*, 2013; Takashima *et al.*, 2014; Theunissen *et al.*, 2014; Chen *et al.*, 2015; Guo *et al.*, 2017). These naïve human iPSCs are competent to generate cross-species chimaeric embryos after microinjection into mouse morula (Gafni *et al.*, 2013; Mascetti and Pedersen, 2016). Preliminary results suggest that a similar conversion could be achieved in rabbits. Partial reprogramming of rabbit iPSCs to a naïve-like state of pluripotency was achieved by overexpressing the human *Oct4* gene (Honda *et al.*, 2013) or both human Klf2 and Klf4 genes (Tapponnier *et al.*, 2017). After injection into the preimplantation embryo, these new cells were capable of colonizing the epiblast of the blastocyst and participating in epiblast expansion at E6.0 pre-gastrula stage. Recently, rabbit ESCs were derived using small molecule inhibitors, namely GSK3βi (CHIR99021, 3 μM), MEKi (PD0325901, 1 μM, and PKCi (Gö6983, 5 μM), in the presence of FGF2 (10 ng/mL), LIF (10^3 U/mL) and feeder cells, and showed typical dome-shaped colonies like naïve mESC (Liu *et al.*, 2019). However, those rabbit naïve-like cells remained dependent on FGF2 and feeders to self-renew, supporting the essential role of the activin/TGFβ/Smad2/3 signalling pathway for the maintenance of rabbit pluripotent stem cells *in vitro*.

16.8 Conclusion

Based on their capacity to produce chimaeras, mESCs and miPSCs are powerful tools for creating transgenic mice with complex genetic modifications. However, the use of rabbit pluripotent stem cells in generating transgenic rabbits is currently limited by the lack of *bona fide* ESCs and iPSCs that self-renew in the naïve state of pluripotency. Thus, future research aims to unravel the molecular mechanisms underlying pluripotency in the rabbit preimplantation embryo, as well as design culture conditions to capture the original state of pluripotency in a culture dish.

References

Afanassieff, M., Gavin-Plagne, L., Perold, F., Voisin, S., Moreira, S. *et al.* (2019) Interest of rabbit induced Pluripotent Stem Cells for species preservation. *Thai Journal of Veterinary Medicine* 49(suppl. 2), 103–107.

Aladjem, M.I., Spike, B.T., Rodewald, L.W., Hope, T.J., Klemm, M. *et al.* (1998) ES cells do not activate p53-dependent stress responses and undergo p53-independent apoptosis in response to DNA damage. *Current Biology* 8, 145–155.

Andrey, P., Kieu, K., Kress, C., Lehmann, G., Tirichine, L. *et al.* (2010) Statistical analysis of 3D images detects regular spatial distributions of centromeres and chromocenters in animal and plant nuclei. *PLOS Computational Biology* 6, e1000853.

Anokye-Danso, F., Trivedi, C.M., Juhr, D., Gupta, M., Cui, Z. *et al.* (2011) Highly efficient miRNA-mediated reprogramming of mouse and human somatic cells to pluripotency. *Cell and Stem Cell* 8, 376–388.

Bao, S., Tang, F., Li, X., Hayashi, K., Gillich, A. *et al.* (2009) Epigenetic reversion of post-implantation epiblast to pluripotent embryonic stem cells. *Nature* 461, 1292–1295.

Berg, D.K., Smith, C.S., Pearton, D.J., Wells, D.N., Broadhurst, R. *et al.* (2011) Trophectoderm lineage determination in cattle. *Developmental Cell* 20, 244–255.

Blackburn, H.D. (2004) Development of national animal genetic resource programs. *Reproduction, Fertility and Development* 16, 27–32.

Blackburn, H.D. (2012) Genetic selection and conservation of genetic diversity. *Reproduction in Domestic Animals* 47 (Suppl. 4), 249–254.

Bodo, S., Gocza, E., Revay, T., Hiripi, L., Carstea, B. *et al.* (2004) Production of transgenic chimeric rabbits and transmission of the transgene through the germline. *Molecular Reproduction and Development* 68, 435–440.

Bodrogi, L., Brands, R., Raaben, W., Seinen, W., Baranyi, M. *et al.* (2006) High level expression of tissue-nonspecific alkaline phosphatase in the milk of transgenic rabbits. *Transgenic Research* 15, 627–636.

Boland, M.J., Hazen, J.L., Nazor, K.L., Rodriguez, A.R., Gifford, W. *et al.* (2009) Adult mice generated from induced pluripotent stem cells. *Nature* 461, 91–94.

Bosze, Z., Hiripi, L., Hoffmann, O., Gocza, E., Kvell, K. *et al.* (2012) Creation and characterization of second generation transgenic rabbit models. *Transgenic Research* 21, 901–925.

Bosze, Z., Major, P., Baczko, I., Odening, K.E., Bodrogi, L. *et al.* (2016) The potential impact of new generation transgenic methods on creating rabbit models of cardiac diseases. *Progress in Biophysics and Molecular Biology* 121, 123–130.

Brunner, M., Peng, X., Liu, G.X., Ren, X.Q., Ziv, O. *et al.* (2008) Mechanisms of cardiac arrhythmias and sudden death in transgenic rabbits with long QT syndrome. *Journal of Clinical Investigation* 118, 2246–2259.

Cao, H., Yang, C.S. and Rana, T. M. (2008) Evolutionary emergence of microRNAs in human embryonic stem cells. *PLOS ONE* 3 e282,0.

Catunda, A.P., Gocza, E., Carstea, B.V., Hiripi, L. *et al.* (2008) Characterization, chromosomal assignment, and role of LIFR in early embryogenesis and stem cell establishment of rabbits. *Cloning Stem Cells* 10, 523–534.

Catunda Lemos, A.P., Cervenak, J., Bender, B., Hoffmann, O.I., Baranyi, M. *et al.* (2012) Characterization of the rabbit neonatal Fc receptor (FcRn) and analyzing the immunophenotype of the transgenic rabbits that overexpresses FcRn. *PLOS ONE* 7, e28869.

Chan, Y.S., Goke, J., Ng, J.H., Lu, X., Gonzales, K.A. *et al.* (2013) Induction of a human pluripotent state with distinct regulatory circuitry that resembles preimplantation epiblast. *Cell Stem Cell* 13, 663–675.

Chen, H., Aksoy, I., Gonnot, F., Osteil, P., Aubry, M. *et al.* (2015) Reinforcement of STAT3 activity reprogrammes human embryonic stem cells to naïve-like pluripotency. *Nature Communication* 6, 7095–7111.

Chesne, P., Adenot, P.G., Viglietta, C., Baratte, M., Boulanger, L. and Renard, J.P. (2002) Cloned rabbits produced by nuclear transfer from adult somatic cells. *Nature Biotechnology* 20, 366–369.

Chevreux, G., Faid, V., Scohyers, J.M. and Bihoreau, N. (2013) N-/O-glycosylation analysis of human FVIIa produced in the milk of transgenic rabbits. *Glycobiology* 23, 1531–1546.

Chevreux, G., Tilly, N., Leblanc, Y., Ramon. C., Faid, V. *et al.* (2017) Biochemical characterization of LR769, a new recombinant factor Viia bypassing agent produced in the milk of transgenic rabbits. *Haemophilia* 23, e324–e334.

Chrenek, P., Makarevich, A.V., Bauer, M. and Jurcik, R. (2008) Developmental rate and allocation of transgenic cells in rabbit chimeric embryos. *Zygote* 16, 87–91.

Christians, E., Rao, V.H. and Renard, J.P. (1994) Sequential acquisition of transcriptional control during early embryonic development in the rabbit. *Developmental Biology* 164, 160–172.

Cladel, N.M., Peng, X., Christensen, N. and Hu, J. (2019) The rabbit papillomavirus model: a valuable tools to study viral-host interactions. *Philosophical Transactions of the Royal Society B: Biological Sciences* 374, 20180294.

Daheron, L., Opitz, S.L., Zaehres, H., Lensch, M.W., Andrews, P.W. *et al.* (2004) LIF/STAT3 signaling fails to maintain self-renewal of human embryonic stem cells. *Stem Cells* 22, 770–778.

Duranthon, V., Beaujean, N., Brunner, M., Odening, K.E., Santos, A.N. *et al.* (2012) On the emerging role of rabbit as human disease model and the instrumental role of novel transgenic tools. *Transgenic Research* 21, 699–713.

Evans, M.J. and Kaufman, M.H. (1981) Establishment in culture of pluripotential cells from mouse embryos. *Nature* 292, 154–156.

Ezashi, T., Telugu, B.P. and Roberts, R.M. (2012) Induced pluripotent stem cells from pigs and other ungulate species: an alternative to embryonic stem cells? *Reproduction in Domestic Animals* 47 (Suppl. 4), 92–97.

Fan, J., Chen, Y., Yan, H., Niimi, M., Wang, Y. and Liang, J. (2018) Principles and applications of rabbit models for artherosclerosis research. *Journal of Atherosclerosis and Thrombosis* 25, 213–220.

Fang, Z.F., Gai, H., Huang, Y.Z., Li, S.G., Chen, X.J. *et al.* (2006) Rabbit embryonic stem cell lines derived from fertilized, parthenogenetic or somatic cell nuclear transfer embryos. *Experimental Cell Research* 312, 3669–3682.

Fluckiger, A.-C., Marcy, G., Marchand, M., Negre, D., Cosset, F. *et al.* (2006) Cell cycle features of primate embryonic stem cells. *Stem Cells* 24, 547–556.

Fox, R.R. (1984) The rabbit as a research subject. *Physiologist* 27, 393–402.

Gafni, O., Weinberger, L., Mansour, A.A., Manor, Y.S., Chomsky, E. *et al.* (2013) Derivation of novel human ground state naive pluripotent stem cells. *Nature* 504, 282–286.

Garrels, W., Ivics, Z. and Kues, W.A. (2012) Precision genetic engineering in large mammals. *Trends in Biotechnology* 30, 386–393.

Gasiunas, G. and Siksnys, V. (2013) RNA-dependent DNA endonuclease Cas9 of the CRISPR system: Holy Grail of genome editing? *Trends in Microbiology* 21, 562–567.

Gendrel, A.V. and Heard, E. (2011) Fifty years of X-inactivation research. *Development* 138, 5049–5055.

Giles, J.R., Yang, X., Mark, W. and Foote, R.H. (1993) Pluripotency of cultured rabbit inner cell mass cells detected by isozyme analysis and eye pigmentation of fetuses following injection into blastocysts or morulae. *Molecular Reproduction and Development* 36, 130–138.

Graves, K.H. and Moreadith, R.W. (1993) Derivation and characterization of putative pluripotential embryonic stem cells from preimplantation rabbit embryos. *Molecular Reproduction and Development* 36, 424–433.

Greber, B., Lehrach, H. and Adjaye, J. (2007) Fibroblast growth factor 2 modulates transforming growth factor beta signaling in mouse embryonic fibroblasts and human ESCs (hESCs) to support hESC self-renewal. *Stem Cells* 25, 455–464.

Greber, B., Wu, G., Bernemann, C., Joo, J.Y., Han, D.W. *et al.* (2010) Conserved and divergent roles of FGF signaling in mouse epiblast stem cells and human embryonic stem cells. *Cell Stem Cell* 6, 215–226.

Gu, P., Reid, J.G., Gao, X., Shaw, C.A., Creighton, C. *et al.* (2008) Novel microRNA candidates and miRNA-mRNA pairs in embryonic stem (ES) cells. *PLOS ONE* 3, e2548.

Guenther, M.G., Frampton, G.M., Soldner, F., Hockemeyer, D., Mitalipova, M. *et al.* (2010) Chromatin structure and gene expression programs of human embryonic and induced pluripotent stem cells. *Cell Stem Cell* 7, 249–257.

Guo, G., Yang, J., Nichols, J., Hall, J.S., Eyres, I. *et al.* (2009) Klf4 reverts developmentally programmed restriction of ground state pluripotency. *Development* 136, 1063–1099.

Guo, G., von Meyenn, F., Rostvskaya, M., Clarke, J., Dietmann, S. *et al.* (2017) Epigenetic resetting of human pluripotency. *Development* 144, 2748–2763.

Hall, J., Guo, G., Wray, J., Eyres, I., Nichols, J. *et al.* (2009) Oct4 and LIF/Stat3 additively induce Kruppel factors to sustain embryonic stem cell self-renewal. *Cell Stem Cell* 5, 597–609.

Hamatani, T., Carter, M.G., Sharov, A.A. and Ko, M.S. (2004) Dynamics of global gene expression changes during mouse preimplantation development. *Developmental Cell* 6, 117–131.

Hanna, J., Cheng, A.W., Saha, K., Kim, J., Lengner, C.J. *et al.* (2010) Human embryonic stem cells with biological and epigenetic characteristics similar to those of mouse ESCs. *Proceedings of the National Academy of Sciences of the USA* 107, 9222–9227.

Hiripi, L., Negre, D., Cosset, F.L., Kvell, K., Czompoly, T. *et al.* (2010) Transgenic rabbit production with simian immunodeficiency virus-derived lentiviral vector. *Transgenic Research* 19, 799–808.

Honda, A. and Ogura, A. (2017) Rabbit models for biomedical research revisited via genome editing approaches. *Journal of Reproduction and Development* 63, 435–438.

Honda, A., Hirose, M., Inoue, K., Ogonuki, N., Miki, H. *et al.* (2008) Stable embryonic stem cell lines in rabbits: potential small animal models for human research. *Reproductive Biomedicine Online* 17, 706–715.

Honda, A., Hirose, M. and Ogura, A. (2009) Basic FGF and Activin/Nodal but not LIF signaling sustain undifferentiated status of rabbit embryonic stem cells. *Experimental Cell Research* 315, 2033–2042.

Honda, A., Hirose, M., Hatori, M., Matoba, S., Miyoshi, H. *et al.* (2010) Generation of induced pluripotent stem cells in rabbits: potential experimental models for human regenerative medicine. *Journal of Biological Chemistry* 285, 31362–31369.

Honda, A., Hatori, M., Hirose, M., Honda, C., Izu, H. *et al.* (2013) Naive-like conversion overcomes the limited differentiation capacity of induced pluripotent stem cells. *Journal of Biological Chemistry* 288, 26157–26166.

Honda, A., Hirose, M., Sankai, T., Yasmin, L., Yuzawa, K. *et al.* (2015) Single-step generation of rabbit carrying a targeted allele of the Tyrosinase gene unsing CRISPR/Cas9. *Experimental Animals* 64, 31–37.

Houdebine, L.M. (2005) Use of transgenic animals to improve human health and animal production. *Reproduction in Domestic Animals* 40, 269–281.

Houdebine, L.M. (2009) Production of pharmaceutical proteins by transgenic animals. *Comparative Immunology, Microbiology and Infectious Diseases* 32, 107–121.

Hsieh, Y.C., Intawicha, P., Lee, K.H., Chiu, Y.T., Lo, N.W. and Ju, J.C. (2011) LIF and FGF cooperatively support stemness of rabbit embryonic stem cells derived from parthenogenetically activated embryos. *Cellular Reprogrammming* 13, 241–255.

Intawicha, P., Ou, Y.W., Lo, N.W., Zhang, S.C., Chen, Y.Z. *et al.* (2009) Characterization of embryonic stem cell lines derived from New Zealand white rabbit embryos. *Cloning Stem Cells* 11, 27–38.

Ivics, Z., Li, M.A., Mates, L., Boeke, J.D., Nagy, A. *et al.* (2009) Transposon-mediated genome manipulation in vertebrates. *Nature Methods* 6, 415–422.

Jouneau, A., Ciaudo, C., Sismeiro, O., Brochard, V., Jouneau, L. *et al.* (2012) Naive and primed murine pluripotent stem cells have distinct miRNA expression profiles. *RNA* 18, 253–264.

Judson, R.L., Babiarz, J.E., Venere, M. and Blelloch, R. (2009) Embryonic stem cell-specific microRNAs promote induced pluripotency. *Nature Biotechnology* 27, 459–461.

Kerekes, A., Hoffmann, O., Iski, G., Liptak, N., Gocza, E. *et al.* (2017) Secretion of a recombinant protein without a signal peptides by the exocrine glands of transgenic rabbits. *PLOS ONE* 12, e0187214.

Kesavan, A.K., Brooks, M., Tufariello, J., Chan, J. and Manabe, Y.C. (2009) Tuberculosis genes expressed during persistence and reactivation in the resistant rabbit model. *Tuberculosis (Edinb)* 89, 17–21.

Kevorkova, O., Ethier-Chiasson, M. and Lafond, J. (2007) Differential expression of glucose transporters in rabbit placenta: effect of hypercholesterolemia in dams. *Biology of Reproduction* 76, 487–495.

Kim, K., Doi, A., Wen, B., Ng, K., Zhao, R. *et al.* (2010) Epigenetic memory in induced pluripotent stem cells. *Nature* 467, 285–290.

Knight, K.L., Spieker-Polet, H., Kazdin, D.S. and Oi, V.T. (1988) Transgenic rabbits with lymphocytic leukemia induced by the c-myc oncogene fused with the immunoglobulin heavy chain enhancer. *Proceedings of the National Academy of Sciences of the USA* 85, 3130–3134.

Koles, K., van Berkel, P.H., Pieper, F.R., Nuijens, J.H., Mannesse, M.L. *et al.* (2004) N- and O-glycans of recombinant human C1 inhibitor expressed in the milk of transgenic rabbits. *Glycobiology* 14, 51–64.

Kominami, T., Ueno, S., Nishida, K., Inooka, D., Kominami, A. *et al.* (2019) Electrically evoked potentials are reduced compared to axon numbers in rhodopsin P347L transgenic rabbits with severe photoreceptor degeneration. *Investigative Ophthalmology and Vision Science* 60, 2543–2550.

Kondo, M., Sakai, T., Komeima, K., Kurimoto, Y., Ueno, S. *et al.* (2009) Generation of a transgenic rabbit model of retinal degeneration. *Investigative Ophthalmology and Vision Science* 50, 1371–1377.

Kreider, J.W. and Bartlett, G.L. (1981) The Shope papilloma-carcinoma complex of rabbits: a model system of neoplastic progression and spontaneous regression. *Advances in Cancer Research* 35, 81–110.

Li, M., Suzuki, K., Kim, N.Y., Liu, G.H. and Izpisua Belmonte, J.C. (2013) A cut above the rest: targeted genome editing technologies in human pluripotent stem cells. *Journal of Biological Chemistry* 289, 4594–4599.

Lillico, S., Vasey, D., King, T. and Whitelaw, B. (2011) Lentiviral transgenesis in livestock. *Transgenic Research* 20, 441–442.

Liu, J., Zhu, X., Li, J., Liu, Z., Liu, Y. *et al.* (2019) Deriving rabbit embryonic stem cells by small molecule inhibitors. *American Journal of Translational Research* 11, 5122–5133.

Lozano, W.M, Arias-Mutis, O.J., Calvo, C.J., Chorro, F.J. and Zarzoso, M. (2019) Diet-induced rabbit models for study of metabolic syndrome. *Animals 9, pii*: E463.

Lu, R., Zhang, T., Song, S., Zhou, M., Jiang, L. *et al.* (2019) Accurately cleavable goat b-lactoglobulin signal peptide efficiently guided translation of a recombinant human plasminogen activator in transgenic rabbit mammary gland. *Bioscience Reports* 39, pii: BSR20190596.

Maraghechi, P., Hiripi, L., Toth, G., Bontovics, B., Bosze, Z. and Gocza, E. (2013) Discovery of pluripotency-associated microRNAs in rabbit preimplantation embryos and embryonic stem-like cells. *Reproduction* 145, 421–437.

Marian, A.J., Wu, Y., Lim, D.S., McCluggage, M., Youker, K. *et al.* (1999) A transgenic rabbit model for human hypertrophic cardiomyopathy. *Journal of Clinical Invest* 104, 1683–1692.

Marson, A., Levine, S.S., Cole, M.F., Frampton, G.M., Brambrink, T. *et al.* (2008) Connecting microRNA genes to the core transcriptional regulatory circuitry of embryonic stem cells. *Cell* 134, 521–533.

Martin, G.R. (1981) Isolation of a pluripotent cell line from early mouse embryos cultured in medium conditioned by teratocarcinoma stem cells. *Proceedings of the National Academy of Sciences of the USA* 78, 7634–7638.

Mascetti, V.L. and Pedersen, R.A. (2016) Human-mouse chimerism validates human stem cell pluripotency. *Cell Stem Cell* 18, 1–6.

Matsuda, J., Takahashi, S., Ohkoshi, K., Kaminaka, K., Kaminaka, S. *et al.* (2002) Production of transgenic chimera rabbit fetuses using somatic cell nuclear transfer. *Cloning and Stem Cells* 4, 9–19.

Miura, K., Okada, Y., Aoi, T., Okada, A., Takahashi, K. *et al.* (2009) Variation in the safety of induced pluripotent stem cell lines. *Nature Biotechnology* 27, 743–745.

Miyoshi, N., Ishii, H., Nagano, H., Haraguchi, N., Dewi, D.L. *et al.* (2011) Reprogramming of mouse and human cells to pluripotency using mature microRNAs. *Cell Stem Cell* 8, 633–638.

Nichols, J. and Smith, A. (2009) Naïve and primed pluripotent states. *Cell Stem Cell* 4, 487–492.

Niwa, H., Burdon, T., Chambers, I. and Smith, A. (1998) Self-renewal of pluripotent embryonic stem cells is mediated via activation of STAT3. *Genes and Development* 12, 2048–2060.

Niwa, H., Toyooka, Y., Shimosato, D., Strumpf, D., Takahashi, K. *et al.* (2005) Interaction between Oct3/4 and Cdx2 determines trophectoderm differentiation. *Cell* 123, 917–929.

Odening, K.E., Bodi, I., Franke, G., Rieke, R., Ryan de Medreiros, A. *et al.* (2019) Transgenic short-QT syndrome 1 rabbits mimic the human disease phenotype with QT/action potential duration shortening in the atria and ventricles and increased ventricular tachycardia/ventricular fibrillation inducibility. *European Heart Journal* 40, 842–853.

Okamoto, I., Patrat, C., Thepot, D., Peynot, N., Fauque, P. *et al.* (2011) Eutherian mammals use diverse strategies to initiate X-chromosome inactivation during development. *Nature* 472, 370–374.

Okita, K. and Yamanaka, S. (2006) Intracellular signaling pathways regulating pluripotency of embryonic stem cells. *Current Stem Cell Research and Therapy* 1, 103–111.

Okita, K., Ichisaka, T. and Yamanaka, S. (2007) Generation of germline-competent induced pluripotent stem cells. *Nature* 448, 313–317.

Osteil, P., Tapponnier, Y., Markossian, S., Godet, M., Schmaltz-Panneau, B. *et al.* (2013) Induced pluripotent stem cells derived from rabbits exhibit some characteristics of naive pluripotency. *Biology Open* 2, 613–628.

Osteil., P., Moulin, A., Santamaria, C., Joly, T., Jouneau, L. *et al.* (2016) A panel of embryonic stem cell lines reveals the variety and dynamic of pluripotent states in rabbits. *Stem Cell Reports* 7, 383–398.

Palinski, W., D'Armiento, F.P., Witztum, J.L., de Nigris, F., Casanada, F. *et al.* (2001) Maternal hypercholesterolemia and treatment during pregnancy influence the long-term progression of atherosclerosis in offspring of rabbits. *Circulation Research* 89, 991–996.

Perez, E.E., Wang, J., Miller, J.C., Jouvenot, Y., Kim, K.A. *et al.* (2008) Establishment of HIV-1 resistance in CD4+ T cells by genome editing using zinc-finger nucleases. *Nature Biotechnology* 26, 808–816.

Quan, L., Chen, Y., Song, J., Yan, Q., Zhang, Q. *et al.* (2014) Establishment of rabbit Oct4 promoter-based EGFP reporter system. *PLOS ONE* 9, e109728.

Ralston, A. and Rossant, J. (2005) Genetic regulation of stem cell origins in the mouse embryo. *Clinical Genetics* 68, 106–112.

Reis Silva, A.R., Adenot, P., Daniel, N., Archilla, C., Peynot, N. *et al.* (2011) Dynamics of DNA methylation levels in maternal and paternal rabbit genomes after fertilization. *Epigenetics* 6, 987–993.

Saito, S., Liu, B. and Yokoyama, K. (2004) Animal embryonic stem (ES) cells: self-renewal, pluripotency, transgenesis and nuclear transfer. *Human Cell* 17, 107–115.

Schoonjans, L., Albright, G.M., Li, J.L., Collen, D. and Moreadith, R.W. (1996) Pluripotential rabbit embryonic stem (ES) cells are capable of forming overt coat color chimeras following injection into blastocysts. *Molecula Reproduction and Development* 45, 439–443.

Sharpe, C.R. and Franco, E.L. (1995) Etiology of Wilms' tumor. *Epidemiologic Reviews* 17, 415–432.

Soler, E., Le Saux, A., Guinut, F., Passet, B., Cohen, R. *et al.* (2005) Production of two vaccinating recombinant rotavirus proteins in the milk of transgenic rabbits. *Transgenic Research* 14, 833–844.

Song, J., Zhong, J., Guo, X., Chen, Y., Zou, Q. *et al.* (2013) Generation of RAG 1- and 2-deficient rabbits by embryo microinjection of TALENs. *Cell Research* 23, 1059–1062.

Sui, T., Lau, Y.S., Liu, D., Liu, T., Xu., L. *et al.* (2018) A novel rabbit model of duchenne muscular dystrophy generated by CRISPR/Cas9. *Disease Model and Mechanisms* 11, dmm032201.

Takahashi, K. and Yamanaka, S. (2006) Induction of pluripotent stem cells from mouse embryonic and adult fibroblast cultures by defined factors. *Cell* 126, 663–676.

Takashima, Y., Guo, G., Loos, R., Nichols, J., Ficz, G. *et al.* (2014) Resetting transcription factor control circuitry toward ground-state pluripotency in human. *Cell* 158, 1254–1269.

Tam, P.P. and Behringer, R.R. (1997) Mouse gastrulation: the formation of a mammalian body plan. *Mechanisms of Development* 68, 3–25.

Tancos, Z., Nemes, C., Varga, E., Bock, I., Rungarunlert, S. *et al.* (2017) Establishment of a rabbit induced pluripotent stem cell (RbiPSC) line using lentiviral delivery of human pluripotency factors. *Stem Cell Research* 21, 16–18.

Tang, F., Kaneda, M., O'Carroll, D., Hajkova, P., Barton, S.C. *et al.* (2007) Maternal microRNAs are essential for mouse zygotic development. *Genes and Development* 21, 644–648.

Tapponnier, T., Afanassieff, M., Aksoy, I., Aubry, M., Moulin A. *et al.* (2017) Reprogramming of rabbit induced Pluripotent Stem cells toward epiblast and chimeric competency using Kruppel-like factors. *Stem Cell Research* 24, 106–117.

Theunissen, T.W., Powell, B.E., Wang, H., Mitalipova, M., Faddah, D.A. *et al.* (2014) Systematic identification of culture conditions for induction and maintenance of naive human pluripotency. *Cell Stem Cell* 15, 471–487.

Thisse, B. and Thisse, C. (2005) Functions and regulations of fibroblast growth factor signaling during embryonic development. *Developmental Biology* 287, 390–402.

Van den Hout, J.M., Reuser, A.J., de Klerk, J.B., Arts, W.F., Smeitink, J.A. and Van der Ploeg, A.T. (2001) Enzyme therapy for pompe disease with recombinant human alpha-glucosidase from rabbit milk. *Journal of Inherited Metabolic Disease* 24, 266–274.

Wan, Y., Guo, R., Deng, M., Liu, Z., Pang, J. *et al.* (2019) Efficient generation of CLPG1-edited rabbits using the CRISPR/Cas9 system. *Reproduction in Domestic Animals* 54, 536–544.

Wang, S., Tang, X., Niu, Y., Chen, H., Li, B. *et al.* (2007) Generation and characterization of rabbit embryonic stem cells. *Stem Cells* 25, 481–489.

Wang, S., Shen, Y., Yuan, X., Chen, K., Guo, X. *et al.* (2008) Dissecting signaling pathways that govern self-renewal of rabbit embryonic stem cells. *Journal of Biological Chemistry* 283, 35929–35940.

Wang, Y., Zhao, S., Bai, L., Fan, J. and Liu, E. (2013) Expression systems and species used for transgenic animal bioreactors. *Biomedical Research International* 2013, 580463.

Xue, F., Ma, Y., Chen, E., Zhang, J., Lin, T.A. *et al.* (2012) Recombinant rabbit Leukemia Inhibitory Factor and rabbit embryonic fibroblasts support the derivation and maintenance of rabbit Embryonic Stem Cells. *Cellular Reprogramming* 14, 364–376.

Yang, C.X., Liu, Z., Fleurot, R., Adenot, P., Duranthon, V. *et al.* (2013a) Heterochromatin reprogramming in rabbit embryos after fertilization, intra-, and inter-species SCNT correlates with preimplantation development. *Reproduction* 145, 149–159.

Yang, D., Zhang, J., Xu, J., Zhu, T., Fan, Y. *et al.* (2013b) Production of apolipoprotein C-III knockout rabbits using Zinc finger nucleases. *Journal of Visualized Experiments* 81, 50957.

Yang, D., Xu, J., Zhu, T., Fan, J., Lai, L. *et al.* (2014) Effective gene targeting in rabbits using RNA-guided Cas9 nucleases. *Journal of Molecular and Cell Biology* 6, 97–99.

Yang, D., Song, J., Zhang, J., Xu, J., Zhu, T. *et al.* (2016) Identification and characterization of rabbit ROSA26 for gene knock-in and stable reporter gene expression. *Scientific Reports* 6, 25161.

Yang, W.J., Yang, D.D., Na, S., Sandusky, G.E., Zhang, Q. and Zhao, G. (2005) Dicer is required for embryonic angiogenesis during mouse development. *Journal of Biological Chemistry* 280, 9330–9335.

Yin, M., Jiang, W., Fang, Z., Kong, P., Xing, F. *et al.* (2015) Generation of hypoxanthine phosphoribosyltransferase gene knockout rabbits by homologous recombination and gene trapping through somatic cell nuclear transfer. *Scientific Reports* 5, 16023.

Yuan, L., Sui, T., Chen, M., Deng, J., Huang, Y. *et al.* (2016) CRISPR/Cas9-mediated GJA8 knockout in rabbits recapitulates human congenital catacracts. *Scientific Reports* 6, 22024.

Zakhartchenko, V., Flisikowska, T., Li, S., Richter, T., Wieland, H. *et al.* (2011) Cell-mediated transgenesis in rabbits: chimeric and nuclear transfer animals. *Biology of Reproduction* 84, 229–237.

17 Biotechnology Applications in the Rabbit

Peter Chrenek[1,2*], Zsuzsanna Bősze[3] and Luca Fontanesi[4]

[1]Department of Biochemistry and Biotechnology, Faculty of Biotechnology and Food Sciences, Slovak University of Agriculture in Nitra, Nitra, Slovak Republic; [2]NPPC—Research Institute for Animal Production Nitra, Lužianky, Slovak Republic; [3]National Agricultural Research and Innovation Centre, Agricultural Biotechnology Institute, Gödöllő, Hungary; [4]Department of Agricultural and Food Sciences, Division of Animal Sciences, University of Bologna, Italy

17.1 Introduction

To understand the biological functions of proteins and all the regulatory elements involved in protein expression, modification, degradation and transport, and their possible roles in the human diseases, it is important to develop many animal models that could help in dissecting all these issues and features (Houldebine and Fan, 2009). Rodents (i.e. mice and rats) offer many examples and possibilities, including the availability of genetically modified models, that are useful to understand the basic functional elements needed in this context but do not cover all possible needs for many reasons. Therefore, other non-rodent models are essential to complete the biological vision of the foundational mechanisms underlying human diseases. Rabbits have a more diverse genetic background than inbred and outbred mouse and rat strains (Bősze and Houdebine, 2006). This can be useful when complex disease models (e.g. for atherosclerosis) are investigated or when it is needed to test and develop therapeutic strategies, since this situation mimics human genetic diversity more accurately (Bősze and Houdebine, 2006). For these reasons and considering that the rabbit is genetically and physiologically closer to humans for several aspects, including, for example, the cardiovascular system and metabolic characteristics like lipid metabolism, this species has been widely used in many biomedical studies. Moreover, another advantage of the rabbit derives from the larger size compared to that of the rodent models, which makes all manipulations and surgical interventions needed to study in more detail organ function and physiology easier. On the other hand, the rabbit has several advantages over other large laboratory animal species: it has a short gestation period that gives a short generation interval; it is a polytocous species; does produce large numbers of embryos with the production of a large litter and are multiparous.

Genetically modified rabbits can be produced using several approaches including microinjection (Chrenek *et al.*, 2005) and other methods such as lentiviral and transposon vectors and, more recently, by using gene-editing methodologies (see Chapter 15). For biotechnology applications, the rabbit is used to produce proteins of pharmaceutical relevance and many polyclonal antibodies that are essential in many research and clinical activities (Houdebine and Fan, 2009; Zhao *et al.*, 2010).

Contact e-mail address: chrenekp@cvzv.sk

DOI: 10.1079/9781780643342.0017

As briefly mentioned above, relevant biological and practical aspects for which the rabbit can be considered a useful model open the possibilities for several biotechnological applications (Chrenek and Makarevich, 2008):

- the litter size and the number of litters that can be produced per year per doe can easily produce large families and populations;
- the does have an induced ovulation with a superovulation outcome, i.e. about 25 embryos are produced per ovulation;
- embryos can provide a very good visualization of pronuclei;
- the short duration of the pregnancy (about 30 days) and the relatively fast maturation and favourable growth rate of the newborn rabbits reduce the generation interval;
- the general low cost of the animals and of the management of this species creates the economic conditions to use the rabbit to generate a large number of transgenic founders;
- the mammary gland is easily accessible for the collection of milk; it is a natural bioreactor for post-translation modification (e.g. glycosylation, sulfation and acylation) of recombinant proteins;
- the quite large amount of milk produced per day in the lactation period (150–250 mL, depending on peak of lactation) makes it possible to consider this biofluid for the production and then the collection of recombinant proteins;
- for some aspects, the rabbit, being genetically more similar to humans than dairy animals, is preferred for the production of the milk of human therapeutic proteins; and
- although the rabbit is no longer classified as resistant to Transmissible Spongiform Encephalopathies (TSEs), it is unlikely that it could transmit prionic diseases (Chianini *et al.*, 2012).

With the improvement of the genetic modification techniques that can be applicable to the rabbit (gene knock-down, gene knock-out, gene knock-in or gene replacement) a few specialized biotech companies (Transgenic Rabbit Models, TRM; PolyGene, www.polygene.ch, accessed 7 March 2021; ImmunoGenes, immunogenes.com) have started to offer genetically modified rabbits for several purposes, mainly as research tools. Therefore, it is expected that genetically modified rabbits could become more widely used animal models for several aspects that cannot be investigated using rodent models. The use of transgenic rabbit models in biomedical research promises to accelerate the development of new human diagnostic and therapeutic treatments (Fan and Watanabe, 2003; Duranthon *et al.*, 2012; Baczkó *et al.*, 2016; see Chapter 18).

17.2 Transgenic Rabbits for Biotechnology Applications

Transgenic rabbits have been used for several purposes and applications as described below. For example, for medical research activities, transgenic rabbits are used to identify the functions of specific factors in complex homeostatic systems through over- or under-expression of a modified gene (the inserted transgene) (Chrenek, 2012). Transgenic rabbits are used in toxicology to detect toxicants as responsive test animals. Transgenic animals are produced to analyse the genetic factors affecting mammalian development and the effect of disruption of gene regulations at the whole animal level. The pharmaceutical industry uses transgenic rabbits for the production of pharmaceutical proteins and drugs and for testing the therapeutic efficacy of drugs. Several uses have been developed and many more are forecast, particularly in a few areas.

17.2.1 Targeted production of pharmaceutical proteins

One of the most relevant uses of transgenic rabbits is for pharmaceutical applications in which the animals are considered bioreactors for the production of valuable human proteins, hormones, enzymes and growth factors (Table 17.1; Chrenek, 2012). These products may be recombinant or modified. The production of the functional protein from the animals is obtained by using tissue-specific regulatory DNA sequences to specifically address the expression in a defined target tissue.

The biotechnology industry usually uses large-scale cell cultures to produce recombinant proteins. Genetically modified eukaryotic

Table 17.1. Production of recombinant proteins by transgenic rabbits.

Recombinant protein[1]	Use	Reference
h alpha1-antitrypsin	Emhysema	Massoud *et al.* (1990)
h IL-2	Viral diseases	Bühler *et al.* (1990)
h tPA	Thrombosis	Reigo *et al.* (1993)
h erythropoietin	Anemia	Rodriguez *et al.* (1995)
h IGF-1	GH deficiency	Brem *et al.* (1994)
h ESDM	Ischaemia	Strömqvist *et al.* (1997)
h plasminogen activator	Acute myocardial infarction	Song *et al.* (2016); He *et al.* (2018); Lu *et al.* (2019)
h GH	GH deficiency	Hammer *et al.* (1985)
h alpha-glucuronidase	Glycogen storage disease	Bijvoet *et al.* (1999)
Salmon calcitonin	Osteoporosis	McKee *et al.* (1998)
eCG	Hormonal treatments	Galet (2001)
h NGF-beta	Neuropathy	Coulibaly *et al.* (1999)
h PC	hPC defficiency	Chrenek *et al.* (2002)
h FVIIa (eptacog beta)	Haemophilia	Chevreux *et al.* (2013, 2017); Biron-Andreani and Schved (2019)
h FVIII	Haemophilia	Chrenek *et al.* (2007)
h FVIII	Haemophilia	Hiripi *et al.* (2003)
Bochymosin	Cheese production	Brem *et al.* (1995)
Low Phe kappa casein	Cheese production	Baranyi *et al.* (2007)
h tissue non-specific alkaline phosphatase	Gram-negative acterial sepsis	Bodrogi *et al.* (2006)

[1] h indicates 'human'.

cells or bacteria that can express the target protein are cultured in nutrient medium which is continually replaced and from which the bioengineered product is then isolated (Chrenek, 2012). The growing medium should be pathogen-free, is buffered, and growth is temperature-regulated. The use of transgenic rabbits, as bioreactors ('pharmaceutical pharming') can be a cost-effective alternative to cell-culture methods for some specific products that could not be obtained in alternative systems, especially when post-translational modifications are needed (Wang *et al.*, 2013). By directing (or targeting) the expression of the pharmaceutically relevant product in the secretory cells of the liver, lactating mammary gland or kidney, their collection can be simplified as the products are available as body fluids, like milks, or in other tissues. The mammary gland is the most promising target tissue because it produces large amounts of proteins in a temperature-regulated fluid that may be collected daily in a non-invasive fashion (Houdebine, 2000; Chrenek, 2012).

17.2.2 The purification steps of recombinant proteins from milk and other tissues

The purification of the recombinant proteins from milk raises generally no particular problems (Houdebine, 2009). Milk contains a low concentration of proteases, which means that recombinant proteins are not degraded or destroyed. Recombinant proteins can be isolated from milk quite easily by non-destructive or expensive procedures. Chromatography may lead, on a case-by-case basis, to a high purity of the proteins (Houdebine, 2009). However, as milk is a relatively complex biological fluid, in some cases it could be quite complicated to eliminate all potential milk-derived contaminants from the targeted protein isolate, particularly when the recombinant protein is similar in weight, structure and properties to abundant milk-constitutive proteins (Houdebine, 2009). For example, the human serum albumin cannot be easily separated from the constitutive albumin protein normally derived by the animal used as bioreactor (Van Cott *et al.*, 1996).

17.2.3 Animal systems and tissues used to produce recombinant proteins

MILK Milk is currently considered the best available biofluid for the production and collection of recombinant proteins (Houdebine, 2009). A large number of studies have shown that this biofluid can be the source of complex recombinant proteins. However, not all attempts were successful, and failures occurred for various reasons (Chrenek *et al.*, 2010). Some of these failures were due to problems in the production of transgenic animals whereas others were related to the effective production and maturation of the final recombinant protein (Houdebine, 2009).

BLOOD Serum, which collects secretion from many tissues, may be the source of recombinant proteins (Houdebine, 2009). However, many proteins are not very stable in serum and their presence in this circulating biofluid could alter the physiology of the animal and, in turn, hamper its health, preventing the possibility to use this system for large-scale production of recombinant proteins. In addition, serum would need repeated bleeding of the animals and this practice may raise welfare concerns.

URINE Urine is another abundant biofluid that has been already used to prepare proteins such as gonadotropins for pharmaceutical use (Houdebine, 2000). The use of urine as a source of recombinant proteins means that these products are expressed and matured in the urothelium. This system can be preferred to the production in milk only if the maturation is more specific in the urothelium or if there are less negative perturbations of the animal physiology, and if the recombinant protein is reserved in the urine instead of in the milk. The collection of urine is also usually more complicated than the collection of milk in all species, including the rabbit.

SEMINAL PLASMA Seminal plasma is a relatively abundant biological fluid in some species and it can be easily collected (Houdebine, 2000). However, in rabbits, this system has some limitations due to the ejaculate volume that is limited compared to that of other livestock species. It is not known how complex proteins are matured and secreted in this biofluid and this is also true for the rabbit (Houdebine, 2000).

17.3 Rabbit Antibodies

The rabbit has been used in immunology and to develop immunologically derived techniques and approaches for more than a century. Therefore, many standard methodologies and protocols on these aspects have been first applied and standardized in this species, including immunization methods and antibody purification systems for an efficient and cost-effective production of antibodies. Thus, nowadays rabbits constitute one of the most relevant sources for a wide spectrum of monoclonal and polyclonal antibodies with broad applications in the biomedical and biotechnological sectors (Weber *et al.*, 2017). The global market share of antibodies used in biomedical research is represented by about 40–45% of rabbit antibodies (Marker Analysis Report, 2020). Polyclonal antibodies are defined as a set of different non-epitope-specific antibodies produced by the organism in response to a specific antigen or pathogen. Monoclonal antibodies have a specifically defined antigen-binding site that typically binds with high specificity and affinity to only one epitope. Rabbit monoclonal antibodies have been produced by applying and adapting hybridoma technology, the phage display technology and other alternative methods (Weber *et al.*, 2017). Rabbit polyclonal antibodies are commonly used in biomedical research as analytical or diagnostic tools in many immunologically based techniques (e.g. western blottings, immunochemistry). Many biotech companies sell rabbit polyclonal and monoclonal antibodies for these purposes.

Some rabbit monoclonal and polyclonal antibodies are used in clinical applications and in food-safety monitoring protocols (Thomas *et al.*, 1984; Campbell *et al.*, 2007). For example, the rabbit anti-thymocyte globulin known as Thymoglobulin, produced by Sanofi Genzyme Inc. (Cambridge, Massachusetts) was approved by the US Food and Drug Administration (FDA) in 1998 as an immunosuppressive drug and another rabbit polyclonal antibody has been approved in a diagnostic assay to evidence gastrointestinal stromal tumours (Weber *et al.*, 2017). Other rabbit monoclonal antibodies have been approved by the same agency for *in vitro* diagnostic methods to detect the expression of several tumour-associated antigens and the infection of *Helicobacter pylori* (Weber *et al.*, 2017).

The attactiveness of the rabbit for the production of antibodies derives from several features of this lagomorph species. Rabbit antibodies complement the rodent antibody recognition potential against human antigens as they can recognize epitopes that are not recognized by the rodent system. In addition, rabbit antibody reactivity against rodent antigens can be valuable to further characterize the rodent models in many biological investigations. A strong immune response is elicited in rabbits against small molecules and haptens that usually do not turn on similar responses in rodents (Weber *et al.*, 2017). The lower level of inbreeding that is usually observed in rabbit populations compared to that present in mice or rat populations facilitates the production of stronger antibody responses in rabbits that, in turn, result in higher sensitivity of rabbit antibodies than rodent antibodies (Weber *et al.*, 2017). Rabbits are better than rodents for the production of monoclonal antibodies that are based on the recovery of B cells from spleen, bone marrow or blood due to the larger size of the rabbits that can provide many more B cells. Another advantage derived from the rabbit is that this species uses different mechanisms to genetically generate and diversify their primary and secondary antibody repertoires compared to humans and mice, effectively creating a complementary set of binders for many different applications that complete what can be available from the rodents (Weber *et al.*, 2017).

17.4 Biotechnology Applications: Cryopreservation of Rabbit Lines

The cryopreservation of rabbit gametes is relevant for the conservation of genetic resources that can be useful for agricultural, medical and veterinary applications. The cryopreservation of ovarian and testicular tissues can allow the simultaneous preservation of thousands of follicles located in the ovarian stock, up to the primary stage, or sperm production cells. Because no stimulation is necessary, it could be used as an emergency preservation method. Immature follicles are better suited to cryopreservation because of their small size and the absence of *zona pellucida* and because they are metabolically quiescent and undifferentiated (Neto *et al.*, 2008). By these properties, frozen ovarian tissue takes advantage over cryopreservation of mature oocytes.

Moreover, the cryopreservation of ovarian tissue could be applied to the preservation of animal genetic resources: endangered wild species, domestic breeds, transgenic animals or biomedical models. Doe rabbits may constitute a model for the human and the animal applications of the ovarian tissue cryopreservation, because of their biological and breeding characteristics (Neto *et al.*, 2008). Because its prolificacy is high and its generation interval is reduced (5–6 months), the rabbit would allow rapid assessment of the long-term deleterious effects of the cryopreservation process over successive generations. The rabbit is also the most used species among the non-rodent species for regulatory teratology studies.

Gene banks have been established to collect and store the genetic material of such animals. This method could represent a new tool for genetic resource storage by the female pathway. After cryopreservation, mature oocytes could be obtained from the ovarian tissue by grafting or *in vitro* folliculogenesis (Neto *et al.*, 2008). Live offspring have been born from cryopreserved ovarian tissue autografted or allografted in mice, rabbits, ewes and women. Large antral follicles developed after xenograft of frozen ovarian tissue from different animal species into immunodeficient mice and also in humans. Although offspring were obtained from cryopreserved ovarian tissue, it is necessary to determine the critical steps of the freezing protocol by *in vitro* evaluation before grafting.

17.5 Concluding Remarks

The rabbit has numerous biotechnological applications. For some of them transgenic rabbits are more suitable and, in the medium to long term, the rabbit is expected to be used more widely as genetic-modification technologies are also maturing in this species. One of the emerging and promising fields of application is the production of recombinant proteins for pharmacological use in the rabbit milk even if some limitations of this approach should still be solved. The production of antibodies is one of the most relevant biotechnology applications in this species that generates a large economic value and is still contributing to advance related biomedical studies.

References

Baranyi, M., Hiripi, L., Szabó, L., Catunda, A.P., Harsányi, I. *et al.* (2007) Isolation of functional low-phenylalanine kappa casein expressed in the milk of transgenic rabbits. *Journal of Biotechnology* 128, 383–392.

Baczkó, I., Jost, N., Virág, L., Bősze, Z. and Varró, A. (2016) Rabbit models as tools for preclinical cardiac electrophysiological safety testing: importance of repolarization reserve. *Progress in Biophysics and Molecular Biology* 121, 157–168.

Bijvoet, A.G., Van Hirtum, H., Kroos, M.A., Van de Kamp, E.H., Schoneveld, O. *et al.* (1999) Human acid alpha-glucosidase from rabbit milk has therapeutic effect in mice with glycogen storage disease type II. *Human Molecular Genetics* 8, 2145–2153.

Biron-Andreani, C. and Schved, J.F. (2019) Eptacog beta: a novel recombinant human factor VIIa for the treatment of hemophilia A and B with inhibitors. *Expert Review of Hematology* 12, 21–28.

Bodrogi, L., Brands, R., Raaben, W., Seinen, W., Baranyi, M. *et al.* (2006) High level expression of tissue non-specific alkaline phosphatase in the milk of transgenic rabbits. *Transgenic Research* 15, 627–636.

Bősze, Zs. and Houdebine, L.M. (2006) Application of rabbits in biomedical research: a review. *World Rabbit Science* 14, 1–14.

Brem, G., Hartl, P., Besenfelder, U., Wolf, E., Zinovieva, N. and Pfaller, R. (1994) Expression of synthetic cDNA sequences encoding human insulin-like growth factor-1 (IGF-1) in the mammary gland of transgenic rabbits. *Gene* 149, 351–355.

Brem, G., Besenfelder, U., Zinovieva, N., Seregi, J., Solti, L. and Hartl, P. (1995) Mammary gland specific expression of chymosin constructs in transgenic rabbits. *Theriogenology* 43(1), 175.

Bühler, T.A., Bruyère, T., Went, D.F., Stranzinger, G. and Bürki, K. (1990) Rabbit beta-casein promoter directs secretion of human interleukin-2 into the milk of transgenic rabbits. *Biotechnology* 8, 140–143.

Campbell, K., Stewart, L.D., Doucette, G.J., Fodey, T.L., Haughey, S.A. *et al.* (2007) Assessment of specific binding proteins suitable for the detection of paralytic shellfish poisons using optical biosensor technology. *Analytical Chemistry* 79, 5906–5914.

Chevreux, G., Faid, V., Scohyers, J.M. and Bihoreau, N. (2013) N-/O-glycosylation analysis of human FVIIa produced in the milk of transgenic rabbits. *Glycobiology* 23, 1531–1546.

Chevreux, G., Tilly, N., Leblanc, Y., Ramon, C., Faid, V. *et al.* (2017) Biochemical characterization of LR769, a new recombinant factor VIIa bypassing agent produced in the milk of transgenic rabbits. *Haemophilia* 23, e324–e334.

Chianini, F., Fernández-Borges, N., Vidal, E., Gibbard, L., Pintado, B. *et al.* (2012) Rabbits are not resistant to prion infection. *Proceedings of the National Academy of Sciences of the USA* 109, 5080–5085.

Chrenek, P. (2012) Rabbit as a suitable biological model for agricultural and biomedical research. *Proceedings of the XXV. konferencia s medzinárodnou účasťou Aktuálne smery v chove brojlerových králikov*. Králik ako produkčné a modelové zviera, 21 November, Nitra, Slovak Republic.

Chrenek, P. and Makarevich, A.V. (2008) Transgenic rabbits – production and application. *Slovak Journal of Animal Science* 41, 113–120.

Chrenek, P., Vasicek, D., Makarevich, A., Uhrin, P., Petrovicova, I. *et al.* (2002) Integration and expression of the WAP-hPC gene in three generations of transgenic rabbits. *Czech Journal Animal Science* 47, 45–49.

Chrenek, P., Vasicek, D., Makarevich, A., Jurcik, R., Suvegova, K. *et al.* (2005) Increased transgene integration efficiency upon microinjection of DNA into both pronuclei of rabbit embryos. *Transgenic Research* 14, 417–428.

Chrenek, P., Ryban, L., Vetr, H., Makarevich, A.V., Uhrin, P. *et al.* (2007) Expression of recombinant human factor VIII in milk of several generations of transgenic rabbits. *Transgenic Research* 16, 353–361.

Chrenek, P., Makarevich, A.V., Pivko, J. and Bulla, J. (2010) Transgenic farm animal production and application. *Slovak Journal of Animal Science* 43, 45–49.

Coulibaly, S., Besenfelder, U., Fleischmann, M., Zinovieva, N., Grossmann, A. *et al.* (1999) Human nerve growth factor beta (hNGF-beta): mammary gland specific expression and production in transgenic rabbits. *FEBS Letters* 444, 111–116.

Duranthon, V., Beaujean, N., Brunner, M., Odening, K.E., Santos, A.N. *et al.* (2012) On the emerging role of rabbit as human disease model and the instrumental role of novel transgenic tools. *Transgenic Research* 21, 699–713.

Fan, J. and Watanabe, T. (2003) Transgenic rabbits as therapeutic protein bioreactors and human disease models. *Pharmacology and Therapeutics* 99, 261–282.

Galet, C., Le Bourhis, C.M., Chopineau, M., Le Griec, G., Perrin, A. *et al.* (2001) Expression of a single betaalpha chain protein of equine LH/CG in milk of transgenic rabbits and its biological activity. *Molecular Cell Endocrinology* 174, 31–40.

Hammer, R.E., Pursel, V.G., Rexroad, C.E. Jr., Wall, R.J., Bolt, D.J. *et al.* (1985) Production of transgenic rabbits, sheep and pigs by microinjection. *Nature* 315, 118–122.

He, Z., Jiang, L., Zhang, T., Zhou, M., Wu, D. *et al.* (2018) Efficient increase of the novel recombinant human plasminogen activator expression level and stability through the use of homozygote transgenic rabbits. *International Journal of Molecular Medicine* 42, 2269–2275.

Hiripi, L., Makovics, F., Halter, R., Baranyi, M., Paul, D. *et al.* (2003) Expression of active human blood clotting Factor VIII in the mammary gland of transgenic rabbits. *DNA and Cell Biology* 22, 41–45.

Houdebine, L.M. (2000) Transgenic animal bioreactors. *Transgenic Research* 9, 305–320.

Houdebine, L.M. (2009) Production of pharmaceutical proteins by transgenic animals. *Comparative Immunology, Microbiology and Infectious Diseases* 32, 107–121.

Houdebine, L.M. and Fan, J. (2009) Introduction. In: Houdebine, L.M. and Fan, J. (eds) *Rabbit Biotechnology, Rabbit Genomics, Transgenesis, Cloning and Models*. Springer, Dordrecht, Heidelberg, London, New York, pp. 1–2.

Lu, R., Zhang, T., Song, S., Zhou, M., Jiang, L. *et al.* (2019) Accurately cleavable goat β-lactoglobulin signal peptide efficiently guided translation of a recombinant human plasminogen activator in transgenic rabbit mammary gland. *Bioscience Reports* 39, pii: BSR20190596.

Market Analysis Report (2020) *Research Antibodies Market Size, Share & Trends Analysis Report by Type (Monoclonal, Polyclonal), by End Use, by Product, by Technology, by Source, by Application (Oncology, Stem Cells), and Segment Forecasts, 2020–2027*. Grand View Research, Report ID: GVR-2-68038-124-5.

Massoud, M., Bischoff, R., Dalemans, W., Pointu, H., Attal, J. *et al.* (1990) Production of human proteins in the blood of transgenic animals. *Central Research Academic Science III* 311, 275–280.

McKee, C., Gibson, A., Dalrymple, M., Emslie, L., Garner, I. and Cottingham, I. (1998) Production of biologically active salmon calcitonin in the milk of transgenic rabbits. *Nature Biotechnolology* 16, 647–651.

Neto, V., Buff, S., Lornage, J., Bottollier, B., Gu, P. and Joly, T. (2008) Effects of different freezing parameters on the morphology and viability of preantral follicles after cryopreservation of doe rabbit ovarian tissue. *Fertility and Sterility* 89, 1348–1356.

Reigo, E., Limonta, J., Aguilar, A., Perez, A., De Armas, R. *et al.* (1993) Production of transgenic mice and rabbit that carry and express the human tissue plasminogen activator cDNA under control of bovine alpha S1 casein promoter. *Theriogenology* 39, 11731–11785.

Rodriguez, A., Castro, F.O., Aguilar, A., Ramos, B., Del Barco, D.G. *et al.* (1995) Expression of active human erythropoietin in the mammary gland of lactating transgenic mice and rabbits. *Biological Research* 28, 141–153.

Song, S., Ge, X., Cheng, Y., Lu, R., Zhang, T. *et al.* (2016) High-level expression of a novel recombinant human plasminogen activator (rhPA) in the milk of transgenic rabbits and its thrombolytic bioactivity in vitro. *Molecular Biology Reports* 43, 775–783.

Strömqvist, M., Houdebine, M., Andersson, J.O., Edlund, A., Johansson, T. *et al.* (1997) Recombinant human extracellular superoxide dismutase produced in milk of transgenic rabbits. *Transgenic Research* 6, 271–278.

Thomas, F.T., Griesedieck, C., Thomas, J., Carver, M., Whitley, T. *et al.* (1984) Differential effects of horse ATG and rabbit ATG on T cell and T cell subset levels measured by monoclonal antibodies. *Transplantation Proceedings* 16, 1561–1563.

Van Cott, K.E., Williams, B., Velander, W.H., Gwazdauskas, F., Lee, T. *et al.* (1996) Affinity purification of biologically active and inactive forms of recombinant human protein C produced in porcine mammary gland. *Journal of Molecular Recognition* 9, 407–414.

Wang, Y., Zhao, S., Bai, L., Fan, J. and Liu, E. (2013) Expression systems and species used for transgenic animal bioreactors. *BioMed Research International* 2013, 580463.
Weber, J., Peng, H. and Rader, C. (2017) From rabbit antibody repertoires to rabbit monoclonal antibodies. *Experimental & Molecular Medicine* 49, e305.
Zhao, S., Wei, K.,, Yu, Q.L.Y., Cheng, F., Wang, Y. *et al.* (2010) Applications of transgenic rabbits in biomedical research – based on the literature search. *World Rabbit Science* 18, 159–167.

18 The Rabbit as a Biomedical Model

Jianglin Fan[1]*, Cesare Castellini[2], Clive P. Page[3] and Domenico Spina[3]
[1]Department of Molecular Pathology, Interdisciplinary Graduate School of Medicine, University of Yamanashi, Japan; [2]Department of Agricultural, Food and Environmental Science, University of Perugia, Italy; [3]Sackler Institute of Pulmonary Pharmacology, Institute of Pharmaceutical Science, King's College London, United Kingdom

18.1 Introduction

Rabbits are one of the most used experimental animals for biomedical research, particularly as a widely used bioreactor for the production of polyclonal and monoclonal antibodies. However, many unique features of the rabbit have also made it an excellent species for investigating a number of aspects of human diseases. Rabbits are phylogenetically closer to humans than rodents, in addition to their relatively proper size, tame disposition and ease of use and maintenance in the laboratory facility. Because of their short lifespan, short gestation period, high number of progeny, low cost and availability of genomic and proteomic tools and information, rabbits usually serve to bridge the gap between smaller rodents (mice and rats) and larger animals (dogs and monkeys) and play an important role in many translational research activities such as preclinical testing of drugs and diagnostic methods for patients. The principle of using rabbits (rather than other animals) as experimental models is very simple. Rabbits should be used for certain research to solve particular problems (e.g. translational research) that are hardly accomplished by other species. In this chapter, we will provide a brief introduction and some guidelines for using the rabbits as models for studying human diseases such as cardiovascular disease, respiratory disease, immune-related diseases, osteoarthritis, ocular research and Alzheimer's disease, along with reproductive physiology. The chapter will not be comprehensive but rather concise; therefore, readers are also encouraged to read other chapters of this book and other references for further details.

18.2 Rabbit Models for Atherosclerosis

Atherosclerosis is the major factor for mortality globally. Disclosing the pathologic mechanisms of atherosclerosis and developing effective therapy for treating atherosclerosis is an important and continuous task. Rabbits have been widely used as experimental models for studying human atherosclerosis, with the first studies being carried out more than a century ago (Ignatowski, 1908). As herbivores, rabbits do not develop atherosclerosis spontaneously on a normal diet; however, they do develop atherosclerotic lesions when they are fed with a diet containing cholesterol or are bred to be genetically abnormal and to have known human risk factors (Fan and Watanabe, 2000; Fan et al., 2015).

Contact e-mail address: jianglin@yamanashi.ac.jp

 The Genetics and Genomics of the Rabbit (ed Luca Fontanesi)
DOI: 10.1079/9781780643342.0018

The rationale for using rabbits for studying human lipid metabolism and atherosclerosis is that:

1. The lipid metabolism features of rabbits are more similar to those of humans than those of rodents as summarized in Table 18.1. One of these examples is that the major plasma lipoproteins, low-density lipoproteins are present in rabbits as in humans, and rabbits also have abundant cholesteryl ester transfer protein in the plasma, an important regulator of lipoprotein metabolism, which is actually absent in mice and rats.
2. Rabbits are sensitive to a diet containing cholesterol and they develop into hypercholesterolaemia and atherosclerosis rapidly.
3. Pathological features of aortic and coronary atherosclerosis in rabbits resemble human atherosclerosis (Fig. 18.1).

There are three kinds of rabbit models, which are commonly used for the investigation of human atherosclerosis and hyperlipidaemia, depending on the research purpose: (i) cholesterol-fed rabbits; (ii) Watanabe heritable hyperlipidaemic (WHHL) rabbits; and (iii) transgenic (including knockout and knock-in) rabbits.

Cholesterol-fed rabbit models are the most well-known models for the study of human atherosclerosis. For such experiments, rabbits are usually fed with a diet containing 0.3–0.8% cholesterol along with 3% vegetable oil for more than three months (Taylor and Fan, 1997; Fan and Watanabe, 2000). Feeding rabbits with a diet containing more than 1% cholesterol, especially without oil for a long time is usually toxic and can cause massive accumulation of foam cells in many parts of the body in addition to extremely 'high' hypercholesterolaemia. These phenomena are often criticized as 'un-physiological in humans', so are not recommended.

WHHL rabbits are genetically deficient in LDL receptors and spontaneously develop hypercholaesterolaemia and atherosclerosis (Watanabe, 1980). WHHL rabbits are the best model for human familial hypercholaesterolaemia and have made a great contribution to discovery of many lipid-lowering drugs including statins (Shiomi and Ito, 2009). WHHL rabbits can also develop coronary atherosclerosis and myocardial infarction (Shiomi *et al.*, 2003; Shiomi and Fan, 2008).

Transgenic rabbits are other powerful models for investigating many gene functions and their relationships with lipoprotein metabolism and development of atherosclerosis (Fan and Watanabe, 2003). During the past years, a large number of transgenes involved in the lipid metabolism and/or atherosclerosis have been introduced into transgenic rabbits including apolipoproteins [AI, AII, (a), B-100, CI, CIII, E2, and E3] (Fan *et al.*, 1995, 1998, 1999; Duverger *et al.*, 1996; Huang *et al.*, 1997; Ding *et al.*, 2011; Wang *et al.*, 2013; Gautier *et al.*, 2019), lipid metabolism-related enzymes (hepatic lipase, lipoprotein lipase, lecithin:cholesterol acyltransferase, apoB mRNA editing protein, phospholipid transfer protein) (Fan *et al.*, 1994, 2001; Yamanaka *et al.*, 1995; Hoeg *et al.*, 1996; Masson *et al.*, 2011), and lesional macrophage-secreting factors (15-lipoxygenase and matrix metalloproteinase-12) (Shen *et al.*, 1996; Liang *et al.*, 2006) or other factors such as C-reactive protein (Koike *et al.*, 2009) and vascular endothelial cell growth factor (Kitajima *et al.*, 2005). Recently, apoCIII and apoE knock-out (KO) rabbits have been made using ZFN or TALEN technologies (Yang *et al.*, 2013, 2014; Niimi *et al.*, 2016), and, hopefully,

Table 18.1. Comparison of human, rabbit and mouse lipoprotein metabolism and susceptibility to atherosclerosis.

Features/traits	Humans	Rabbits	Mice
Major plasma lipoproteins	Low density lipoproteins	Low density lipoproteins	High density lipoproteins
Cholesteryl ester transfer protein	Abundant	Abundant	None
Response to a cholesterol diet	Sensitive	Sensitive	Resistant
Aortic atherosclerosis	Abdominal aorta	Aortic arch and thoracic aorta	Aortic root
Coronary atherosclerosis	yes	yes	no
Cerebral atherosclerosis	yes	yes	no

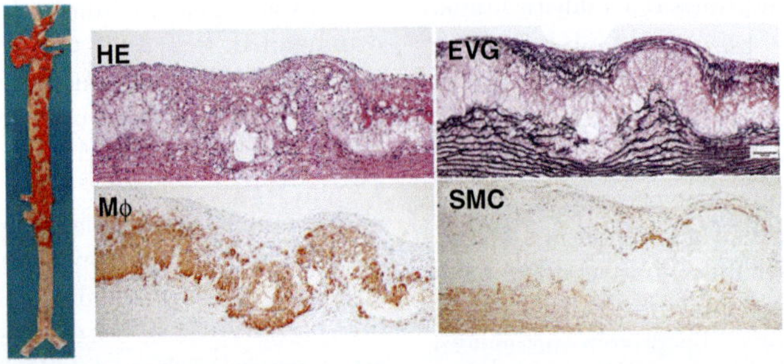

Fig. 18.1. Cholesterol diet-induced aortic atherosclerosis in rabbits. The left figure shows aortic gross atherosclerosis visualized by Sudan IV staining. The right figure (4 panels) shows microscopic features of the lesions stained by hematoxylin and eosin (HE), elastica van Gieson (EVG), and immunohistochemically stained with antibodies against rabbit macrophages (Mϕ) and smooth muscle cells (SMC).

these KO rabbits will provide a novel valuable model for the study of human atherosclerosis in the future.

18.3 Rabbit Models for Heart Disease

Because of its intermediate size, the rabbit also offers several potential advantages over other species in terms of heart disease studies (Pogwizd and Bers, 2008). The rabbit heart is large enough to perform surgical and catheter-based interventions at a much lower cost (5–15 times less expensive than dogs). Surgical interventions are still easier than microsurgical approaches in rodents. More importantly, rabbit cardiac physiology is more similar to human cardiac physiology than with mice or rats. First, cellular electrophysiology and Ca^{2+} transport in rabbits are much more like those in humans than is the case for either rats or mice. This is particularly relevant for the study of heart failure (HF) and arrhythmia, because alterations in ion channel and Ca^{2+} transporter function or expression are thought to contribute directly to depressed contractile performance and arrhythmogenesis (Pogwizd and Bers, 2004). Second, the rabbit can be considered an excellent model for ventricle myocardial hypertrophy because the rabbit myocardium is similar to the adult human myocardium as it expresses almost entirely β-myosin heavy chain, whereas ventricles from adult mice and rats normally express mainly the fast α-myosin isoform (Gupta, 2007) (which allows faster cross-bridge cycling and muscle shortening than β-myosin), but during hypertrophy and HF there is isoform switching to β-myosin (Gupta, 2007). These various differences are not arguments against other large animal models but are crucial when extrapolating results from mouse and rat hearts in the context of human disease. Thirdly, rabbit coronary arteries can be used for investigating ischemic heart disease such as myocardial infarction (Fan *et al.*, 2015). The coronary ischaemia can be induced by ligation or occlusion methods and coronary atherosclerosis (which is almost impossible to produce in hypercholesterolaemic rodents) can occur in both cholesterol-fed rabbits (Liang *et al.*, 2006) and WHHL rabbits (Shiomi *et al.*, 2003). Due to these unique cardiac characteristics in the rabbit, transgenic rabbits were also created to study long QT syndrome (Brunner *et al.*, 2008; Lang *et al.*, 2016; Odening *et al.*, 2019), hypertrophic cardiomyopathy (Marian *et al.*, 1999; Sanbe *et al.*, 2005; Lowey *et al.*, 2018), the G protein G_{sa} (Nishizawa *et al.*, 2006) and phospholamban (Pattison *et al.*, 2008).

18.4 Rabbit Models for Alzheimer's Disease

Hypercholesterolaemic rabbits are not only useful for the study of atherosclerosis, as described above, but have also emerged as new models for the study of human Alzheimer's disease (AD) (Sparks, 2008). It has been reported that hypercholesterolaemic rabbits exhibit 16 neuropathologic features observed in human AD. For example, brains from hypercholesterolaemic rabbits show increased levels of cholesterol along with amyloid β (Aβ) and decreased levels of acetylcholine, tau, and apoE immunoreactivity, Aβ plaques in the extracellular space, a breakdown in the blood-brain barrier, and an increase in microglial and decrease in neuronal cell population (Woodruff-Pak, 2008). More importantly, rabbits with hypercholesterolaemia-induced AD show age-dependent deficits in learning and memory which are often observed in human AD patients. Homologous similarity between human and rabbit Aβ protein sequences are 97%, so rabbit AD models may be superior to other AD animal models. Anatomically, rabbit brains are large enough for pathological study and also cognitive observations. For generating AD pathology, rabbits are usually fed a diet with 2% cholesterol alone or with a trace of copper in the drinking water for 8–10 weeks (Sparks *et al.*, 1994). It seems that there is a sex-dependent difference in terms of neuropathological severity in cholesterol-fed rabbits.

18.5 Rabbit Models for Ophthalmological Research

Big eyeballs of rabbits relative to their body size have made them a valuable experimental model for ophthalmological research, including surgical interventions and drug-delivery experiments. Rabbit eyes have also been used for the study of cataracts, herpes simplex virus, keratitis, glaucoma, uveitis, proliferative vitreoretinopathy, retinoblastoma and retinitis pigmentosa. However, the histology of rabbit eyes is not completely similar to that of humans. For example, the cornea of rabbits is thinner than that for humans, and the rabbit's lens is larger. As a result, the effects of drugs on the retina of a rabbit are not easily translated to humans, and rabbits are infrequently used to study the pathogenesis of ocular diseases (Burkholder *et al.*, 2012). In spite of this, transgenic rabbits expressing rhodopsin P347L have been generated as models of retinitis pigmentosa (Kondo *et al.*, 2009; Kominami *et al.*, 2017, 2019).

18.6 Rabbit Models for Osteoarthritis

The rabbit knee is grossly similar to that of humans so rabbits are an excellent model for the study of human osteoarthritis (OA). In humans, OA develops idiopathically and progresses slowly with age. Rabbit OA models have many advantages over other animal models. The rabbit knee is large enough to harvest adequate amounts of tissue for pathological analysis, a major limitation of smaller animals. In adult humans, the growth plates in the long bones no longer retain the capacity for growth. This is also the case in skeletally mature rabbits. By contrast, mouse and rat growth plates do not normally close completely and longitudinal bone growth can be reinitiated in mature animals. This is a complicating issue for data interpretation and limits the translation of research performed using mice and rats. OA in rabbits can be induced using surgical intervention (such as meniscectomy, or anterior cruciate ligament transection, mechanical manipulation or chemical injection of inflammatory stimuli (IL-1β, collagenase and carrageenan) (e.g. Gregory *et al.*, 2012; Liu *et al.*, 2016).

18.7 Rabbit Models for Infectious and Autoimmune Disease

The rabbit is sensitive to infections and has been used as an animal model for many infectious diseases such as tuberculosis (see below), syphilis (Fitzgerald, 1985), anthrax (Zaucha *et al.*, 1998), tularemia, pox-virus diseases, hepatitis E and several other infectious diseases (e.g. Cheng *et al.*, 2012; Lu *et al.*, 2019; Tran *et al.*, 2020). It is hard or impossible to perform these experiments in other species. When it is unethical to conduct clinical trials of therapeutic strategy against infectious diseases such as anthrax or pneumonic plague because of the severity of the disease, the US Food and Drug Administration's (FDA) animal efficacy rule requires testing in two different species of animals as a part of the approval process. The first therapy to be approved under this 'Animal Rule' was a fully human monoclonal antibody for protection against or treatment of inhalation anthrax.

Raxibacumab was tested in three studies using rabbits while one study was with monkeys.

In addition, rabbits are used for investigations of autoimmune diseases such as systemic lupus erythematosus (SLE) (Rai *et al.*, 2010). Most animal studies of SLE are conducted in mice; however, more and more evidences show that the differences in gene expression between mice and humans are huge and lots of results obtained in mouse models were followed by failure of drug clinical trials in humans. Therefore, alternative human-close models such as rabbits are required. In this regard, the rabbit SLE models are now being applied to studies of human SLE patients (Mage and Rai, 2012). Another example of these is human immunodeficiency virus (HIV) infection. HIV can infect humans through CD4 lymphocytes and when the human CD4 transgene was introduced into transgenic rabbits, but not transgenic mice, CD4 transgenic rabbits can be successfully infected by HIV as a possible AIDS model (Yamamura *et al.*, 1991; Speck *et al.*, 1998).

18.8 Rabbit Models for Studying Reproductive Physiology

In the last decades, male infertility has become a very important concern (Kajihara *et al.*, 2006). The main causes of male infertility are thought to be secondary to environmental exposure to chemical compounds and/or infections/inflammation of the male reproductive system. Mice are widely used as models for physiological and toxicological studies because of the massive amount of information on their development and functions. However, the rabbit is the smallest species in which almost all the reproductive and toxicological endpoints of humans can be measured. The rabbit semen can be easily collected by the use of an artificial vagina and the fertility of sperm tested (Foote and Carney, 2000). Furthermore, the rabbit permits longitudinal studies on ejaculated sperm instead of the investigation of testicular sperms in rats or mice. Accordingly, rabbits are excellent models for research on reproductive traits and they are particularly attractive for studying sperm alterations because the collection of semen does not require killing the animal and the sperm can be used for toxicological studies. The mature sperm is a cell that eliminates almost all the normal biological functions and retains only those dedicated to provide motility and lead to fertilization. Therefore, any damage to one of these basic functions caused by an environmental or toxic chemical is easily detectable as ejaculated sperm is unable to activate a repairing process.

Following these considerations, it is sound to use the sperm as a good target for metabolic and toxicological studies (D'Cruz *et al.*, 2000; Castellini *et al.*, 2019). This approach also lowers the use of laboratory animals that is very advantageous for ethical and economic reasons. Some researchers (Young *et al.*, 1992) have suggested that rabbit spermatozoa are a good model for *in vitro* assessment of cytotoxicity and have proposed motion-based indices as endpoints. However, it is quite evident that there is a need to develop a robust, standardized *in vitro* sperm model to reduce variability factors (Seed *et al.*, 1996) and to increase the discriminating power of the model. Rabbit spermatozoa have been used as a model of spermiotoxicity for some metal ions (Castellini *et al.*, 2009) with the kinetic traits and the integrity of acrosome, which is very sensitive for detecting metal ion damage and toxicity (Fig. 18.2).

Rabbit sperms have also been used to assess the toxicity of nanomaterials (1–100 nm in size). Indeed, professional and environmental exposure to nanomaterials has greatly increased because of the great expansion of nanotechnology. As a consequence of this widespread use, nanoparticles are likely to be ubiquitous in the environment and can enter the human body by different ways (inhalation, ingestion, skin adsorption and biomedical applications). Nanoparticles differ from bulk materials in their chemical and physical properties and generally have different toxic profiles (Bernardini *et al.*, 2011).

Many studies have shown that particles that are not toxic in the micrometer range may become toxic in the nanometer range (Warheit *et al.*, 2007). Moreover, some nanoparticles can diffuse across biological barriers such as the blood-testis barrier, reducing the sperm viability. A preliminary research showed that sperm is very sensitive to some nano-components of environmental pollution (Moretti *et al.*, 2009; Castellini *et al.*, 2014), and to other nanometals

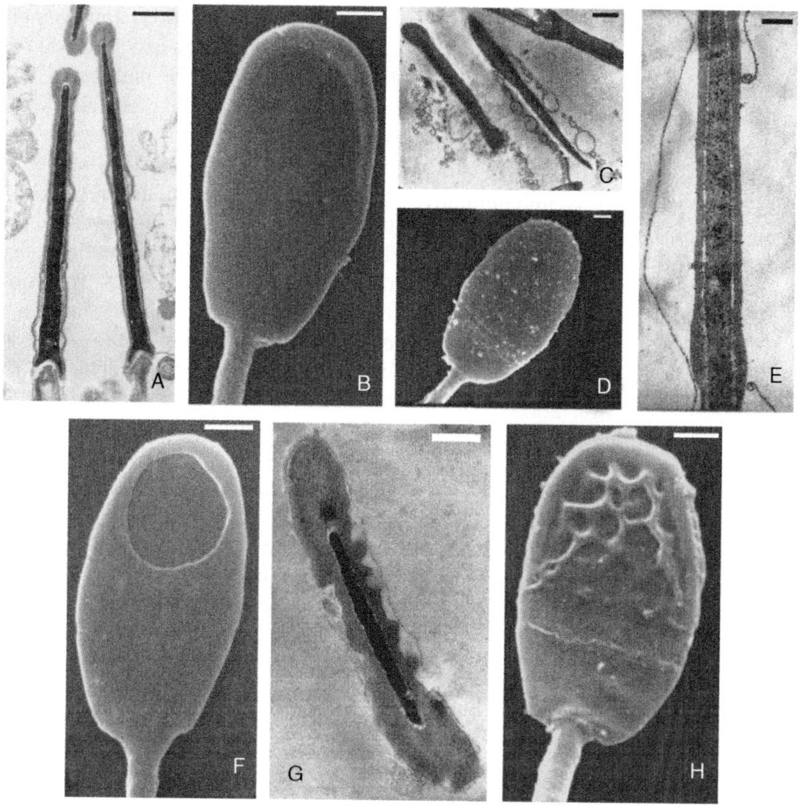

Fig. 18.2 Transmission Electron Microscope (TEM) and Scanning Electron Microscope (SEM) views of ejaculated rabbit sperm. (A) TEM (10,000×) and (B) SEM views (10,400×) of control sperm heads; (C) SEM (8400×) and (D) TEM (6800×) views of sperm heads after 4 hours of incubation with Pt and Hg compounds; (E) TEM (14,200×) and (F) SEM views (10,000×) of a sperm head after 4 hours of incubation with Ch compound; (G) TEM (10,000×) and (H) SEM views (31,500×) of a sperm after 4 hours of incubation with Vn compound. (Modified from Castellini *et al.*, 2009.)

that are often used in biomedical applications (Moretti *et al.*, 2013).

As previously stated, infection and inflammation of the genitourinary tract is one of the most important causes of male infertility in humans and in domestic animals (Kauffold *et al.*, 2007). However, the exact mechanisms by which such conditions impair the male reproductive system are not fully understood. Lipopolysaccharide (LPS)-induced animal models are widely used to highlight the mechanism of inflammation and provide useful information on the whole process of sperm production to better comprehend this complex process.

Buck rabbits treated with a low LPS dose (50 mg/kg body weight *E. coli*) could be a useful model for studying the short- and long-term effect of an induced systemic inflammation on spermatogenesis (Brecchia *et al.*, 2010). Sperm-membrane integrity and the number of necrotic sperm are affected for a long period after the LPS challenge, reaching a maximum at the end of the spermatogenic cycle. These models confirm that a sub-acute inflammation may cause infertility by compromising the testicle cell structure cells and sperm membrane integrity (Collodel *et al.*, 2012) shedding new light on the effect of systemic inflammation on spermatogenesis. Gene-expression analyses in rabbits could be important to better understand all these mechanisms leading to male infertility.

18.9 Respiratory System

The rabbit has been used to increase our understanding of biological processes in the respiratory tract, including nerve reflexes involved in normal physiology and in pathological situations such as cough, anaphylaxis and pulmonary embolism (Karczewski and Widdicombe, 1969; Mills *et al.*, 1969). Additionally, there has been a considerable body of work using the rabbit to understand various biological processes involved in a range of respiratory diseases such as asthma, chronic obstructive pulmonary disease (COPD) (Keir and Page, 2008), cough (Adcock *et al.*, 2003) and tuberculosis (Dharmadhikari and Nardell, 2008). Furthermore, recent work exploiting the rabbit genome is leading to the development of transgenic rabbits, which will be increasingly invaluable in understanding various biological mechanisms and could also aid in the development of new medicines.

18.9.1 Asthma and COPD

A comparison of the anatomical features of the pulmonary system between humans, rabbits and mice demonstrates a number of differences, although in some respect the rabbit is a better model to study airway respiratory physiology (Table 18.2). The mouse is used in greater frequency to study respiratory disease because of the availability of knock-out and transgenic strains and the immunology is also well characterized in this species compared with others. However, the rabbit offers a number of notable advantages over the mouse. For example, the large intrapulmonary airways in the rabbit are supplied by the bronchial circulation (like in humans) whilst in the mouse, blood supply to the intrapulmonary airways is supplied by the pulmonary circulation. The type of mucus-secreting cells also differs and there are fewer airway branches in the mouse compared with rabbits and humans in order to afford the high respiratory rate in this species (Table 18.2). The sensory innervation also differs, and whilst the mammalian lung is innervated by both un-myelinated C-fibres and myelinated Aβ fibres, the density of substance P containing sensory nerves is absent in man, sparse in rabbit (Laitinen *et al.*, 1983; Canning and Spina, 2009) and promotes bronchodilation in the mouse (Table 18.2). This is consistent with the paucity of contractile response of human and rabbit airways to capsaicin, an agent which stimulates sensory neuropeptide release via activation of the transient receptor potential vanilloid receptor (TRPV1) (Spina *et al.*, 1998). In the mouse, activation of TRPV1 promotes a prostaglandin E_2-dependent relaxation (Manzini, 1992). The mouse airways are responsive to a small range of bronchoconstrictor stimuli, and lack the ability to cough, unlike humans. Rabbits, on the other hand, can be induced to cough to a tussive agent following exposure to pollutants like ozone (Table 18.2).

The other major difference lies in the ability to undertake precise measures of changes in respiratory lung mechanics over time in order to understand the temporal aspects of inflammatory disease on airway physiology. Various techniques have been developed to monitor pulmonary lung mechanics in the mouse using various invasive and non-invasive techniques. The latter involve the measurement of penH, a dimensionless parameter which reflects changes in the pattern of breathing but which has been highly criticized for its lack of correlation with total lung resistance (Irvin and Bates, 2003). Hence, more invasive techniques are therefore required but suffer from the disadvantage that this can only be undertaken once in an animal. The main advantage of using the rabbit is that its large size permits the repeated non-invasive measurement of pulmonary lung mechanics in anaesthetized animals (Minshall *et al.*, 1993; Keir *et al.*, 2011).

Like the mouse, the rabbit has been useful for the study of allergic pulmonary inflammation and offers several advantages over other species. It was recognized in the 1970s that rabbits immunized at birth to allergens developed an IgE allergic phenotype that was maintained into the adult animal (Meng *et al.*, 1973; Pinckard *et al.*, 1977), characterized by anaphylaxis to allergen, and early- and late-phase pulmonary responses to inhaled antigen (Shampain *et al.*, 1982). In contrast, animals immunized seven days after birth produced both IgE and IgG but had blunted physiological responses to antigen, highlighting the importance of neonatal immunization for the development of long-lived IgE responses (Meng *et al.*, 1973; Pinckard *et al.*,

Table 18.2. Comparative respiratory physiology and pharmacology between humans, rabbits and mice.

Features of interest	Humans	Rabbits	Mice	References
Anatomy				
• Airway generations	23	32	13–17	Irvin and Bates (2003); Shapiro (2007); Kamaruzaman *et al.* (2013)
• Branching pattern	Symmetrical	Less symmetrical	Less symmetrical	
• Respiratory rate	12 breaths/min	100–200 breaths/min	250–300 breaths/min	Verloop (1949)
• Respiratory bronchioles	Present	Absent	Absent	
• Mucus-producing cells	Submucosal and goblet cells	Goblet cells	Clara cells	Mills *et al.* (1969); Laitinen *et al.* (1983); Manzini (1992); Matsumoto *et al.* (1997); Adcock *et al.* (2003); Canning and Spina (2009)
• Bronchial circulation	extra- and intrapulmonary	large intrapulmonary	extrapulmonary airways	
• Sensory innervation				
• C- and A-delta fibres	present	present	present	
• Neuropeptide containing C-fibres	absent	sparse	sparse (dilator)	
Physiology				
Bronchoconstrictors	Yes	Yes	Yes	Canning (2003)
• Muscarinic agonists	Yes	Not determined	No	
• Leukotrienes	Yes	Yes	No	
• Histamine	Yes	No	No	
• Neurokinins	No	No	Yes	
• Serotonin				
Bronchodilators				
• Beta-agonists	Yes (β2-adrenoceptor)	(Yes β1-receptor)	(Yes β1-receptor)	
• Prostaglandin E2	YES	YES	YES	
Allergic inflammation				Canning (2003)
• Early-phase response	Yes	Yes	Yes	
• Late phase	Yes	Yes	Yes	
• Eosinophilia	Yes	Yes	Yes	
• Bronchial hyper-responsiveness	Yes	Yes	Yes	
Cough (to citric acid)	Yes	Yes (after ozone)	No	Canning and Spina (2009)
Cigarette smoke-induced inflammation				
• Emphysema	Yes	Yes	Yes	Canning (2003)
• Neutrophil recruitment	Yes	Yes	Yes	
Tuberculosis				
• Caseating cavitation	Yes	Yes	No	Dharmadhikari and Nardell (2008)
• Susceptibility to Mtb	Yes	Yes (Bovis and NTG strain)	Low	

1977; Shampain *et al.*, 1982). The pulmonary recruitment of eosinophils (Coyle *et al.*, 1990; Minshall *et al.*, 1993) and heightened airway responsiveness to adenosine and histamine (Coyle *et al.*, 1990; Minshall *et al.*, 1993; el-Hashim *et al.*, 1996), features observed in human asthma (Martinez and Vercelli, 2013), are reproduced in the rabbit. Furthermore, the pharmacology of anti-inflammatory drugs can also be investigated in the rabbit. Glucocorticosteroids suppressed pulmonary eosinophilia and bronchial hyperresponsiveness in the allergic rabbit (Gozzard *et al.*, 1996) and phosphodiesterase 4 inhibitors, including CDP840, which was the first PDE4 inhibitor to be shown to be effective at inhibiting the late-onset response in patients with allergic asthma (Harbinson *et al.*, 1997), was previously shown to suppress the airway inflammatory response in allergic rabbits (Gozzard *et al.*, 1996). This effect was later confirmed with the more active PDE4 inhibitor Roflumilast in allergic asthmatics (Gauvreau *et al.*, 2011) that is currently licenced for the treatment of COPD.

The role of the immune system in regulating the allergic inflammatory response is actively being pursued in attempts to discover novel therapeutic targets, and the mouse has been used widely for this purpose because of the wide availability of transgenic, knock-out and monoclonal antibodies directed at numerous immunological targets (Lloyd and Hessel, 2010). Therefore, it is encouraging that the genome sequence of the rabbit Th2 region is available and identification of transcription factor binding sites similar to those described in man and the mouse highlight the potential of manipulating the rabbit genome to study immunological responses in this species (Gertz *et al.*, 2011).

COPD represents a major global health issue, principally caused by exposure to cigarette smoke or smoke from burning wood for cooking. This chronic smoke exposure is thought to trigger airway inflammation, small airway fibrosis and emphysema (Decramer *et al.*, 2012). Some features of COPD can be investigated using guinea pigs and mice where inflammation and emphysematous changes to the lung can be measured and genes of interest manipulated, particularly in mice, to ascribe function to genotype (Shapiro, 2008; Barnes, 2013). However, the physiological changes accompanying such changes in these models are difficult to measure in mice. In contrast, the rabbit is large enough to permit the recording of lung function repeatedly in the same animal and our laboratory has measured lung function in the same animals over a one-year period (Minshall *et al.*, 1993), and repeatedly following exposure, for example, to porcine pancreatic elastase, as a model of neutrophil elastase, to induce emphysematous changes and compromised lung function (Nishi *et al.*, 2003; Baila *et al.*, 2012). Reports are emerging of the nature of the signalling pathways that are activated following chronic cigarette smoke in rabbits, and in one study there was a significant protein expression of Toll-like receptor (TLR)4 and matrix metalloproteinase (MMP)-1 in the lungs of rabbits exposed to cigarette smoke with the implication that cigarette smoke stimulates TLR4 via activation of oxidant-signalling pathways, which leads to the elevation of MMP-1 (Geraghty *et al.*, 2011). This shows the potential for studying molecular events in the rabbit and, if coupled to measures of lung function, offer a powerful model to study the temporal effects of cigarette smoke exposure on lung inflammation, tissue damage and lung function.

18.9.2 Cough

Chronic cough is a symptom of various inflammatory diseases including asthma, COPD, gastroesophageal reflex, idiopathic pulmonary fibrosis and other unidentified causes, and is a debilitating condition for which there are currently only limited pharmacological options. Chronic cough remains a significant unmet clinical need in respiratory medicine (Spina and Page, 2013). Mammals, including guinea pigs and rabbits, but not rodents such as mice or rats, possess the physiological reflex of cough and have for a long time been used to study the basic physiology of cough (Table 18.2). Sub-populations of myelinated Aβ fibres with terminations within and beneath airway epithelium of extrapulmonary and large intrapulmonary airways are responsible for initiating cough caused by various insults including low pH and mechanical stimuli (e.g. mucus). In contrast, un-myelinated slow-conduction C-fibres with terminals found within and beneath the airway epithelium throughout the respiratory tract can be activated by a variety of chemical stimuli thought to participate in the sensitization of the cough reflex (Canning and Spina, 2009). Of particular interest is that neuropeptide containing C-fibres are sparse in the rabbit and mediate a weak contractile response to capsaicin similar to that observed in human airways (Spina *et al.*, 1998). In contrast, the airways of the guinea pig, a species that is also used to study the cough reflex, contracts robustly to capsaicin reflecting the dense network of neuropeptide containing afferent nerves in the airways (Watanabe *et al.*, 2006). Hence, the sensory innervation pattern in the rabbit more closely reflects that seen in man compared with other species and therefore is a useful model to study the cough reflex. Rabbits have also been used to study the mechanisms leading to hypertussive responses. For example, following exposure to an environmental pollutant, like ozone, the cough sensitivity to citric acid is significantly enhanced in the rabbit and this has been used as a model to study both the physiology and pharmacology of this heightened cough response (Adcock *et al.*, 2003). This heightened tussive response is thought to better mimic chronic cough syndromes seen in human disease (Canning and Spina, 2009). It is of further interest that this heightened cough response can be suppressed by drugs that are known to be effective anti-tussive in the clinic such as codeine and local anaesthetics (Adcock *et al.*, 2003). The size of these animals also permits the administration and study of the pharmacological action of substances injected into specific regions of the central nervous system known to regulate cough and breathing. For example, angiotensin converting enzyme (ACE) inhibitors like linosopril cause dry cough and are thought to be due to the peripheral sensitizing actions of bradykinin on the cough reflex in humans. When injected into the *nucleus tractus solitarii* (NTS) of anaesthetized rabbits, linosopril enhanced cough number in response to stimulation of lung afferents either mechanically (rubbing the surface of the trachea) or pharmacologically (citric acid). Hence, formation of bradykinin within the NTS can augment the cough reflex thereby uncovering a novel mechanism to explain dry cough in patients taking ACE inhibitors (Cinelli *et al.*, 2015). In other examples of the utility of this model to study central mechanisms regulating the cough reflex, the same group has reported that activation of alpha2-adrenoceptors and galanin receptors within the NTS can suppress cough (Cinelli *et al.*, 2013; Mutolo *et al.*, 2014). The potential use of genetically modified rabbits to study the cough reflex would be invaluable in the discovery of novel drug targets.

18.9.3 Tuberculosis

Tuberculosis (TB) is a global disease which resulted in 1.5 million deaths and 10 million new cases in 2018, with global incidence rates beginning to fall since 2001 with the rate of decline in world incidence of tuberculosis of between 1% and 8.5% per annum, yet there remains the issue concerning the susceptibility of HIV patients to TB and multi-drug resistance (Glaziou *et al.*, 2013). Hence, murine, rabbit, guinea pig and non-human primates have been used to study the underlying mechanism of this disease and to evaluate novel treatments (Dharmadhikari and Nardell, 2008; Myllymaki *et al.*, 2015). The rabbit offers several advantages as a disease model which exhibits similar histopathological features to those observed in human infection, particularly the observation of a caseating granuloma cavitation, which is also observed in non-human primates, but not in mice (Table 18.2). Outbred strains of New Zealand White rabbit are relatively resistant to *Mycobacterium tuberculosis* but are susceptible to *M. bovis* infection and have therefore been used to model the human disease (Dharmadhikari and Nardell, 2008). However, an inbred strain of New Zealand White rabbit (the 'Thorbecke strain'), which is no longer available, was susceptible to *M. tuberculosis* infection. These rabbits demonstrated a diminished impairment in macrophage function compared with outbred rabbits and developed histological lesions consistent with *M. tuberculosis* infection compared with the outbred counterparts (Dorman *et al.*, 2004; Mendez *et al.*, 2008). The reason for this susceptibility might be explained by differences in the genome sequence of Th2 cytokines between the inbred and outbred strains of New Zealand White rabbit (Gertz *et al.*, 2011). Further refinements in this animal model appear to have been made with the reporting of susceptibility of outbred New Zealand White rabbit to *M. tuberculosis* HN878

(Beijing strain) resulting in granulomatous pulmonary disease (Subbian *et al.* 2011a, 2011b, 2013). These studies also described the molecular determinants of this infectious disease and have investigated gene expression in the diseased lung using microarray and/or quantitative PCR to study the expression of specific genes involved in macrophage activation over the course of this disease (Subbian *et al.*, 2011b, 2013). These studies documented the potential for studying gene expression, gene pathways and the temporal aspects of the evolution of the disease in a rabbit model with a human relevant infectious agent.

New methods to monitor the progression of infection in rabbits using positive emission tomography-computed tomography (PET-CT) imaging have been described that permit the study of temporal aspects of the infection and the monitoring of the efficacy of antimicrobial agents in a non-invasive manner (Via *et al.*, 2012; Luna *et al.*, 2015). Finally, the pharmacokinetic modelling of antimicrobial agents in non-involved and involved regions of the rabbit lung can enhance the understanding of the distribution of these drugs in the pulmonary system aiding understanding of drug disposition in humans (Kjellsson *et al.*, 2012).

References

Adcock, J.J., Douglas, G.J., Garabette, M., Gascoigne, M., Beatch, G. *et al.* (2003) RSD931, a novel anti-tussive agent acting on airway sensory nerves. *British Journal of Pharmacology* 138, 407–416.

Baila, B., Ohno, Y., Nagamoto, H., Kotosai, K., Yabuuchi, Y. *et al.* (2012) Tetomilast attenuates elastase-induced pulmonary emphysema through inhibition of oxidative stress in rabbits. *Biological and Pharmaceutical Bulletin* 35, 494–502.

Barnes, P.J. (2013) Development of new drugs for COPD. *Current Medicinal Chemistry* 20, 1531–1540.

Bernardini, G., Cattaneo, A.G., Sabbioni, E., Di Gioacchino, M., Chiriva-Internati, M. and Gornati, R. (2011) Toxicology of engineered metal nanoparticles. In: Sahu, S.C. and Casciano, D. (eds) *Handbook of System Toxicology*. Wiley & Sons Ltd, Chichester, UK, pp. 729–742.

Brecchia, G., Cardinali, R., Mourvaki, E., Collodel, G., Moretti, E. *et al.* (2010) Short- and long-term effects of lipopolysaccharide-induced inflammation on rabbit sperm quality. *Animal Reproduction Science* 118, 310–316.

Brunner, M., Peng, X., Liu, G.X., Ren, X.Q., Ziv, O. *et al.* (2008) Mechanisms of cardiac arrhythmias and sudden death in transgenic rabbits with long QT syndrome. *Journal of Clinical Investigation* 118, 2246–2259.

Burkholder, T.H., Linton, G., Hoyt, J.R.F. and Young, R. (2012) The rabbit as an experimental model. In: Suckow, M.A., Steven, K.A. and Wilson, R.P. (eds) *The Laboratory Rabbit, Guinea Pig, Hamster, and Other Rodents*. Elsevier, pp. 529–560.

Canning, B.J. (2003) Modeling asthma and COPD in animals: a pointless exercise? *Current Opinion in Pharmacology* 3, 244–250.

Canning, B.J. and Spina, D. (2009) Sensory nerves and airway irritability. *Handbook of Experimental Pharmacology* (194), 139–183.

Cinelli, E., Bongianni, F., Pantaleo, T. and Mutolo, D. (2013) Suppression of the cough reflex by alpha 2-adrenergic receptor agonists in the rabbit. *Physiological Reports* 1, e00122.

Cinelli, E., Bongianni, F., Pantaleo, T. and Mutolo, D. (2015) The cough reflex is upregulated by lisinopril microinjected into the caudal nucleus tractus solitarii of the rabbit. *Respiratory Physiology and Neurobiology* 219, 9–17.

Castellini, C., Mourvaki, E., Sartini, B., Cardinali, R., Moretti, E. *et al.* (2009) In vitro toxic effects of metal compounds on kinetic traits and ultrastructure of rabbit spermatozoa. *Reproductive Toxicology* 27, 46–54.

Castellini, C., Ruggeri, S., Mattioli, S., Bernardini, G., Macchioni, L. *et al.* (2014) Long-term effects of silver nanoparticles on reproductive activity of rabbit buck. *Systems Biology in Reproductive Medicine* 60, 143–150.

Castellini, C., Mattioli, S., Dal Bosco, A., Collodel, G., Pistilli, A. *et al.* (2019) In vitro effect of nerve growth factor on the main traits of rabbit sperm. *Reproductive Biology and Endocrinology* 17, 1–11.

Cheng, X., Wang, S., Dai, X., Shi, C., Wen, Y. *et al.* (2012) Rabbit as a novel animal model for hepatitis E virus infection and vaccine evaluation. *PLOS ONE* 7, e51616.

Collodel, G., Castellini, C., del Vecchio, M.T., Cardinali, R., Geminiani, M. *et al.* (2012) Effect of a bacterial lipopolysaccharide treatment on rabbit testis and ejaculated sperm. *Reproduction in Domestic Animals* 47, 372–378.

Coyle, A.J., Spina, D. and Page, C.P. (1990) PAF-induced bronchial hyperresponsiveness in the rabbit: contribution of platelets and airway smooth muscle. *British Journal of Pharmacology* 101, 31–38.

D'Cruz, O.J., Vassilev, A. and Uckun, F.M. (2000) Evaluation of boar sperm as a model system to study the mechanism of spermicidal activity of vanadocenes. *Biochemical and Biophysical Research Communications* 270, 826–830.

Decramer, M., Janssens, W. and Miravitlles, M. (2012) Chronic obstructive pulmonary disease. *Lancet* 379, 1341–1351.

Dharmadhikari, A.S. and Nardell, E.A. (2008) What animal models teach humans about tuberculosis. *American Journal of Respiratory Cell and Molecular Biology* 39, 503–508.

Ding, Y., Wang, Y., Zhu, H., Fan, J., Yu, L. *et al.* (2011) Hypertriglyceridemia and delayed clearance of fat load in transgenic rabbits expressing human apolipoprotein CIII. *Transgenic Research* 20, 867–875.

Dorman, S.E., Hatem, C.L., Tyagi, S., Aird, K., Lopez-Molina, J. *et al.* (2004) Susceptibility to tuberculosis: clues from studies with inbred and outbred New Zealand white rabbits. *Infection and Immunity* 72, 1700–1705.

Duverger, N., Kruth, H., Emmanuel, F., Caillaud, J.M., Viglietta, C. *et al.* (1996) Inhibition of atherosclerosis development in cholesterol-fed human apolipoprotein A-I-transgenic rabbits. *Circulation* 94, 713–717.

el-Hashim, A., D'Agostino, B., Matera, M.G. and Page, C. (1996) Characterization of adenosine receptors involved in adenosine-induced bronchoconstriction in allergic rabbits. *British Journal of Pharmacology* 119, 1262–1268.

Fan, J. and Watanabe, T. (2000) Cholesterol-fed and transgenic rabbits for the study of atherosclerosis. *Journal of Atherosclerosis and Thrombosis* 7, 26–32.

Fan, J. and Watanabe, T. (2003) Transgenic rabbits as therapeutic protein bioreactors and human disease models. *Pharmacology and Therapeutics* 99, 261–282.

Fan, J., Wang, J., Bensadoun, A., Lauer, S.J., Dang, Q. *et al.* (1994) Overexpression of hepatic lipase in transgenic rabbits leads to a marked reduction of plasma high density lipoproteins and intermediate density lipoproteins. *Proceedings of the National Academy of Sciences of the USA* 91, 8724–8728.

Fan, J., McCormick, S.P., Krauss, R.M., Taylor, S., Quan, R. *et al.* (1995) Overexpression of human apolipoprotein B-100 in transgenic rabbits results in increased levels of LDL and decreased levels of HDL. *Arteriosclerosis Thrombosis and Vascular Biology* 15, 1889–1899.

Fan, J., Ji, Z.S., Huang, Y., de Silva, H., Sanan, D. *et al.* (1998) Increased expression of apolipoprotein E in transgenic rabbits results in reduced levels of very low density lipoproteins and an accumulation of low density lipoproteins in plasma. *Journal of Clinical Investigation* 101, 2151–2164.

Fan, J., Araki, M., Wu, L., Challah, M., Shimoyamada, H. *et al.* (1999) Assembly of lipoprotein (a) in transgenic rabbits expressing human apolipoprotein (a). *Biochemical and Biophysical Research Communication* 255, 639–644.

Fan, J., Unoki, H., Kojima, N., Sun, H., Shimoyamada, H. *et al.* (2001) Overexpression of lipoprotein lipase in transgenic rabbits inhibits diet-induced hypercholesterolemia and atherosclerosis. *Journal of Biological Chemistry* 276, 40071–40079.

Fan, J., Kitajima, S., Watanabe, T., Xu, J., Zhang, J. *et al.* (2015) Rabbit models for the study of human atherosclerosis: from pathophysiological mechanisms to translational medicine. *Pharmacology and Therapeutics* 146C, 104–119.

Fitzgerald, T.J. (1985) Experimental congenital syphilis in rabbits. *Canadian Journal of Microbiology* 31, 757–762.

Foote, R.H. and Carney, E.W. (2000) The rabbit as a model for reproductive and developmental toxicity studies. *Reproductive Toxicology* 14, 477–493.

Gautier, T., Deckert, V., Aires, V., Le Guern, N., Patoli, D. *et al.* (2019) Human apolipoprotein C1 transgenesis to inhibit Cetp and to attenuate atherosclerosis in hypercholesterolemic rabbits. *Atherosclerosis* 287, e81.

Gaureau, G.M., Boulet, L.P., Schmid-Wirlitsch, C., Côté, J., Duong, M. *et al.* (2011) Roflumilast attenuates allergen-induced inflammation in mild asthmatic subjects. *Respiratory Research* 12, 140.

Geraghty, P., Dabo, A.J. and D'Armiento, J. (2011) TLR4 protein contributes to cigarette smoke-induced matrix metalloproteinase-1 (MMP-1) expression in chronic obstructive pulmonary disease. *Journal of Biological Chemistry* 286, 30211–30218.

Gertz, E.M., Agarwala, R., Mage, R.G. and Schaffer, A.A. (2011) Comparative analysis of genome sequences of the Th2 cytokine region of rabbit (*Oryctolagus cuniculus*) with those of nine different species. *Immunol Immunogenet Insights* 3, 59–82.

Glaziou, P., Falzon, D., Floyd, K. and Raviglione, M. (2013) Global epidemiology of tuberculosis. *Seminars in Respiratory and Critical Care Medicine* 34, 3–16.

Gozzard, N., el-Hashim, A., Herd, C.M., Blake, S.M., Holbrook, M. *et al.* (1996) Effect of the glucocorticosteroid budesonide and a novel phosphodiesterase type 4 inhibitor CDP840 on antigen-induced airway responses in neonatally immunised rabbits. *British Journal of Pharmacology* 118, 1201–1208.

Gregory, M.H., Capito, N., Kuroki, K., Stoker, A.M., Cook, J.L. and Sherman, S.L. (2012) A review of translational animal models for knee osteoarthritis. *Arthritis* 2012,764621.

Gupta, M.P. (2007) Factors controlling cardiac myosin-isoform shift during hypertrophy and heart failure. *Journal of Molecular and Cellular Cardiology* 43, 388–403.

Harbinson, P.L., MacLeod, D., Hawksworth, R., O'Toole, S., Sullivan, P.J. *et al.* (1997) The effect of a novel orally active selective PDE4 isoenzyme inhibitor (CDP840) on allergen-induced responses in asthmatic subjects. *The European Respiratory Journal* 10, 1008–1014.

Hoeg, J.M., Vaisman, B.L., Demosky, S.J. Jr., Meyn, S.M., Talley, G.D. *et al.* (1996) Lecithin:cholesterol acyltransferase overexpression generates hyperalpha-lipoproteinemia and a nonatherogenic lipoprotein pattern in transgenic rabbits. *Journal of Biological Chemistry* 271, 4396–4402.

Huang, Y., Schwendner, S.W., Rall, S.C.J., Sanan, D.A. and Mahley, R.W. (1997) Apolipoprotein E2 transgenic rabbits. *Journal of Biological Chemistry* 272, 22685–22694.

Ignatowski, A.C. (1908) Influence of animal food on the organism of rabbits. *S Peterb Izviest Imp Voyenno-Med Akad* 16, 154–173.

Irvin, C.G. and Bates, J.H. (2003) Measuring the lung function in the mouse: the challenge of size. *Respiratory Research* 4, 4.

Kajihara, T., Okagaki, R. and Ishihara, O. (2006) LPS-induced transient testicular dysfunction accompanied by apoptosis of testicular germ cells in mice. *Medical Molecular Morphology* 39, 203–208.

Kamaruzaman, N.A., Kardia, E., Kamaldin, N., Latahir, A.Z. and Yahaya, B.H. (2013) The rabbit as a model for studying lung disease and stem cell therapy. *Biomedical Research International* 2013, 691830.

Karczewski, W. and Widdicombe, J.G. (1969) The role of the vagus nerves in the respiratory and circulatory reactions to anaphylaxis in rabbits. *Journal of Physiology* 201, 293–304.

Kauffold, J., Henning, K., Bachmanna, R., Hotzel, H. and Melzer, F. (2007) The prevalence of chlamydiae of bulls from six bull studs in Germany. *Animal Reproduction Science* 102, 111–121.

Keir, S. and Page, C. (2008) The rabbit as a model to study asthma and other lung diseases. *Pulmonary Pharmacology and Therapeutics* 21, 721–730.

Keir, S.D., Spina, D., Douglas, G., Herd, C. and Page, C.P. (2011) Airway responsiveness in an allergic rabbit model. *Journal of Pharmacological and Toxicological Methods* 64, 187–195.

Kitajima, S., Liu, E., Morimoto, M., Koike, T., Yu, Y. *et al.* (2005) Transgenic rabbits with increased VEGF expression develop hemangiomas in the liver: a new model for Kasabach-Merritt syndrome. *Laboratory Investigation* 85, 1517–1527.

Kjellsson, M.C., Via L.E., Goh, A., Weiner, D., Low, K.M. *et al.* 2012. Pharmacokinetic evaluation of the penetration of antituberculosis agents in rabbit pulmonary lesions. *Antimicrobial Agents and Chemotherapy* 56(1), 446–457.

Koike, T., Kitajima, S., Yu, Y., Nishijima, K., Zhang, J. *et al.* (2009) Human C-reactive protein does not promote atherosclerosis in transgenic rabbits. *Circulation* 120, 2088–2094.

Kominami, T., Ueno, S., Okado, S., Nakanishi, A., Kondo, M. and Terasaki, H. (2017) Contributions of second- and third-order retinal neurons to cone electroretinograms after loss of rod function in rhodopsin P347L transgenic rabbits. *Investigative Ophthalmology and Visual Science* 58, 1417–1424.

Kominami, T., Ueno, S., Nishida, K., Inooka, D., Kominami, A. *et al.* (2019) Electrically evoked potentials are reduced compared to axon numbers in rhodopsin P347L transgenic rabbits with severe photoreceptor degeneration. *Investigative Ophthalmology and Visual Science* 60, 2543–2550.

Kondo, M., Sakai, T., Komeima, K., Kurimoto, Y., Ueno, S. *et al.* (2009) Generation of a transgenic rabbit model of retinal degeneration. *Investigative Ophthalmology and Visual Science* 50, 1371–1377.

Laitinen, L.A., Laitinen, A., Panula, P.A., Partanen, M., Tervo, K. and Tervo, T. (1983) Immunohistochemical demonstration of substance P in the lower respiratory tract of the rabbit and not of man. *Thorax* 38, 531–536.

Lang, C.N., Koren, G. and Odening, K.E. (2016) Transgenic rabbit models to investigate the cardiac ion channel disease long QT syndrome. *Progress in Biophysics and Molecular Biology* 121, 142–156.

Liang, J., Liu, E., Yu, Y., Kitajima, S., Koike, T. *et al.* (2006) Macrophage metalloelastase accelerates the progression of atherosclerosis in transgenic rabbits. *Circulation* 113, 1993–2001.

Liu, Z., Hu, X., Man, Z., Zhang, J., Jiang, Y. and Ao, Y. (2016) A novel rabbit model of early osteoarthritis exhibits gradual cartilage degeneration after medial collateral ligament transection outside the joint capsule. *Scientific Reports* 6, 34423.

Lloyd, C.M. and Hessel, E.M. (2010) Functions of T cells in asthma: more than just T(H)2 cells. *Nature Review Immunology* 10, 838–848.

Lowey, S., Bretton, V., Joel, P.B., Trybus, K.M., Gulick, J. *et al.* (2018) Hypertrophic cardiomyopathy R403Q mutation in rabbit β-myosin reduces contractile function at the molecular and myofibrillar levels. *Proceedings of the National Academy of Sciences of the USA* 115, 11238–11243.

Lu, G., Wang, C., Wu, C., Yan, L. and Tang, J. (2019) Identification of early biomarkers in a rabbit model of primary Candida pneumonia. *BMC Infectious Diseases* 19, 698.

Luna, B., Kubler, A., Larsson, C., Foster, B., Bagci, U. *et al.* (2015) In vivo prediction of tuberculosis-associated cavity formation in rabbits. *Journal of Infectious Diseases* 211, 481–485.

Mage, R.G. and Rai, G. (2012) *A Rabbit Model of Systemic Lupus Erythematosus, Useful for Studies of Neuropsychiatric SLE*. InTech-Open Access Publisher, Croatia.

Manzini, S. (1992) Bronchodilatation by tachykinins and capsaicin in the mouse main bronchus. *British Journal of Pharmacology* 105, 968–972.

Marian, A.J., Wu, Y., Lim, D.S., McCluggage, M., Youker, K. *et al.* (1999) A transgenic rabbit model for human hypertrophic cardiomyopathy. *Journal of Clinical Investigation* 104, 1683–1692.

Martinez, F.D. and Vercelli, D. (2013) Asthma. *The Lancet* 382, 1360–1372.

Masson, D., Deckert, V., Gautier, T., Klein, A., Desrumaux, C. *et al.* (2011) Worsening of diet-induced atherosclerosis in a new model of transgenic rabbit expressing the human plasma phospholipid transfer protein. *Arteriosclerosis Thrombosis and Vascular Biology* 31, 766–774.

Matsumoto, S., Takeda, M., Saiki, C., Takahashi, T. and Ojima, K. (1997) Effects of tachykinins on rapidly adapting pulmonary stretch receptors and total lung resistance in anesthetized, artificially ventilated rabbits. *Journal of Pharmacology and Experimental Therapeutics* 283, 1026–1031.

Mendez, S., Hatem, C.L., Kesavan, A.K., Lopez-Molina, J., Pitt, M.L. *et al.* (2008) Susceptibility to tuberculosis: composition of tuberculous granulomas in Thorbecke and outbred New Zealand White rabbits. *Veterinary Immunology and Immunopathology* 122, 167–174.

Meng, A.L., Pinckard, R.N. and Halonen, M. (1973) Ontogeny of the release and reactivity to the mediators of IgE-induced acute allergic reactions in the rabbit: the homologous PCA reaction. *Journal of Immunology* 110, 1554–1561.

Mills, J., Sellick, H. and Widdicombe, J.G. (1969) The role of lung irritant receptors in respiratory responses to multiple pulmonary embolism, anaphylaxis and histamine-induced bronchoconstriction. *Journal of Physiology* 200, 79P–80P.

Minshall, E.M., Riccio, M.M., Herd, C.M., Douglas, G.J., Seeds, E.A. *et al.* (1993) A novel animal model for investigating persistent airway hyperresponsiveness. *Journal of Pharmacological and Toxicological Methods* 30, 177–188.

Moretti, E., Dal Bosco, A., Mourvaki, E., Cardinali, R., Collodel, G. *et al.* (2009) In vitro effects of tyre debris organic extract on the kinetic and morphologic traits of rabbit spermatozoa. *World Rabbit Science* 17, 213–220.

Moretti, E., Terzuoli, G., Renieri, T., Iacoponi, F., Castellini, C. *et al.* (2013) In vitro effect of gold and silver nanoparticles on human spermatozoa. *Andrologia* 45, 392–396.

Mutolo, D., Cinelli, E., Bongianni, F. and Pantaleo, T. (2014) Inhibitory control of the cough reflex by galanin receptors in the caudal nucleus tractus solitarii of the rabbit. *American Journal of Physiology-Regulatory, Integrative and Comparative Physiology* 307, R1358–R1367.

Myllymaki, H., Niskanen, M., Oksanen, K.E. and Ramet, M. (2015) Animal models in tuberculosis research – where is the beef? *Expert Opinion on Drug Discovery* 10, 871–883.

Niimi, M., Yang, D., Kitajima, S., Ning, B., Wang, C. *et al.* (2016) ApoE knockout rabbits: a novel model for the study of human hyperlipidemia. *Atherosclerosis* 245, 187–193.

Nishi, Y., Boswell, V., Ansari, T., Piprawala, F., Satchi, S. and Page, C.P. (2003) Elastase-induced changes in lung function: relationship to morphometry and effect of drugs. *Pulmonary Pharmacology and Therapeutics* 16, 221–229.

Nishizawa, T., Vatner, S.F., Hong, C., Shen, Y.T., Hardt, S.E. *et al.* (2006) Overexpressed cardiac Gsalpha in rabbits. *Journal of Molecular and Cellular Cardiology* 41, 44–50.

Odening, K.E., Bodi, I., Franke, G., Rieke, R., Ryan de Medeiros, A. *et al.* (2019) Transgenic short-QT syndrome 1 rabbits mimic the human disease phenotype with QT/action potential duration shortening in the atria and ventricles and increased ventricular tachycardia/ventricular fibrillation inducibility. *European Heart Journal* 40, 842–853.

Pattison, J.S., Waggoner, J.R., James, J., Martin, L., Gulick, J. *et al.* (2008) Phospholamban overexpression in transgenic rabbits. *Transgenic Research* 17, 157–170.

Pinckard, R.N., Halonen, M., Palmer, J.D., Butler, C., Shaw, J.O. and Henson, P.M. (1977) Intravascular aggregation and pulmonary sequestration of platelets during IgE-induced systemic anaphylaxis in the rabbit: abrogation of lethal anaphylactic shock by platelet depletion. *Journal of Immunology* 119, 2185–2193.

Pogwizd, S.M. and Bers, D.M. (2004) Cellular basis of triggered arrhythmias in heart failure. *Trends in Cardiovascular Medicine* 14(2), 61–66.

Pogwizd, S.M. and Bers, D.M. (2008) Rabbit models of heart disease. *Drug Discovery Today, Disease Models* 5, 185–193.

Rai, G., Ray, S., Milton, J., Yang, J., Ren, P. *et al.* (2010) Gene expression profiles in a rabbit model of systemic lupus erythematosus autoantibody production. *Journal of Immunology* 185, 4446–4456.

Sanbe, A., James, J., Tuzcu, V., Nas, S., Martin, L. *et al.* (2005) Transgenic rabbit model for human troponin I-based hypertrophic cardiomyopathy. *Circulation* 111, 2330–2338.

Seed, J., Chapin, R.E., Clegg, E.D., Dostal, L.A., Foote, R.H. *et al.* (1996) Methods for assessing sperm motility, morphology, and counts in the rat, rabbit, and dog: a consensus report. ILSI Risk Science Institute Expert Working Group on Sperm Evaluation. *Reproductive Toxicology* 10, 237–244.

Shampain, M.P., Behrens, B.L., Larsen, G.L. and Henson, P.M. (1982) An animal model of late pulmonary responses to Alternaria challenge. *American Review of Respiratory Disease* 126, 493–498.

Shapiro, S.D. (2007) Transgenic and gene-targeted mice as models for chronic obstructive pulmonary disease. *The European Respiratory Journal* 29, 375–378.

Shapiro, S.D. (2008) The use of transgenic mice for modeling airways disease. *Pulmonary Pharmacology and Therapeutics* 21, 699–701.

Shen, J., Herderick, E., Cornhill, J.F., Zsigmond, E., Kim, H.S. *et al.* (1996) Macrophage-mediated 15-lipoxygenase expression protects against atherosclerosis development. *Journal of Clinical Investigation* 98, 2201–2208.

Shiomi, M. and Fan, J. (2008) Unstable coronary plaques and cardiac events in myocardial infarction-prone Watanabe heritable hyperlipidemic rabbits: questions and quandaries. *Current Opinion in Lipidology* 19, 631–636.

Shiomi, M. and Ito, T. (2009) The Watanabe heritable hyperlipidemic (WHHL) rabbit, its characteristics and history of development: a tribute to the late Dr. Yoshio Watanabe. *Atherosclerosis* 207, 1–7.

Shiomi, M., Ito, T., Yamada, S., Kawashima, S. and Fan, J. (2003) Development of an animal model for spontaneous myocardial infarction (WHHLMI rabbits). *Arteriosclerosis Thrombosis and Vascular Biology* 23, 1239–1244.

Sparks, D.L. (2008) The early and ongoing experience with the cholesterol-fed rabbit as a model of Alzheimer's disease: the old, the new and the pilot. *Journal of Alzheimer's Disease* 15, 641–656.

Sparks, D.L., Scheff, S.W., Hunsaker, J.C. III, Liu, H., Landers, T. and Gross, D.R. (1994) Induction of Alzheimer-like beta-amyloid immunoreactivity in the brains of rabbits with dietary cholesterol. *Experimental Neurology* 126, 88–94.

Speck, R.F., Penn, M.L., Wimmer, J., Esser, U., Hague, B.F. *et al.* (1998) Rabbit cells expressing human CD4 and human CCR5 are highly permissive for human immunodeficiency virus type 1 infection. *Journal of Virology* 72, 5728–5734.

Spina, D. and Page, C.P. (2013) Regulating cough through modulation of sensory nerve function in the airways. *Pulmonary Pharmacology and Therapeutics* 26, 486–490.

Spina, D., Matera, G.M., Riccio, M.M. and Page, C.P. (1998) A comparison of sensory nerve function in human, guinea-pig, rabbit and marmoset airways. *Life Sciences* 63, 1629–1642.

Subbian, S., Tsenova, L., O'Brien, P., Yang, G., Koo, M.S. *et al.* (2011a) Phosphodiesterase-4 inhibition alters gene expression and improves isoniazid-mediated clearance of Mycobacterium tuberculosis in rabbit lungs. *PLOS Pathogens* 7, e1002262.

Subbian, S., Tsenova, L., Yang, G., O'Brien, P., Parsons, S. *et al.* (2011b) Chronic pulmonary cavitary tuberculosis in rabbits: a failed host immune response. *Open Biology* 1, 110016.

Subbian, S., Bandyopadhyay, N., Tsenova, L., O'Brien, P., Khetani, V. *et al.* (2013) Early innate immunity determines outcome of Mycobacterium tuberculosis pulmonary infection in rabbits. *Cell Communation and Signaling* 11, 60–76.

Taylor, M.J. and Fan, J. (1997) Transgenic rabbit models for the study of atherosclerosis. *Frontiers in Bioscience* 2, 298–308.

Tran, V.G., Venkatasubramaniam, A., Adhikari, R.P., Krishnan, S., Wang, X. *et al.* (2020) Efficacy of active immunization with attenuated α-hemolysin and Panton-Valentine Leukocidin in a rabbit model of *Staphylococcus aureus* necrotizing pneumonia. *Journal of Infectious Diseases* 221, 267–275.

Verloop, M.C. (1949) On the arteriae bronchiales and their anastomosing with the arteria pulmonalis in some rodents; a micro-anatomical study. *Acta Anatomica (Basel)* 7, 1–32.

Via, L.E., Schimel, D., Weiner, D.M., Dartois, V., Dayao, E. *et al.* (2012) Infection dynamics and response to chemotherapy in a rabbit model of tuberculosis using [^{18}F]2-fluoro-deoxy-D-glucose positron emission tomography and computed tomography. *Antimicrobial Agents and Chemotherapy* 56, 4391–4402.

Wang, Y., Niimi, M., Nishijima, K., Waqar, A.B., Yu, Y. *et al.* (2013) Human apolipoprotein AII protects against diet-induced atherosclerosis in transgenic rabbits. *Arteriosclerosis Thrombosis and Vascular Biology* 33, 224–231.

Warheit, D.B., Borm, P.J.A., Hennes, C. and Lademann, J. (2007) Testing strategies to establish the safety of nanomaterials: conclusions of an ECETOC workshop. *Inhalation Toxicology* 19, 631–643.

Watanabe, N., Horie, S., Michael, G.J., Keir, S., Spina, D. *et al.* (2006) Immunohistochemical co-localization of transient receptor potential vanilloid (TRPV)1 and sensory neuropeptides in the guinea-pig respiratory system. *Neuroscience* 141, 1533–1543.

Watanabe, Y. (1980) Serial inbreeding of rabbits with hereditary hyperlipidemia (WHHL-rabbit). *Atherosclerosis* 36, 261–268.

Woodruff-Pak, D.S. (2008) Animal models of Alzheimer's disease: therapeutic implications. *Journal of Alzheimer's Disease* 15, 507–521.

Yamamura, Y., Kotani, M., Chowdhury, M.I., Yamamoto, N., Yamaguchi, K. *et al.* (1991) Infection of human CD4+ rabbit cells with HIV: the possibility of the rabbit as model for HIV-1 infection. *International Immunology* 3, 1183–1187.

Yamanaka, S., Balestra, M.E., Ferrell, L.D., Fan, J., Arnold, K.S. *et al.* (1995) Apolipoprotein B mRNA-editing protein induces hepatocellular carcinoma and dysplasia in transgenic animals. *Proceedings of the National Academy of Science of the USA* 92, 8483–8487.

Yang, D., Zhang, J., Xu, J., Zhu, T., Fan, J. and Chen, Y. (2013) Production of Apolipoprotein C-III knockout rabbits using Zinc Finger Nucleases. *Journal of Visualized Experiments* 81, e50957.

Yang, D., Xu, J., Zhu, T., Fan, J., Lai, L. *et al.* (2014) Effective gene targeting in rabbits using RNA-guided Cas9 nucleases. *Journal of Molecular Cell Biology* 6, 97–99.

Young, R.J., Bodt, B.A., Iturralde, T.G. and Starke, W.C. (1992) Automated-analysis of rabbit sperm motility and the effect of chemicals on sperm motion parameters. *Molecular Reproduction and Development* 33, 347–356.

Zaucha, G.M., Pitt, L.M., Estep, J., Ivins, B.E. and Friedlander, A.M. (1998) The pathology of experimental anthrax in rabbits exposed by inhalation and subcutaneous inoculation. *Archives of Pathology and Laboratory Medicine* 122, 982–992.

Index

Note: Page numbers in **bold** type refer to **figures**
Page numbers in *italic* type refer to *tables*

Printed and bound by CPI Group (UK) Ltd, Croydon, CR0 4YY

15/09/2026

14962382-0001